Polymer Nanocomposites

Handbook

Polymer
Nanocomposites
Handbook

Rakesh K. Gupta
Elliot Kennel
Kwang-Jea Kim

CRC Press
Taylor & Francis Group
Boca Raton   London   New York

CRC Press is an imprint of the
Taylor & Francis Group, an **informa** business

CRC Press
Taylor & Francis Group
6000 Broken Sound Parkway NW, Suite 300
Boca Raton, FL 33487-2742

Printed in the United States of America on acid-free paper
10 9 8 7 6 5 4 3 2 1

International Standard Book Number: 978-0-8493-9777-6 (Hardback)

Library of Congress Cataloging-in-Publication Data

Polymer nanocomposites handbook / editors, Rakesh K. Gupta, Elliot Kennel.
    p. cm.
  Includes bibliographical references and index.
  ISBN 978-0-8493-9777-6 (alk. paper)
  1. Polymeric composites--Handbooks, manuals, etc. 2. Nanostructured materials--Handbooks, manuals, etc. I. Gupta, Rakesh K. (Rakesh Kumar), 1954- II. Kennel, Elliot. III. Title.

TA418.9.C6P6487 2008
620.1'92--dc22
                                         2008048861

Visit the Taylor & Francis Web site at
http://www.taylorandfrancis.com

and the CRC Press Web site at
http://www.crcpress.com

Dedication

For my wife, Gunjan, and for our daughters, Deepti and Neha.

Rakesh K. Gupta

To the memory of Rick Smalley, Roger Bacon, Leonard Singer, and the other pioneers of nanomaterial development, without whom carbon nanocomposites would not exist; and to Max Lake who got us started on nanomaterials in the first place in 1985.

Elliot Kennel

For my wife, Hyekyong, and for my daughter, Carol Tongyon.

Kwang-Jea Kim

The universe is created by the association of "DRAVYA," an ultrafine entity from which matter, energy, time, directions and mind emanate.

—**Maharishi Kannad**
Vaishesik Darshan
written around 2000 B.C.

Contents

About the Editors

Rakesh K. Gupta has been teaching at West Virginia University (WVU) since 1992. He holds B.Tech. and Ph.D. degrees in chemical engineering from the Indian Institute of Technology, Kanpur, and the University of Delaware, respectively. Before coming to WVU, he taught at the State University of New York at Buffalo for 11 years. He has also worked briefly for the Monsanto and DuPont Companies, and he serves as technical advisor to the Polymers Alliance Zone of West Virginia. His research focuses on polymer rheology, polymer processing, and polymer composites. He is the author of *Polymer and Composite Rheology* and the co-author of *Fundamentals of Polymer Engineering*.

Elliot B. Kennel holds an M.S. degree in nuclear engineering from The Ohio State University and a B.S. in physics from Miami University (Ohio). He is a co-founder of Nanographite Materials, Inc., and Pyrograf Products, Inc. While working for the U.S. Air Force Research Laboratory in the early 1980s, he helped sponsor some of the early work in the creation of p-type doped carbon nanotubes and nanocomposites as a means of achieving high-temperature electrical conductors. Currently, Kennel is developing coal-based feedstocks for low-cost nanomaterials and other carbon products such as pitches and cokes.

Previously, Kennel served as vice president and director of research and development at Applied Sciences, Inc. (Cedarville, Ohio), and prior to that he served in the U.S. Air Force Research Laboratory as an officer and then civil servant in the area of aerospace materials and energy conversion.

Kwang-Jea Kim is currently at the University of Akron as a research faculty member. He obtained a Ph.D. degree in polymer engineering from the University of Akron after receiving an M.S. degree in surfactant synthesis at Inha University in South Korea where he continued his research work as a postdoctoral fellow. He worked for Struktol Company of America as a research scientist and project manager for more than five years and later at the chemical engineering department of West Virginia University as a research assistant professor for one year. His research focuses on polymer composites, interfacial science, rheology, reactive processing, chemical additives, nanomaterials, organic and inorganic hybrid materials, and rubbers and plastics. He is the co-author of *Thermoplastic and Rubber Compounds: Technology and Physical Chemistry*. He is currently serving as a guest editor of *Interfaces of Cellulose Polymer Composites*, which specializes in the wood–plastic composite area, and represents special issues of *Composite Interfaces*.

Contributors

Sushant Agarwal
Department of Chemical Engineering
West Virginia University
Morgantown, West Virginia

Paul Andersen
Coperion Corporation
Ramsey, New Jersey

Henry C. Ashton
Schneller Corporation
Kent, Ohio

Enrique V. Barrera
Department of Mechanical
  Engineering and Materials Science
Rice University
Houston, Texas

Vinod K. Berry
Department of Chemical Engineering
West Virginia University
Morgantown, West Virginia

Sati Bhattacharya
The Rheology and Materials
  Processing Centre
School of Civil, Environmental and
  Chemical Engineering
RMIT University, Melbourne
Victoria, Australia

Subhendu Bhattacharya
The Rheology and Materials
  Processing Centre
School of Civil, Environmental and
  Chemical Engineering
RMIT University, Melbourne
Victoria, Australia

D.J. Burton
Applied Sciences, Inc.
Cedarville, Ohio

Erica Corral
Sandia National Laboratories
Albuquerque, New Mexico

Daniel De Kee
Department of Chemical and
  Biomolecular Engineering
Tulane Institute for Macrololecular
  Engineering and Science
Tulane University
New Orleans, Louisiana

Kyle J. Frederic
Department of Chemical and
  Biomolecular Engineering
Tulane Institute for Macrololecular
  Engineering and Science
Tulane University
New Orleans, Louisiana

D. Gerald Glasgow
Applied Sciences, Inc.
Cedarville, Ohio

Rahul K. Gupta
The Rheology and Materials
  Processing Centre
School of Civil, Environmental and
  Chemical Engineering
RMIT University, Melbourne
Victoria, Australia

Rakesh K. Gupta
Department of Chemical Engineering
West Virginia University
Morgantown, West Virginia

Chang Kook Hong
Center for Functional Nano Fine Chemicals
Chonnam National University
Gwangju, South Korea

Avraam I. Isayev
Institute of Polymer Engineering
The University of Akron
Akron, Ohio

Elliot B. Kennel
Chemical Engineering Department
West Virginia University
Morgantown, West Virginia

Kwang-Jea Kim
Chemical Engineering Department
West Virginia University
Morgantown, West Virginia

Max L. Lake
Applied Sciences, Inc.
Cedarville, Ohio

Jue Lu
Department of Chemical Engineering
  and Center for Composite Materials
University of Delaware
Newark, Delaware

Meisha Shofner
School of Polymer, Textile & Fiber
  Engineering
Georgia Institute of Technology
Atlanta, Georgia

Daneesh Simien
National Institute of Standards and
  Technology
Gaithersburg, Maryland

Chang H. Song
Colortech, Inc.
Morristown, Tennessee

Tatsuhiro Takahashi
Department of Polymer Science and
  Engineering
Yamagata University
Yonezawa, Japan

Gary G. Tibbetts
Applied Sciences, Inc.
Cedarville, Ohio

Leszek A. Utracki
Industrial Materials Institute
National Research Council Canada
Boucherville, Canada

James L. White
Institute of Polymer Engineering
The University of Akron
Akron, Ohio

Charles A. Wilkie
Department of Chemistry
Marquette University
Milwaukee, Wisconsin

Richard P. Wool
Department of Chemical
  Engineering and Center for
  Composite Materials
University of Delaware
Newark, Delaware

Koichiro Yonetake
Department of Polymer Science and
  Engineering
Yamagata University
Yonezawa, Japan

Jin Zhu
YTC America, Inc.
Camarillo, California

1

Overview of Challenges and Opportunities

Rakesh K. Gupta, Elliot B. Kennel, and Kwang-Jea Kim

CONTENTS

1.1 Introduction to the Book

Nanomaterials, and, in particular, nanoreinforcements for polymer composites have in recent years been the subject of intense research, development, and commercialization. A remarkable 1959 talk by Nobel Laureate Richard Feynman at the meeting of the American Physical Society at Caltech is recognized by many scientific historians as a salient event in the history of nanotechnology.[1] In his talk, Feynman foresaw the development of nanomaterials, nanolithography, nanoscale digital storage, molecular electronics, and nanomanufacturing methods. Among other things, Feynman famously offered two prizes, for a thousand dollars apiece, in which he asked for a working motor smaller than 1/64 of a cubic inch; or to anyone who could reduce text to the size such that only an electron microscope could read it (i.e., nanolithography). Both prizes were awarded within a few years.

Nanomaterials are an important subset of nanotechnology. Feynman was interested not only in the small dimensions that might be created, but also in the special attributes of materials whose size might be controlled to only a few atomic layers in thickness. These attributes, taken together, help to more precisely define the concept of nanomaterials. That is, nanomaterials of interest should not only have very small physical dimensions, but should also exhibit some unusual properties by virtue of their small size; and moreover, the producers of these materials should have control over the dimensions of the materials and hence the resultant property enhancements.

On this basis, it might be argued that a tire made of rubber compounded with carbon black was one of the earliest primitive nanocomposites. As early as the 1860s, the ability of

1

carbon black to enhance the mechanical properties of vulcanized rubber was recognized by researchers who experimented with adding different materials to the basic rubber formulation. By virtue of its high surface area, surface energy, and mechanical properties, carbon black is able to significantly enhance the properties of rubber. Other well-known nanoscale reinforcements available in the early twentieth century included fumed silica and precipitated calcium carbonate.

Today, industrial applications of nanomaterials can be found in a wide variety of industries. Most readers would be familiar with applications in the field of electronics and in health care. These, however, are not all. Synthetic textiles incorporating nanopowders that endow the fabrics with antibacterial, flame retardant, non-wetting, or self-cleaning properties are becoming common. Thick coatings composed of nanoparticulate metal oxides find use in waterfast ink-jet media with photo-parity, while thin coatings can be used for optical amplifying systems for light-emitting diodes. Thermal spray of nanopowders allows for the coating of flight- and land-based turbines where corrosion or erosion must be prevented. Other applications may be found in buildings and construction, in automotive and aerospace components, and in environmental remediation and energy storage technologies. In the United States, the National Nanotechnology Initiative that was begun about a decade ago has aided these efforts.

As the name suggests, polymer nanocomposites are polymer-matrix composites but contain materials having at least one dimension below about 100 nm, wherein the small size offers some level of controllable performance that is different from the expectations developed in the macroworld. The notion is that there must be some advantage in achieving the nanoscale whether it is for mechanical reinforcement or for the enhancement of another desirable property. If the nanocomposite (e.g., a polypropylene carbon nanotube composite) were to have identical properties to its macroscopic counterpart (e.g., the same polypropylene reinforced with chopped carbon fiber derived from a polyacrylonitrile precursor), then there would be little point in developing nanocomposites in the first place. The challenge, then, is to create structures either by whittling down large features or by engineering atoms such that materials and devices have novel mechanical, chemical, electrical, magnetic, or optical properties.

The first nanoclay composite, in which silicates were used as a means of influencing the macroscopic properties of the composite, was described in a patent from the National Lead Company in 1950, which describes the use of clays to reinforce elastomers.[2] Yet, very little commercial activity proceeded from this patent, and it would be nearly four decades later before Toyota (Okada et al.) would patent a nanoclay-polyamide system in 1988, which was probably the first time that the molecular layering of silicates was recognized to be a key in transferring nano-properties to the macrocomposite.[3] Thus, this was a true polymer nanocomposite and can be regarded as a key milestone in the modern nanocomposite era. Following this development, Toyota launched the first commercial automotive application of polymer nanoclay composites, with a Nylon-6 timing belt cover in 1993.[4-6] By 2001, Toyota was producing body panels and bumpers containing nanoclays. Similarly, General Motors began using nanoclay composites for step assists on its GM Safari and Chevrolet Astro models in 2002.

Special mention should be made of carbon-based nanomaterials. Prior to 1980, it was thought that only two allotropes of carbon existed: (1) the diamond lattice and (2) the graphite lattice. A single plane of densely packed sp2-bonded carbon atoms arranged in a hexagonal close-packed configuration is referred to as graphene. Graphite, then, consists of multiple layers of graphene. However, by 1996 when the Nobel Prize in Chemistry was awarded to Robert F. Curl Jr., Sir Harold W. Kroto, and Richard E. Smalley, it was clear that at least a third allotrope existed—the so-called "buckyball" or C-60 atom. The nickname derives from the similarity between C-60 of the geodesic dome structures designed by the architect Buckminster Fuller. This can be considered a watershed event in the history of nanomaterials, as it catalyzed a flurry of research on nanotechnology in various fields, not only materials science. Later, other carbon molecules were discovered, such as C-72, C-76, C-84, and even as high as C-100.[7] Later, as a thought experiment at least, it was suggested that single-walled carbon nanotubes (SWNTs or SWCNTs) might be considered as very large elongated fullerenes.[8] A SWCNT consists of a single graphene layer wrapped into a tubular shape. In practice, however, these nanomaterials are grown with the aid of a metal catalyst particle, and thus the identification of nanofilaments with fullerene molecules is mainly a theoretical one.

Multiwalled carbon nanotubes (MWNTs or MWCNTs) consist of several layers of graphene wrapped into tubular shapes. Carbon nanofibers (CNFs) are considered to consist of layers of truncated conic sections or "stacked cups" of graphene.[9] Some variants of such nanomaterials had semiconductor characteristics, whereas others were electrically and thermally conductive, similar to metals. Potentially, these materials could have an enormous range of applications, including nanoelectronic devices, pharmaceuticals, and catalyst supports. Putting aside for the moment the practical objections of cost (about a million dollars per pound in the early days, except that no one could produce as much as a pound), the reported mechanical properties led materials scientists to wonder if practical composites could be made with nanoreinforcements. Individual nanotubes were said to offer strength, modulus, and strain values many times greater than those of steel.

Thus, from the standpoint of compounding polymer nanocomposites, filamentary carbons seemingly offer the prospect of significantly influencing mechanical properties. Initial attempts at fabricating polymer nanocomposites, however, did not result in the expected level of performance. Indeed, the nanocomposite properties were often inferior to the neat polymer, causing materials scientists to revisit their textbooks and to study the problems such as the material interface at the nanoscale as well as agglomeration and dispersion of nanoscale additives. In particular, polymer matrix materials must bond to the graphene surface of nanotube fillers. However, the surface energy of nanotubes is usually very low (e.g., analogous to Teflon®). Functionalizing the surface may provide a means to improve bonding to nanotubes, although the functional groups would presumably damage the graphene lattice. Consequently, the mechanical properties of carbon nanocomposites have not necessarily proven exceptional as of the first decade of the twenty-first century. Yet other properties may be just as important.

Electrically and thermally conductive polymer composites can be of interest for many niche applications. For example, even their weak electrical conductivity can render polymers suitable for electrostatic paint spraying, resulting in a less-expensive and an environmentally attractive process due to less wasted paint and elimination of the need for a primer coat. Alternatives such as chopped microfiber composites are not always viable due to the effect of the fibers on surface finish. Surface finish is often very important for automotive applications, obviously. Hyperion Catalysis, Inc., was one of the first producers of multiwalled nanotubes (MWNT) to bring nanocomposites to commercial status, based on the electrostatic paint spray application, as Ford Motor Company introduced MWNT nanocomposites in mirror housings on the 1998 Ford Taurus.[10] Thus, like their nanoclay cousins, carbon nanotube composites found early commercial success in the automotive industry, at a time when many scholarly researchers were unaware that these materials had been reduced to commercial practice. This may be partly due to the innocuous trade name (Fibrils™) used by Hyperion to market its material.

1.2   Organization of the Book

This book is divided into five main sections:

Section 1: Overview
Section 2: Nanomaterials and Surface Treatment
Section 3: Processing
Section 4: Structure Characterization
Section 5: Properties

1.2.1   Section 1: Overview

Chapters 1 and 2 are of an introductory nature. Chapter 2 reviews the history of carbon nanofilaments of different types from Gary Tibbetts, who at General Motors Research was an early pioneer in nanosynthesis, with the intention of producing low-cost commercial reinforcements for the automotive industry.

1.2.2   Section 2: Nanomaterials and Surface Treatment

Chapters 3 through 5 are concerned with surface treatment. In Chapter 3, Henry Ashton of Schneller Corporation provides an introduction to the incorporation of nanomaterials of different types into polymer media. In the case of the hydrophilic nanoclays, one needs to coat the clay surface with alkylammonium or other organic cations that are compatible with hydrophobic polymers. The coatings, however, can be thermally unstable and can decompose at the processing temperatures of some polymers. Solving this problem remains a challenge for the future. Chapter 4, authored by Kwang-Jea Kim and James L. White of the University of Akron, discusses the key issues of nanoparticle dispersion and reinforcement, including the effect of surface modification and additives such as chemical coupling of silanes on silica surfaces and its effect on the reduction of silica agglomerate size. Coupling conditions such as temperature, moisture level in the silica, and conditions affecting the particle dispersion are also introduced. Chapter 5 is particularly devoted to the modification of carbon nanoreinforcement surfaces to promote bonding with polymer matrix materials, and it is written by Max Lake and Jerry Glasgow, two pioneers in the development of carbon polymer nanocomposites.

1.2.3   Section 3: Processing

Chapters 6 through 11 deal with issues of processing nano-additives into polymer melts. Although the right chemistry is very helpful in facilitating the dispersion of nanofillers into polymers, extruders and other mixers must necessarily be used for the manufacture of polymer nanocomposites. The practical issues of compounding of layered silicate nanocomposites with thermoplastics using a twin-screw extruder are introduced by Paul Anderson of Coperion in Chapter 6. In Chapter 7, authored by Kwang-Jea Kim and James L. White of the University of Akron, a parallel treatment is provided to the processing of various nano-sized particles into elastomers using various processing instruments with various processing conditions. Chapter 8, on nanocomposite rheology and written by Subhendu Bhattacharya, Rahul Gupta, and Sati Bhattacharya of RMIT University in Australia, reviews what is known about the flow behavior, especially the dynamic mechanical behavior of different polymer and filler combinations. This information is needed for the successful conduct of subsequent shaping operations.

Chapter 9 introduces the fundamentals of polymer carbon nanocomposites, written by Enrique Barrera of Rice University. Barrera was one of the first persons to investigate the use of true nanotubes in polymer composites. Chapter 10, by Tatsuhiro Takahashi and Koichiro Yonetake of Ymagata University, recognizes the importance of controlling the physical alignment of nanotubes, and thus provides specialized insight into the use of electromagnetic fields to align carbon nanotubes within polymer nanocomposites.

Chapter 11, authored by Chang H. Song and Avraam I. Isayev of the Institute of Polymer Engineering at the University of Akron, explores the formation of nanofibrillar structured liquid crystal polymer (LCP) dispersed in polyester matrices during processing of unidirectional sheets and fibers. Polyester blends and operating conditions were examined to determine the main factors affecting the tensile strength and Young's modulus.

1.2.4 Section 4: Structure Characterization

Although a variety of techniques are employed to characterize the structure of polymer nanocomposites, x-ray diffraction and transmission electron microscopy (TEM) are the most common ones encountered. TEM, in particular, is the only method that actually allows one to see the microstructure. It is, however, a very tedious and labor-intensive technique. Vin Berry of West Virginia University explains the fundamentals of electron microscopy and its different variations in Chapter 12.

1.2.5 Section 5: Properties

Chapters 13 through 18 are devoted to the specialized properties attained from both clay-based and carbon-based nanocomposites. First, Chapter 13, authored by Leszek Utracki of the National Research Council of Canada, discusses the mechanical properties obtained from nanoclay polymer composites. Note that nanocomposites having the theoretical strength and stiffness of ideal, defect-free nanocomposites are now a reality.[11] Chapter 14, by Daniel De Kee and Kyle Frederic of Tulane University, describes mass transport issues through polymer nanocomposites. Controlling mass transport is one of the main drivers for considering nanoclay additives.

Similarly, flammability reduction is another objective with great commercial importance. Properties related to flammability and their potential enhancement with nanoclay additives are discussed in Chapter 15 by Charles Wilkie of Marquette University and Jin Zhu of YTC America, Inc.

Chapter 16, written by Elliot Kennel of West Virginia University, discusses the enhanced electrical properties of carbon nanocomposites. Chapter 17, which discusses the enhancement of thermal conductivity using nanoadditives, was written by Sushant Agarwal and Rakesh Gupta of West Virginia University.

Chapter 18 discusses specific niche nanocomposites. This chapter, by Chang-Kook Hong of Chonnam National University in Korea, and by Jue Lu and Richard P. Wool of the University of Delaware, delves into bio-nanocomposites from plant oil.

1.3 The Challenge

When dealing with blends and composites, one finds that improvement in one property comes at the expense of another property. Thus, the addition of an elastomer to a glassy polymer improves the impact strength but reduces the stiffness. In the case of polymer nanocomposites, however, one has the possibility of improving several properties simultaneously without degrading any key property and that too at only 3 to 5 wt% of added nanofiller. To fulfill this promise, however, one must be able to control both the dispersion of the filler particles and the interfacial interaction between the filler and the matrix. The current inability to accomplish these twin tasks is the reason for the slow progress in the commercialization of polymer nanocomposites. Rapid advancements are likely to occur only if scientific understanding is used to guide the future development of the technology. We hope that this handbook will contribute to this effort.

References

1. Richard P. Feynman, There's Plenty of Room at the Bottom, *American Physical Society Meeting*, Pasadena, CA, December 1959. *See also* Anthony J.G. Hey and Richard P. Feynman, *Feynman and Computation*, Westview Press, 2002.
2. Lawrence W. Carter, John G. Hendricks, and Don S. Bolley, U.S. Patent 2,531,396, November 28, 1950; submitted March 29, 1947.
3. Akane Okada, Yoshiaki Fukushima, Masaya Kawasumi, Shinji Inagaki, Arimitsu Usuki, Shigetoshi Sugiyama, Toshio Kurauchi, and Osami Kamigaito, Composite Material and Process for Manufacturing Same, U.S. Patent 4,739,007, April 19, 1988.
4. A. Osaka and A. Usuki, The chemistry of polymer-clay hybrids, *Mater. Sci. Eng.*, C3, 109, 1995.
5. J.S. Shelley and K.L. Devries, Degradation of Nylon-6/Clay Nanocomposites in NO(x), United States Air Force Report, National Technical Information Service ADA409288, April 24, 2000.
6. A. Okada and A. Usuki, Twenty years of polymer-clay nanocomposites, *Macromol. Mater. Eng.*, 291, 1449, 2006.
7. Y.L. Voytekhovsky and D.G. Stepenshchikov, C72 to C100 fullerenes: combinatorial types and symmetries, *Acta Cryst.*, A59, 283, 2003.
8. Sumio Iijima, Helical microtubules of graphitic carbon, *Nature*, 354, 56, 1991.
9. Yoong-Ahm Kim, Takuya Hayashi, Satoru Naokawa, Takashi Yanagisawa, and Morinobu Endo, Comparative study of herringbone and stacked-cup carbon nanofibers, *Carbon*, 43, 3005, 2005.
10. Gary S. Vasilash, Automotive Design and Production, Product & Process Improvement Because It's Right, August 1998.
11. P. Podsiadlo et al., Ultratrong and stiff layered polymer nanocomposites, *Science*, 318, 80, 2007.

2

History of Carbon Nanomaterials

Gary G. Tibbetts

CONTENTS

2.1   Carbon Nanofilaments

Carbon nanofilaments may be formed over a broad range of conditions by carburizing atmospheres in contact with metallic impurities. In fact, carbon nanofilaments may even predate the formation of the Earth, as they seem to be a minor constituent of interplanetary dust! Chondritic porous (unmelted) micrometeorites believed to be representative of the primordial cloud were collected from the stratosphere and examined by Bradley et al.[1] TEMs (transmission electron micrographs) clearly showed carbon nanofibers resembling those produced under laboratory conditions growing from iron-nickel alloy particles (Figure 2.1).

Carburizing gases in contact with metallic particles have been utilized by many human technologies, but documented observations of nanofiber-related structures began in the nineteenth century. One clear example was the 1889 U.S. patent of Hughes and Chambers, describing "hair-like carbon filaments" grown at high temperature by flowing a hydrogen/methane feedstock through iron crucibles.[2] This is a good description of carbon filaments thickened with vapor-deposited carbon. However, unraveling this multi-step growth process was too formidable a task for nineteenth-century technology, so this process was not practicable for producing its intended product: electric light bulb filaments.

In 1890, Schultzenberger observed a felt-like material comprised of very fine, long filaments deposited in a red-hot porcelain tube after the decomposition of cyanogen, $(CN)_2$, over "Cryolith" powder, a sodium-aluminum fluoride glass comprised of AlF_2, NaF, and Al.[3] The material, probably a mat of slightly thickened nanofibers, deposited a pencil-like line when rubbed against paper.

Over the next 60 years, what we know about carbon nanofiber growth was inferred by the study of similar structures: carbon nanofibers thickened to visible dimension by vapor deposition of carbon. By 1948, Iley and Riley could study the production of carbon fibers of centimeter length and a diameter of many micrometers from pure hydrocarbon gases such as methane.[4]

But the invention of the transmission electron microscope gave Davis, Slawsen, and Rigby in 1953 the first view of the truly microscopic nature of the carbon nanofibers

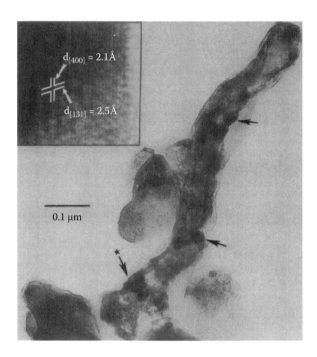

FIGURE 2.1
TEM showing a carbon nanofilament with included magnetite particles (arrows) obtained from a chondritic meteorite by Bradley et al. (ref 1) . The insert shows a lattice fringe image of an embedded crystal.

themselves.[5] In studies on CO decomposition on blast furnace brickwork, they observed twisted bundles of 10- to 200-nm nanofibers formed at 450°C (Figure 2.2). Their work first spotlighted the role of iron catalyst particles, "formed originally as specks on the surface of the iron oxide-each speck giving rise to a thread of carbon…. We suggest that the particles of catalyst are located on the growing ends of the thread."

At this time, studies of carbon nanofibers bifurcate into two strands of research. One group studied the growth of "filamentous carbon," viewing it as a pesky corrosion product associated with the decomposition of carbon-containing gases. Perhaps because it is easier to avoid CVD (chemical vapor deposition) using CO as a feedstock rather than hydrocarbons, these researchers concentrated on the growth of nanofibers from CO. Meanwhile, other groups continued studying the macroscopic CVD-thickened fibers, not always recognizing that a carbon nanofiber or "filament" was the core and template for each macroscopic fiber.

A large literature describing the growth of carbon filaments developed after 1953. For example, Hofer, Sterling, and McCartney (1955) compared the morphologies of filaments grown on Fe, Ni, and Co.[6] They introduced themes that would be developed by many authors over the next 50 years: determining whether the filaments were hollow or solid, twisted or straight, and what was the composition (i.e., metal or carbide), morphology, and position of the catalyst particles within the filament.

Mayer (1957) described the growth of "filamentary graphite crystals" about 0.1 mm thick on a carbon fiber heated to 2000°C in acetylene.[7] Influenced by the then-popular study of metal whiskers (Hardy, 1956), Meyer attributed the growth of 1-mm long "whiskers" to

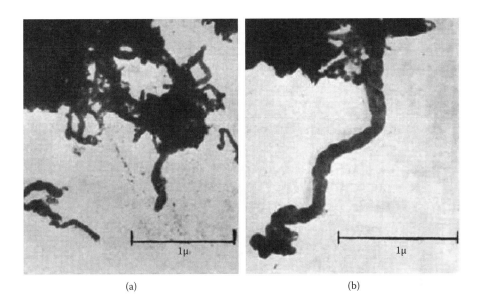

FIGURE 2.2
TEM of nanofibers observed by Davis et al. (ref 5) growing on blast furnace brickwork.

"nucleation of faults" at the base leading to diffusion of carbon to the high-energy boundary between the "whisker" and matrix (Figure 2.3).[8]

By 1958, Hillert and Lange had developed a better understanding of how these macroscopic fibers grew.[9] "Graphite filaments were *unintentionally* produced from an atmosphere of N_2 saturated with n-heptane…led through a quartz tube containing some pieces of iron." A large number of different structures grew at 1000°C, including submicron hollow filaments of various morphologies that "attached to dense, conical particles, assumed to be

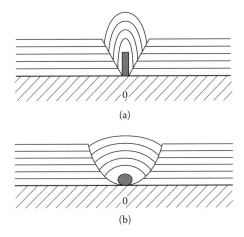

FIGURE 2.3
Diagram from Mayer (ref 7) showing growth of a carbon whisker from fault nucleus at its base.

catalyst particles." The macroscopic threads "were formed by a two-stage process. Initially, very thin filaments were formed and grew length-wise with a constant cross section. Later on, this growth was succeeded by a radial growth...."

Also inspired by work on whisker growth, Roger Bacon introduced an utterly novel nanofilament production method in 1960.[10] Studying the deposits that grew on the negative graphite electrode of a high-pressure, direct-current discharge in argon, he found imbedded carbon fibers of diameters ranging from 1 to 5 μm in diameter and of lengths up to 3 cm. His motivation for this technique was to elevate the carbon vapor pressure to near its equilibrium vapor pressure in the high-temperature arc, conditions that would favor whisker formation. Electrical data showed that the fibers were very well crystallized, with room-temperature resistivities less than a factor of 2 above that of single-crystal graphite. X-ray and electron diffraction showed that the exterior planes of the fibers were graphene planes, and examination of an electrically exploded fiber led Bacon to favor the rolled-up scroll morphology rather than the alternative nested coaxial cylindrical morphology (Figure 2.4). The scroll morphology resembled the screw dislocation whisker growth mechanism favored at that time for metallic whiskers.[11] Bacon speculated that fiber growth proceeds when "a very thin graphite sheet or ribbon coils itself up...in order to reduce its surface energy. If the resulting scroll or cylinder is properly oriented, with its axis parallel to the general growth direction, then it can grow rapidly in the direction of increasing length, while thickening by tangential growth in spiral fashion."

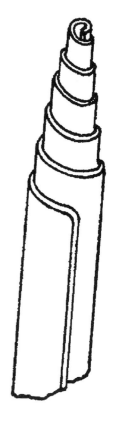

FIGURE 2.4
Bacon's diagram (ref 10) showing scroll growth morphology of carbon filament.

It seems clear that the ability of these early workers to find a suitable growth mechanism had been limited by the available theory. Burton, Cabrera, and Frank described growth at a screw dislocation in 1951; it provides a self-perpetuating nucleation site of lower energy than simple two-dimensional nucleation could.[11] It provided a viable explanation in cases where no catalytic particle was visible, and, because these particles can be of nanometer size, they are frequently difficult to discern.

In 1965, Wagner and Ellis introduced the VLS (vapor liquid solid) mechanism describing crystal growth from *vapor*-phase constituents dissolving in a *liquid* "catalyst" before precipitating as a *solid*.[12] Their article showed that they had mastered the growth of Si crystals from $SiCl_4$ gas using Au, Ag, Ni, Cu, or Pt catalyst particles (Figure 2.5). The mechanism seems amenable to explaining nanofiber growth as it can deposit long and slender crystals resembling whiskers, but it requires a liquid catalyst droplet and was capable of growing crystals more than 100 μm in diameter. Neither of these observations seemed to be characteristic of carbon nanofiber growth.

For workers at the Harwell Atomic Energy Research Establishment in the United Kingdom, filamentous carbon growth in nuclear fuel cladding might be avoided by altering the metallic composition so that active catalyst particles would not be formed. Of course, that required extensive comparative studies of how particles of different metals and sizes grew filaments. By the 1970s, R.T.K. Baker was showing conference audiences vivid *in situ* controlled atmosphere TEM video recordings of filament growth, which undoubtedly

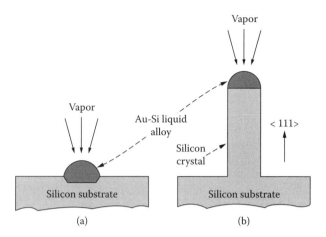

FIGURE 2.5
Vapor Liquid Solid growth mechanism from Wagner and Ellis (ref 12) showing catalyzed growth of Si crystals from the vapor phase.

promoted interest in filamentous carbon. This method allowed direct measurement of growth rates that could be used to compare the effectiveness of different catalytic particles. In a 1973 article, Baker et al. showed that the activation energies for filament growth on Fe, Co, and Cr were close to those for the diffusion of carbon through these metals.[13] Furthermore, the dependence of growth rate on the square root of catalyst diameter was also consistent with diffusion-controlled growth. This work, along with an extensive historical review of filamentous carbon formation, is nicely described in Baker's 1978 book chapter.[14]

Baker et al. (1972) introduced an influential conceptual model of how carbon filaments grew in acetylene atmospheres (Figure 2.6).[15] The large enthalpic energy released on acetylene decomposition could establish a thermal gradient driving carbon atoms across the catalytic particle toward the filament face where they precipitate. Much later (1987), Tibbetts et al. calculated that the temperature gradient across the small, conductive particle would be far less effective than the carbon concentration gradient in driving diffusion.[16]

In a rather ironic twist, Baker and colleagues began working in the 1980s on promoting rather than inhibiting the growth of filaments as it became clear that these structures had a promising future as a reinforcement and conductive filler for polymers. He designed an optimal binary Cu-Ni catalyst material effective at growing graphitic filaments at near 600°C. Other workers thought that the 7-μm macroscopic fibers might be useful in composites and directed their studies toward practicable production of these "vapor-grown" carbon fibers. Ground-breaking work was performed in the Soviet Union, Japan, France, and the United States. Although these fibers are generally at the 10-μm diameter level, and not, strictly speaking, nanomaterials, those who wished to grow these vapor-grown carbon fibers (VGCFs) eventually realized that nanoscale filamentous carbon was the critical template for the formation of the micron-scale fiber, and were able to gain considerable information about the conditions under which these filaments grew.

In 1970, Tesner et al. published kinetic data on the growth of carbon fibers on nichrome wires heated to 450 to 600°C in acetylene/hydrogen or acetylene/nitrogen mixtures.[17] They directly measured the mass increase of these wires as a function of temperature and

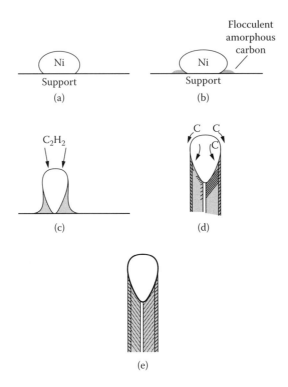

FIGURE 2.6
Model of Baker et al. (ref 15) for the growth of a carbon filament from a Ni catalyst particle driven by the enthalpic energy of the decomposing acetylene.

reaction time and showed that yields were largest at high temperatures and high acetylene concentration. The rate of mass deposition peaked just 1 to 3 minutes into growth, indicating to the authors that a "catalytic activation of the surface with a creation of active centers" was necessary for filament growth.

In Japan in 1972, Koyama grew good yields of 3- to 80-µm fibers up to 25 cm long from benzene/hydrogen mixtures (Figure 2.7), with a clear nested cylindrical structure (Figure 2.7A).[18] The article does not mention catalyst particles, so they appear to be adventitious. He was able to show that thinner fibers had strengths up to 3 GPa. Koyama and Endo were able to produce large amounts of vapor-grown carbon fibers by thermal decomposition of benzene near 1200°C.[19] Being an electrical engineer by training, Endo studied the electrical conductivity and magnetoresistance of these fibers, and showed that they are extremely graphitizable, approaching the electrical conductivity of graphite after they were heat treated to over 2800°C.[20,21]

In 1976, Oberlin, Endo, and Koyama delineated in more complete fashion the relationship between carbon filaments and VGCF.[22] They used TEM studies of incompletely thickened VGCF to show that the central tube varied from 2 to 50 nm in diameter. The exterior layer could be initially deposited in rather unexpected forms, including spindle-shaped initial layers. Ultimately, the exterior layer thickened to become turbostratic carbon, similar to vapor-deposited carbon with poorly crystallized graphene planes deposited cylindrically around the central axis. The central tubes were more crystalline and stronger than

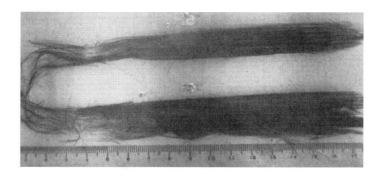

FIGURE 2.7
Photograph of skeins of vapor-grown carbon fibers produced by Koyama (ref 18) showing a length of many cm.

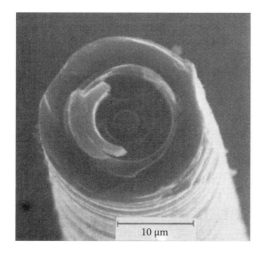

FIGURE 2.7A
SEM shows nested cylindrical structure of the fibers.

the vapor-deposited material (Figure 2.8). Although not recognized at the time, that article's Figure 11 may be the first published photo of a single-walled nanotube. Much of that article concerns the morphology and composition of the catalytic particles, which their post mortem showed to be Fe_3C. This was naturally a topic of great interest in promoting lavish growth of VGCF. As for the growth mechanism of the hollow cored filament, Oberlin and Endo searched for and rejected the supposition of a specific (carbide) catalyst orientation whose anisotropic carbon deposition could account for the hollow core. They found Baker's hypothesis of temperature-driven diffusion of carbon to the cool side of the catalytic particle less convincing than Baird et al.'s notion that carbon islands migrate to the catalyst particle/filament interface where they are fixed in position by contact angle energetics (Figure 2.9).[23]

At General Motors Research Labs in the United States, studies on the production of VGCF began with the accidental growth of these materials in an apparatus designed to measure the diffusion of carbon through steels near 1000°C.[24] Hoping to move toward

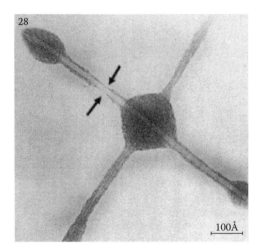

FIGURE 2.8
TEM from Oberlin et al. (ref 22) of a central carbon filament or nanotube in the initial stages of CVD thickening to become a vapor-grown carbon fiber

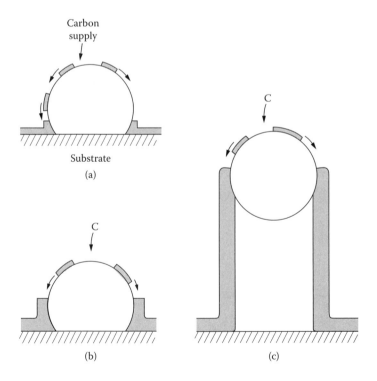

FIGURE 2.9
Filament growth model proposed by Baird et al. (ref 23) showing surface diffusion of active hydrogen-carbon species across the catalyst particle before the carbon is incorporated in the walls.

industrialization of VGCF, this group published data that showed how the lengthening and thickening process under methane atmospheres could be made disparate.[25] Studies of filament lengths achieved at different methane concentrations on iron particles indicated that the rate of methane decomposition controlled the filament growth rate. For iron-based particles, it seemed likely that filament growth proceeded from the austenetic phase supersaturated with carbon (Tibbetts et al., 1987), and that transition to the carbide phase results in particle deactivation.[16] Because active catalyst particles are generally in a liquid-like state and only crystallize on cooling after their catalytic effect is lost, the search for the most active catalyst crystal orientation is chimerical.

By carefully studying histograms of VGCF lengths, further work clarified the random mechanism that limited the length attained by the fibers before poisoning the growth catalyst by pyrolysis products.[26] More complete optimization studies for this system showed that the number of fibers grown was also limited by the sintering of the 12-nm magnetite particles used.[27] As this VGCF research matured, it became clear that industrial production of very long CVD-thickened nanotubes was impractical due to the tendency of the nanofibers to submerge in the CVD layers deposited on the tube and substrate during CVD thickening. Calculations showed that vastly more nanofibers were initially formed than ultimately survived as macroscopic VGCF. In a 1984 publication entitled "Why are Carbon Filaments Tubular?," Tibbetts published a diagram of a hollow-cored MWNT (Figure 2.10) along with a free energy calculation showing why the elastic energy required to precipitate structures comprised of coaxial graphene tubes mandates a hollow core.[28]

Benissad et al. in 1988 used design-of-experiment methods to extract some optimum conditions for fiber growth and related them to the melting points of small iron particles to argue that filament growth proceeded from liquid catalytic particles.[29, 30]

A more practical approach for bulk production was to concentrate on *in situ* production of catalyst nanoparticles and coax them into growing nanotubes so efficiently that the nanofibers themselves might be harvested for composite fabrication. This approach was being pioneered in the United States by Howard Tennant and colleagues at Hyperion Catalysis (fibrils.com). Hyperion had developed a two-step process based on nanometer-sized catalyst particle preparation, followed by growth on the nanoparticles; this work was published in the patent literature rather than the scientific literature.[31] Their work utilized alloy nanoparticles that were frequently comprised of three metals, including Mo, which had generally been neglected in previous literature.

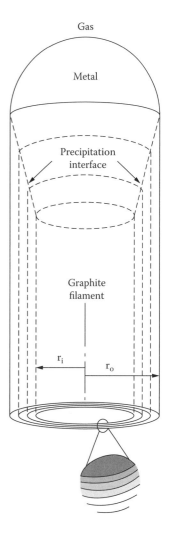

FIGURE 2.10

Tibbetts' diagram (ref 28) of a multi-walled nanotube used in a calculation showing how a hollow core is energetically preferred to avoid interior highly strained cylinders.

In 1985, Endo et al. published a prescription for a reactor that infusion pumped a ferrocene/benzene mixture into a 1100°C reactor to create 0.1-μm fibers "deposited like a sponge" near the reactor outlet.[32] Essential to fiber production was that the ferrocene create numerous 5-nm Fe particles to efficiently catalyze this filament growth. In a 1988 review, Endo diagrammed a vertical tubular reactor that was "more efficient than substrate seeding."[33]

However, there are many problems in really making a commercially feasible reactor for producing kilogram quantities of nanofibers. Utilizing liquid solutions of metallo-organics as in the Endo articles meant that the heat of vaporization of the solvent must be provided within the very volume where sudden heating of the feedstock must occur. This heating is vital to the creation of very small particles that must grow nanofibers before they agglomerate. Tibbetts et al. tried to cope with this problem of using liquid feedstocks in 1991 by fabricating a wire mesh heat exchanger to rapidly evaporate ferrocene/hexane feedstock mixtures and utilizing a nested tube configuration to allow the feedstocks to be injected from the bottom of a vertical reactor.[34] Yet, really efficient fiber production seemed to require another catalytic boost, and this group, inspired by a publication of Katsuki et al. in 1980 showing how the addition of S made VGCF growth on substrates markedly more efficient, utilized a small quantity of H_2S to vastly increase the proportion of catalytic Fe particles that actually grew fibers.[35]

By 1994, Tibbetts et al. had completely abandoned the use of liquid feedstocks and were using $Fe(CO)_5$ as a source of catalyst particles.[36] This liquid has substantial vapor pressure at room temperature and can be incorporated into the feedstock by bubbling gaseous constituents through it. The increased efficiency of fiber formation allowed a closer examination of the role of sulfur (S), and these workers speculated that the S dissolved in Fe to produce FeS, whose 950°C eutectic allowed VLS production of fibers at lower temperatures. However, this vastly improved efficiency brought yet another set of problems in ushering many grams per minute of solid material out of a high-temperature reactor without clogging. Nevertheless, Applied Sciences, Inc., forged a commercial relationship with General Motors and began improving and marketing PYROGRAF carbon nanofibers in kilogram quantities produced with this approximate prescription in 1991.

Because of his extensive work on growth and applications of carbon nanofibers, Endo became the spokesman of this community and frequently was invited to lecture on CNF applications at international meetings.

Meanwhile, the world of carbon nanomaterials had dramatically changed in 1985 due to the discovery of carbon fullerenes.[37] The value of the discovery of the C_{60} molecule of exceptional stability was affirmed by the awarding of a Nobel Prize to Curl, Kroto, and Smalley for this work in 1996. Later developments, including the discovery of stable trans-C_{60} compounds, along with ideas about how pentagons and heptagons could stabilize more complex structures primarily formed from hexagons, gave a new legitimacy to the search for geometrically simple nanostructures.[38] This work was ably supported by the efforts of numerous theoreticians.[39]

Another novelty re-introduced by the fullerene work was the use of arc discharges with carbon electrodes in making new materials. Arc discharge production and the atomic-scale investigative methods used for fullerenes were applied in 1991 by Iijima to produce and examine carbon nanotubes of wall thicknesses as small as two atomic layers.[40] Because catalytic growth was not evident in these nanotubes and TEM indicated that the structure of the MWNT (multiwalled nanotube) was that of coaxial cylinders whose individual layers were rolled up in helical fashion, Iijima speculated that "individual tubes themselves

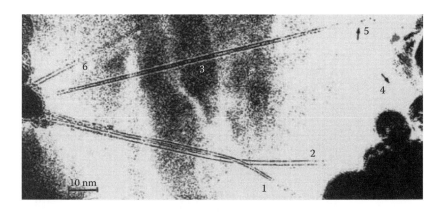

FIGURE 2.11
Iijima and Ichihashi's (ref 41) TEM of single walled carbon nanotubes produced with iron as a catalyst.

can have spiral growth steps at the tube ends." In 1993, Iijima and Ichihashi improved the discharge method by doping the cathode with iron to produce single-walled nanotubes (SWNTs) as thin as 0.7 nm (Figure 2.11).[41] Although they "speculate that the iron in the present experiments acts as a homogenous catalyst in the vapor phase," they neverthe-less believed that "the tubule ends are open so that the carbon atoms are easily captured at dangling bonds, and the multi-shell tubules grow in the direction of the tube axis, and also perpendicular to it." The article used electron diffraction to determine the helicity of the SWNTs and cited Hamada et al.'s band structure calculations anticipating that SWNTs could show semiconducting behavior.[42]

A simultaneous 1993 article by Bethune et al. reported the "formation of carbon nano-tubes which all have very small diameters (about 1.2 nm) and walls only a single atomic layer thick" in an arc discharge using a Co-doped carbon electrode in an atmosphere of He of pressure below 500 Torr.[43] The individual fibers were several micrometers long and formed a "web-like deposit" that "could be peeled off {the walls} in long strips." Nanofiber growth was found to be unique to the 20-nm and larger Co catalyst particles that the dis-charge formed, and was not observed with Fe, Ni, or Ni/Cu doped electrodes.

At Rice University, Smalley's group developed the technique of laser vaporization of a Ni-Co-doped graphite susceptor to produce SWNTs of 1.4-nm diameter with 70% yield. By 1996, Thess et al. had shown how, under these efficient production conditions, up to 500 SWNTs can organize into ropes with a triangular lattice cross-section of spacing 1.7 nm (Figure 2.12).[44] Direct measurement showed these ropes had a metallic conductivity near 10^{-4} Ω-cm. Accounting for the profuse SWNT growth was a difficult problem because suf-ficient metallic clusters of such uniform diameter cannot be produced during the vapor-ization step. But just a single metallic atom might be sufficient to hold a nanotube open as it "scooted" around the periphery of an open tube "helping to anneal away any carbon structures which are not energetically favored." The energetically favored tubes are the (10,10) SWNTs primarily observed.

By 1996, Journet et al. had demonstrated that similar yields and crystalline rope struc-tures could be achieved using a simple He arc discharge with the anode drilled out and doped with powders of Ni, Co, and Y.[45] The best yields, about 2 g including soot, were obtained with a 4:1 mixture of Ni:Y. While it was important to have a two-component catalyst, Y was found to most strongly favor the production of NTs. XRD (x-ray diffraction)

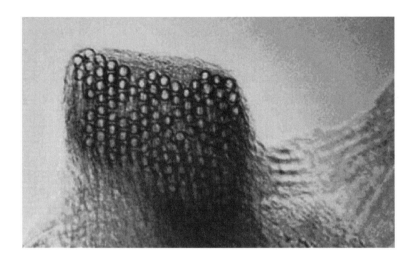

FIGURE 2.12
Profuse production of single walled carbon nanotubes can allow them to crystallize into crystalline ropes, as shown in this TEM from Thess et al. (ref 44).

studies showed that the fiber diameters were near 1.4 nm and compared closely with the laser ablation products generated by Thess et al.[44]

For many applications, such as field emission and the fabrication of papers, producing aligned nanotubes was even more important than efficiency. This problem has been attacked in several ways. In 1996, Li et al. showed that NTs grown on iron particles embedded in mesoporous silica could produce an array of 50-μm long NTs spaced about 100 nm apart.[46] They demonstrated that the aligned channels in the silica actually imposed vertical orientation on the nanofibers by comparing with growth on a flat quartz plate coated with similar Fe particles (Figure 2.12).

Another alignment method was described by Ren et al in 1998.[47] By using hot-filament CVD in plasma, these workers were able to grow dense arrays of carbon nanofibers on Ni coated glass substrates at only 666 C with excellent alignment. Presumably, the nanofibers assume positions along the electric field lines as they grow and are frozen into position as they thicken and become more rigid.

Thus by the late 1990's the fundamental scientific picture of the growth of carbon nanofilaments was becoming clearer. Metal catalysts particles were desirable for abundant fiber production, yet not always visible in the nanofilament itself. Many different papers have recommended many different catalysts, making us suspect that the most desirable metal depends not only on the chemical identity of the metal, but on particle size, or the specific vapor phase conditions. Particle morphology (in the sense of a preferred face on which the nanofilament forms) is not likely to be a factor, as very small particles at high temperatures behave more like liquids. In SWNT growth, there is hardly room for a macroscopic particle to be present: a single atom may be sufficient. The morphology of the nanofilament, whether SWNT, MWNT, or stacked-cup nanofiber depends on the size and identity of the catalyst particle and the local supersaturation of carbon in the gas phase. Further, non-catalytic deposition of carbon from the vapor phase onto a SWNT, MWNT, or carbon nanofilament can create a complex array of different structures from the sub-nanometer to the mm in diameter range, with lengths ranging up to a meter.

References

1. Bradley, J.P., Brownlee, D.E., and Fraundorf, F., Carbon compounds in interplanetary dust: evidence for formation by heterogeneous catalysis, *Science,* 223, 56–58, 1984.
2. Hughes, T.V. and Chambers, C.R., Manufacture of Carbon Filaments. U.S. Patent No. 405,480 (1889).
3. Schulzenberger, P.L., *C.R. Acad. Sci., Paris, Ser.,* 11, 774, 1890.
4. Iley, R. and Riley, H.L., The deposition of carbon on vitreous silica, *J. Chem. Soc. London II,* p. 1362, 1948.
5. Davis, W.R., Slawson, J., and Rigby, G.R., An unusual form of carbon, *Nature,* 171, 756, 1953.
6. Hofer, L.J.E., Sterling, E., and McCartney, J.T., Structure of the carbon deposited from CO on iron, cobalt, and nickel, *J. Phys. Chem.,* 59 1153–1155, 1955.
7. Meyer, L., Graphite whiskers, *Zeit. Kristal.,* 109, 61–67, 1957.
8. Hardy, H.K., The filamentary growth of metals, *Progr. Metal Phys.,* 6, 45, 1956.
9. Hillert, M. and Lange, N., The structure of graphite filaments. *Zeit Krist.,* 111, 24–34, 1958.
10. Bacon, R.J., Growth, structure, and properties of graphite whiskers, *J. Appl. Phys.,* 31(2), 283–290, 1960.
11. Burton, W.K., Cabrera, N., and Frank, F.C., *Phil. Trans. Roy. Soc. London, Ser. A,* 243, 299–358, 1951.
12. Wagner, R.S. and Ellis, W.C., The vapor-liquid-solid mechanism of crystal growth and its application to silicon, *Trans. Met. Soc. AIME,* 233, 1053–1064, 1965.
13. Baker, R.T.K., Harris, P.S., Thomas, B.B., and Waite, R.J., Formation of filamentous carbon from iron, cobalt, and chromium catalyzed decomposition of acetylene, *J. Catal.,* 30, 86–95, 1973.
14. Baker, R.T.K. and Harris, P.S., The formation of filamentous carbon. In Walker, P.L. and Thrower, P.A., Eds., *Chemistry and Physics of Carbon 14.* New York: Dekker, 1978, p. 83–162.
15. Baker, R.T.K., Barber, M.A., Harris, P.S., Feates, F.S., and Waite, R.J., Nucleation and growth of carbon deposits from the Ni catalyzed decomposition of acetylene, *J. Catal.,* 26 51–62, 1972.
16. Tibbetts, G.G., Devour, M.G., and Rodda, E.J., An adsorption-diffusion isotherm and its application to the growth of carbon filaments on iron catalyst particles, *Carbon,* 25, 367, 1987.
17. Tesner, P.A., Robinovich, E.Y., Rafalkes, I.S., and Arefieva, E.f., Formation of carbon fibers from aceetylene, *Carbon,* 8, 435–442, 1970.
18. Koyama, T., Formation of carbon fibers from benzene, *Carbon,* 10, 757, 1972.
19. Koyama, T. and Endo, M., Structure and growth processes of vapor-grown carbon fibers (in Japanese), *O. Buturi,* 42, 690–696, 1973.
20. Koyama, T. and Endom M., Electrical resistivity of carbon fibers prepared from benzene, *Japan J. Appl. Phys.,* 13, 1175–11176, 1974
21. Endo, M., Koyama, T., and Hishiyama, Y., Structural improvement of carbon fibers prepared from benzene, *Japan J. Appl. Phys.,* 15(11) 2073–2076, 1976.
22. Oberlin, A., Endo, M., and Koyama, T., filamentous growth of carbon through benzene decomposition, *J. Cryst. Growth,* 32, 335–349, 1976.
23. Baird, T., Fryer, J.R., and Grant, B., Carbon formation on iron and steel foils by hydrocarbon pyrolysis reactions at 700°C, *Carbon,* 12, 591–602, 1974.
24. Tibbetts, G.G., Carbon fibers produced by pyrolysis of natural gas in stainless steel tubes, *Appl. Phys. Lett.,* 42(8), 666–668, 1983.
25. Tibbetts, G.G., From catalysis to chemical vapor deposition: graphite fibers from natural gas, *Materials Research Society Symposia Proceedings,* 20, Intercalated Graphite, Boston, p. 196, 1984.
26. Tibbetts, G.G., Lengths of carbon fibers grown from iron catalyst particles in natural gas, *J. Cryst. Growth,* 73, 431–438, 1986.
27. Tibbetts, G.G., Growing carbon fibers with a linearly increasing temperature sweep: experiments and modeling, *Carbon,* 1992; 30, 399–406, 1992.

28. Tibbetts, G.G., Why are carbon filaments tubular?, *J. Cryst. Growth,* 66, 632, 1984.
29. Benissad, F., Gadale, P., Coulon, M., and Bonnetain, L., Formation de fibres de carbone a partir du methane. I. Croissance catalytique et epaississment pyrolytique, *Carbon,* 61,. 26, 1988.
30. Benissad, F., Gadale, P., Coulon, M., and Bonnetain, L., Formation de fibres de carbone a partir du methane: germination du carbone et fusion des particules catalytiques, *Carbon,* 61, 425–432, 1988.
31. Tennant, H.G., Barber, J.J., and Hoch, R., U.S. Patent No. 5,165,909, November 24, 1992, and references therein.
32. Endo, M., Shikata, M., Momose, T., and Shiraishi, M., Vapor grown carbon fibers obtained by fluid ultra fine catalytic particles, *17th Biennial Carbon Conference Proceedings,* Lexington, KY, American Carbon Society, 1985, p. 295–296.
33. Endo, M., Grow carbon fibers in the vapor phase, *Chemtech,* September, 568–576, 1988.
34. Tibbetts, G.G., Gorkiewicz, D.W., and Alig, R.L., A new reactor for growing carbon fibers from liquid and vapor phase hydrocarbons, *Carbon,* 31(5), 809–814, 1993.
35. Katsuki, H., Matsunaga, K., Egashira, M., and Kawasumi, S., Formation of carbon fibers on some sulfur-containing substrates, *Carbon,* 19, 148–150, 1981.
36. Tibbetts, G.G., Bernardo, C.A., Gorkiewicz, D.W., and Alig, R.L., Role of sulfur in the production of carbon fibers in the vapor phase, *Carbon,* 32(4), 569–576, 1994.
37. Kroto, H.W., Heath, J.R., O'Brian, S.C., Curl, R.F., and Smalley, R.E., *Nature,* 318, 1652, 1985.
38. Ebbesen, T.W., Carbon nanotubes, *Physics Today,* June, 26–31, 1996.
39. Charlier, J.-C., De Vita, A., Blasé, X., and Car, R., Microscopic growth mechanisms for carbon nanotubes, *Science,* 275, 646–649, 1997.
40. Iijima, S., Helical microtubules of graphitic carbon, *Nature,* 354, 56–58, 1991.
41. Iijima, S. and Ichihashi, Y., Single-shell carbon nanotubes of 1 nm in diameter, *Nature,* 363, 603–605, 1993.
42. Hamada, N., Sawada, S., and Oshiyama, S., New one-dimensional conductors: Graphitic microtubules, *Phys. Rev. Lett.,* 68, 1579–1581, 1992.
43. Bethune, D.S., Klang, C.H., deVries, M.S., Gorman, G., Savoy, R., Vazquez, J., and Beyers, R., Cobalt-catalyzed growth of carbon nanotubes with single-atomic layer walls, *Nature,* 363, 605–607, 1993.
44. Thess, A., Lee, R., Nikolaev, P., Dai, H., and Smalley, R.E., Crystalline ropes of metallic carbon nanotubes, *Science,* 273, 483–487, 1996.
45. Journet, C., Maser, W.K., Bernier, P., Loiseau, A., Lamy de la Chapelle, M., Lefrant, S., Deniard, P., Lee, R., and Fischer, J.E., Large-scale production of single-walled nanotubes by the electric-arc technique, *Nature,* 388, 756–758, 1997.
46. Li, W.Z., Xie, S.S., Quian, L.X., Chang, B.H., Zou ,B.S., Zhou, W.Y., Zhao, R.A., and Wang, G., Large scale synthesis of aligned carbon nanotubes, *Science,* 274, 1701–1703, 1996.
47. Ren, Z.F., Huang, Z.P., Xu, J.W., Wang, J.H., Bush, P., Siegal, M.P., and Provencio, P.N., Synthesis of large arrays of well-aligned carbon nanotubes on glass, , *Science,* 282, 1105–1107, 1998.

3

The Incorporation of Nanomaterials into Polymer Media

Henry C. Ashton

CONTENTS

3.1 Introduction

3.1.1 Introduction

The advent and increasing availability of nanomaterials offers the promise of developing new polymer composite materials with hitherto unseen properties. Nevertheless, the development of these materials has been frustrated by difficulties in incorporation and dispersion of these materials into polymer media. This chapter addresses some of the underlying concepts and offers an insight into this problem.

With the increasing development and availability of nanomaterials has come the promise of new and enhanced materials that can find use in a wide variety of applications that include electronics, pharmaceuticals, medicine, and materials science. One area of particular interest is the development of polymer nanocomposites. Materials formed from a combination of inorganic nanomaterials and organic polymers are a merger between organic and inorganic materials and offer properties that are representative of both components. In this way they have often been referred to as hybrid materials. Nevertheless, traditional composite materials often exhibit localized heterogeneity that can lead to the loss of desired properties. It is fair to say that even with inorganic particles of the order of a cubic micron in size, the concept of true hybridization at the molecular level is not obtained.

With the advent of nanomaterials comes the possibility of increasing the interaction between the organic and inorganic phases by several orders of magnitude.

To put this statement in perspective, the diameter of a carbon atom is about 140 picometers (1 picometer (pm) = 10^{-12} m) or 14Å (Angstroms). Therefore, seven carbon atoms side by side would describe a length of approximately 1 nm (nanometer). This is considerably smaller than any linear dimension that would tend to be associated with a polymer molecule. Hence, a case can be made that the interaction of nano-sized particles with a polymer molecule is truly a molecular-level interaction. A consequence of this fact is that the chemical surface reactivity as well as bulk mixing dynamics would be expected to influence the incorporation of nanomaterials into a polymer medium.

3.1.2 Polymer Nanocomposites

Nanocomposites are composites in which the interphases dominate the composite properties because of the very small size of the components—100 nm or less.

3.2 The Polymer-Nanomaterial Interface

The polymer-nanomaterial interface is a key determinant of nanocomposite properties. To address the subject of the polymer-nanomaterial interface, some simple calculations are useful:

- For the sake of simplicity, consider a 1-μ cube. The surface area equals 6 μ^3.
- Expressed in nanometers, this equals 6×10^6 nm^2.
- If the cube were to fracture into cubes of 1-nm length, then the total surface area of one of the cubes generated would be 6 nm^2.

- The total number of cubes generated would be 10^9. Therefore, the total surface area of the generated cubes equals 6×10^9 nm^2, representing an increase of the total surface area by a factor of 10^3.

The factor for the real-life situation may be somewhat less than 10^3, but the important point is that the available surface area increases by orders of magnitude, not incrementally, as we move from a conventional or micron-sized material to an equivalent weight of nano-sized material. Given such an increase in the interfacial area, it is reasonable to expect that there would be significant increases in the interfacial drag upon incorporation of these materials into polymer media, resulting in difficulties in mixing. This is, in fact, what is observed when nanoclay precursors are added to a polymer matrix. As the nanomaterial precursors wet-out and separate into tactoids, there is an increase in viscosity. As these tactoids separate or exfoliate into individual platelets, the viscosity can increase by orders of magnitude.[1]

[*Note:* previous reports have indicated that nanomaterials incorporate into polymer materials in no different a way than into conventional materials. A likely explanation for this observation is that there has been little or no exfoliation and the material is essentially dispersed into micron-sized particles. In such a case, incremental property increases would be expected but not the anticipated high level changes that would be expected with exfoliation.]

Definition:

Exfoliation is a term generally used to describe the dispersion of individual clay platelets derived from nanoclay materials into another medium—usually an organic polymer. The term *nanoclay* is often loosely used when what is in fact meant is a nanoclay precursor that under appropriate conditions will delaminate into the individual constituent plates. The term *nanoclay* is applied to these platelets as at least one of the dimensions of the clay is on the nanoscale. In the case of Montmorillonite, the clay platelets are approximately 1 nm in thickness. Other dimensions can be of the order of 150 nm, up to as large as 1 or 2 μ.[2] This property of exfoliation is unique to nano-sized clays and must be considered along with dispersion and distribution when characterizing the clay-based nanocomposites.

The preceding calculations do not take into account the chemistry that may take place at the interfacial surfaces of the polymer and nanomaterial. Chemical surface modification of conventional fillers to achieve compatibility with a polymer matrix is a well-documented art. The same concepts can be applied to nanomaterials, but there are some challenges that are unique to these materials. The actual process of functionalization itself is far from facile. It makes sense to look for surface chemical reactions that are thermodynamically and kinetically favored in order to accomplish this.

3.3 Why Nanomaterials?

3.3.1 Mechanical Properties

The challenge to produce stronger, tougher, lighter-weight materials continues apace, driven by demands for property improvements, economics, and material availability. The advantages that have been cited to date are that nanocomposites offer similar or improved properties at significantly lower filler loading levels than would be the case with conventional filler materials. The addition of nanomaterials to polymers to produce composite materials significantly impacts mechanical and barrier properties, flame retardancy, and electrical conductivity.[3]

In applications where structural integrity is required, nanocomposites show increases in tensile strength, heat deflection temperature, stiffness, and toughness. Nanocomposites can slow the transmission of gases and moisture vapor as a consequence of their exceedingly high surface area-to-volume ratio. This renders them able to create a tortuous path for gas molecules. Nanoclays and nanotalcs are favored in such cases because of their platy structure. Applications such as food packaging, beer bottles, and tire innerliners have made use of nanomaterials to improve impermeability to gases. The potential exists to improve vapor barrier properties, leading to lower emissions in applications such as automotive fuel system components such as hoses, seals and gaskets, PET bottles, and packaging film.[4]

3.3.2 Fire Retardant Applications

In fire-retardant applications, the use of nanomaterials is becoming increasingly more prevalent. While there are a relatively small number of commercialized fire-resistant materials containing nano-sized materials, there is a considerable amount of work being done. Work is underway at the Building Fire and Research Laboratory (BFRL) at the National Institute of Standards and Technology (NIST) to develop materials with unidirectional heat properties that take advantage of the fact that the thermal conductivity of carbon nanotubes (CNTs) is at least three orders of magnitude greater along the tube axis than other materials used to make protective fabrics for clothing.[5] Furthermore, CNT materials are highly reflective so that they will absorb only a small fraction of the radiation from a fire. This congruence of favorable thermal properties was previously thought to occur only in metals, which are clearly not suitable for clothing.[6, 7]

There are numerous accounts of the above in the literature. The references cited herein are generally relevant to property enhancement, either directly or indirectly.

BFRL researchers have shown that a new class of nano-based hydrotalcite-type additives for polymers acts as an effective fire retardant at levels that do not adversely affect the polymer properties to the same degree as additives currently in widespread use. The nano-flame retardant, layered double hydroxide (LDH, see Figure 3.1) gives intumescent, self-extinguishing properties when combined with conventional flame retardants. This material can increase the volume of intumescent char formed when nano-dispersed with ammonium polyphosphate in epoxy coatings. This new nano-flame retardant is also useful in fire-resistant coatings on steel and in flame-retarding thermoplastics. Future work will focus on the use of LDH in these applications and in polyurethane (PU) flexible foams and polyurea blast coatings.

Due to environmental concerns related to the halogenated flame retardants widely used today, new flame retardant nano-additives such as these are of great interest to industry because they are inherently non-halogenated. In addition, they impart improvements in the modulus of the polymers with which they are combined, in contrast to most non-halogenated flame retardants that usually reduce the modulus and flexibility of the polymer.

Another area of research interest is the effectiveness of nano-sized flame retardants in reducing the flammability of flexible foams used in furniture. Deaths resulting from burning polyurethane foam products are a significant fraction of those incurred in home fires. More stringent requirements for flammability protection of residential furniture are under consideration by the CPSC (Consumer Product Safety Commission). A ban was also placed on the use of penta-BDE (pentabromodiphenyl ether) as a flame retardant in

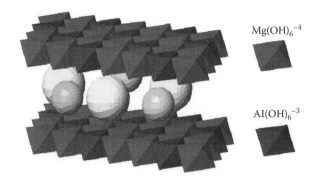

$Mg(OH)_6^{-4}$

$Al(OH)_6^{-3}$

FIGURE 3.1
Drawing shows the layer structure of the LDH formed by the combination of two metal hydroxides along with the "gallery" anions (carbonate, CO_3^{2-}) and water located between the layers.

PU foam. As a consequence, there is a need to develop new flame retardant systems for polyurethanes. Nano-sized materials such as carbon nanotubes (CNTs), layered double hydroxides (LDHs), and poly-oligo silsesquioxanes (POSSs) have been shown to influence polymer properties, including oxidation resistance and flammability. The effects of these materials remain to be characterized in PUs.[8, 9]

The challenge of introducing these materials to a PU system is far from trivial. As a consequence, the mode of combining nanomaterials with PUs to produce composite materials will require careful study. The manufacturing of PU foams is a complex process in which polymer cure rates, bubble formation, coalescence to form cell structures, and phase separation are controlled at carefully balanced rates. Nano-sized additives would be expected to affect bubble nucleation, viscosity, and surface properties. Seemingly minor changes in foam formulations can have major effects in foam structure and properties, including flammability. The work described above is underway at NIST under the "Fire Retarded Polyurethane Foam Flammability Project" (866-5237).[10]

Other workers have reported the introduction of organically modified nanoclays at 5% loading into polymers such as film-grade poly-L-lactic acid (PLLA) using a variety of masterbatches based on either semi-crystalline or amorphous poly-(lactic acid), as well as biodegradable aromatic aliphatic polyester. The PLLA masterbatches and compounds were prepared using a twin screw extruder, while the films were prepared using a single-screw cast film extruder. The use of a masterbatch appeared to improve the overall results. Tensile moduli and elongation showed increases of over 30%. As the balance of properties is improved by the use of nanomaterials, the range of applications in which these new composites may find use is broadened.[11–13]

Fire retardant (FR) effects of nano-sized hydrotalcite and magnesium hydroxide were studied in an ethylene-vinyl acetate (EVA) copolymer. Loading levels used were high—up to 150 phr (60%), somewhat unusual in the case of nano-sized materials. Laser Raman spectroscopy identified the presence of graphitic-like charred layers formed during combustion. It is reported that the FR effect of nano-hydrotalcite is better than that of nano-magnesium hydroxide based on LOI, UL-94, and heat release rate (HRR) data. This is not the case with conventionally sized versions of these materials. However, a key difference is the ability of the nano-sized hydrotalcite to form a platy char unlike conventionally sized hydrotalcite.[14–19]

Wang et al. describe the preparation of a modified nano-hydrotalcite inorganic flame-retardant filler for poly (ethylene terephthalate) (PET) polymers. This was prepared by dispersing hydrotalcite in brominated polystyrene solution. After solvent evaporation, an FR compound was obtained. A flame-retardant PET composite was prepared by melt-compounding the flame-retardant compound of PBS/LDH and PET. Improvement in the fire retardancy of the nano-flame-retardant PET composite obtained was found by measuring the oxygen index. The nanostructure of the flame-retardant PET composite was characterized by scanning electron microscopy (SEM) of the flame-retardant PET composite. The mechanical properties of the flame-retardant PET nano-composite were also characterized.[20]

3.3.3 Conductivity

There are numerous reports of enhancement of conductivity through the use of nano-materials, in particular carbon nanotubes. This topic is treated below in the section on nanotubes.

3.4 Methods of Preparation of Polymer Nanocomposites

3.4.1 *In Situ* Polymerization

The earliest reports of nanocomposite development came out of Japan in the 1970s. Unitika Ltd. and Toyota central research and development laboratories reported that mechanical and barrier properties could be enhanced by using low levels (5% or less) of nanoclays in nylon-6 nanocomposites.[21, 22]

The process described involved the addition of a nanoclay (precursor) to the caprolactam monomer prior to polymerization. The monomer intercalates into the clay galleries and, under appropriate conditions, polymerizes. The platelets are expanded by the action of the polymerization, resulting in exfoliation of the clay to become an integral part of the bulk polymer.[23, 24] A variant of the above process involves the use of a solvent, which is subsequently removed from the system, to carry the monomer into the clay galleries prior to polymerization and exfoliation.

U.S. Patent No. 4,739,007 teaches a process for manufacturing composite material with high mechanical strength and excellent high-temperature characteristics. The patent describes the steps of (a) bringing a swelling agent into contact with a clay mineral having a cation-exchange capacity of 50 to 200 mEq/100 g to form a complex capable of being swollen by a polyamide monomer at a temperature higher than the melting point of said monomer; (b) mixing said complex with said polyamide monomer; and (c) heating the mixture obtained in said mixing step to a prescribed temperature to effect polymerization. The clay mineral may be a smectite or vermiculite.[25]

U.S. Patent No. 4,810,734 describes a process for producing a composite material from a polymer and a layered silicate clay mineral that is linked to the polymer through ionic bonding. The process consists of (a) contacting a layered clay mineral with a swelling agent in the presence of a dispersion medium and thereby forming a complex; (b) mixing the complex and dispersion medium with a monomer; and (c) polymerizing the monomer contained in the resultant mixture. The process permits the economical and efficient production of a composite material in which the layered silicate is uniformly dispersed.[26]

The above processes are not always universally applicable. For some vinyl compounds, it is possible to prepare polymers of narrow molecular weight distribution by living anion polymerization or by group-transfer polymerization, but these techniques are not applicable to polyamide.

Not all polymerization processes are unaffected by the presence of nanomaterials. Condensation polymerization is least affected but processes involving supported heterogeneous catalysts or emulsion-based processes could easily be significantly disrupted or poisoned by the addition of the nanomaterials to the polymerization process.

3.4.2 Surface Chemistry

Understanding the role of interfacial chemistry between a nano-sized material and a polymer medium is key to understanding the formation and subsequent stabilization of nanocomposite materials. Intuitively, this makes sense, as the interfacial area increases by orders of magnitude as the nano-scale is approached.

3.4.2.1 *Chemical Functionalization of Clays*

Many clay minerals have the unique property of being able to change volume by absorbing water, often from other polar ions, into their structures. Many nanoclays are based on Montmorillonite, a hydrated swellable, smectite-type clay of general formula (Na, Ca)$(AlMg)_6(Si_4O_{10})_3(OH)_6 \cdot XH_2O$. Silica is a major component of these clays but alumina is also essential. Montmorillonite may be converted into a nanoclay compatible with organic polymers by surface treatment with an organic cation such as a quaternary ammonium ion. The reaction that takes place is an ion exchange whereby the cations on the clay surface, usually sodium, are exchanged with the quaternary ammonium ion. Formation of a sodium salt by combination with the anion of the quaternary ammonium salt, usually a chloride, also helps to drive the surface treatment reaction. The quaternary ammonium cations are of the general formula $[R_1R_2R_3R_4N]^+$, where the chains R_1 through R_4 may be combinations of aliphatic chains, methyl or benzyl groups. In addition to the use of quaternary ammonium ions, the use of phosphonium, sulfonium, or oxonium ions has also been reported.[27, 28]

Nanoclay precursor materials thus treated may be added to polymers in the melt phase, affording exfoliation of the clay to yield a nanocomposite. These precursor clays often have particle sizes typical of what would be expected for more conventional filler clays.[29] For example, the Cloisite® Montmorillonite materials data sheet describe the particle size distributions as in Table 3.1.

The preparation of delaminated kaolins, for example, is well described. However, in the case of non-swellable clays such as kaolins, the delamination involves a breaking of the stacked platelets of the clay under shear into smaller stacks of platelets. However, there is a point in the process at which the clay stacks will not break any further, a point that is well short of attaining any nano-sized dimensons. The ability of the Montmorillonite clays to

TABLE 3.1

Dry Particle Sizes

Weight Fraction	10%	50%	90%
Particle Diameter	<2 μm	<6 μm	<13 μm

Source: Product data sheets for Cloisite ® -10A, -15A, -20A, -25A, 30B, -93A, -NA, Southern Clay Products, Gonzalez, TX.

intercalate polymers and swell is a key determinant in achieving the fracturing of the clays into individual layers.

3.4.2.2 Nanotalcs

In addition to platy nanoclays, there are also developments in the field of nano-sized talcs. Fine talcs are usually considered to have dimensions of 1 to 12 µ with particle thickness in the range of 0.2 to 0.6 µ. There have been recent market introductions of nanotalcs. These materials have a significantly higher surface area than conventional talc—250 m^2/g compared to conventional talc platelets that range from 8 to 18 m^2/g.[30, 31]

- Nanocomposites: Enhancing Value in the Global plastics industry, 2005, Principia Partners, Exton, PA.[32]

- Nanova uses a proprietary, patent-pending mechano-chemical synthesis process to reduce the size and increase the surface area of conventional talc. The process maintains talc's high aspect ratio, which gives improved physical properties such as impact resistance and rigidity to the modified polymer.[33] Talc has a high surface activity, and thus would be expected to agglomerate on the nano-size scale. Chemical surface modification can help prevents this and increase compatibility with polymer matrices. Nanotalcs provide better dimensional stability than conventional talcs. Optical properties such as scratch and mar resistance are also improved optics.

3.4.2.3 Chemical Functionalization of Nanocarbon Materials

Graphene structures have a relatively low chemical reactivity.[34] This can lead to a lower than expected interaction between a nanotube and a polymer matrix in which it is embedded. One way to enhance and strengthen this interaction is to chemically change the nature of the nanotube surface.

There have been several published accounts of chemical modification of the surface of carbon nanotubes. These often involve chemical modification of the surface under vigorous or forcing conditions.[35–40] Functional groups such as fluorine and hydrogen can be generated that must be removed, and this can often require the use of aggressive chemicals such as alkyl-lithiums in the case of fluorine. Carboxyl groups generated in an oxidizing medium can also be used as sites for further chemistry, such as amidation or transesterification type reactions. These reactions can assist in the solubilization of nanotubes.[41] Sulfonate groups can be attached to multiwalled nanotubes (MWNTs).[42] These functionalized materials were compounded into polyaniline to form composites with enhanced electrical conductivity arising from the doping of the polymer by the sulfonated MWNTs. Another approach to solubilization of nanotubes in water is achieved by wrapping the nanotubes with polyvinylpyrollidone (PVP) or polytheylene imine (PEI).[43–45]

Another approach to derivatization of nanotubes involves chemical interactions with defects in the graphitic structure and/or at the ends of the tubes. These sites are reactive as a result of their chemical isolation and are susceptible to oxidation reactions.[46–49]

One approach that has been well reported is the use of diazonium salts to functionalize carbon nanotubes. One advantage of this approach is that it is essentially a one-step reaction that can be carried out by electrochemically induced reactions, thermally induced reactions (via *in situ* generation of diazonium compounds or via preformed diazonium compounds), and photochemically induced reactions. Up to one in twenty carbons on the nanotube surface may be functionalized in this way. The derivatization causes significant changes in the spectroscopic properties of the nanotubes. The estimated degree of functionality is ca. one out of every twenty to thirty carbons in a nanotube bearing a functionality moiety.[50–53] While this approach is perhaps the most benign chemically of those reported, it is a process that has not yet been scaled to industrial production.

3.4.2.4 Reagglomeration

A key feature of nanomaterials that is unique to them is that following exfoliation and separation in a polymer matrix into their individual components, there is a strong driving force toward reagglomeration. This phenomenon can be dealt with by suitable surface functionalization of the materials to make them more chemically compatible with the host polymer, which helps in maintaining the distribution and nano-scale dispersion of the materials.

In the case of conventionally sized materials, chemical surface treatments are often employed to assist in dispersion. Once good dispersion is obtained, there is little likelihood that the material will reagglomerate. This suggests the possibility that surface treatments that work well with a given material in the micron-sized domain may not work as well or at all in the nano-sized domain.

There is some evidence of this. Materials that have been typically used fall into three categories based on reactivity:

1. Encapsulating or nonreactive with the substrate
2. Reactive with the substrate and noncoupling with the host medium
3. Reactive and coupling

The preponderance of organic functional groups used are alkyl, phenyl, amino, epoxy, isocyanate, vinyl, and methacryl. Examples are shown in Table 3.2. These materials can modify the organophilic/hydrophobic character of the material being dispersed. They can improve the compatibility and wetting of the material being dispersed with the host medium, and in so doing they provide reinforcement or mechanical strength, in some cases through cross-linking. This approach may appear to work with nanomaterials, but there is a serious caveat. Addition of some of these materials, such as metal stearates, may actually assist in coating the nanomaterial in such a manner that the material easily incorporates with little rise in viscosity of the medium. This is generally desirable with micron-sized materials; but with nano-sized materials, it may simply mean that micron-sized

TABLE 3.2

Chemical Surface Treatments for Clay

Chemical Reactivity	Material
Encapsulating or nonreactive	Dispersants, surfactants (ionic or non-ionic)
Reactive non-coupling	Metal stearates, alkylphenyls, polysiloxanes
Reactive and coupling	Methacryl-, vinyl-, S-containing-, amino- and epoxy-silanes

agglomerates of the nanomaterials are being incorporated into the host material with poor dispersion or exfoliation on the nano-scale. In short, the nanomaterials behave like conventionally sized materials and their potential remains unrealized.

There are numerous reports in the literature of materials, such as quaternary ammonium salts, being successfully used to enable the incorporation of nanoclays into polymer matrices leading to stable composites. These treatments are rarely used in the case of more conventionally sized clays where metal stearates and silanes are the norm. There are also reports of the use of diazonium salts, fluorine, and lithium alkyls (cited already) to surface-treat carbon nanotubes and facilitate their incorporation into polymer media.

3.4.2.5 Nanocomposite Commercialization

Anybody wishing to produce nanocomposites on a commercial scale needs to bear in mind that the incorporation and subsequent dispersion and exfoliation of nanomaterials into polymeric materials is governed by both physical and mechanical dynamics. Also bear in mind that the interfacial chemistry is, in turn, governed by both chemical kinetics and thermodynamics (chemical potential). Workers at Case Western Reserve University reported that the interactions between polymer–surfactant, solvent–surfactant, and polymer–solvent play an important role for the solution intercalation of polymers into a nanoclay to form a composite.[54] To successfully disperse nanomaterials into polymers, it is often required to exceed the yield stress of the materials in question. To do this often requires a sufficiently high melt viscosity or polymer molecular weight.[55] The challenge here is that as the nanomaterials disperse, the melt viscosity further rises, making processing more difficult.

3.4.3 Melt-Compounding

An area of research and development that has received considerable attention is the study of the incorporation of nanomaterials into polymers via melt-compounding.[56, 57]

In melt-compounding, the objective is to achieve distributive mixing, in essence the homogeneous distribution of an additive in a substrate medium. The objective is to surround each individual particle with a coating of polymer. In practice with many materials, this is not always achieved. However, if a sufficient degree of randomness on a sufficiently small scale is achieved, the desired composite properties may be attained.

3.4.3.1 Laminar vs. Chaotic Mixing

It is possible to overcome inertial torquing forces, especially with high aspect ratio materials, by laminar incorporation. The key here is to induce flow in the direction of the longest axis of the material to be incorporated during mixing and allow laminar flow to surround the material with a sufficient quantity of the substrate medium to effect good dispersion or distribution. This might require sufficient residence time.

Chaotic mixing tends to randomize. During chaotic mixing, density rather than viscosity tends to dominate (see Hildebrandt's text); as in a truly chaotic system, all resistance to flow has been overcome.

Melt-compounding of nanoclays into polymers is often achieved using single- or twin-screw extruders. This is more of a laminar mixing regime. The work that is being reported describes methodologies that are a combination of chemical and mechanical techniques designed to incorporate, distribute, and exfoliate the nanoclays.

3.4.3.2 Plastics

Nano-sized materials are used in thermoplastics, such as polyethylene, polypropylene, TPOs, polyesters, polycarbonate, polyetherimides, and nylon. Nanocomposites have the potential to provide property enhancements such as reinforcement, heat resistance, dimensional stability, stiffness, flame retardancy, and electrical conductivity. Areas of application include automotive, electronics (hard drive components), packaging (films and PET bottles), controlled release of dyes, and additives such as biocides. While progress is being made in growing commercial markets for thermoplastic nanocomposites, the ability to successfully incorporate and disperse materials to the nano-dimension still remains a significant challenge.

Several workers have described the use of extrusion techniques to generate a polymer melt phase to which nanoclays are directly added. Under conditions of high shear and elevated temperatures, exfoliation of the clay into the polymer takes place.[58–61]

U.S. Patent No. 6,864,308 describes a method of making polyolefin nanocomposites by melt-blending via twin-screw extrusion of a polyolefin and a smectite clay in the presence of an intercalating agent such as a hydroxy-substituted carboxylic ester, amides, hydroxyl-amides, and oxidized polyolefins where the ratio of the intercalating agent to clay is 1:3 based on the ash content of the clay.[62]

In another U.S. Patent (Patent No. 6,887,931), the preparation of inorganic clay dispersions and their use is described. More specifically, these are inorganic clay nanodispersions comprising an inorganic clay treated *in situ* with an intercalating agent and a nonaqueous chemically reactive organic intercalating facilitating agent. These dispersions are used to prepare thermosetting nanocomposite articles.[63]

The hydrophilic clay is treated with water to swell it by expanding the d-spacing between the inorganic layers. The swollen clay is then treated with an intercalating agent such as a quaternary ammonium salt. This renders the clay organophilic and further increases the d-spacing by cation exchange between the inorganic clay and the intercalating agent. The intercalated clay is recovered and dried. The dried clay is then mixed with a thermoplastic or thermosetting resin or monomer that causes exfoliation of some or all of the layers of the inorganic clay.

Fillers such as silica, talc, alumina, and calcium carbonate are often mixed with the exfoliated clay prior to forming a finished article that may be molded or thermoformed into a finished article.

3.4.3.3 Elastomers

Mixing: The incorporation of nanomaterials into elastomer materials has received somewhat less attention to date than in the case of thermoplastics. However, the problems that one encounters are generally of the same nature. While melt-processing is seldom an option, published work has described the use of medium to high shear mixers as at least a part of the process of making elastomer nanocomposites. The use of conventional methods of mixing rubber is also reported in several instances.

Ansarifar et al. report a method for preparing rubber (SBR, NR, and BR) formulations using a pretreated silanized silica nanofiller and mixing in a laboratory Haake Rheocord 90 mixer equipped with counter-rotating rotors. A safe-processing delayed-action accelerator (TBBS N-*t*-butyl-2-benzothiazole sulfenamide) was used to help in controlling the state of cure in the composite. Results indicated that this method of processing helped to substantially reduce the use of curing agents without compromising the mechanical properties of the vulcanized material, which are necessary for maintaining good performance and long service life of the compounds. It remains to be seen if such a process can be commercialized but the approach holds promise.[64]

U.S. Patent No. 5,840,796 teaches the use of elevated temperature and mechanical shear such as ball milling or pebble milling to form composites by intercalation of a fluoroelastomer into layered silicate structures to yield a composite. The silicates are pretreated by a quaternary ammonium salt to enhance wetting of the silicate by the polymer. The fluoroelastomer molecules penetrate the MTS (mica-type silicate) laminae, causing each lamina to be surrounded by polymer chains of the fluoroelastomer as the MTS exfoliates. The exfoliated nanocomposite may be formed with or without going through the intermediate stage of intercalation.[65]

Also, in the same patent, the examples describe the production of a composite of a fluoroelastomer with a pretreated clay on a two-roll mill with a tight nip. When the rollers were cooled to keep the temperature of the composition below 100°F during milling x-ray diffraction (XRD), analysis showed that no intercalation had taken place. When the same procedure was repeated heating the mill rolls to 120°F, XRD indicated that some intercalation had taken place as the interlayer spacing appeared to have increased from 28.8 to 33 Å. Further raising the temperature to 150°F afforded material that exhibited no peaks in the x-ray spectra. There was no evidence of layer spacing or crystal ordering, indicating that exfoliation had taken place to such an extent that the individual layers of the clay were peeled apart and distributed randomly within the fluoroelastomer.

The above examples are illustrative of the fact that several factors can control exfoliation: temperature, which influences kinetics and thermodynamics of the reaction, and chemical surface treatment, which improves chemical compatibility and the mechanical dynamics of the milling. Controlling these input parameters is key to controlling the dispersion process and achieving the desired composite properties.

3.4.4 Dispersion Destabilization

As true melt-processing is seldom an option with elastomers, another approach is to mix dispersions of nanomaterials and polymers, including lattices, under various conditions of shear to afford nano-scale dispersion of the nanomaterials into a polymer medium. The balance between stability of the dispersions must be controlled, as it is not unusual for the addition of one dispersion to destabilize another dispersion. While ultimately this is required, it is not desirable to have destabilization or flocculation prior to obtaining the required dispersion of the nanomaterials.

U.S. Patent No. 6,849,680 B2 (registered February 1, 2005) describes the preparation of polymer nanocomposites by dispersion destabilization. First, dispersions of nanoclays and polymers are mixed. After mixing, the dispersions are destabilized by the addition of flocculating agents. The elastomer dispersions are prepared in a liquid carrier at up to 80% concentration (the concentration required is viscosity dependent) in organic or aqueous media. Polymers may be polyester, PU, PVC, SBR, acrylic rubber, chlorosulfonated PE rubber, fluoroelastomer, polyisoprene, PC, Nylon, PO, or thermoplastic.[66] The clay dispersion can be prepared by adding 1 to 10% by weight of clay to a carrier liquid; this can be further processed in a high shear mixer such as a Manton-Gaulin mixer.[67]

U.S. Patent No. 6,861,462 B2 describes the formation *in situ* of a nanocomposite by an elastomer matrix and a dispersion of at least partially exfoliated nanoclay particles. The composite is formed by blending the above nanocomposite with additional elastomer(s), additional reinforcing fillers, and/or a coupling agent. This patent illustrates that an aqueous cationic latex of SBR was slowly added to a preheated and aggressively agitated water/clay dispersion. The cationic surfactant in the latex displaced the sodium ions of the multilayered Montmorillonite clay, and thereby expanded the interlayer spacing, allowing intercalation by the polymer with the latex coagulating as a result of the destabilizaton of the latex by removal of the cationic surfactant.[68]

3.5 Nanotubes

3.5.1 Characterization

Hitherto we have described the exfoliation of clays. Another interesting class of nanomaterials are nanotubes based on carbon. The discovery of the new allotropes of carbon, the "buckyballs" C_{60} and C_{72}, led to renewed interest in the structures of carbon. In 1985, Smalley and Kroto reported the isolation of another allotropic form of carbon, the C-60 and C-72 spherical carbon clusters known as fullerenes.[69, 70]

Carbon nanotube are long, thin cylinders of carbon. Iijima first reported their discovery in 1991.[71] They differ from clay in that they are not flat-sheet platelets but are rod-like in structure.

Carbon nanotubes are, in fact, a class of allotropes of carbon. A single-walled carbon nanotube can be considered as simply a graphite-type sheet folded into a cylinder that is capped at either end by a "buckyball" hemisphere. Typically, they have a large aspect ratio, often of the order of 1000 or more. The wall thickness can range from a single layer of carbon to multiple walls, in excess of 20 or more.

These are large macromolecules that are unique for their size, shape, and remarkable physical properties. They can be thought of as a sheet of graphite (a hexagonal lattice of carbon) rolled into a cylinder.[72] These intriguing structures have sparked much excitement in recent years, and a large amount of research has been dedicated to their understanding. Currently, the physical properties are still being discovered and disputed. What makes it so difficult is that nanotubes have a very broad range of electronic, thermal, and structural properties that change depending on the different kinds of nanotube (defined by its diameter, length, and chirality, or twist). For example, these materials have a range of inherent conductivities that depend on how the carbon sheets are folded. Nanotubes may be a single cylindrical wall (SWNTs) or can have multiple walls (MWNTs)—cylinders inside the other cylinders.

Carbon nanotubes may be thought of as molecular-scale tubes of graphitic carbon. They are among the stiffest and strongest fibers known. Literature reports show that carbon nanotubes have a Young's modulus of 1.4 TPa.[37] They have an expected elongation to failure of 20 to 30%, which, combined with the stiffness, projects to a tensile strength well above 100 GPa (possibly higher), by far the highest known. For comparison, the Young's modulus of high-strength steel is around 200 GPa, and its tensile strength is 1 to 2 GPa.[73–76]

As a result of their seamless cylindrical graphite structure, carbon nanotubes are expected to have high stiffness and axial strength.[77–81] Because of their small size, they are not amenable to direct measurement of physical properties. However, some workers have estimated the Young's modulus of isolated nanotubes using TEM (transmission electron microscopy) to measure the amplitude of intrinsic thermal vibrational modes. Results indicated that the Young's moduli were extremely high, in the terapascal range (TPa).[82]

3.5.2 Incorporation into Polymers

The challenge in incorporating carbon nanotubes into polymers arises from several perspectives. Because of their high aspect ratio, they are capable of entanglements and as a consequence they will be expected to give very high increases in the system viscosity as the individual nanotubes un-bundle, one from the other. A second consideration is that the individual tubes are difficult to separate from each other as a result of Van der Waals forces that then tend to make them clump together.

Work by Islam et al. shows that at a volume fraction of 0.0026, aqueous suspensions of single-walled nanotubes (SWNTs) exhibit the onset of solid-like elasticity. Development of a master curve for shear-dependent stresses enables derivation of an inter-nanotube interaction energy of 40 k_BT per bond. The authors postulate that SWNTs form interconnected networks in suspension with bonds that freely rotate and resist stretching. The observed elasticity originates from bonds between the nanotubes rather than from stiffness or stretching of the individual tubes.[83]

Current research is focusing on the CNT (carbon nanotube) surface and the interface between the CNT and surrounding polymer matrix. It is through shear stress build-up at this interface that stress is mechanically transferred from the surrounding polymer matrix to the CNT. Numerous researchers have attributed lower-than-predicted CNT-polymer composite properties to a lack of interfacial bonding.[84, 85]

The surface of a CNT is essentially an exposed graphene sheet. The weak inter-planar interaction of graphite provides its solid lubricant quality, but this in turn can lead to a weakening of the polymer-NT interaction and resistance to matrix adhesion. In the case of MWNTs, this could be expected to lead to weakening of the polymer-NT interaction. Furthermore, the relatively chemically inert nature of graphene structures does not favor a strong polymer-NT interaction in the case of both MWNTs and SWNTs.[86]

The high aspect ratio of CNTs is generally a barrier to facile incorporation of nanotubes into polymer media. The high surface energies of the nanotubes render separation into individual tubes difficult but also increase the difficulty of maintaining separated tubes in a polymer media. Several workers have reported methods to chemically functionalize the surface of nanotubes but these methods are generally on a very small scale and have not been scaled up to any scale suitable for commercial activity.

Incorporation of nanotubes (NTs) into polymer media is rendered difficult by

- High aspect ratio
- Small size
- Tendency to agglomerate or clump.

However, when NTs can be dispersed effectively, significant toughening of polymer matrices, such as ultrahigh molecular weight polyethylene (UHMWPE), HDPE, PAN, PVA, and iPP, has been reported at relatively low loading levels of 1 to 5%.[87] Nanotubes can induce changes in the crystallization and morphology of polymers.[88–91] Modifications of morphology and also measurements of the energy required for debonding of NTs suggest that the NTs may increase the energy absorption or toughness of the composite and lead to prolonged fatigue-life.[92, 93] Potschke et al. report results that show the rigid rod-like nanotubes in PA 12 enhance load-bearing and anti-wear properties of thin films. Nanotubes that were pre-encapsulated with SMA were melt-mixed with PA12 matrix in a conical twin-screw extruder. The process of encapsulation by SMA copolymer leads to a finer dispersion of SWNT and enhanced interfacial adhesion between PA12 and the SMA modified

SWNT. This leads to enhanced mechanical properties, which are manifested by tensile and other measurements.[94]

This would clearly indicate that the transfer of mechanical energy at the nanotube-polymer interface is critical. The mechanisms that may be at work are prolongation of the formation of crazing or microcracking as a result of the nanoparticle acting as a stress concentrator. This mechanism is well known in fracture mechanics. Another possible mechanism is simply physical bridging of a developing crack by a nanotube to prevent further crack propagation. However, further work is needed to fully understand the reinforcing mechanisms at work in carbon nanotube-polymer composites. The nature of these materials indicates that sophisticated analytical techniques will have to be used to understand the mechanisms at work.[95, 96]

3.5.3 Percolation Threshold

Percolation is a statistical concept concerned with the formation of long-range clusters of connected particles or pathways in random systems. One definition of percolation relates to the slow flow of fluids through a porous medium, but more theoretically it refers to lattice models of random systems and their internal connectivity. In such systems, there is a critical concentration at which the connectivity appears. This conceptual model applies well to dispersions of nanotubes in polymers. As the concentration of nanotubes progressively increases, there is a concentration where there is a jump in conductivity. This is correlated with the onset of connectivity and this concentration is usually defined as the percolation threshold. Zou et al. showed that for the dispersion of MWNTs in a polymer matrix by a twin-screw extruder, there is a critical MWNT concentration of 1.0 wt% where a fine network of filler is formed above which the composite possesses improved mechanical properties.[97] Pötschke et al. report a study in which a polycarbonate (PC) masterbatch was progressively diluted by PC, in a conical twin-screw extruder, to afford different nanotube concentrations. Electrical measurements indicated percolation of MWNT between 1.0 and 1.5 wt%.[98]

Nanocomposites comprised of double-wall carbon nanotubes (DWCNTs) and an epoxy matrix were produced by a standard calandering technique. A very good dispersion of both the DWCNT and carbon black (CB) in an epoxy resin was obtained. Fracture-mechanical properties resulted in an increase of strength, Young's modulus, and strain to failure at a nanotube content of only 0.1 wt%.[99]

Lowering the percolation threshold is a critical consideration in achieving low loading levels (lower cost) of a conducting medium in an insulator (conducting composites). This has relevance in applications where charge dissipation and electrical conductivity is desired. Rod-shaped particles tend to have lower percolation thresholds rather than lower aspect-ratio particles. Altering the interaction potential between particles also has a significant impact. In the case of nanotubes, there are different measures of percolation thresholds, ranging from concentrations below 1% to concentrations in the 4 to 5% range and above. One interpretation of these results is that the use of dispersants and surfactants can have a significant impact.[100] The correct balance of attraction between individual particles combined with dispersion and distribution in the polymer medium is what is required. In practice, there is large variability of experimental results in the literature dealing with the electrical behavior of composites loaded with conducting rods. Control of the percolation threshold is difficult as numerous factors have an impact.[101]

It is clear that the dispersion of nanomaterials is a combination of chemical and mechanical effects. While it is desirable to have a certain attraction between individual nanotubes

Brabender or Haake Torque mixing	Single screw extrusion	Twin screw extruder	Gear pump extruder	Buss Kneader
Can be considered as a preliminary method of distributive mixing in a polymer system. Good for evaluation on a lab. scale but the question of scalability to larger processing equipment has to be considered.	In Single screw extruder-conveying behavior is affected by frictional drag of solid phase and viscous drag in the melt. Back-mixing can be accomplished by use of a Maddock or pineapple mixing element.	More positive displacement in a closely intermeshing counter rotating geometry. Back mixing can be designed into the screw configuration. Has all the advantage of single screw plus more. Laminar flow can be set up. This is a type of flow in which the surface area between the two phases can be increased.	Generally there is leakage around the intermeshing gears. Breakage of NT's might be a problem. Temperature rise can be a problem and they are not energy efficient. Over 50% (50 to 80%) of applied energy is lost. In the case of very abrasive materials this is not a recommended approach.	A single screw type machine. The screw shaft oscillates axially once per revolution in sinusoidal motion generated by a synchronized drive. Generally introduces low shear into the polymer and provides uniformity of pressure and good distribution.
Kenics static mixer	**Banbury mixer**	**Farrell continuous mixer**	**Double arm kneading mixers (Sigma mixer)**	**High shear blenders**
Distributive mixing is usually excellent. However dispersive mixing may be poor especially when viscosity ratios are high.	Generally used for rubber or materials that exhibit viscoelastic behavior. Counter rotating rotors at different speeds. The degree of distribution is usually a function of the mixing time and specific energy input.	Similar to the Banbury mixer except that it can be run in a continuous mode.	Ideal for mixing, kneading of highly viscous mass, sticky and dough like products. Mixing of pastes, rubber, and heavy plastic masses.	Household type blenders. These are good for solid/liquid dispersions. They can damage the solid being dispersed. The Draiswerke Gelimat mixer is a good example of this kind of blender.

FIGURE 3.2
Polymer processing methodologies.

to achieve percolation and other property enhancements, it is not desirable that this interaction be so strong as to cause clumping or non-dispersing behavior and thus defeat the very purpose of using nanotubes.

Figure 3.2 provides several examples of equipment used to process polymers. It is difficult to recommend one particular piece of equipment for the processing of nanomaterials. The user will have to take into account the desired properties and the degree of work and shear that the nanomaterial can withstand.

To date, the majority of successful production of nanocomposites have involved extrusion melt-compounding.

3.6 Commercial Nanomaterials

3.6.1 Nanocomposites

- General Motors has used nanocomposites since 2002 in parts such as body-side molding. A nanoclay-TPO composite is in use in the 2005 Hummer.

- Nanocomposite concentrates are being evaluated in films not only for enhancing barrier, but also to control the release and migration of additives such as biocides and dyes.

- PolyOne commercialized nanocomposite products in the Nanoblend™ family of concentrates and compounds for polyolefin resins. These composites are based on nanoclays and can reduce the amount of mineral fillers, often used as fire retardants that are required in many compounds.

- RTP offers nylon-based themoplastics that are reinforced with nanoclays. They have also developed nanotube-based thermoplastic composites for the electronics industry in hard disk driven and wafer handling equipment.

- Bayer Polymers Durethan® polyamide 6 grade contains nano-sized, chemically modified layered silicates that are incorporated during the polymerization of the resin. The application is in barrier film, paper, and coatings.

- Honeywell Polymer has developed Aegis TM OX for high-barrier beer bottles and Aegis NC (Nylon 6/barrier nylon) for medium barrier bottles and films.

- Basell USA developed PP-based nanocomposites using Cloisite® nanoclay products from Southern Clay Products.

- Nobel Polymers—Forte™ nanocomposite. This is a polypropylene-based nanocomposite produced by Noble Polymers. It is one of the first commercializations of nanocomposites in the automotive industry and can be used in a variety of applications replacing talc-filled, glass-filled, or virgin polypropylene.

- Polykemi AB offers a PP-based nanocomposite called Scancomp®.

- Nanocyl recently launched NC9000, a nanotube-filled HDPE.

- Mitsubishi gas chemical company has developed M9, a high barrier property nanocomposite for the multi-layer films, juice and beer bottles, and containers.

- The use of multiwalled carbon nanotubes as a flame retardant in EVA has been reported.[102, 103]

- Solar cells: the U.S. Department of Energy at the Lawrence Berkeley National Laboratory and The University of California-Berkeley have developed a hybrid semiconductor-polymer photovoltaic device using nanometer-sized cadmium-selenide rods.[104]

- Thermoplastic nanocomposite foams: workers at Ohio State University have made dense plastic foams using nanocomposites made from polymer blends such as PMMA/PS and a nanoclay. These foams are lighter weight than solid plastics and have the potential to replace them in furniture and packaging applications.[105]

- Putsch GmbH used nanofil to produce Elan XP, which is a blend of PP/PS.

- Chemtura's Polybound® X5104 is a maleated PP used as a coupling agent to improve physical and thermal properties of nanoclays, natural fibers, and glass in filled PP composites.

- Dyneon has developed hydrocarbon-based block copolymers as compatibilizers and coupling agents for polyolefin and styrenic nanocomposites produced by melt compounding nanoclays. These copolymers contain amine, epoxy anhydride, and acid, and are more efficient in enhancing the exfoliation and dispersion of organically modified clays.

- Nycoa has developed a nylon nanocomposite to meet CARB and U.S. EPA requirements for gas permeation in fuel tanks and fuel systems such as fuel hoses.[106]

- U.S. Patent 6,942,897 describes a nanoparticle barrier-coated substrate and a method for making the same. The pigment nanoparticles may be talc, calcium carbonate, clay, silica, or plastic. This development may find application in paper, wood, wallboard, fiberglass, metal, and ceramics.

- U.S. Patent 6,646,026 B2 describes a method to enhance the ability to dye polymers by dispersion of nanocomposites into a polymer and reacting the nanocomposites with a dye. The nanomaterial may be a clay, silica, or an oxide of a metal such as silver or zinc.

3.6.2 Nano-Sized Materials

- Sud Chemie has introduced two new Nanofil® products. Nanofil SE 3000 is designed for applications in EVA, PE, PP, and polyamide. Nanofil SE 3010 is suited for nylon, polyamide, PS, ABS, and polycarbonate alloys.

- Elementis Specialties manufactures Bentone® nanoclays and has recently filed a patent covering a method for measuring the degree of exfoliation and alignment of nanoclay additives into plastic composites.

3.7 Summary

While some industries such as the automotive industry have been somewhat slow in adopting nanocomposite technology, there is an increasing level of activity. The drivers of recycling, performance, fuel efficiency aligned with product stewardship are forcing the polymer producers and OEMs and fabricators to improve performance in this area.

The fact that there is a steadily increasing rate of commercialization of nanomaterials and polymer composites of these materials suggests that workers in diverse fields are coming to grips with the challenges of incorporating nanomaterials into polymers. However, the inherent challenges arising from the unique physical and chemical properties of these materials remain. As a consequence, the methodologies employed in the case of conventionally sized materials will have to be significantly modified in practice.

In a practical sense, the speed of commercialization is faster if nanocomposites can be produced without the need to use or develop highly sophisticated dispersion and mixing equipment. To date, the majority of commercial polymer nanocomposite materials have been produced by melt-processing of the nanomaterial, usually using a twin-screw extruder. Companies such as Bayer and Honeywell have developed nylon nanocomposites for the automotive market. Basell and Poly One have developed polymer nanocomposites. Another factor to consider is that the use of chemical functionalization methods will be hampered if the reagents used are difficult to handle or present disposal and environmental difficulties.

The difficulties in producing the nanomaterials themselves have been a key determinant in the relatively slow rate of commercialization of polymer nanocomposites. There has been a lot of investment in patenting but the majority of the ideas patented have not to date yielded commercial products. Nevertheless, the potential to generate new and innovative products remains if the developed technologies can be exploited commercially.

While it is difficult to make predictions about the markets in which nanocomposites may be rapidly commercialized, it will be critical that property enhancements will be sufficient to overcome the added development, material, and processing costs. At present, the most likely type of materials to see commercialization in polymers are nanoclays. Future growth in these areas may benefit from the improved properties obtained at lower weight—a fact that in an age of increasing energy costs may be beneficial as lower-weight materials can mean lower transportation costs.

References

1. Kamena, K., Nanoclays and their emerging markets, in *Functional Fillers for Plastics*, Wiley-VCH VerlagGmbH & Co KgaA, 2005, p. 172.
2. Kamena, K., Nanoclays and their emerging markets, in *Functional Fillers for Plastics*, Wiley-VCH VerlagGmbH & Co KgaA, 2005, p. 163.
3. Manolis Sherman, L., Nanocomposites—a little goes a long way, in *Plastics Technology Online,* June 1999, http://www.ptonline.com/articles/199906fa4.html.
4. Leaversuch, R., Nanocomposites broaden roles in automotive, barrier packaging, in *Plastics Technology Online,* October 2001, http://www.ptonline.com/articles/200110fa3.html.
5. See the Web site of NIST http://www.bfrl.nist.gov/Annual/2004-2005/BFRL06.pdf.
6. Report of the Activities, Accomplishments and Recognitions of the Building and Fire Research Laboratory (BFRL) of the NIST, 2006, SP, 838-19.
7. Carbon Nanotube Fabric for Fire Fighter Protective Clothing, Report of the Activities, Accomplishments and Recognitions of the Building and Fire Research Laboratory (BFRL) of the NIST, 23, 2006.
8. Dodiuk, H., Belinski, I., Dotan, A., and Kenig, S., Polyurethane adhesives containing functionalized nanoclays, *J. Adhesion Sci. Technol.,* 20, 2006.
9. Efrat, T., Dodiuk, H., Kenig, S., and McCarthy, S., Nanotailoring of polyurethane adhesive by polyhedral oligomeric silsesquioxane (POSS), *J. Adhesion Sci. Technol.,* 20, 2006.
10. Gilman, J., Fire Retarded Polyurethane Foam Flammability Project as published by NIST (866-5237).
11. Lewitus, D., McCarthy, S., Ophir, A., and Kenig, S., The effect of nanoclays on the properties of plla-modified polymers. 1. Mechanical and thermal properties, *J. Polym. Environ.,* 14, 1566, 2006.
12. Dodiuk, H., Belinski, I., Dotan, A., and Kenig, S., Polyurethane adhesives containing functionalized nanoclays, *J. Adhesion Sci. Technol.,* 20, 2006.
13. Efrat, T., Dodiuk, H., Kenig, S., and McCarthy, S., Nanotailoring of polyurethane adhesive by polyhedral oligomeric silsesquioxane (POSS), *J. Adhesion Sci. Technol.,* 20, 2006.
14. Jiao, C.M., Wang Z.Z., Ye, Z., Hu, Y., and Fan, W.C., Flame retardation of ethylene-vinyl acetate copolymer using nano magnesium hydroxide and nano hydrotalcite, in *2005 International Conference on Advanced Fibers and Polymer Materials (ICAFPM 2005)*, ICAFPM 2005 Shanghai, 2006.

15. Anhui, State Key Laboratory of Fire Science, University of Science and Technology of China, 230026, PR China, in *2005 International Conference on Advanced Fibers and Polymer Materials*, China, p. 1867, 2006.
16. Beyer, G., Flame retardant properties of EVA-nanocomposites and improvements by combination of nanofillers with aluminium trihydrate, *Fire Mater.*, 25, 193, 2001.
17. Riva, A., Canimo, G., Fomperie, L., and Amiqouet, P., Fire retardant mechanism in intumescent ethylene vinyl acetate compositions, *Polym. Degrad. Stab.*, 82, 341, 2003.
18. Qiu, L.Z., Xie, R.C., Ding, P., and Qu, B.J., Preparation and characterization of $Mg(OH)_2$ nanoparticles and flame-retardant property of its nanocomposites with EVA , *Compos. Struct.*, 62, 391, 2003.
19. Yang, W.S., Kim, Y., Liu, P.K.T., Sahimi, M., and Tsotsis, T.T., A study by *in situ* techniques of the thermal evolution of the structure of a Mg-Al-CO_3 layered double hydroxide, *Chem. Eng. Sci.*, 57, 2945, 2002.
20. Wang, M., Zhu, Meifang, and Sun, Bin, A new nano-structured flame-retardant poly (ethylene terephthalate), *J. Macromolec. Sci.: Pure Appl. Chem.*, 43, 1867, 2006.
21. Fujiwara, S. and Sakamoto, T., Japanese Patent No. JPA51-109998, 1976.
22. Kato, M. and Usuki, A., Polymerization of organic monomers and biomolecules on hectorite, in *Polymer-Clay Nanocomposites*, Pinnavia, T.J. and Beal, G.W., Eds., John Wiley & Sons, New York, 2000.
23. Usuki et al., Synthesis of nylon 6-clay hybrid, *J. Mater. Res.*, 8, 1179, 1993.
24. Usuki et al., Synthesis of nylon 6-clay hybrid, *J. Mater. Res.*, 8, 1179, 1993.
25. U.S. Patent. No. 4,739,007.
26. U.S. Patent. No. 4,810,734.
27. Qian, G., Cho, J.W., and Lan, T., U.S. Patent No. 6,632,868.
28. Qian, G., Lan, T., Fay, A., and Tomlin, A., U.S. Patent No. 6,462,122.
29. Product data sheets for Cloisite® –10A, –15A, –20A, –25A, 30B, –93A, –NA, Southern Clay Products, Gonzalez, TX.
30. Flaris, V., Talc, in *Functional Fillers for Plastics*, Wiley-VCH VerlagGmbH & Co KgaA, 2005, p. 211.
31. Harris, P., *Industrial Minerals*, 443, p. 60–63, October 2003.
32. *Nanocomposites—Enhancing Value in the Global Plastics Industry, 2005*, Principia Partners, Exton, PA.
33. *Nanocomposites—Enhancing Value in the Global Plastics Industry, 2005*, Principia Partners, Exton, PA.
34. Aihara, J., Lack of superaromaticity in carbon nanotubes, *J. Phys. Chem.*, 98, 9773, 1994.
35. Chen, J. et al., Solution properties of single-walled carbon nanotubes, *Science*, 282, 95, 1998.
36. Mickelson et al., Fluorination of single-wall carbon nanotubes, *Chem. Phys. Lett.*, 296, 188, 1998.
37. Bahr, J.L. et al., Functionalization of carbon nanotubes by electrochemical reduction of aryl diazonium salts: a bucky paper electrode, *J. Am. Chem. Soc.*, 123, 6536, 2001.
38. Pekker, S. et al., Hydrogenation of carbon nanotubes and graphite, in Liquid Ammonia, *J. Phys. Chem.*, 105, 7938, 2001.
39. Lin., Y. et al., Characterization of fractions from repeated functionalization reactions of carbon nanotubes, *J. Phys. Chem.*, 107, 914, 2003.
40. Sun, Y.P. et al., Functionalized carbon nanotubes: properties and applications, *Acc. Chem. Res.*, 35, 1096, 2002.
41. Sun, Y. P. et al., Soluble dendron-functionalized carbon nanotubes: preparation, characterization, and properties, *Chem. Mater.*, 13, 2864, 2001.
42. Dai, L., Mau, A.W.H., Controlled synthesis and modification of carbon nanotubes and C_{60}: carbon nanostructures for advanced polymeric composite materials, *Adv. Mater.*, 13, 899, 2001.
43. Star, A., Trong-Ru, H., Joshi,V., and Gruner, G., Polymer coatings of carbon nanotube sensors, *Polymer Preprints*, 44(2), 201, 2003.
44. Shim, M. et al., Charge-tunable optical properties in colloidal semiconductor nanocrystals, *J. Am. Chem. Soc.*, 123, 11512, 2001.

45. Chen, J. et al., Solution properties of single-walled carbon nanotubes, *Science*, 282, 95, 1998.
46. Iqbal, Z. and Goyal A., carbon nanotubes/nanofibers and carbon fibers, in *Functional Fillers for Plastics*, Wiley-VCH VerlagGmbH & Co KgaA, 181, 2005.
47. Chen, J. et al., Chemical attachment of organic functional groups to single-walled carbon nanotube material, *J. Mater. Res.*, 13, 2423, 1998.
48. Rao et al., Functionalized carbon nanotubes from solutions, *Chem. Commun.*, 1525, 1996.
49. Allongue, P. et al., Covalent modification of carbon surfaces by aryl radicals generated from the electrochemical reduction of diazonium salts, *J. Am. Chem. Soc.*, 119, 201, 1997.
50. Saby, C. et al., Electrochemical modification of glassy carbon electrode using aromatic diazonium salts. 1. Blocking effect of 4-nitrophenyl and 4-carboxyphenyl groups, *Langmuir*, 13, 6805, 1997.
51. Delamar, M. et al., Modification of carbon fiber surfaces by electrochemical reduction of aryl diazonium salts: application to carbon epoxy composites, *Carbon*, 35, 801, 1997.
52. Delamar, M. et al., Covalent modification of carbon surfaces by grafting of functionalized aryl radicals produced from electrochemical reduction of diazonium salts, *J. Am. Chem. Soc.*, 114, 5883, 1992.
53. U.S. Patent No. 7,250, 147, Tour et al.
54. Li, Y. and Ishida, H., Novel thermosetting resins based on 4-(N-maleimidophenyl)glycidylether I. Preparation and characterization of monomer and cured resins, *Polymer*, 44, 6571, 2003.
55. Vlasveld, D.P.N., Vaidya, S.G., Bersee, H.E.N., and Picken, S.J., A comparison of the temperature dependence of the modulus, yield stress and ductility of nanocomposites based on high and low MW PA6 and PA66, *Polymer*, 46, 3452, 2005.
56. Vaia, R.A. et al., Synthesis and properties of two-dimensional nonostructures by direct intercalation of polymer melts in layered silicates, *Chem. Mater.*, 5, 1694, 1993.
57. Vaia, R.A. et al., New polymer electrolyte nanocomposites: melt intercalation of poly (ethylene oxide) in mica-type silicates, *Adv. Mater.*, 7, 154, 1995.
58. Vaia, R., Polymer clay nanocomposites, in *Polymer-Clay Nanocomposites*, Pinnavia, T.J. and Beal, G. W., Eds., John Wiley & Sons, New York, 2000.
59. Utracki, L.A. and Cole, K.C. Eds., Preface, *Proceedings of the 2nd International Symposium on Polymer Nanocomposites, 2004.*
60. Utracki, L.A. and Cole, K.C. Eds., Preface, *Proceedings of the 2nd International Symposium on Polymer Nanocomposites*, 8, 2004.
61. Stewart, R., Nanomaterials, *Plastics Engineering (SPE)*, 62, 12, 2006.
62. U.S. Patent No. 6,864,308 B2, Rosenthal, J.S., Wolkowicz, M.D., (Assigned to Basell Italia S.p.A.), 2005.
63. Twardowska, H. and Dammann, L.G., U.S. Patent No. 6,887,931.
64. Ansarifar, A., Wang, L., Ellis, R.J., and Kirtley, S.P., Method for preparing rubber formulations using silanized silica nanofiller, *Rubber World*, 236, 24, 2007.
65. U.S. Patent No. 5,840,796.
66. U.S. Patent No. 6,849,680 B2 2005.
67. U.S. Patent No. 4,664,842.
68. U.S. Patent No. 6,861,462 B2.
69. Kroto, H.W., et al., C-60—Buckminsterfullerene, *Nature*, 318, 162, 1985.
70. Smalley, R.E., Doping the fullerenes, in *Fullerenes: Synthesis, Properties, and Chemistry of Large Carbon Clusters*, Hammond, G.S. and Kuck, V.J., Eds., Am. Chem. Soc. Symp. Series No. 481, Washington, D.C., 1992, p. 141.
71. Iijima, S. et al., Helical microtubules of graphitic carbon, *Nature*, 354, 56, 1991.
72. Harris, P., *Carbon Nanotubes and Related Structures*, Cambridge University Press, Cambridge, 1999.
73. Yu, M.F. et al., Tensile loading of ropes of single wall carbon nanotubes and their mechanical properties, *Phys. Rev. Lett.*, 84, 5552, 2000.
74. Khare, R. and Bose, S., Carbon Nanotube Based Composites—A Review, *J. Minerals Mater. Characterization Eng.*, 4, 31, 2005.

75. Wong, E.W., Sheehan, P.E., and Lieber, C.M., Nanobeam mechanics: elasticity, strength, and toughness of nanorods and nanotubes, *Science*, 277, 1971–1975, 1997.
76. Liew, K.M., He, X.Q., and Wong, C.H., On the study of elastic and plastic properties of multi-walled carbon nanotubes under axial tension using molecular dynamics simulation, *Acta Materiala*, 52, 2521, 2004.
77. Ebbesen, T.W., Carbon nanotubes, *A. Rev. Mater. Sci.*, 24, 235, 1994.
78. Overney, G., Zhong, W., and Tomanek, D., Structural rigidity and low frequency vibrational modes of long carbon tubules, *Z. Phys.*, D27, 93, 1993.
79. Robertson, D.H., Brenner, D.W., and Mintmire, J.W., Energetics of nanoscale graphitic tubules, *Phys. Rev.*, B45, 12592, 1992.
80. Calvert, P., Strength in disunity, *Nature*, 357, 365, 1992.
81. Dresselhaus, M.S. et al., Chapter 6 in *Graphite Fibers and Filaments*, Gonser, U., Mooradian, A., Muller, K.A., Panish, M.B., and Sakaki, H., Eds., Springer, New York, 1988, p. 120.
82. Treacy, M.M.J. et al., Exceptionally high Young's modulus observed for individual carbon nanotubes, *Nature*, 381, 678, 1996.
83. Hough, L.A. et al., Viscoelasticity of single wall carbon nanotube suspensions, *Phys. Rev. Lett.*, 93, 2004.
84. Weisenberger, M.C. et al., Enhanced mechanical properties of polyacrylonitrile/multiwall carbon nanotube composite fibers, *J. Nanosci. Nanotechnol.*, 3, 39, 2003.
85. Nan, C.W., Shi, Z., and Lin, Y., Capillary-induced filling of carbon nanotubes, *Chem. Phys. Lett.*, 375, 666, 2003.
86. Cadek, M. et al., Mechanical and thermal properties of multiwalled carbon nanotube reinforced polymer composites, in *Conference Proceeding of 9th International Conference of Composite Engineering*, San Diego, CA, 2002.
87. Dalton, A.B. et al., Super-tough carbon-nanotubes—these extraordinary composite fibres can be woven into electronic textiles, *Nature*, 423, 703, 2003.
88. Assouline, E., Nucleation ability of multiwall carbon nanotubes in polypropylene composites, *Polym. Sci.*, Part B 41, 520, 2003.
89. Coleman, J.N. et al., Effect of nanotube inclusions on polymer morphology in composite systems, in *AIP Conferece Proceedings*, New York, 633, 557, 2002.
90. Cadek, M. et al., Mechanical properties and morphology of carbon nanotube reinforced semi crystalline and amorphous polymer composites, *Appl. Phys. Lett.*, 81, 5123–5125, 2002.
91. Ryan, K.P. et al., Photo-luminescence quenching and degradation studies to determine the effect of nanotube inclusions on polymer morphology in conjugated polymer-carbon nanotube composites, in *Conference Proceeding of SPIE*, Galway, 4876, 2002.
92. Cadek, M., Investigations on Toughening of Novel Epoxy Resins for Solid Insulations, Diploma thesis, University of Applied Science Darmstadt, 2000.
93. Cadek, M., Mechanical and thermal properties of CNT and CNF reinforced polymer composites, Austria, in *AIP Conference Proceedings*, New York, 633, 562, 2002.
94. Bhattacharyya, A.R. et al., Effect of encapsulated SWNT on the mechanical properties of melt mixed PA12/SWNT composites, *Chem. Phys. Lett.*, 392, 28, 2004.
95. Ryan, K.P. et al., Photo-luminescence quenching and degradation studies to determine the effect of nanotube inclusions on polymer morphology in conjugated polymer-carbon nanotube composites, in *Conference Proceeding of SPIE*, Galway, 4876, 2002.
96. Ren, Y. et al., Tension–tension fatigue behavior of unidirectional single-walled carbon nanotube reinforced epoxy composite, *Carbon*, 41, 2159, 2003.
97. Zou, Y. et al., Processing and properties of MWNT/HDPE composites, *Carbon*, 42, 271, 2004.
98. Pötschke, P., Bhattacharyya, A.R., and Janke, A., Melt mixing of polycarbonate with multi-walled carbon nanotubes: microscopic studies on the state of dispersion, *Eur. Polymer J.*, 40, 137, 2004.
99. Gojny, F.H. et al., Carbon nanotube-reinforced epoxy-composites: enhanced stiffness and fracture toughness at low nanotube content, *Compos., Sci. Technol.*, 64, 2363, 2004.

100. Gojny, F.H. et al., Carbon nanotube-reinforced epoxy-composites: enhanced stiffness and fracture toughness at low nanotube content, *Compos., Sci. Technol.*, 64, 2363, 2004.
101. Vigolo, B. et al., An experimental approach to the percolation of sticky nanotubes, *Science*, 309, 920, 2005.
102. Beyer, G., Short communication: carbon nanotubes as flame retardants for polymers, *Fire Mater.*, 26, 291, 2002.
103. Kashiwagi, T. et al., Thermal degradation and flammability properties of poly(propylene)/carbon nanotube composites, *Macromol. Rapid Commun.*, 23, 761, 2002.
104. Huynh, W.U., Dittmer, J.J., and Paul Alivisatos, A.P., Hybrid nanorod-polymer solar cells, *Science*, 29, 2425, 2002.
105. Xiangmin, H., Jiong, S., and Yong, Y., Thermoplastic Nanocomposite Foam—online reference—http://www.chbmeng.ohio-state.edu/state.edu/facultypages/-leeresearch/2/ThermoplasticNano.htm.
106. Molinars, H. (Ed.), Nylon nanocomposite resin: Transparent nylons, *Plastics Engineering*, 62(9), 6, 2006.

Appendix 3.A

Mixing Terminology

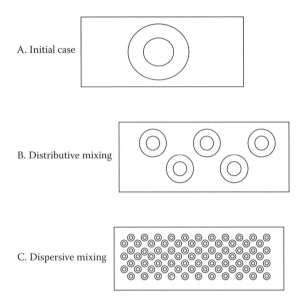

A. Initial case

B. Distributive mixing

C. Dispersive mixing

Consider the above three cases: the problem is to disperse a donut-shaped particle into a box. Let us assume that as the particle breaks, it breaks into progressively smaller particles of similar morphology (has fractal properties). Case A is the initial case. Case B is an example of good distributive mixing. A good case could be made that over the whole area of the box, the donuts are well distributed. However, if we were to divide the box into five or six equal-sized cells, then within each cell there would not be good distribution of the donut. This is where dispersive mixing comes in. Case C shows what would happen if good dispersive mixing were applied.

The key difference in dispersive versus distributive mixing is that there is a characteristic yield point in dispersive mixing that is not seen in distributive mixing. This can be characterized in extrusion or mixing studies. This yield point can correspond to the de-aggregation of an aggregate to form primary particles that are dispersed or, for example, in the case of some synthetic silica, it can correspond to de-agglomeration to yield individual aggregates composed solely of chemically bonded primary silica particles of colloidal size.

4

Nanoparticle Dispersion and Reinforcement by Surface Modification with Additives for Rubber Compounds

Kwang-Jea Kim and James L. White

Contents

4.1 Particle Dispersion and Agglomeration[1]

Aggregates are particles combined by covalent bonds. *Agglomerates* are particles held together by van der Waals attraction and polar bonds. However, the literature is very large and often uses contradictory terminology. Prior to establishing the above terminology, the term "aggregation" was used to include gelling, coagulation, flocculation, and coacervation (liquid precipitation).[1]

Small particles tend to agglomerate. This is due to inter-particle attractive forces. As the particle sizes decrease, the surface area becomes large relative to mass. The large surface areas include agglomeration. This tendency is much greater for polar particles such as oxides (SiO_2, TiO_2, etc.) and carbonates ($CaCO_3$) than for nonpolar particles such as carbon black. When polar particles are suspended in polar media containing ions, they tend to accumulate ions on their surfaces, making them electrically charged. At the negative charged colloidal particle surface, counter-ions accumulate and form an electrical double-layer, called the Stern layer and the diffuse layer. The electrostatic potential is highest at the particle surface and decreases as distance from it increases. Charged particles with counter-ions repel each other, while uncharged particles attract each other.

4.1.1 Particle Agglomeration[2-6]

Nano-sized particles (commonly less than 0.1 µm or 100 nm) are increasingly being developed for the tire, electronic, ceramics, pharmaceutical, biomedical, energy, and catalyst industries. When these particles are charged, they repel each other; uncharged particles are free to collide and agglomerate as shown in Figure 4.1.

To keep each particle separate and prevent particles from gathering into larger agglomerates, the attractive forces between individual particles must be reduced. When colloid particle sizes are very small, the surface forces at the interface of particles are large. One of the major surface effects is electrokinetic. When particles hold the same electrical charge (positive or negative), they produce a force of mutual electrostatic repulsion between adjacent particles. When the degree of charge is high, the particles will remain separate, diffuse, and in suspension, as shown in Figure 4.1a. When the degree of charge is low, it will create the opposite effect. This results in agglomeration as shown in Figure 4.1b. Charge can be modified and

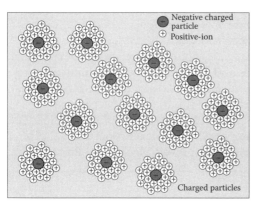
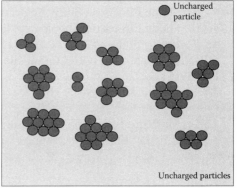

(a) Charged Particles   (b) Uncharged Particles

FIGURE 4.1
Collision behavior of (a) charged particles and (b) uncharged particles.

controlled by changing the liquid's pH or changing the ionic species in solution.[2-6] The most direct technique to prevent agglomeration is the use of surfactants. The surfactants adsorb or are chemically bonded on the particle surface and change its characteristics.

4.1.2 Electrokinetic Potential and Electrical Repulsion

Consider isolated particles suspended in a liquid where the particles have identical charges. The electrokinetic potential is highest at the surface of the particle and decreases as the distance from the surface increases in the suspending liquid. The strength of the electrical barrier of a suspended particle represents the electrical potential, which is called the Zeta potential (ζ).

The Zeta potential is the electrostatic potential generated by the accumulation of ions at the surface of a colloidal particle, which is surrounded by an electrical double layer (the Stern layer and the diffuse layer) as shown in Figure 4.2.[7] The concentration of counter-ions, which consist of a double layer, depends on the type. As the counter-ion concentration in solution increases, the thickness of the double layer increases. The Stern layer is the layer of positive ions that attach to a charged surface. These ions are temporarily bound, and they shield the surface charge. Charge density is greatest near the particle and gradually decreases toward zero as the concentration of positive and negative ions merge together.

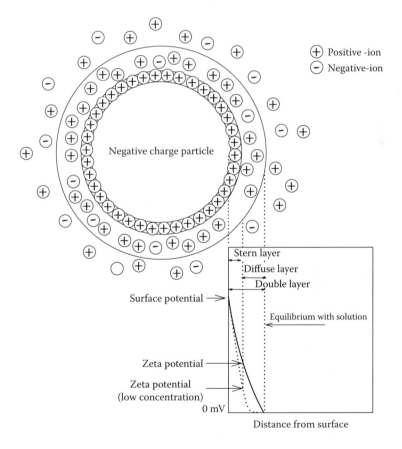

FIGURE 4.2
Particle in a fluid, surrounding double layer (Stern layer and diffuse layer), surface potential, and Zeta potential with respect to distance from the particle surface.

Formation of the double layer neutralizes the charged particle. This double layer leads an electrokinetic potential difference between the surface of the particle and any point in the mass of the suspending fluid. This voltage difference is on the order of millivolts and is referred to as the surface potential. The magnitude of the surface potential is related to the surface charge and the thickness of the double layer. As the distance from the particle surface increases, the surface potential decreases roughly linearly in the Stern layer, and then exponentially through the diffuse layer, approaching zero or equilibrium with the solution at the imaginary boundary of the double layer.

The electrokinetic potential curve indicates the strength of the electrical force between particles and the distance at which this force arises. A charged particle will move at a fixed velocity under an applied voltage. This phenomenon is called electrophoresis. The particle's mobility is related to the dielectric constant, the viscosity of the suspending liquid, and the electrical potential at the boundary between the moving particle and the liquid. This boundary is called the slip plane and is usually defined as the point where the Stern layer and the diffuse layer meet. The Stern layer is considered rigidly attached to the particle, while the diffuse layer is not. The electrical potential at this junction is related to the mobility of the particle and is called the Zeta potential. Although the Zeta potential is an intermediate value, it is sometimes considered more significant than the surface potential as far as electrostatic repulsion is concerned. The Zeta potential can be measured by tracking colloidal particles with a microscope as they migrate under a known applied voltage.

4.1.2.1 Attractive Inter-Particle Forces

More important to the rubbers and plastics are attractive forces between small particles. Nonpolar particles such as carbon blacks (with minimal surface oxidation) have attractive van der Waals forces. This causes them to agglomerate. Polar particles such as calcium carbonate, silica, titanium dioxide, and zinc oxide have dipole-dipole attractive forces, which are much stronger and form more difficult-to-disperse agglomerates.

4.1.2.2 Aliphatic Carboxylic Acid Surface Modified Particles

Aliphatic carboxylic acids first became known when technologists hydrolyzed fats and used the products as soaps and for the creation of emulsions. The chemical structures of these fatty carboxylic acids came to be known in the nineteenth century. By the end of the early twentieth century, it was known that fatty carboxylic acids would form monolayers on the surface of water and reduce its surface tension. The -COOH molecular ends would be immersed in the water and the hydrocarbon chains would extend vertically upward. This technology has become highly advanced.[8]

Aliphatic carboxylic acids were found by researchers at the New Jersey Zinc Company in 1925.[9, 10] They found that introducing small amounts of aliphatic carboxylic acids (e.g., stearic acid, propionic acid) with fine zinc oxide particles clearly enhanced its incorporation into natural rubber. Similar behavior was found to exist with other polar particles such as calcium carbonate and titanium dioxide. By the 1970s, various researchers found that carboxylic acids added to these particulates, significantly reduced the compound viscosity.[11–14]

4.1.3 Clay Dispersion

4.1.3.1 Kaolin Clay[15–20]

In the rubber industry, silicate layer minerals represent the largest volume of filler second to carbon black. The primary silicate mineral used is Kaolin clay. Kaolin is naturally

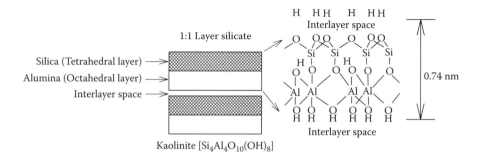

FIGURE 4.3
Crystal structure of kaolinite clay. (Source: From White, J.L. and Kim, K.J., *Thermoplastic and Rubber Compounds: Technology and Physical Chemistry*, Hanser Publisher, 2008.)

occurring and is harvested by mining. Subsequently, it is pulverized and refined by air and water separation processing to remove impurities (mainly quartz and mica). Both "soft" and "hard" clays are used. The classification is based on the rubber compounds produced. Hard clays possess small particles (<2 μm) and soft clays contain larger particles. The name "clay" itself implies that particles of the material are very fine.

Clay minerals are classified into kaolinite, illite, smectite (montmorillonite), chlorite, halloysite, and the vermiculite group. The most important commercial clay minerals are kaolinite and montmorillonite. Kaolin or kaolinite, known as China clay, has the basic chemical formula $Si_2Al_2O_5(OH)_4$. Montmorillonite is $AlSi_2O_5(OH) \cdot xH_2O$. Kaolinite is formed by various hydrothermal alterations or weathering of feldspars, and other silicate minerals.

Comparing the structure of kaolin and montmorillonite, kaolin has 1:1 layer lattice and montmorillonite has 2:1 lattice structure. Kaolin consists of successive layers of alumina and silicate mineral, whereas montmorillonite has two layers of silicate with a layer of alumina sandwiched between the layers.[15–19] Kaolinite, first given by Gruner in 1932, has alternating octahedral alumina and tetrahedral silica sheets (1:1 layer silicate) (see Figure 4.3).[20]

4.1.3.2 Montmorillonite Clay

Montmorillonite is a clay similar to kaolinite but differs in its detailed structure. Montmorillonite clay is a hydrous alumina silicate mineral whose lamellae are constructed from an octahedral alumina sheet sandwiched between two tetrahedral silicate sheets (2:1 layer silicate) (see Figure 4.4). The crystal structure of montmorillonite was worked out by Hoffmann et al.[21, 22] Montmorillonite is similar but different from kaolinite in that the silicate surface of montmorillonite exhibits a negative charge, and these surfaces adsorb cations such as Na^+ or Ca^{2+}. It was found that montmorillonite clay can absorb large amounts of water and polar liquids, which separate the silicate layers. Indeed, montmorillonite can absorb 20 to 30 times (or more) its volume in water. Hofmann et al.[21] found that the swollen clays were able to continue to diffract x-rays ((001) spacing) up to high water contents.

In 1934, Smith[23] found that montmorillonite absorbs varying amounts of polar organic liquids, especially amines. This finding was confirmed in subsequent studies by several researchers.[24–29]

4.1.3.3 Clay Dispersion with Polar Polymers

Generally, polar polymers can be absorbed into montmorillonite. This can result in the breakup of the layered structure effectively dispersing it. The earliest efforts were by

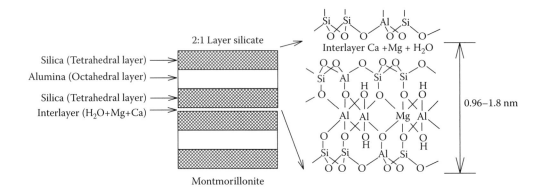

FIGURE 4.4
Crystal structure of montmorillonite clay. (Source: From White, J.L. and K.J. Kim, *Thermoplastic and Rubber Compounds: Technology and Physical Chemistry*, Hanser Publisher, 2008.)

Unitika and Toyota researchers with highly polar polyamides.[30–33] It was only in later years that it became clear that this ability was related to polymer polarities.[34, 35] Various halogenated polyolefins could be used to produce "nanocomposites," but not poly-ethylene or polypropylene. Polystyrene can only be used to produce nanocomposites with difficulty but this readily occurs with styrene-acrylonitriles copolymers. Kim and White[36] sought to correlate the ability of polymers to absorb into the clays in terms of their dielectric constants. As early as 1997, Kawasumi et al.[37] found it necessary to graft maleic anhydride onto polypropylene for it to be absorbed into montmorillonite. There have been many investigations of the absorption of maleated polypropylene into mont-morillonite since that time.

The first important factor to achieve the exfoliated and homogeneous dispersion of the layers is the intercalation capability of polymers between the layers.

4.2 Silica Particle Dispersion with Organosilanes

4.2.1 Background

It has been found that very small polar particles are difficult to disperse through process-ing equipment. With SiO_2 at dimensions on the order of 0.01 to 0.05 µm or 10-50 nm, this is especially true.[38–40] Very polar particles are more difficult to disperse than nonpolar particles due to increased dipole-dipole forces, as seen in silica particles as opposed to car-bon black particles.[39] Silica particles tend to easily agglomerate and re-agglomerate, even after mixing. Silica particles are a polar mineral due to the differences in electronegativity between the silicon (1.90) and oxygen (3.44) atoms, while carbon blacks are nonpolar due to the same electronegativity between carbon atoms (2.55). The silica particle in suspension is negatively charged at a certain pH level because of the loss of protons from H_3O^+ mol-ecules located in the spaces between the oxygen atoms of the SiO_2 structure.[1, 41]

Silica-filled tires, which have been called "green tires," show a lower rolling resistance and higher traction than traditional carbon black (CB) tires. This improves gasoline mileage and gives a good grip on road surfaces.[42–44] A high level of wear-resistance with low rolling resistance was made possible using silica in the rubber matrix.[43–46] However, silica particles were well known to be hard to process and known to consume significant energy during processing due to their polar character.[39, 47–50] This is because silica particles are agglomerated not only with van der Waals, but also from dipole-dipole forces.

4.2.2 Theory of Silane Hydrolysis and Condensation with Silica Surface

The chemical structures of various silane coupling agents, which include bis (triethoxysilylpropyl)tetrasulfide (TESPT or SN069), hexadecyltrimethoxysilane (trimethoxysilylhexadecane or SN116), propyltriethoxysilane (triethoxysilylpropane or SN203), and octyltriethoxysilane (triethoxysilyloctane or SN208), are summarized in Table 4.1. These coupling agents contain ethoxy groups (SN069, SN203, SN208) and methoxy groups (SN116), which react chemically with the hydroxyl groups on silica surfaces. SN069 has a chemical structure of four sulfur atoms located in between the triethoxysilyl propyl groups. The general purpose of the sulfur atoms in TESPT is to chemically bond with double bonds in rubber chains. The purpose of the alkoxy groups is to chemically couple with hydroxyl groups in silica surfaces via a hydrolysis reaction. Thus, TESPT chemically bridges the silica surfaces and the rubber chains.

Hexadecyltrimethoxysilane (trimethoxysilylhexadecane; SN116) consists of three methoxy functional groups and a sixteen-carbon alkyl chain attached to a silicon atom. Propyltriethoxysilane (triethoxysilylpropane; SN203) consists of three ethoxy functional groups and a three-carbon alkyl chain attached to a silicon atom. Octyltriethoxysilane (triethoxysilyloctane; SN208) consists of three ethoxy functional groups and an eight-carbon alkyl chain attached to a silicon atom. McNeish and Byers[45] have described coupling methods of silanes with silica particles.

As the silica particle size decreases to nanometer dimensions, its surface area per unit volume increases both the van der Waals force and the accumulated hydrogen-bonding force between silica particles. To disperse agglomerated silica particles, the inter-particle forces should be reduced by coating the particle surfaces with additives that make the surfaces more compatible with the surrounding polymer.

The surface of silica can be modified with surfactant chemicals such as the silanes of Table 4.1. Silanes help silica dispersion in two ways. First, the chemical coupling of a silane on the silica surface changes the hydrophilic silica surface to a hydrophobic surface, which makes the surfactant-treated silica surface compatible with hydrophobic polymer chains. The compatibilized silica particle agglomerates are easily broken down during the mixing

TABLE 4.1

Chemical Structures of Silanes Used and Codes

Name	Chemical Conformation	Code (Degussa)
Bis(triethoxysilylpropyl)tetrasulfide (TESPT)	$[(EtO)_3Si-(CH_2)_3]_2-S_4$	SN069
Hexadecyltrimethoxysilane(trimethoxysilylhexadecane)	$(MeO)_3Si-C_{16}H_{33}$	SN116
Propyltriethoxysilane (triethoxysilylpropane)	$(EtO)_3Si-C_3H_7$	SN203
Octyltriethoxysilane(triethoxysilyloctane)	$(EtO)_3Si-C_8H_{17}$	SN208

Note: MeO= CH_3O, EtO= C_2H_5O.

TABLE 4.2

Surface area of particles and rubber

Material	Supplier Product Name	Nitrogen Desorption (BET) Surface Area(m²/g) (Diameter:nano meter)	Code
Silica	PPG		
	*Hi-Sil(R)255LD (untreated)	185 (18)	S185
	*Ciptane 255LD (treated)	170 (18)	S170T
	*Silene 732D (untreated)	35 (75)	S35
Carbon Black	Cabot		
	* Vulcan 9 (untreated)	143 (21)	CB143 (N110)
	* Sterling NS-1 (untreated)	29 (89)	CB29 (N762)

stage as compared to untreated silica particles. The silane-bonded silica surface, which now has improved compatibility with polymer chains, will rather easily "wet out" (peel off) silica particles from the surface of silica aggregates during mixing. Second, the coupling of silane on silica surface reduces the number of hydroxyl groups on the silica surface per unit volume, which implies the reduction of the hydrogen bonding forces between silica particles. Thus, the coupling of silane on the silica surface improves silica particle dispersion. To chemically couple alkoxysilanes onto silica surfaces, the silane should be first hydrolyzed and then the hydroxyl group in the hydrolyzed silane should be condensed with the hydroxyl group on the silica surface. This method somewhat reduces the polarity of the silica surface. The surface areas of typical silicas and carbon blacks that have been investigated are summarized in Table 4.2.

Alkoxysilanes undergo hydrolysis by both base- and acid-catalyzed mechanisms as shown in Figure 4.5.[51, 52] The acid-catalyzed hydrolysis mechanism is an S_N2-Si type mechanism in which the incoming nucleophile and outgoing leaving group are further from the sp^2 hybridized silicone atom than the nucleophile and the leaving group in an sp^3d hybridized silicone atom of a pentacoordinate intermediate as shown in Figure 4.5a. This mechanism is a rapid equilibrium protonation of the substrate, followed by a bimolecular S_N2-type displacement of the leaving group by water, which has one transition stage, while the base-catalyzed mechanism, which has two transition stages, is slow as shown in Figure 4.5b.[52, 53] The hydrolyzed silanes undergo a condensation reaction with hydroxyl group on the silica surface. Figure 4.6 represents the coupling mechanism between the silica surface and the silane, which includes a silane condensation mechanism.

Allen[54] described a hydrolysis and condensation model of trialkoxysilanes derived from Kay and Assink[55] model as shown in Figure 4.7. The rate of the acid hydrolysis reaction is significantly greater than that of the base hydrolysis and is minimally affected by other carbon-bond substituents. Hydrolysis and condensation reactions are affected by the pH level of the solution. The slowest rate of hydrolysis occurs at an approximately neutral pH.[56, 57] The condensation reaction is also controlled by pH, with a minimum at pH 4. Alcohols reverse the hydrolysis reaction, stabilizing the solution for extended periods; thus, removal of alcohol in the compound eliminates the possible reverse reaction in the compound.

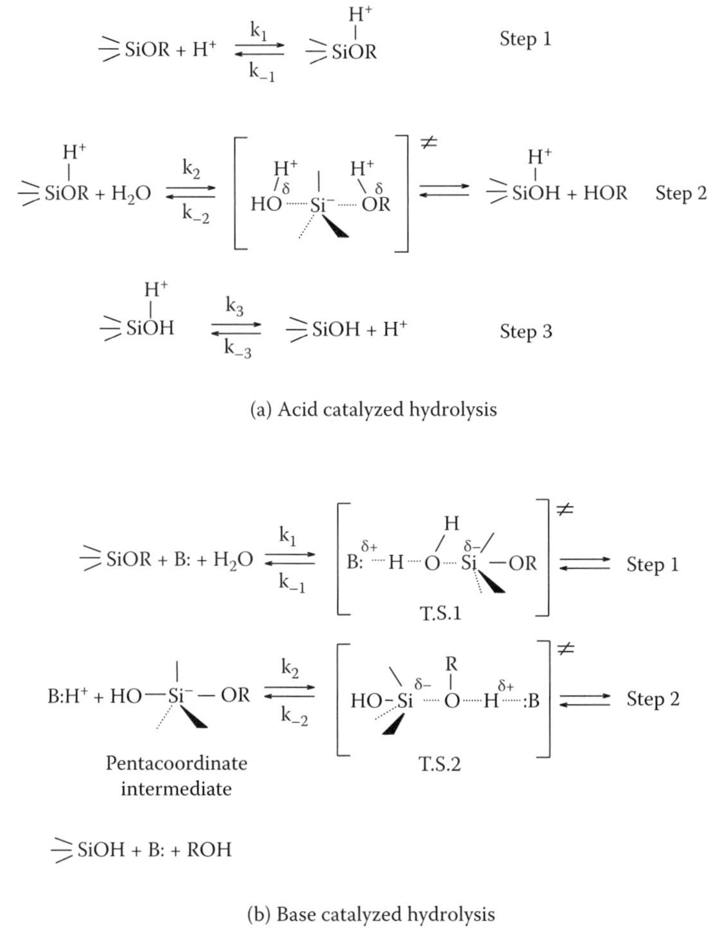

(a) Acid catalyzed hydrolysis

(b) Base catalyzed hydrolysis

FIGURE 4.5
Hydrolysis reaction mechanism: (a) acid-catalyzed system and (b) base-catalyzed system. (Source: Mittal, K.L., Ed., *Silane Coupling Agents*, VSP, Utrecht, 1992; McNeil, K.J., J.A. DiCapri, D.A. Walsh, and R.F. Pratt, *J. Am. Chem. Soc.*, 102, 1859, 1980.)

The increased level of condensation interferes with bond formation. The completion of hydrolysis thus takes longer at room temperature.[56] Moisture affects the vulcanization time and mechanical properties of the silane-silica compound.[58, 59] This appears to be due to improved silane coupling on the silica surface.

The theories on silane coupling mechanism are well established.[52–55] However, there has been little research on the effects of large amounts of moisture in silane-silica coupling. More recently, a new mechanism for large amounts of moisture effects on silica-silane coupling was introduced.[60–64] Enough water molecules need to be supplied to achieve a hydrolysis reaction of an alkoxysilane. In moisture-starved conditions (i.e., in less than three moles of water molecules), the hydrolysis reaction stops before all the alkoxy groups are hydrolyzed (see Figure 4.7). The condensation reaction then takes over. The fully hydrolyzed silane would finish 100% coupling on silica surface (bottom right) via the condensation reaction; however, the partially hydrolyzed silane (top middle), due to moisture

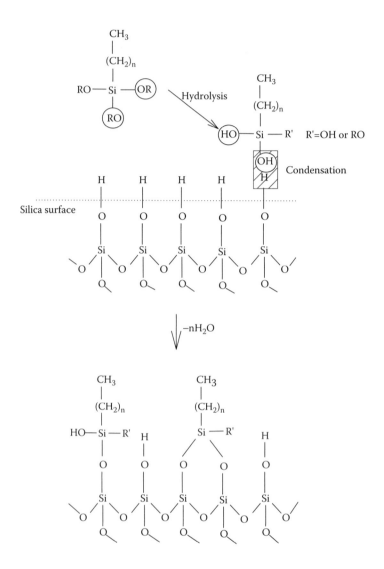

FIGURE 4.6
Coupling mechanism between silica surface and silane. (Source: Redrawn from Kim, K.J. and J.L. White, *J. Ind. Eng. Chem.*, 7, 50, 2001.)

starvation, would react only with one or two hydroxy groups in the silane, resulting in one or two alkoxy groups remaining in the silane (bottom middle).

Figure 4.8 presents the various silica particles as normal, condensed, and water molecule emulsified forms.[60-64] Normal silica particles are agglomerated with hydrogen/polar bonding (Figure 4.8, top). Under hot storage or shipping conditions, the nano-sized silica particles are chemically aggregated easily due to a self-condensation reaction. Once they are chemically bonded to each other, it is difficult to break down during processing (Figure 4.8, middle). These types of silica particles were investigated by Schaal et al.[65] Sufficient moisture on adsorbed silica particles causes water molecules to react with alkoxy groups in a silane molecule (Figure 4.8, bottom). Thus, when an alkoxy silane approaches the silica surface, the water molecules, which are emulsifying on silica surface, react with the

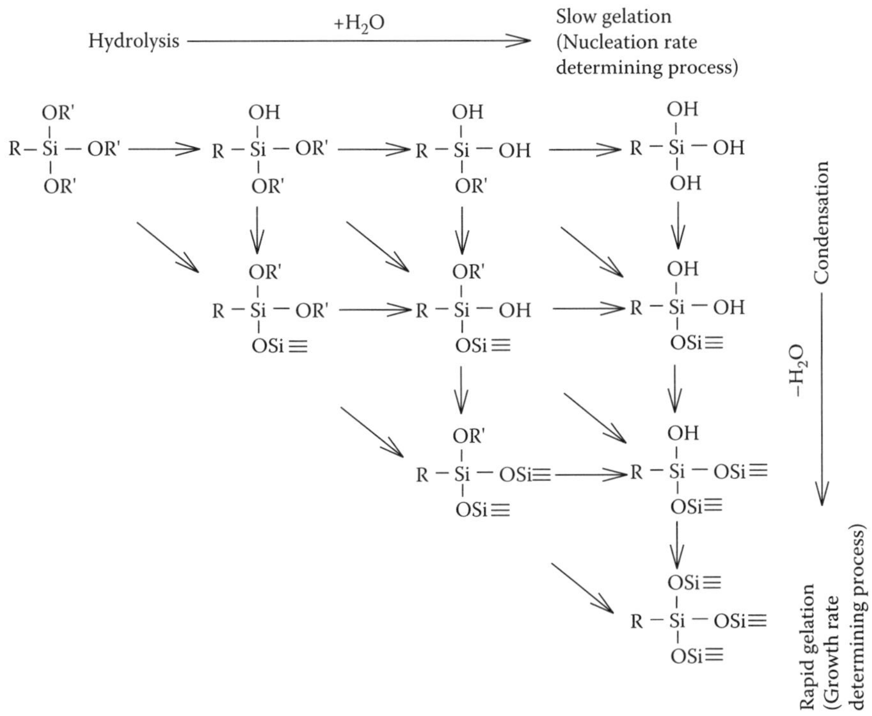

Hydrolysis ——————— $+H_2O$ ——————→ Slow gelation (Nucleation rate determining process)

$$\begin{array}{ccc}
\text{OR'} & \text{OH} & \text{OH} & \text{OH}\\
| & | & | & |\\
\text{R}-\text{Si}-\text{OR'} \longrightarrow & \text{R}-\text{Si}-\text{OR'} \longrightarrow & \text{R}-\text{Si}-\text{OH} \longrightarrow & \text{R}-\text{Si}-\text{OH}\\
| & | & | & |\\
\text{OR'} & \text{OR'} & \text{OR'} & \text{OH}
\end{array}$$

$$\begin{array}{ccc}
\text{OR'} & \text{OR'} & \text{OH}\\
| & | & |\\
\text{R}-\text{Si}-\text{OR'} \longrightarrow & \text{R}-\text{Si}-\text{OH} \longrightarrow & \text{R}-\text{Si}-\text{OH}\\
| & | & |\\
\text{OSi}\equiv & \text{OSi}\equiv & \text{OSi}\equiv
\end{array}$$

$$\begin{array}{cc}
\text{OR'} & \text{OH}\\
| & |\\
\text{R}-\text{Si}-\text{OSi}\equiv \longrightarrow & \text{R}-\text{Si}-\text{OSi}\equiv\\
| & |\\
\text{OSi}\equiv & \text{OSi}\equiv
\end{array}$$

$$\begin{array}{c}
\text{OSi}\equiv\\
|\\
\text{R}-\text{Si}-\text{OSi}\equiv\\
|\\
\text{OSi}\equiv
\end{array}$$

Condensation — $-H_2O$ —

Rapid gelation (Growth rate determining process)

FIGURE 4.7
Hydrolysis and condensation cascade reaction mechanism of trialkoxy silanes. (Source: Allen, K.W., *Silane Coupling Agents*, K. L. Mittal, Ed., , P.93, VSP, Utrecht, 1992; Kay, B.D. and R.A. Assink, *J. Non-Cryst. Solids*, 104, 112, 1988.)

alkoxy group (see Figure 4.8, bottom). This provides an improved hydrolysis reaction to approaching alkoxy silanes compared to the moisture-starved conditions (see Figure 4.8, top and middle).

During mixing, the polar-bonded silica agglomerates disperse easily via the presence of large amounts of polar water molecules, because hydrogen bonding is about five times stronger than polar bonding. A large amount of water molecules loosens the agglomeration force between silica particles. This facilitates dispersion of the silica particles in a polymer matrix. Well-dispersed silica particle compounds produce less heat build-up (HBU) and show better mechanical properties.[61]

The steric hindrance theory, in which a third alkoxy group in a trialkoxysilane does not hydrolyze (i.e., 33% alkoxy group remain in the silane) due to steric hindrance, has been accepted in the silane hydrolysis process as a mechanism.[68–70] The unreacted alkoxy group remains and later undergoes a secondary hydrolysis reaction and produces alcohols in the final stage. However, water-molecule-treated silica systems have shown additional hydrolyzation of the alkoxy group, remained in a silane (alkoxy group remains in the silane 12%, 19%, 29%), and produced fewer alcohols than the steric hindrance theory value. It also generates less volatile organic chemicals (VOCs) and improves silica particle dispersion as well as the mechanical properties of the compound.[61–64, 66, 67] Well-dispersed silica particles reduce filler-filler interactions (such as the Payne effect) and give better mechanical properties.[71]

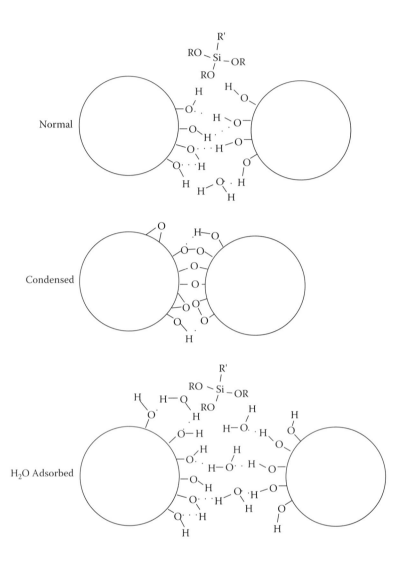

FIGURE 4.8
Schematic presentation of various silica particles existing in normal, condensed, and H$_2$O-adsorbed form. (Source: Kim, K.J. and J. Vanderkooi, *Int. Polym. Process.*, 19, 364, 2004; Kim, K.J. and J. Vanderkooi, *Composite Interfaces*, 11, 471, 2004.)

Once the silica particles are treated with silanes, the particle interactions decrease due to reduced inter-particle reactions and polarity, and reduced hydrogen bonding on the silica surface.[72, 73]

4.2.3 Silane Effects on Silica Dispersion

The silica surface is usually chemically modified to improve its performance in a rubber matrix. It is possible to chemically bond a silica surface with organic chains known as silanes.[74–80] These compounds react with silanol groups on the surface of the silica via a hydrolysis reaction. The organic chains are chemically bonded to the silica surface. The chemicals should react with each other at an optimum temperature and distribute well in the compound. To improve the uniformity of distribution and temperature control, which also prevent silane degradation during processing and longer mixing cycles with decreased numbers of mixing stages, an intermeshing rotor was applied.[81, 84]

The treatment of a sulfur-containing bifunctional organosilane coupling agent on a silica surface increases the tensile modulus, tear strength, hysteresis, and abrasion resistance properties of the compounds.[76] The levels of sulfur atoms in a silane chain affect the stability of silane during processing.[85] Silica grafted with rubber was previously studied and compared with silane-treated systems in a rubber-silica system.[74] The surface energies of the silicas and their interactions were also studied.[86, 87] Kim and White presented the effects of different chain length silanes, and their ability to enhance the dispersion of silica agglomerates during mixing.[88, 89] Silica particles and silanes were compared.[88] This work was extended to materials including a treated silica particle and an additional silane.[89] This was intended for the investigation of chain-length effects on silica dispersion and for the investigation of commercially treated silica particle dispersions.

4.2.4 Silica Coupling with Bifunctional Organosilanes

Bifunctional silane, bis(triethoxysilylpropyl)tetrasulfide (TESPT), which contains two triethoxysilylpropyl groups and four sulfur atoms, has been studied since the 1970s in an effort to improve the chemical bonding between the silica surface and unsaturated elastomer chains.[90, 91] It was introduced by Michelin in tires in the 1990s.[43, 44] Later, in an effort to overcome the pre-scorching problem of TESPT, the bis(triethoxysilylpropyl)disulfide (TESPD), which contains two triethoxysilylpropyl groups and two sulfur atoms, was introduced.[85] These bifunctional silanes have been widely used in the "green tire" manufacturing industry as a chemical bonding agent as well as a processing aid.[90, 91]

Good dispersion of particles is proportionally related to a good reinforcement of the compound. The primary silica particle sizes range from several to hundreds of nanometers. The dispersion of agglomerated primary silica particles has been a big issue in the "green tire" manufacturing industry. There have been numerous efforts to disperse the agglomerated silica particles.[38, 40, 74, 81, 88, 89, 92] Examples include *in situ* processing, improved mixer rotor design, application of various different silanes, ultrasound treatment on silica particles, etc.[38, 40, 81, 88, 89] The polymer chain was chemically grafted onto the silica surface during the precipitation stage of silica *in situ* processing and compared with silane-treated systems by Kohjiya and Ikeda.[74] The change in rotor design from tangential to intermeshing was originally introduced to increase dispersive mixing by providing for mixing between rotors. It was subsequently found to control the temperature in the being mixed stock and to prevent the thermal degradation of silane in the compound during mixing and to decrease the mixing cycle by Heiss.[81] Silane chain-length effects on improved dispersion and processability of the silica filled compounds have been studied by Kim and White.[88, 89] Ultrasound treatment on silica agglomerates was applied to break down the silica agglomerates in the elastomer matrix and was found to provide improved silica agglomerate dispersion.[38, 40] Different silane applications on magnetic particles and their effects on dispersion and rheological properties were reported in an effort to replace the acrylate binder.[92] Silanes can change the hydrophilic surface of a material to hydrophobic. Monolayer coating of trichlorosilane solutions on the glass wall changed the micro-channel to a hydrophobic surface, suggesting control of microfluidic systems.[93]

4.2.5. Silane Effects on Silica Agglomerate Size Reduction

The treatment of silane coupling agents on silica agglomerate further reduced their size during processing.[39, 88, 89]

Typical SEM photomicrographs for various coupling agent treated agglomerates at 20 vol% and 30 RPM are shown in Figure 4.9.[39] The S185/propyltriethoxysilane(triethoxysi-lylpropane) (S185/SN203) system exhibits the smoothest surface compared to the other systems. In previous work (coupling of silanes with silica particles), the current

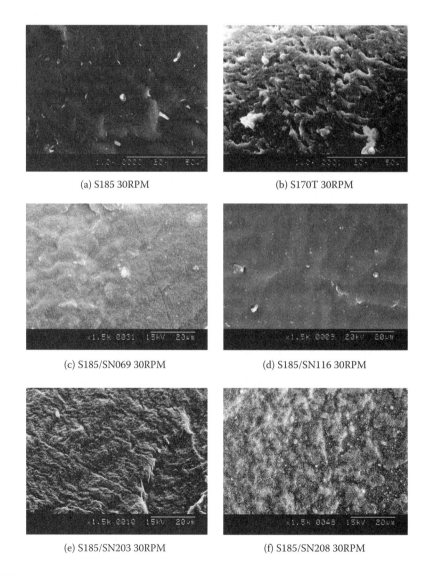

(a) S185 30RPM   (b) S170T 30RPM

(c) S185/SN069 30RPM   (d) S185/SN116 30RPM

(e) S185/SN203 30RPM   (f) S185/SN208 30RPM

FIGURE 4.9
SEM photographs of (a) S185, (b) S170T, (c) S185/SN069, (d) S185/SN116, (e) S185/SN203, and (f) S185/SN208 in EPDM matrix at 20 vol% and 30 RPM.

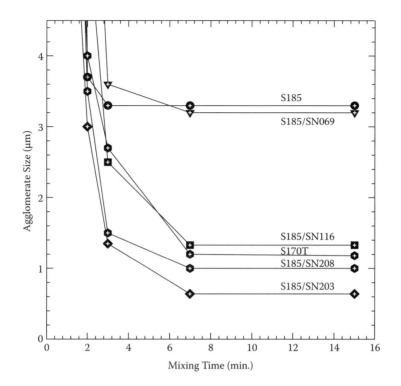

FIGURE 4.10
Agglomerate size as a function of mixing time for silane-treated silica in EPDM at 20 vol%, 60 RPM.

authors presented SEM photomicrographs of silane-treated silica agglomerate at 20 vol% and 60 RPM, and the S185/SN203 system exhibited the smoothest surface compared to the other treated systems. In Figure 4.10, the different coupling agent-treated silica agglomerate sizes and S170T compound are plotted as a function of the mixing time with a 60-RPM rotor speed and 20 vol% loading.[39] The data for the different coupling agent-treated silica varied relative to the mixing time. The SN069-treated silica agglomerates were the largest and closest to the untreated silica, while the agglomerate size of the S170T compound was closest to those of the hexadecyltrimethoxysilane(trimethoxysilylhexadecane) (SN116) and octyltriethoxysilane(triethoxysilyloctane) (SN208) treated compounds. The agglomerate size decreased exponentially and reached dimensions on the order of 0.6 to 4 m. The S185/SN203 system exhibited the smallest agglomerate particle size.

From the above observations, the following conclusion can be made: first the silica agglomerates of the propyltriethoxysilane(triethoxysilylpropane) (SN203) treated system were smaller than those of the other treated systems and the untreated silica systems. When small-chain silane bonded on the silica surface, it formed a brush-like shape, as illustrated in Figure 4.11a. The probability of short-chain silane bonding on a silica surface is higher than that of long-chain silane. At the same weight, SN203 has more molecules than SN116, which implies that more functional groups exist in SN203 to bond to the silica surface. This would seem to reduce the silica agglomerate sizes. The second trend was that

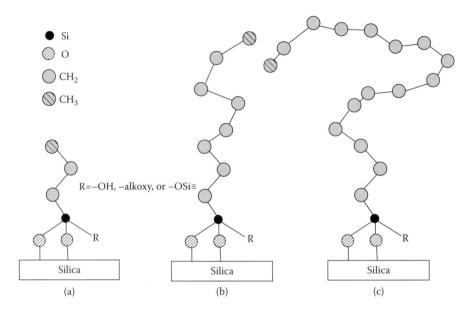

FIGURE 4.11
Schematic presentation of different chain length silanes bonded on silica surface (a) SN203, (b) SN208, and (c) SN116.

as the chain length of the coupling agents decreased, the agglomerate size also decreased. This appears to be related to steric hindrance of the long-chain silane and a polarity reduction on the silica surface. When a long aliphatic chain silane bonded on the silica surface, it causes steric hindrance to the next chain approaching to the silica surface as shown in Figure 4.11b.[88] Similar explanations for other systems can be found from the research of Mizoguchi and Hamada.[94]

The effect of the rotor speed on the agglomerate size was investigated.[137] In Figure 4.12, the agglomerate breakdown for each coupling-agent-treated system at 30 and 60 RPM was compared. The experiments at 60 RPM produced smaller agglomerate sizes compared to the 30-RPM experiments. An increasing rotor speed reduced the agglomerate size of the untreated and treated S185 compounds; however, their agglomerate size was not significantly reduced at 60 RPM. The size of the silica agglomerate was more significantly reduced by the treatment of silane on the silica surface rather than by increasing the rotor speed, except for the S185/SN069 compound.

The number of ultimate particles in an agglomerate is plotted as a function of the particle BET surface area in Figure 4.13.[40, 89] For carbon black CB143, the number of ultimate particles per agglomerate was about 12. For the untreated S185, the number of ultimate particles per agglomerate was about 116. For the compounds of SN069, S185/SN116, S170T, S185/SN208, and S185/SN203 added systems, the numbers of ultimate particles per agglomerate decreased to 110, 84, 66, 48, and 40, respectively. The commercially treated S170T appeared between S185/SN116 and S185/SN208.

Clearly, the silane coupling agents help the dispersion of silica agglomerates during mixing.

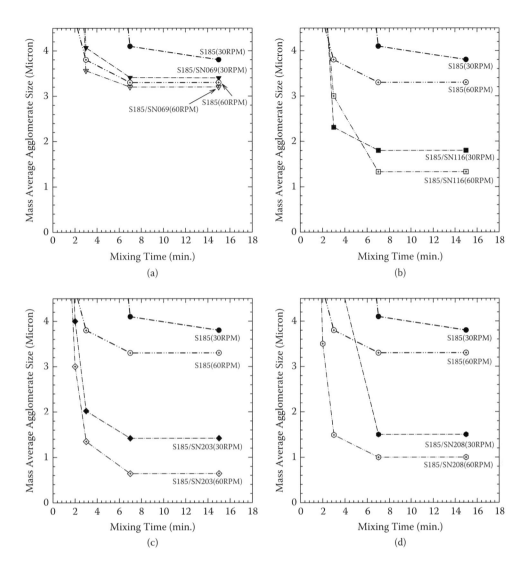

FIGURE 4.12
Agglomerate size as a function of mixing time for (a) S185/SN069, (b) S185/SN116, (c) S185/SN203, and (d) S185/SN208 in EPDM system at 20 vol%, and 30 and 60 RPM.

4.2.6 Silane Effects on Rheological Properties

Figure 4.14 presents the shear viscosity of all the different compounds, including silane-treated systems and untreated systems in EPDM. The S185/SN116, S185/SN069, and S185/SN208 systems exhibited lower viscosities relatively, while the S185, S170T, and S185/SN203 systems exhibited the highest viscosities at 20 vol%. It should be noted that S185/SN069 and S185/SN116 showed the most significant viscosity reduction. The viscosities of the S185/SN069 and S185/SN116 compounds were close to that of the carbon black CB143 compound (Table 4.2).

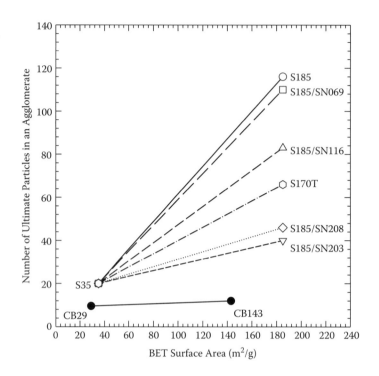

FIGURE 4.13
Number of ultimate particles in agglomerate as a function of BET surface area of silane-treated silica in EPDM matrix at 20 vol%, 60 RPM.

The carbon black filled systems exhibited a lower viscosity than silica or any treated silica system.

4.2.7 Interpretation

Clearly, the silane coupling agents are playing the role of surfactants in dispersing the silica agglomerates during mixing.

In Figure 4.13 we plot agglomerate size for silica and all the different surface treatment systems after 7 minutes of mixing as a function of BET surface area. There are clear correlations. First, the silica agglomerates of the SN203-treated system are smaller than those of the other treated systems and the untreated silica system. The second trend is that as the chain length of coupling agents decreases, the agglomerate diameters also decrease.

We can convert the particle's BET surface area to the size of the ultimate particle and divide the agglomerate size by this number. This gives an estimate of the number of ultimate particles in an agglomerate. We plot this as a function of BET particle surface area in Figure 4.13. For carbon black CB143, we find that the number of ultimate particles per agglomerate is about 12. For untreated S185, the number of ultimate particles per agglomerate is approximately 180. For surface-treated S185 with SN116 and SN203, the numbers of ultimate particles per agglomerate decreased to the range 40 to 50.

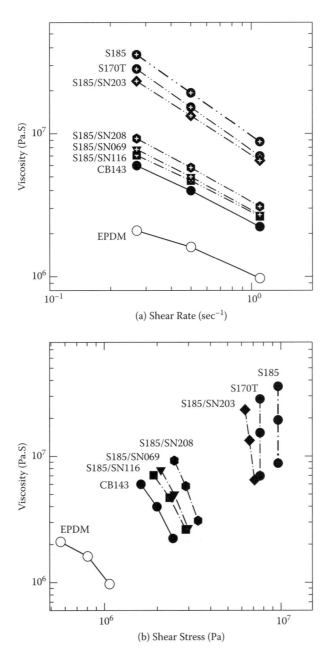

FIGURE 4.14
Viscosity as a function of (a) shear rate and (b) shear stress of EPDM filled with 20 vol% of silane-treated silica.

The roughly spherical ultimate particles of carbon black and silica are fused to other ultimate particles through covalent chemical bonds to form "aggregates." This gives carbon black and silica what is called "structure." We can also estimate the number of aggregates rather than the number of ultimate particles in an agglomerate. The question is: how many ultimate particles are there per carbon black or silica aggregate? This can be judged

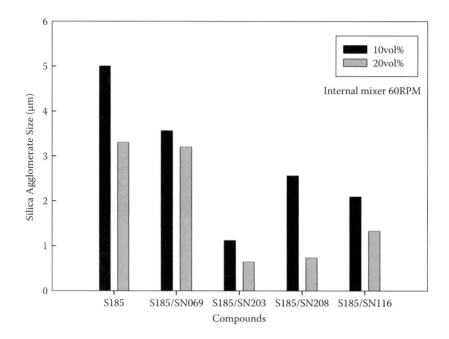

FIGURE 4.15
Silica size reductions by the treatment of silane coupling agents processed through an internal mixer at 60 RPM (10 vol% and 20 vol%).

from studying transmission electron photomicrographs. It is clearly seen that the number of ultimate particles per aggregate is higher for carbon black than for silica, but it varies from grade to grade. The numbers observed for carbon black vary from about 7 to 35. Here, 7 is used as an estimate for carbon black and 5 for silica. This suggests that small carbon black particle (CB143) agglomerates have about 2 aggregates. For S185, an agglomerate of untreated particles has roughly 37 aggregates, and SN203-treated S185 has about 8 aggregates.

Figure 4.15 summarizes the silane effect on silica agglomerate dispersion observed from 10 vol% and 20 vol% compounds (Figure 4.10 and Figure 4.12) processed through an internal mixer at 60 RPM. Monofunctional silane-treated compounds (S185/SN116, S185/SN203, and S185/SN208) showed smaller agglomerate sizes compared to bifunctional silane-treated compounds (S185/SN069) at 10 vol% and 20 vol%, respectively. It seems related to low-molecular-weight silanes having more functional groups per unit weight to chemically react with silica surface, thus reducing particle interaction between silica particles (as described above in the section on Zeta potential).

Processing of those compounds with two-roll mill and ultrasound further reduced silica agglomerate size.[38, 40]

The viscosity level of an untreated S185-filled system exhibited the highest viscosity level compared to treated S185-filled systems. Carbon black-filled systems exhibited a lower viscosity level than silica or any treated silica system.

The capillary extrudate swell ratios of each compound were compared. Untreated silica compounds exhibited significant reductions in extrudate swell. Presumably, strong

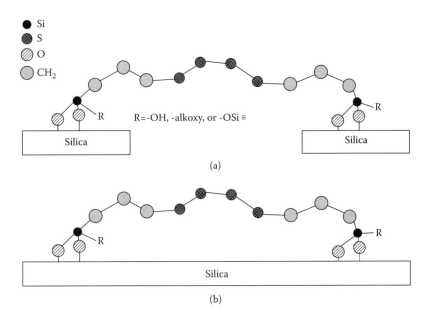

FIGURE 4.16
Schematic presentation of TESPT bonded on silica surface (a) between silicas and (b) silica itself.

silica-polymer networks were formed. The compounds containing silanes exhibited large extrudate swell. Presumably the surfactant action of the silane reduces the tendency to form silica-polymer chain networks.

The TESPT (SN069)-treated silica system did not show significant agglomerate particle size reduction compared to the other silane coupling agent treated silicas; however, they exhibited good processability during processing close to that of the SN116-treated system. The ethoxy groups on both ends of the SN069 chain probably bonded to adjacent silica particles (see Figure 4.16a) or bonded to an individual silica surface itself (see Figure 4.16b).[89] This (Figure 4.16b) shape seemed to reduce the viscosity of the S185/SN069 compound. The bonding types shown in Figure 4.16 did not seem to affect the bonding between the silica and the rubber in the end because the sulfur-sulfur bonds easily split by shear and heat during the mixing procedure, and then bond to rubber at 160°C and form a silica-silane-sulfur-rubber bond at the curing stage, as previously pointed out by others.[81]

The silica compounds of 60 RPM, which experienced higher shear stress than at 30 RPM in the internal mixer, exhibited smaller agglomerates than the 30-RPM-treated silica compounds. There seems to exist a critical stress breakdown of silica agglomerates.

The silane-treated silica compounds exhibited smaller agglomerates than the untreated silica compounds; however, they still exhibited larger agglomerates than compounds based on carbon black of the same particle size.

Among the agglomerate sizes of the treated compounds, the short-chain silane coupling agent (SN203) treated compounds exhibited a more significant agglomerate particle size reduction than the other silane coupling agent (SN069, S170T, SN116, SN208) treated compounds. The effect of polarity reduction seems to be greater on a short-chain silane because it seems to have more functional groups bonded to the silane. The reduced polarity of the silica agglomerates treated with silane meant that it split more easily with less shear stress than the untreated agglomerates.

The addition of silane-treated silicas to elastomers reduced the viscosity level of the S185 compounds in all systems. Among them, the SN116 and TESPT (SN069) treated systems exhibited the lowest viscosity compared with the other silane-treated systems as shown in Figure 4.14. Emulsified silica with silane made the flow easier, as shown in Figure 4.14.

In summary, various organosilane coupling agents with different alkyl chain lengths were treated on silica particles and compounded in EPDM matrix using an internal mixer, and their agglomerate particle sizes and viscosity were investigated. The treated silica compounds showed a smaller agglomerate size and lower viscosity than the untreated silica compound after equivalent mixing times. The TESPT-treated silica compound exhibited a lower viscosity than the other treated compounds but exhibited a large agglomerate size, whereas the short-chain silane- (SN203-) treated compounds exhibited a smaller agglomerate size with a higher viscosity. The long-chain silane- (SN116-) treated compound exhibited a lower viscosity with better dispersion than the TESPT- (SN069-) treated compound. The TESPT (SN069) acted as a processing aid in the silica/EPDM compounds.

4.3 Silica Reinforcement with Organosilanes and Their Effects on Mechanical Properties

The addition of bifunctional silanes such as TESPT and bis(triethoxysilylpropyl)disulfide (TESPD) improves the chemical bonding between the silica surface and unsaturated elastomer chain. One end of the bifunctional silane (the alkoxy group) reacts with hydroxyl group on the silica surface via a hydrolysis reaction, and the other end of the bifunctional silane (the sulfur group) reacts with the double bond in the rubber chain and chemically bonds to the rubber chain.[66, 68–70, 95–98] Practically, the tire manufacturing industry has made extensive use of these sulfur-containing bifunctional silanes as a chemical bonding agent as well as a processing aid.[90, 91] They reinforced silica-containing elastomer compounds by chemically bonding and improving the hardness and toughness of the elastomer compounds, resulting in improved gas mileage and improved traction on the road surface. During chemical reaction between the silane and the silica surface, experimental conditions such as temperature and moisture level are very important parameters in determining the degree of coupling.

4.3.1 Temperature Effects on Silica Surface Modification and Mechanical Properties

Temperature effects were investigated on silica surface modification with a bifunctional silane with respect to the vulcanization characteristics, the physical properties, and the alcohol residues. Kim and Vanderkooi[61] investigated mixing temperature effects on the moisture (distilled water (H_2O) treatment on silica particles at 20 phr) treated silica compounds, silica/TESPD/carbon black (CB)/SBR. They showed the moisture treatment on the silica surface increased the hydrolysis reaction with the silane and improved the degree of coupling on the silica surface, which resulted in a strong 3-dimensional network structure.

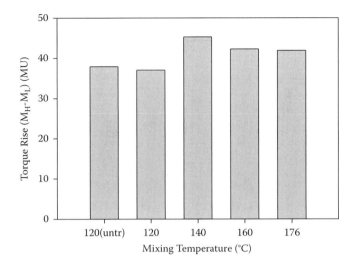

FIGURE 4.17
Torque rise (M_H-M_L) of various mixing-temperature compounds measured from a cure rheometer at 160°C.

This resulted in higher torque rises (M_H-M_L) (see Figure 4.17), less heat build-ups (HBU) (see Figure 4.18), and equal or less alcohol residues of the compounds (see Figure 4.19) compared to moisture untreated compound. As the mixing temperatures of the compound were increased, it increased the tensile moduli (see Figure 4.20) and lowered the tanδ values (see Figure 4.21) in the compound.

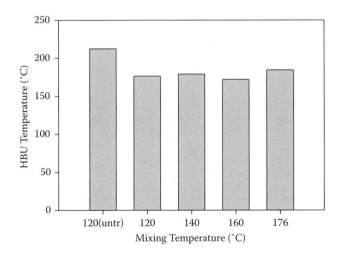

FIGURE 4.18
Heat build-up (HBU) temperature of various mixing-temperature compounds measured by Firestone flexometer.

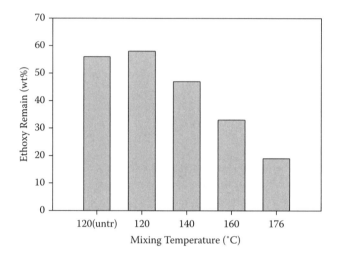

FIGURE 4.19
Alcohol (ethoxy) % remaining in moisture-treated compounds, as measured by a headspace/gas chromatographic technique.

4.3.2 Moisture Effects on Silica Coupling and Mechanical Properties

The moisture level also affects silica-silane bond formation.[60, 72] Goerl et al. showed that the rate of ethanol evolution increased as the moisture content increased the moisture level range between 2.6% and 9.0% in the silica-TESPT system.[72.]

Kim and vanderkooi[60] investigated moisture level effects on silica/TESPD/carbon black (CB)/SBR compounds with respect to vulcanization characteristics, physical properties, and alcohol residues. The effects of moisture concentration were investigated with

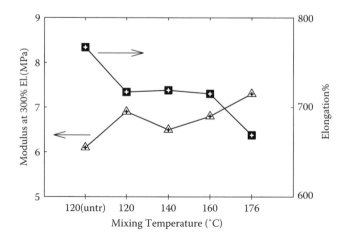

FIGURE 4.20
Modulus (MPa) at 300% elongation and elongation % of various compounds.

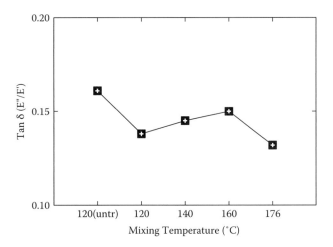

FIGURE 4.21
Viscoelastic behavior (tanδ) of various compounds measured by mechanical energy resolver (MER) at 100°C.

respect to the processability and mechanical properties. It was observed that the addition of water molecules improved the silane reaction with silica surface via improved hydrolysis and resulted in an increased level of cross-linking.[62, 63] This resulted in lower viscous heat generation during mixing (see Figure 4.22), increased torque rise (M_H-M_L) (see Figure 4.23), lower the tanδ (E"/E') (see Figure 4.24), increased elongation modulus (see Figure 4.25), lower heat build-up (HBU), and increased blowout (BO) time (see Figure 4.26) of the compounds.[60]

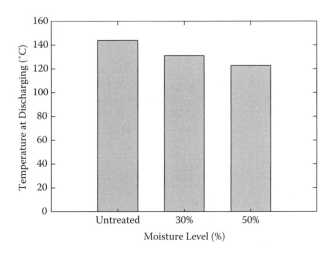

FIGURE 4.22
Probe temperature of master batch compounds after internal mixer mixing.

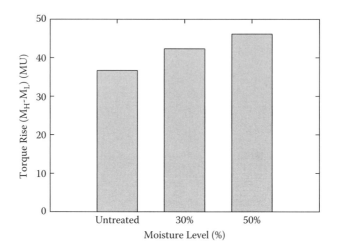

FIGURE 4.23
Torque rise (M_H-M_L) of each compound measured at 160°C.

4.3.3 Zinc Ion Effects on Organosilanes and Improved Mechanical Properties

Zinc ions not only improve reaction kinetics of volcanization, but also improve the mechanical properties of elastomer compounds. Addition of a zinc ion containing fatty acid into bifunctional organosilane (TESPD and TESPT) containing elastomers improved their mechanical properties.[99–104] The effect is significant in natural rubber compounds.[101] Kim and Vanderkooi[101] showed that a zinc ion containing fatty acids aided the formation

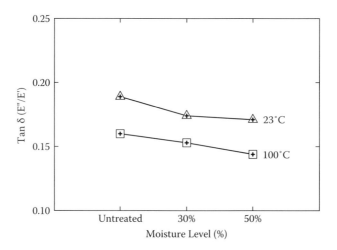

FIGURE 4.24
Viscoelastic property (tanδ) of each compound measured by MER.

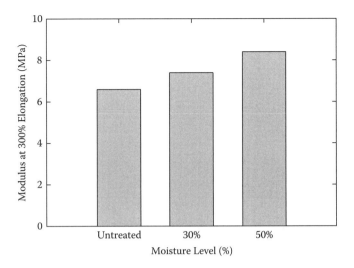

FIGURE 4.25
Modulus (MPa) at 300% elongation of each compound.

of a crosslink between the sulfur group in a silane and the double bond in a polymer chain due to a contraction and expansion mechanism of the zinc ion. This increased the cross-linking density of the compounds and resulted in a lower tanδ (E″/E′), increased elongation modulus, lower heat build-up, and significantly increased blowout time (see Figure 4.27).[90, 100]

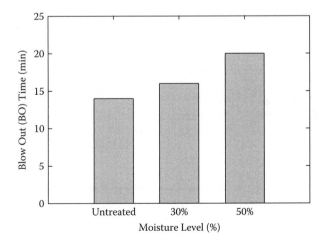

FIGURE 4.26
Blowout (BO) time of each compound measured on Firestone flexometer.

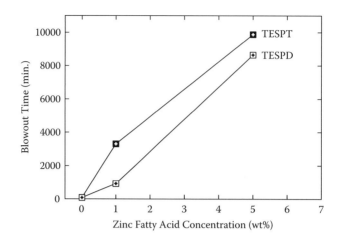

FIGURE 4.27
Zinc fatty acid concentration effects on blowout (BO) time for the TESPT- and TESPD-filled natural rubber compounds.

References

1. Iler, R.K., *The Chemistry of Silica: Solubility, Polymerization, Colloid and Surface Properties,* Wiley-Interscience Publication, New York, 1979, p. 357–364.
2. Schwer, C. and E. Kenndler, "Electrophoresis in fused-silica capillaries: The influence of organic solvents on the electroosmotic velocity and ζ potential,' *Anal. Chem.,* **63**, 1801–1807 (1991)
3. Dickens, J. E., Gorse, J., Everhart, J. A., Ryan, M., 'Dependence of Electroosmotic flow in Capillary Electrophoresis on Group I and II metal Ions', *J. Chromatogr. B,* **657**, 401–407 (1994)
4. Caslavska, J., Thormann, W., 'Electrophoretic seprations in PMMA capillaries with uniform and discontinuous buffers'. *J. Microcol Sep.,* **13**, 69–83(2001)
5. Gaudin, A. M., Fursteneau, D. W., *Trans. ASME,* **202**, 66–72(955)
6. Atamna, I. Z., issaq, H. J., Muschik, G. M., 'optimization of resolution in capillary zone electrophoresis: combined effect of applied voltage and buffer concentration'. *J. Chromatogr.,* **588**, 315–320 (1991)
7. Anonymous, 'Zeta Potential: A Complete Course in 5 Minutes'. Zeta-Meter, Inc. (2007)
8. Lange, K.R., Ed., *Detergents and Cleaners,* Hanser, Munich, 1994.
9. Mathews, W.C. and G.S. Haslam, Some factors involving the Banbury mixing of zinc oxide in rubber stocks, *Rubber Age,* 32, 206, 1932.
10. Jones, H.C. and E.G. Snyder, Banbury mixing of zinc oxide, *Ind. Eng. Chem.,* 43, 2602, 1951.
11. Chapman, F.M. and T.S. Lee, *Soc. Plast Eng. J.,* 26(1), 37, 1970.
12. Tanaka, H. and J.L. White, Experimental investigations of shear and elongational flow properties of polystyrene melts reinforced with calcium carbonate, titanium dioxide and carbon black, *Polym. Eng. Sci.,* 20, 949, 1980.
13. Suetsugu, Y. and J.L. White, The influence of particle size and surface coating on the rheological properties of its suspensions in molten polystyrene, *J. Appl. Polym. Sci.,* 28, 1481, 1983.
14. Kim, K.J., J.L. White, S. Shim, and S. Choe, Effects of stearic acid coated talc, $CaCO_3$ and mixed talc/$CaCO_3$ particles on the rheological properties of polypropylene compounds, *J. Appl. Polym. Sci.,* 93(5), 2105, 2004.
15. Pauling, L., *General Chemistry,* 3rd ed., Dover, New York, 1988.
16. Cooper, B.M. Ed., Kaolin: A review of production and processing, *Ind. Minerals,* Jan. 1979.

17. Bailey, S.W., The status of clay mineral structures, *Clays and Clay Minerals*, 14, 1, 1966.
18. Brown, G., Significance of recent structure determinations of layer silicates for clay studies: *Clays and Clay Minerals*, 6, 73, 1965.
19. Moeller, T., *Inorganic Chemistry*, Wiley, New York, 1952.
20. Gruner, J.W., The crystal structure of kaolinite, *Z. Krystal Mineral*, 83, 75, 1932.
21. Hoffmann, U., K. Endell, and D. Wilm, Krystall strukturund Quellung vo montmorillonit, *Z. Kryst.*, 86, 340, 1933.
22. Hoffmann, U., K. Endell, and D. Wilm, Rontgenographische und colloid chemische untersachungen über ton, *Angew. Chem.*, 47, 539, 1934.
23. Smith, C. R., Base-exchange reactions of bentonite and salts of organic bases, *J. Am. Chem. Soc.*, 56, 1561, 1934.
24. Gieseking, J.E., The mechanism for cation exchange in the montmorillonite-beidellite-nontronite type of clay minerals, *Soil Science*, 47, 1, 1939.
25. Hendricks, S.B., Base exchange of the clay mineral montmorillonite for organic cations and its dependence upon adsorption due to van der Waals forces, *J. Phys. Chem.*, 45, 65, 1941.
26. McEwan, D.M.C., Complexes of clays with organic compounds. I. Complex formation between montmorillonite and halloysite and certain organic liquids, *Trans. Faraday Soc.*, 44, 349, 1948.
27. Barshad, I., Factors affecting the interlayer expansion of vermiculite and montmorillonite with organic substances, *Proc. Soil Sci. Soc. Amer.*, 16, 176, 1952.
28. Jordan, J.W., Organophilic bentonites, *J. Phys. Coll. Chem.*, 53, 284, 1949.
29. Weiss, A., Organic derivatives of mica-type layer-silicates, *Angew Chem. Int.*, 2, 134, 1963.
30. Usuki, A. et al., *J. Mater. Res.*, 8(5), 1179, 1993.
31. Kojima, Y., A. Usuki, M. Kawasumi, A. Okada, T. Kurauci, and O. Kamigaito, *J. Polym. Sci.: Part A: Polym. Chemistry*, **31**, 983 (1993); ibid, Kojima, Y., A. Usuki. M. Kawasumi, A. Okada, T. Kurauvi, and O. Kamigaito, *J. Appl. Polym. Sci.*, **49**, 1259 (1993)
32. Kojima, Y., A. Usuki, M. Kawasumi, A. Okada, Y. Fukushima, T. Kurauci, and O. Kamigaito, *J. Mater. Res.*, **8**(5), 1185 (1993)
33. Kojima, Y., A. Usuki, M. Kawasumi, A. Okada, T. Kurauci, and O. Kamigaito, *J. Polym. Sci.: Part A: Polym. Chemistry*, **31**, 1755 (1993)
34. Kim, Y. and J.L. White, Melt intercalation nanocomposites with chlorinated polymers, *J. Appl. Polym. Sci.*, 90, 1581, 2003.
35. Kim, Y. and J.L. White, Melt intercalation nanocomposites with fluorinated polymers, *J. Appl. Polym. Sci.*, 92, 1061, 2004.
36. Kim, Y. and J.L. White, Formation of polymer nanocomposites with various organoclays, *J. Appl. Polym. Sci.*, 96, 1888, 2005.
37. Kawasumi, M., N. Hasegawa, M. Kato, A. Usuki, and A. Okada, Preparation and mechanical properties of polypropylene-clay hybrids, *Macromolecules*, 30, 6333, 1997.
38. Isayev, A.I., C.K. Hong, and K.J. Kim, Continuous mixing and compounding of polymer/filler and polymer/polymer mixtures with the aid of ultrasound, *Rubber Chem. Technol.*, 76(4), 923, 2003.
39. Kim, K.J. and J.L. White, Breakdown of silica agglomerates and other particles during mixing in an internal mixer and their processing character, *J. Ind. Eng. Chem.*, 6(4), 262, 2000.
40. Kim, K.J. and J.L. White, Silica agglomerate breakdown in three-stage mix including a continuous ultrasonic extruder, *J. Ind. Eng. Chem.*, 6(6), 372, 2000.
41. Iler, R. K., *Colloid Chemistry of Silica and Silicates*, Cornell University Press, Ithaca, New York, 1955 p. 36, 87.
42. Waddell, W.H., J.H. O'Haver, L.R. Evans, and J.H. Harwell, Organic polymer-surface modified precipitated silica, *J. Appl. Polym. Sci.*, 55, 1627, 1995.
43. Rauline, R. (Michelin), U.S. Patent (filed Feb. 20, 1992) 5,227,425, 1993.
44. Rauline, R. (Michelin), Europe Patent EP0501 227, 1991.
45. McNeish, A. and J.T. Byers, Low Rolling Resistance Tread Compounds-Some Compounding Solutions, in *ACS Rubber Division Meeting*, Anaheim, CA, Degussa Corporation, 1997.

46. Cruse, R.W. et al., Effects of polysulfidic silane sulfur on rolling resistance, *Rubber and Plastics News*, Witco, 1997.
47. Ismail, H. and P.K. Freakley, The effect of multifunctional additive on filler dispersion in carbon black and silica filled natural rubber compounds, *Polym. Plast. Technol. Eng.*, 36, 873, 1997.
48. Stone, C.R, M. Hensel, and K.H. Menting, The processability of "green tyre" tread compounds based on the new inversion carbon blacks, *Kautsch. Gummi Kunstst.*, 51, 568, 1998.
49. Hockley, J.A. and B.A. Pethica, Surface hydration of silicas, *Trans Faraday Soc.*, 57, 2247, 1961.
50. Bassett, R., E.A. Boucher, and A.C. Zettlemoyer, Adsorptive studies on hydrated and dehydrated silicas, *J. Colloid Interface Sci.*, 27, 649, 1968.
51. Mittal, K.L., Ed., Silane Coupling Agents, VSP, Utrecht, 1992.
52. McNeil, K.J., J.A. DiCapri, D.A. Walsh, and R.F. Pratt, Kinetics and mechanism of hydrolysis of a silicate triester, tris(2-methoxyethoxy)phenylsilane, *J. Am. Chem. Soc.*, 102, 1859, 1980.
53. Vorokonov, M.G., V.P. Meleshkevisk, and Y.A. Yuzkelvski, *The Siloxane Bond*, Plenum Press, New York, 1978, p. 375–380.
54. Allen, K.W., *Silane Coupling Agents*, K. L. Mittal, Ed., P.93, VSP, Utrecht, 1992.
55. Kay, B.D. and R.A. Assink, Sol-gel kinetics 11: chemical speciation modeling, *J. Non-Cryst. Solids*, 104, 112, 1988.
56. Iler, R. K., "The chemistry of silica", Wiley-Interscience Publication, N. Y., 1979
57. Osterholtz, F.D. and E.P. Pohl, *Silane Coupling Agents*, K. L. Mittal, Ed., P.93, VSP, Utrecht, 1992
58. Kim, K.J. and J. Vanderkooi, Effects of zinc fatty acids on improved polymer processing for critical extrusions, *Int. Polym. Process.*, 18, 156, 2003.
59. Kim, K.J. and J. Vanderkooi (Struktol Co. America), Method for Improved Mixing of Coupled Silica Filled Rubber, U.S. Patent 2005/0080179 A1 (Apr. 14, 2005) (filed Oct. 14, 2003).
60. Kim, K.J. and J. Vanderkooi, Moisture effects on TESPD-silica/CB/SBR compounds, *Rubber Chem. Technol.*, 78, 84, 2005.
61. Kim, K.J. and J. Vanderkooi, Temperature effects of silane coupling on moisture treated silica surface, *J. Appl. Polym. Sci.*, 95(3), 623, 2005.
62. Kim, K.J. and J. Vanderkooi, Reactive batch mixing for improved silica-silane (TESPD) coupling, *Int. Polym. Process.*, 19, 364, 2004.
63. Kim, K.J. and J. Vanderkooi, moisture effects on improved hydrolysis reaction for TESPT and TESPD silica compounds, *Compos. Interfaces*, 11, 471, 2004.
64. Kim, K.J. and J. Vanderkooi, Moisture level effects on hydrolysis reaction in TESPD/Silica/CB/S-SBR compound, 166th *ACS: Rubber Division Meeting*, Columbus, OH, October 2004, Paper No. 57.
65. Schaal, S., A.Y. Coran, and S.K. Mowdood, The effects of certain recipe ingredients and mixing sequence on the rheology and processability of silica- and carbon black-filled tire compounds, *Rubber Chem. Technol.*, 73, 240, 2000.
66. Wolff, S., Reinforcing and vulcanization effects of silane Si69 in silica-filled compounds, *Kautsch. Gummi Kunstst.*, 34, 280, 1981.
67. Wolff, S., Vernetzung von 1,5-dien-kautschuken mittels bis-(3-trialkoxysilylpropyl)-tetrasulfid, *Kautsch. Gummi, Kunstst*, 30, 516, 1977.
68. Culler, S.R., H. Ishida, and J.L. Koenig, Structure of silane coupling agents adsorbed on silicon powder, *J. Colloid Interface Sci.*, 106, 334, 1985.
69. Ishida, H., A review on the recent progress in the studies of molecular and microstructure of silane coupling agents in composites materials, adhesive joints, and coatings, in *Adhesion Aspects of Polymeric Coatings*, Mittal, K.L., Ed., Plenum Press, New York, 1983, p. 45.
70. Ishida, H., A review on the recent progress in the studies of molecular and microstructure of silane coupling agents in composites materials, adhesive joints, and coatings, *Polym. Composites*, 5, 101, 1984.
71. Payne, A.R., Effect of dispersion on dynamic properties of filler-loaded rubbers, *Rubber Chem. Technol.*, 39, 365, 1966.
72. Goerl, U., A. Hunsche, A. Mueller, and H.G. Koban, Investigations into the silica/silane reaction system, *Rubber Chem. Technol.*, 70, 608, 1997.

73. Lin, C., W.L. Hergenrother, and A.S. Hilton, Mooney viscosity stability and polymer filler interactions in silica filled rubbers, *Rubber Chem.Technol.*, 75(2), 215, 2002.
74. Kohjiya, S. and Y. Ikeda, Reinforcement of general-purpose grade rubbers by silica generated *in situ*, *Rubber Chem. Technol.*, 73, 534, 2000.
75. Krysztafkiewicz, A., Modified silica precipitated in the medium of organic solvents—an active rubber filler, *Colloid Polym. Sci.*, 267, 399, 1989.
76. Vondracek, P., M. Hradec, V. Chvalovsky, and H.D. Khanh, The effect of the structure of sulfur containing silane coupling agents on their activity in silica-filled SBR, *Rubber Chem. Technol.*, 57, 675, 1984.
77. Wagner, M.P., Reinforcing silicas and silicates, *Rubber Chem. Technol.*, 49, 703, 1976.
78. Dannenberg, E.M., The effects of surface chemical interactions on the properties of filler-reinforced rubbers, *Rubber Chem. Technol.*, 48, 410, 1975.
79. Dannenberg, E.M., Reinforcement of silicone rubber by particulate silica, *Rubber Chem. Technol.*, 48, 558, 1975.
80. Bachmann, J.H., J.W. Sellers, and M.P. Wagner, Fine particle reinforcing silicas and silicates in elastomers, *Rubber Chem. Technol.*, 32, 1286, 1959.
81. Heiss, G., Mixing Silica Compounds, Krupp Rubber Machinery Inc., ITEC-2000 (No 5B), Akron, OH, September 2000.
82. White, J.L., *Rubber Processing: Technology, Materials, and Principles,* Hanser Publishers, New York, 1995.
83. White, J.L., "Principles of Polymer Engineering Rheology", Wiley Inter-Science, New York Chichester Brisbane Toronto Singapore, 1990
84. White, J.L., and K.J.Kim, "Thermoplastic and Rubber Compounds: Technology and Physical Chemistry", *Hanser Publisher*, Munich (2007)
85. Luginsland, H., Reactivity of the sulfur functions of the disulfane silane TESPD and the tetrasulfane silane TESPT, Degussa-Hülls, *ACS Rubber Division Meeting*, Chicago, April 1999, Paper No. 74.
86. Wang, M., S. Wolff, and J. Donnet, Filler-elastomer interactions. Part I: Silica surface energies and interactions with model compounds, *Rubber Chem. Technol.*, 64, 559, 1991.
87. Wang, M., S. Wolff, and J. Donnet, Filler-elastomer interactions. Part III. Carbon-black-surface energies and interactions with elastomer analogs, *Rubber Chem. Technol.*, 64, 714, 1991.
88. Kim, K.J. and J.L. White, Silica surface modification using different aliphatic chain length silane coupling agents and their effects on silica agglomerate size and processability, *Compos. Interfaces*, 9, 541 2002.
89. Kim, K.J. and J.L. White, TESPT and different aliphatic silane treated silica compounds effects on silica agglomerate dispersion and on processability during mixing in EPDM, *J. Ind. Eng. Chem.*, 7, 50, 2001.
90. Wolff, S., Fullstoffentwicklung heute unt morgen, *Kautsch. Gummi Kunstst.*, 32, 312, 1979.
91. Thurn, F. and S. Wolff, Neue organosilane fur die reifenindustrie, *Kautsch. Gummi Kunstst.*, 28, 733, 1975.
92. Huh, J.Y., T. Woo, and D.E. Nikles, Effect of silane coupling agents on rheological properties of solventless magnetic ink, *Polym. Mat.: Sci. & Eng.*, 85, 387, 2001.
93. Zhao, B., J.S. Moore, and D.J. Beebe, Surface-directed liquid flow inside microchannels, *Science*, 291, 1023, 2001.
94. Mizoguchi, M. and H. Hamada, Evaluation of interfacial properties through microscopic fractography, Advanced Fibro Science, Kyoto Institute of Technology, *International Conference on Composite Interfaces VIII (ICCI-VIII)*, Cleveland, OH, October 2000.
95. Hunsche, A., U. Görl, A. Müller, M. Knaack, and T. Göbel, Investigations concerning the reaction silica/organosilane and organosilane/polymer, *Kautsch. Gummi Kunstst.*, 50, 881, 1997.
96. Hunsche, A., U. Görl, H.G. Koban, and T. Lehmann, *Kautsch. Gummi Kunstst.*, 51, 525, 1998.
97. Wolff, S., The Role of Rubber-to-Silica Bonds in Reinforcement, paper presented at the first *Franco-German Rubber Symposium*, Obernai, France, November 1985.

98. Wolff, S., 'optimizatiion of silane-silica OTR compounds. Part 1: Variations of mixing temperatue and during the modification of silica with bis-(3-triethoxysilylpropyl)-tetrasulfide', *Rubber Chem. Technol.*, **55**, 967 (1980).

99. Kim, K.J.[2] and J. Vanderkooi, "TESPT and TESPD Treated Silica Compounds on Rheological Property and Silica Break Down in Natural Rubber", *Kautschuk Gummi Kunststoffe*, **55**, 518 (2002).

100. Kim, K.J.[3] and J. Vanderkooi, "Effects of Zinc Icon Containing Surfactant on Bifunctional Silane Treated Silica Compounds in Nature Rubber", *J. Ind. Eng. Chem.*, **8**, 334 (2002)

101. Kim, K.J.[4] and J. Vanderkooi, "Rheological Effects of Zinc Suractant on the TESPT-Silane Mixture in NR and S-SBR compounds", *Int. Polym. Process.*, **17**, 192–200(2002)

102. Kim, K.J.[5] and J. Vanderkooi, "Effects of Zinc Soap on TESPT and TESPD Treated Silica Compounds on processing and Silica Dispersion in Polyisoprene Rubber", *Rubber World*, **226**, 39–83 (2002).

103. Kim, K.J.[4] and J. Vanderkooi, "Zinc Surfactant Effects on NR/TESPD/Silica and SBR/TESPD/Silica Compounds", *Elastomer*, **39**, 263 (2004)

104. Kim, K.J.[5] and J. Vanderkooi, "Zinc Surfactant Effects on Processability and Mechanical Properties of Silica Filled natural Rubber Compounds", *J. Ind. Eng. Chem.*, **10**, 772 (2004)

5

Surface Modification of Carbon Nanofibers

Max L. Lake, D. Gerald Glasgow, Gary G. Tibbetts, and D.J. Burton

CONTENTS

5.1   Introduction

Carbon nanofibers (CNFs) are vapor-grown carbon fibers grown from gaseous hydrocarbons using metallic catalyst particles. CNFs are envisioned for a broad spectrum of applications, including reinforcement or adding conductivity to polymer composites. Good fiber dispersion and adequate adhesion between fibers and matrix resins are essential for the performance of CNF-based composites; consequently, the properties of CNF composites are strongly influenced by fiber surface morphology and chemistry. Methods of surface treatments available include modification of the surface energy and surface area to improve wetting and physical bonding, addition of functional groups to enhance chemical bonding, and control of the graphitization of the CNF surface. Partial graphitization of CNFs by high-temperature heat treatment can give improved composite properties. The intrinsic electrical conductivity of the CNFs is maximized by giving the fibers an initial heat treatment at 1500°C. Similarly, for carbon nanofiber/polypropylene composites having up to 12 vol% fiber, initial fiber heat treatments near 1500°C give tensile modulus and strength superior even to composites made from fibers graphitized at 2800°C. However, optimum composite conductivity is obtained with a somewhat lower heat treatment temperature, near 1300°C. Transmission electron microscopy (TEM) explains these results, showing that high-temperature heat treatment of the fibers alters the exterior planes from continuous, coaxial, and poorly crystallized to discontinuous nested conical crystallites inclined at about 25° to the fiber axis.

Carbon nanofibers (CNFs) are discontinuous hollow filaments having a diameter of approximately 100 nm (nanometer). The promise of these filaments, produced by a catalytic process in which metal particles are exposed to free carbon at elevated temperatures, derives from the fact that the filaments display physical properties approaching those of single-crystal graphite, including high thermal and electrical transport values and high tensile strength and modulus.[1] These properties can improve engineered polymers for structural and conducting applications such as static discharge, electrostatic painting, and radio frequency interference shielding. These discontinuous fibers lend themselves naturally to fabricating composite materials that utilize their short length.[2–4] Because short-fiber-reinforced composites can be fabricated without the expensive limitations of textile processing, they offer advantages in ease of manufacture and low cost. For example, CNF/polymer composites may be continuously fabricated by extrusion or injection molding, allowing for both high volume production and recycling. A review of CNF composites was recently published by Tibbetts et al.[5] Previous work has shown that CNFs in a polypropylene thermoplastic matrix can yield considerably improved composite strength and stiffness.[6–8] Similarly, CNF has been shown to hold promise as an electrical conductivity additive.[9, 10] The sub-micron dimensions of CNFs and consequent high surface area require novel procedures to capture the properties of graphite in polymer composites. Among the important factors is an appropriate interphase between the graphitic filaments and the matrix polymer. The optimal interphase for a given polymer would provide fiber-matrix adhesion to maximize mechanical properties such as tensile strength and would allow for ease of dispersion and good electron transport between the fibers. A large body of work describes surface treatment of now-common continuous carbon fibers.[11] In comparison, the science and technology of surface modification for CNFs is only beginning to emerge.

Modification of the surface state of CNFs can be accomplished by altering the surface area or surface energy, by modification of the degree of graphitization of the CNF surface, or by the addition of functional groups intended to enhance covalent bonding. The work presented here examines the fiber-matrix adhesion problem as influenced by CNF surface area and energy. Also, methods of functionalizing the CNF to render a more reactive surface are discussed. Recent efforts to add surface functional groups, particularly oxygen groups, have demonstrated benefits for interphase development. Carboxyl and phenolic groups contributing to a total surface oxygen concentration in the range from 5 to 20 atom% have been added to CNFs used to fabricate epoxy-based composites, providing improved flexural strength and flex modulus. The effect of similarly functionalized CNFs in bismaleimide (BMI)-based composites is also good. Data for polypropylene, epoxy, and BMI/nanofiber reinforced composites show that higher CNF volume loadings can play a significant role in structural composite properties. Finally, the impact of the degree of graphitization on the role of CNFs as a reinforcement is discussed.

5.2 Experimental

5.2.1 Materials

The CNFs used in this study were produced at Applied Sciences, Inc. CNFs of several types and post-production treatments were produced. Their average diameters ranged from approximately 100 to 200 nm with an aspect ratio exceeding 100. These CNFs were formed by iron-based catalyst particles in a pyrolyzing methane atmosphere. The CNFs are formed in a two-stage process, frequently referred to as a "catalytic phase" and a "CVD phase." The initial catalytic phase filament is generated as a hollow graphitic tube from the growth of a graphitic lattice formed at one side of a metal catalyst from free carbon deposited on the other surface of the metal catalyst. The catalytic phase component for the CNFs produced for this work has predominantly a herringbone orientation of well-formed graphene planes canted at an angle of approximately 25° with respect to the fiber axis. The CVD phase is turbostratic carbon, which is later deposited on the core filament. The turbostratic layer tends to be oriented parallel to the fiber axis. The thickness of the CVD phase can be controlled from near zero to about two thirds of the total fiber diameter.

Montel Pro-Fax 6301 in powder form was used as the polypropylene (PP) resin. The bismaleimide (BMI) used in this work was RTM-651 purchased from Hexcel Corp.

In some cases, thermoplastic resin tensile strength and modulus test specimens were prepared by ball-milling fiber for 2 min (minute) using a Spex 8000 mixer mill, mixing the fiber with powdered polypropylene or polyamide 66 and injection molding in a benchtop MiniMax Molder (Custom Scientific Instruments, Inc.). Other specimens were prepared by first mixing fiber and powdered polymer in a small Brabender, followed by injection molding in the MiniMax Molder. The mold was held at room temperature.

5.2.2 Oxidation Treatments

Some samples of fibers were oxidized after production, using different procedures to increase the surface area, surface energy, and reactivity. Air oxidation was the simplest procedure, while more complex procedures for oxidation of the fiber surface were achieved using protocols taken from organic synthesis methodology for oxidation of aromatic compounds.

Oxidation using ulfuric/nitric acid mixtures was conducted by adding 1 part fiber to 20 parts by volume of a 1:1 mixture of concentrated sulfuric and nitric acids. This mixture was refluxed 1 hr (hour), cooled, and filtered through a sintered glass filter to isolate the fiber. The fiber was washed on the filter with copious amounts of deionized water to remove the remaining acid and then dried at 120°C. Surface oxygen levels as high as 25 atom% oxygen can be obtained.

Air oxidation was accomplished by passing air through the fiber contained in a horizontal tube furnace heated at 400°C. The time at temperature varied somewhat but was 30 min for the best results. Surface oxygen levels of approximately 4 to 5 atom% oxygen were obtained.

The fiber could also be oxidized using peracetic acid at room temperature. In this case, 1 part fiber (by weight) was added to a 200 parts (by volume) mixture of 3:1 glacial acetic acid:30% hydrogen peroxide and allowed to stand at room temperature for 5 days. The fiber was separated by filtration and washed on the filter with copious amounts of deionized water to remove the remaining acid and then dried at 120°C. Surface oxygen levels as high as 10 atom% oxygen could be obtained.

5.2.3 Heat Treatments

For heat-treated CNFs, the fibers were loaded into graphite crucibles. The crucibles were opened when placed in a heat-treatment furnace and were purged by cyclic evacuation and backfilling with argon. The crucibles containing the sample fibers were heated to the target temperatures at a rate of 5°C/min and were held at the target temperature, determined by optical pyrometry, for 1 hr. The samples were then allowed to cool to room temperature over a 20-hr period.

5.2.4   Composites Processing

A benchtop CS-183 MiniMax Molder (Custom Scientific Instruments, Inc., Cedar Knolls, NJ) was used for fabricating CNF/PP composite specimens.[12] The injection molder was equipped with a cylindrical mixing cup (12.7 mm diameter and 25.4 mm height) and a rotating and vertically moving rotor for mixing and injection. The cup temperature was 230°C for PP, while the specimen mold was held at room temperature. Initially, polymer pellets were loaded into the mixing cup and melted. The nanofibers were then gradually loaded into the mixing cup filled with polymer melt and mixed by rotational and vertical motions of the rotor. For composite samples with high fiber concentrations, a dry mixture of fiber and the PP powder were loaded into the mixing cup to facilitate mixing and reduce the degradation of polymer during mixing.

Ultimately, the composite mixture was injected into the mold by opening the valve connecting the mixing cup and the mold while vigorously pushing the lever attached to the rotor downward. The mold was immediately submerged in cold water for at least 5 min to minimize additional crystal growth during solidification.

Composite specimens of BMI loaded with CNFs were made in the form of square panels, approximately 32 cm on a side and from 2 to 5 mm thick (depending on the mold used). The composites were cured and post cured as per the manufacturer's directions. The final cured density of this material was 1.25 g/cc.

Because of the high aspect ratio and surface area of the nanofibers, the nanofibers will tend to increase the viscosity of mixtures to which they are added; thus, it was necessary to use a solvent to extend the BMI during formulation. A solution of the resin in acetone was prepared. Separately, a dispersion of nanofibers in acetone was also prepared. The nanofibers, which enter the process as entangled bundles, were dispersed in the acetone using a Misonix 3000 sonicator equipped with a solid-tip probe for 30 min at a setting of 78 Watts. Sonication tends to reduce the average bundle size of nanofibers and separate some individual nanofibers from bundles. After sonication, the BMI solution was added to the nanofiber dispersion. The mixture was then mechanically stirred to blend the components. During stirring, the mixture was warmed to 90°C to drive off the solvent. The concentrated mixture was then degassed in a vacuum oven at 120°C and ~2 Torr for about 1 hr to complete solvent removal prior to casting the panels.

To form panels, the degassed, nanofiber-filled resin was poured into a silicone rubber mold of appropriate dimensions and placed under pressure in a temperature-programmable press for pre-curing. The pre-cure process consisted of a temperature ramp (5°C/min) up to 135°C, holding at temperature for 20 min, then a similar ramp to 190°C, followed by a hold at temperature of 4 hr. After pre-cure, the panel was then cooled to 30°C. The entire pre-cure temperature cycle was carried out under 10,000 lb. (44,000 N) total pressure, which equates to approximately 64 psi (440 kPa).

The post-cure process consisted of simply heating the panels at 232°C under nitrogen for 16 hr.

5.2.5   Electrical Resistivity Measurements

A rectangular bar (12.7 × 70 × 0.33 mm) obtained by injection molding was lightly sanded in order to smooth the surface and expose the fibers for measuring the electrical resistivity. The volume resistivity was measured using a Keithley 2000 electrometer (Keithley Instruments, Inc., Cleveland, OH) at room temperature. The resistivities in the flow direction were measured by connecting electrodes attached by silver paint to the surfaces of the specimens. For higher resistivities over 10^4 Ω-cm, the two-point measurement was accurate, while four-point measurement was performed for lower resistivities.

The intrinsic electrical resistivity of the fibers could also be measured by putting the neat fibers in a cylindrical die as per ASTM D257-99. A stainless steel pin of diameter 12.94 mm compresses the fibers in a mating Teflon cylinder so that electrical resistivity measurements may be made while the fibers are compressed with a maximum of 1000 lb. force. The die was mounted in an Instron universal testing machine providing controlled pressures and precise distance measurements between the upper and lower probes. The intrinsic resistivity was determined as a function of volume fraction at a certain pressure using the intrinsic density of the fibers, which we previously measured as 2.02 g/cc.[13]

5.2.6 Tensile Measurements

Tensile specimens (ASTM D638 Type V) were mounted in the grips of the Instron universal testing machine and stretched at 1 mm/min until break or yield occurred. Tensile strength was obtained from the ultimate load before separation of the two sections of the specimen, while the tensile modulus was determined from the slope of the initial section of the stress-strain curve.

5.2.7 CNF Surface Area and Surface Energy

The surface area and surface energy values were determined at Laval University, Quebec, Canada, by Drs. Hans Darmstadt, Naizhen Cao, and associates.

5.3 Results and Discussion

5.3.1 Surface Treatments on CNF/Polypropylene Composites

High tensile strength in a fiber-reinforced composite requires good adhesive forces between the fiber and matrix resin. Tensile strength increases as the strength of the interfacial bond increases, and a strong bond can form only when the matrix resin wets the fiber. Fiber wetability, in turn, is related to the surface energy of the fiber and the available surface area.

The surface areas and surface energies (dispersive) of eight CNFs are presented in Table 5.1. These fibers were produced by variations of the fiber production process

TABLE 5.1

Fiber Surface Areas and Energies

	Surface Area (m²/g)		
Sample No.	Total	External	Surface Energy (mJ/m²)
PR-18	15	10	33
PR-1	14	10	48
PR-1A	34	16	57
PR-11-4	123	46	79
PR-11-2	83	28	106
PR-11-5	26	15	156
PR-11-7	91	53	248
PR-11-0	191	57	470

TABLE 5.2

Composite Tensile Strengths and Moduli

Sample No.	Tensile Strength (MPa)	Tensile Modulus (GPa)
PR-18	37	3.2
PR-1	48	3.4
PR-1A	63	3.4
PR-11-4	65	4.2
PR-11-2	66	4.1
PR-11-5	78	4.1
PR-11-7	58	4.1
PR-11-0	53	3.2

used by Applied Sciences, Inc., for the preparation of CNFs. The first two fiber samples represent fibers directly from production. The latter six fibers were obtained by post-production oxidation. Both total and external surface area values are given. The external surface area is that area which is not in micropores (i.e., pores having diameters of less than 2 nm) and comes directly from the measurement method. The external surface area is the area of the fiber considered to be available for bonding with a matrix resin. It is assumed that little if any resin would flow into 2-nm micropores. The data in Table 5.1 data show that fibers with a wide range of surface areas and energies can be produced.

These fibers were used to prepare polypropylene composites containing 15 vol% nanofibers, which then were used for the determination of tensile strength and modulus. The values obtained are shown in Table 5.2. The tensile strengths obtained for composites containing fibers with widely differing surface energies show the influence of the fiber interfacial properties. This effect can be seen more clearly when the tensile strengths are plotted vs. external surface area and surface energy, as shown in the surface plot in Figure 5.1. It is obvious that there is an optimum fiber surface area and energy combination that results in the highest composite tensile strength. The interpretation of these results is that the oxidative process used to produce the higher surface area fibers results in increasing fiber damage at higher oxidation levels. This, in turn, weakens the fiber and results in a lower composite tensile strength. However, at lower levels of surface area, improved fiber wetting due to the higher surface energy results in significant increases in composite tensile strength. The optimum combination of surface area and surface energy is a compromise between the improvement due to higher surface energy and fiber damage from excessive oxidation.

The data in Table 5.2 and Figure 5.1 were used as a guide to engineer an improved fiber production process. The goal was to produce fiber having an external surface area between 10 and 30 m^2/g and a surface energy value between about 150 and 180 mJ/m^2. The results are summarized in Table 5.3. The values for sample PR-21A from this study are in the desired surface area range and just below the targeted surface energy values. However, the fibers from the process that produces PR-21A are wet by a wide range of liquids, including water. This indicates that the surface energy of PR-21A is sufficient to allow wetting by a large variety of matrix resins and produce correspondingly high levels of fiber-matrix adhesion. A continuous fiber production process was developed and fiber from this process was evaluated for various composite applications, but the results were

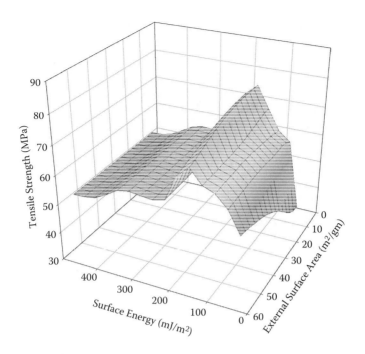

FIGURE 5.1

Surface plot of tensile strength (MPa) vs. external surface area and surface energy.

not encouraging enough to merit production of the fiber. These evaluations occurred prior to the realization of the impact of fiber dispersion on composite properties. Perhaps, in retrospect, the benefits of fiber produced by this process may have been overshadowed by the dispersion problems.

Maleic anhydride modified PP is frequently used as an adhesion promoter for glass fiber based composites, but its use in carbon fiber composites is unknown. Preliminary experiments revealed that maleic anhydride modified PP containing 1% maleic anhydride when used at 4 wt% relative to the PP increased the tensile strength and modulus of the composite by about 30%, even though the CNF has essentially no functional groups on the fiber surface. The data are shown in Figures 5.2 and 5.3.

TABLE 5.3

Properties of Fibers from Alternate Production Process

| Sample No. | Surface Area (m²/g) | | Surface Energy (mJ/m²) |
	Total	External	
PR-11	18	12	26
PR-21A	25	19	125
PR-21B	27	19	
PR-21C	31	22	104

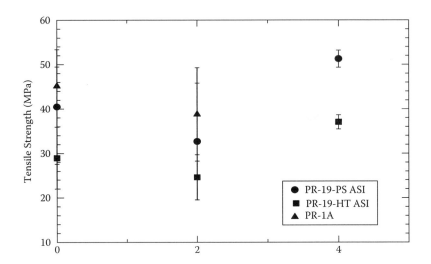

FIGURE 5.2
Effect of maleic anhydride-modified PP addition on tensile strength of 9 vol% carbon nanofiber/PP composites.

The CNF oxidation results prompted further study of continuous fiber production methods for preparation of high surface area and surface energy fibers. The results of these studies were the identification of a series of three fiber types and three fiber grades. These fibers are shown in Table 5.4 with corresponding surface areas and energies.

The electrical resistivities of several of these composites were determined. The data obtained are shown graphically in Figure 5.4, where resistivity is plotted vs. fiber loading for five different resins. The resistivity of composites based on the fully graphitized fiber is significantly better than that obtained using as-grown fiber, probably due to the cleaning associated with graphitization. These resistivity values are two orders of

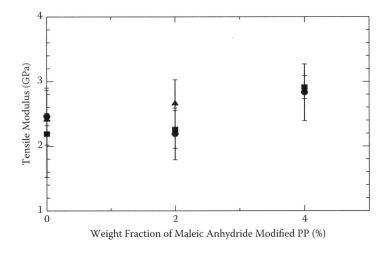

FIGURE 5.3
Effect of maleic anhydride-modified PP addition on tensile modulus of 9 vol% carbon nanofiber/PP composites.

TABLE 5.4

Characteristics of Pyrograf-III® Fiber Grades

Fiber	Grade	Surface Area (m²/g) Total	External[e]	Surface Energy[d] (mJ/m²)	Approximate Fiber Diameter (nm)
PR-19	AG[a]	10–20	9–15	25–40	200
	PS[b]	(10–20)[f]		133	200
	HT[c]	21	18	275	200
PR-21	AG	25–35	19–26	125	200
	PS	(25–35)		136	200
PR-24	PS	53	50		100

[a] AG denotes pelletized as-grown fiber.
[b] PS denotes clean (free of organic surface contaminates) version of the AG grade.
[c] HT denotes the graphitized version of the AG grade.
[d] Dispersive (non-specific) surface energy.
[e] Surface area not in micropores.
[f] Values in parenthesis are estimated, from assumption that the process used to produce these grades would not significantly change the fiber surface area.

magnitude higher than calculated for a simple network of fibers, implying that there may be high-resistance regions between the individual fibers in the composites. It should be noted that the resistivities of these composites would allow them to be used in various applications requiring electrical conductivity, including those requiring static dissipation, electrostatic painting, or electromagnetic interference (EMI) shielding. Achieving the resistivity required for each can be accomplished by adjusting the fiber type and loading.

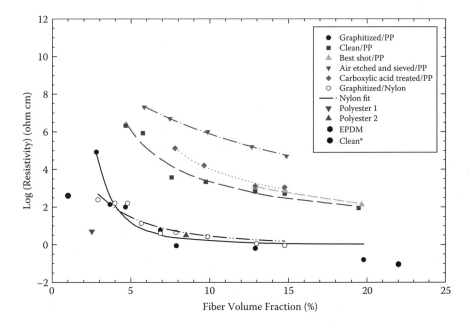

FIGURE 5.4
Electrical resistivities of carbon nanofiber composites.

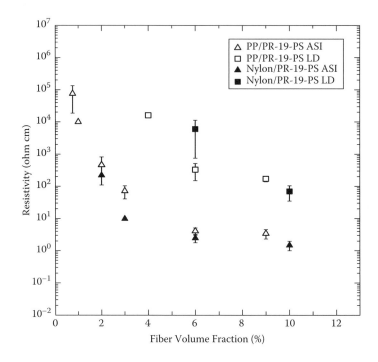

FIGURE 5.5
Electrical resistivity of carbon nanofiber-based PP and nylon composites.

The CNFs produced by ASI are relatively bulky in the as-grown state but can be debulked to various low bulk densities to facilitate handling in compounding or composite synthesis. One process used for debulking results in the generation of small fiber bundles and some reduction in fiber aspect ratio. This, in turn, impacts the resultant electrical resistivity and mechanical properties in a composite. Moreover, the process used to incorporate fiber into a specific matrix resin also affects the fiber aspect ratio. Both of these effects must be considered when choosing a method for incorporating carbon nanofibers into composites. The process that exposes the fiber to the least amount of shear forces will produce the composite with the highest electrical conductivity. Mechanical properties can benefit more from the use of processes that reduce the fiber aspect ratio during fabrication.

The effect of fiber aspect ratio on electrical resistivity can be observed in Figure 5.5. The ASI and LD designations refer to two different debulking methods. The ASI method produces fiber with higher aspect ratio than the LD method. The effect on electrical resistivity is obvious for composites based on either PP or Nylon 66.

Injection molded composites based on CNFs exhibit fiber alignment as a result of the injection molding process. This is demonstrated by the electrical resistivity data in Figure 5.6, which shows data for electrical resistivity in all three axes.

The effect of fiber-matrix adhesion on the coefficient of thermal expansion (CTE) was also investigated. Figure 5.7 shows that neat polypropylene has an inconveniently large isotropic coefficient of thermal expansion. A very significant decrease in CTE was obtained in composites having good fiber-matrix adhesion. This is demonstrated by the data shown

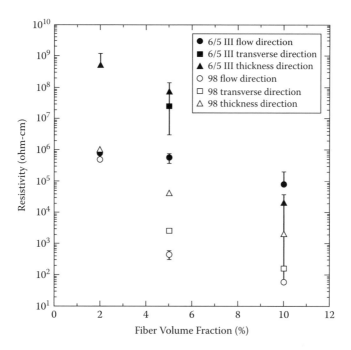

FIGURE 5.6
Anisotropic electrical resistivity of carbon nanofiber based PP composites.

in Figures 5.8 and 5.9. Good adhesion between the CNF and matrix results in a significant decrease in the CTE in the longitudinal direction, but not in the transverse direction. This effect is believed to result from orientation of the CNF in the direction of flow that occurs during injection molding. These data are shown in Figure 5.8. Figure 5.9 presents the CTE data for a composite containing the fully graphitized version of the nanofiber.

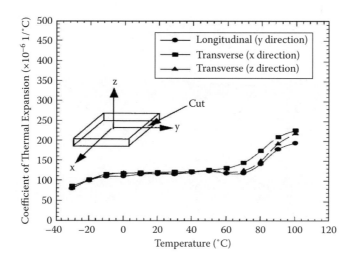

FIGURE 5.7
CTE vs. temperature for polypropylene.

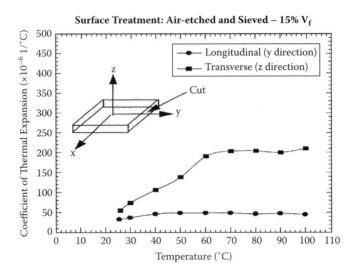

FIGURE 5.8
CTE vs. temperature for 15% air-etched CNF/PP.

This composite has a CTE very similar to the pure polymer up to about 40°C, after which the CTE increases more rapidly with temperature than does the homopolymer. These data suggest that there is relatively little bonding between the matrix and the heat-treated, fully graphitized CNF.

5.3.2 Oxidation Treatments for Functionalization

CNFs are intrinsically discontinuous because they are produced by a catalytic process. Figure 5.10 shows a transmission electron micrograph (TEM) of a typical PR 24-PS fiber

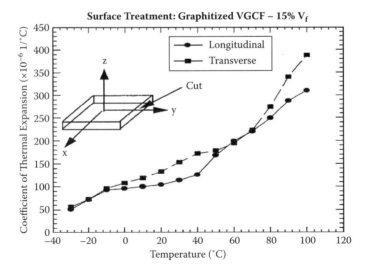

FIGURE 5.9
CTE vs. temperature for graphitized CNF/PP.

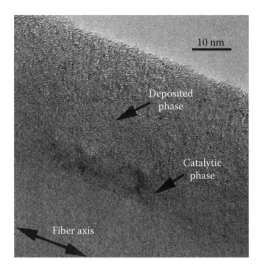

FIGURE 5.10
Transmission electron micrograph of typical PR-24 PS carbon fiber, showing the fiber axis in the center of a hollow core region, surrounded by the catalytic phase formed by precipitation from a Fe-based particle. This region is covered, in turn, by a vapor-deposited phase of less well-crystallized material.

grown at a process temperature of approximately 1100°C. As the image clearly shows, this fiber has a hollow core covered with a layer of vapor deposited carbon and the interior catalytically formed layers are highly ordered, whereas the vapor-deposited layers on the surface are turbostratic carbon. The disordered morphology of the exterior planes, expected to be more reactive than a fully graphitized surface, may be advantageous for functionalizing the surface to bond to polymeric matrices.

A typical x-ray photoemission (XPS) scan of PR-19-Ps-Ox fibers (fibers oxidized with a proprietary process) is shown in Figure 5.11. The oxygen 1s peak is broadened due to the presence of several different types of oxygen bonded to the fiber's surfaces. The deconvoluted peak at smaller binding energies is attributed to doubly bonded oxygen, while that at larger binding energy is attributed to single bonded oxygen. The magnitude of each peak indicates the number of surface atoms of each type; note that they are approximately

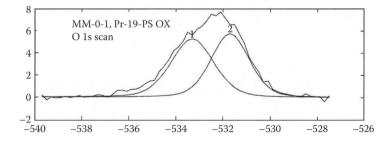

FIGURE 5.11
X-ray photoemission scan of the oxygen 1s peak of PR-19-PS Ox fibers, treated with a proprietary oxidation process. The deconvoluted peak at the lower binding energies represents surface doubly bonded oxygen, while that at larger binding energies is attributed to carboxylic acid.

TABLE 5.5

Level and Type of Oxygenated Fiber Surface Functional Groups on Carbon Nanofibers

Sample Identification	Atom % O	C–O, Acid	C–O, Phenolic	C=O, Acid	C=O, Quinone
PR-24-PS	3	1.5	—	1.5	—
PR-24-LHT-Ox[a]	5.7	2.2	—	2.2	1.3
PR-24-HHT-Ox[a]	5.2	2.3	—	2.3	0.6
PR-24-PS-Ox[a]	12.4	4.7	—	4.7	3
PR-19-PS	2.5	1.5	—	1.5	—
PR-19-LHT-Ox[a]	6.1	2.4	—	2.4	1.3
PR-19-PS-Air[b]	4	2.0	—	2.0	—
PR-19-PS-Ox[a]	14.5	6.9	—	6.9	0.7
PR-19-PS-Acid[c]	21	9.0	3.0	9.0	—
PR-19-PS-Peracid[d]	9.6	4.8	—	4.8	—
PR-19 in-situ	8.7	2.9		2.9	1.3

[a] Oxidant X (oxidant not identified pending patent decision).
[b] Fiber oxidized with air at 400–450°C.
[c] Oxidized with sulfuric/nitric acid mixture.
[d] Peracetic acid oxidation

equal. These data may be converted into percentage of surface atoms. These calculations give up to 14.5% surface oxygen atoms and are tabulated in Table 5.5.

Table 5.5 shows XPS results tabulated for a variety of surface-treated fibers. The pyrolytically stripped fibers have a minimum surface oxygen concentration of 2 to 3%, while air oxidation increases surface oxygen to 4 to 6%. Strong acid oxidation (nitric/sulfuric acids) is a somewhat cumbersome procedure but can raise surface oxygen levels to 21%. A more practical procedure with the proprietary oxidizer is labeled "Ox." It is very encouraging that this procedure is able to raise the surface oxygen on PR-19-PS-Ox to the 15% level; perhaps this shows that the vapor-deposited carbon is a more readily oxidizable structure. Moreover, the oxygen present as carboxylic acid is raised to nearly 7%.

Preliminary results testing the efficacy of these very highly oxidized CNFs in PP matrices gave mixed results. Figure 5.12 shows that while modest amounts of surface oxidation of the nanofibers did improve the tensile strength of composites fabricated with 10 vol% nanofibers, larger amounts of surface oxygen decreased composite tensile strength down to the level of the neat matrix.

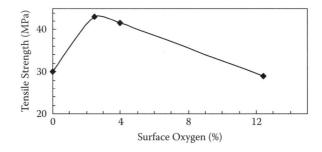

FIGURE 5.12
Tensile strengths of polypropylene composites fabricated with 10 vol% carbon nanofibers surface treated to achieve varying fractions of surface oxygen.

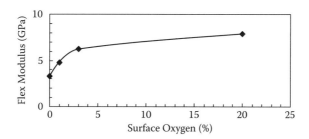

FIGURE 5.13
Improvement in flex modulus with strength for 4 wt% nanofiber/epoxy composites as a function of surface oxygen concentration of carbon nanofibers. (Data of Lafdi and Matzed, Reference 14.)

It may be possible that highly oxidized CNFs form a poor bond with the low-surface-energy, low-polarity polypropylene. This would be the case if increasing the CNF surface oxygen beyond a certain critical level increases the polarity of the CNF surface, rendering it less compatible with the PP.

Epoxy-based composites (Epon 862 cured with Curing Agent W) fabricated from these oxidized fibers have been studied by Lafdi and Matzed.[14] Their data for composites containing 4 wt% nanofibers seem to imply a direct improvement in both flex strength and flex modulus with increasing surface oxygen concentration, as shown in Figures 5.13 and 5.14.

These data indicate that in the selected epoxy system, at least a 5% oxygen concentration on the nanofiber surface is needed to develop optimum flex modulus and flex strength. Further increase in oxygen concentration results in further improvement of these properties, but at a lower marginal rate. Nevertheless, the decrease in tensile strength observed in PP at high oxygen levels is not evident in the epoxy composites.

Preliminary results with BMI as the matrix polymer were similarly encouraging. Figure 5.15 shows how the tensile modulus of BMI composites improves after the addition of only 2 vol% of two types of differently oxidized fibers. The BMI data are again consistent with the observation in epoxy that more highly oxidized CNFs result in composites with better tensile properties.

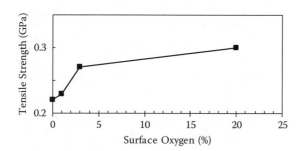

FIGURE 5.14
Improvement of flex strength for 4 wt% nanofiber/epoxy composites as a function of surface oxygen concentration of carbon nanofibers. (Data of Lafdi and Matzed, Reference 14.)

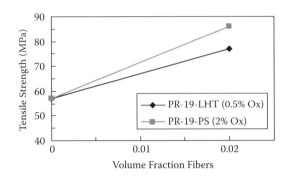

FIGURE 5.15
Improvement in tensile strength observed in BMI: composites with the addition of two different types of oxidized nanofibers.

5.3.3 Heat Treatment for Morphology Modification

Figure 5.10 showed the layered morphology for CNF in the as-grown state. In contrast, Figure 5.16 is a TEM view of a similar nanofiber heat-treated to a temperature of 2900°C, a point at which all the disordered carbon is seen to recrystallize into a highly graphitic structure.

Heat-treating fibers with such a complex morphology results in a complex sequence of changes, recently described by Howe et al.[15] Figure 5.16 shows a fiber in which all traces of the duplex structure have disappeared; the fiber is now comprised of conical sections about 20 nm in size canted at about 25° from the longitudinal fiber axis. These crystallites are apparently nested cones of much-improved graphitization.

Heat-treating the fibers to intermediate temperatures results in partial graphitization of both the catalytically formed interior carbon planes and the vapor-deposited material. These morphological changes may be seen to affect composite properties. Figure 5.17 shows the tensile

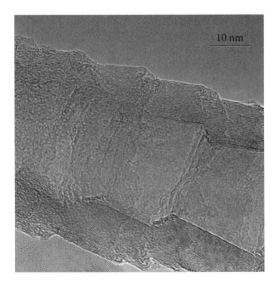

FIGURE 5.16
TEM of a nanofiber heat-treated to 2900°C.

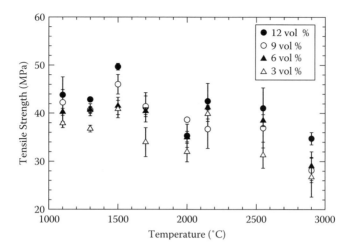

FIGURE 5.17
Tensile strengths of composites as a function of nanofiber heat-treatment temperature. Values for volume fraction of nanofiber in polypropylene vary from 3 to 12 vol%.

strengths measured for PP composites made from the nanofibers heat-treated to temperatures between 1100°C and 2900°C. The four data sets correspond to composites made with nanofiber volume fractions ranging from 3 to 12 vol%. Three samples were tested at each data point, and the resulting error bars are also shown. The PP samples with no reinforcing fibers averaged around 30-MPa tensile strength. The data show a small increase in composite tensile strength with heat treatment temperature between 1100°C and 1500°C, followed by a secular decrease extending up to 2900°C. The peak near 1500°C is only modestly larger than the average error bars.

Young's modulus values for the same samples are plotted in Figure 5.18. These values may be determined with more precision than tensile strengths and show smaller errors.

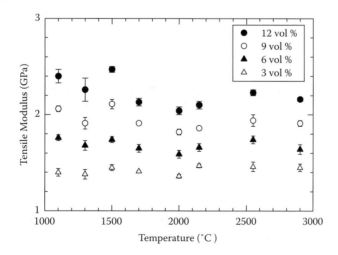

FIGURE 5.18
Tensile modulus of composites fabricated from nanofibers heat-treated at different temperatures. Nanofiber volume fraction in polypropylene varies from 3 to 12 vol%.

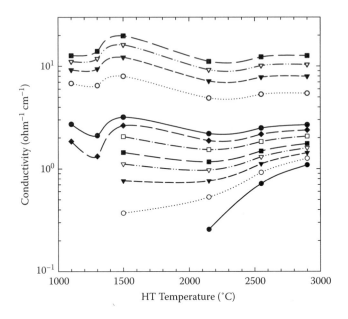

FIGURE 5.19
Intrinsic conductivity of carbon nanofiber bundles compressed in an air-filled pressure die at room temperature.

Nevertheless, the behavior is similar to that of tensile strength, with modulus perhaps increasing marginally in going from 1100°C heat-treatment temperature to 1500°C, and then dropping with higher heat-treatment temperatures. Therefore, the data again indicate that fibers heat-treated to 1500°C give optimum composite modulus values. The tensile modulus of unreinforced polypropylene for an average of six samples studied was 1.25 GPa.

The intrinsic mechanical and electrical properties of CNFs are dauntingly difficult to measure but the electrical conductivity of a simple mass of fibers can be measured easily and is a useful guide related to the intrinsic resistivity of the CNF. Following the methods outlined in Section 5.2.4, the conductivity of a loose mass of fibers confined within an aluminum die was determined. Figure 5.19 shows collected data for conductivity of the heat-treated samples of CNFs as a function of the volume fraction of fibers present in an air-filled pressure die at room temperature. Each separate run on a batch of nanofibers provides information on the complete volume fraction range as the batch is compressed. Experiments were unable to be performed with the later batches of fibers heat-treated to 1700°C and 2000°C.

Electrical conductivity again seems to peak near 1500°C for all volume fractions above 5.5 vol% fibers. Measurements made at very low volume fractions suffer from contact problems and sensitivity to the details of loading, and are not as significant as measurements at higher volume fractions.

Figure 5.20 shows the conductivity vs. volume fraction data for polypropylene composites made from these heat-treated fibers. The conductivity values are lower than those measured in the intrinsic conductivity experiment because the polymer makes fiber-to-fiber electrical contact more difficult. However, the variation in the conductivities with temperature shows similar trends, with conductivity now approaching its peak nearer 1300°C than 1500°C. As with the intrinsic conductivity, composite conductivity diminishes with lower fiber volume fraction. Above 1300°C, conductivity shows a secular decrease with increasing heat-treatment temperature.

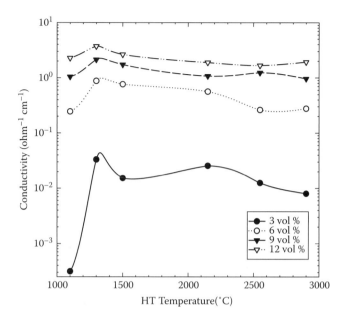

FIGURE 5.20
Conductivities of heat-treated carbon nanofiber/polypropylene composites. The composites were made with nanofiber volume fractions varying from 3 to 12 vol%.

5.4 Conclusions

Processes for modifying the surfaces of carbon nanofibers (CNFs) have been shown to be useful in improving the properties of CNF/polymer composites. In CNF/polypropylene (PP) composites, increasing the surface area and surface energy of the CNF leads to significant increases in the tensile strength and modulus. Addition of oxygen up to 4% surface oxygen also increases mechanical properties. In these CNF/PP composites, further increases in oxygen concentration appear to decrease composite tensile strength, an observation that is attributed to the less polar behavior of the low-energy surface of polypropylene, which may have difficulty bonding to a highly oxygenated nanofiber surface. However, composites made with epoxy give improved tensile properties with strongly oxidized CNF surfaces (where the surface of the CNF is up to 21% oxygen by XPS measurement). Composites made with BMI also show improved performance with oxidized nanofiber surfaces.

Heat treatment of CNF above 1100°C results in graphitization, which improves the crystallinity of the carbon from which they are formed, but also induces gross changes in morphology. CNFs having a filamentary core of conically nested graphene planes graphitize into discontinuous conical crystallites after heat treatment to 2900°C or higher. This discontinuous structure does not give optimum mechanical or electrical properties to composites in which they are used. Intermediate heat treatment to 1500°C yields an apparent optimum heat treatment temperature for tensile strength and modulus, as well as electrical conductivity. This observation resembles the optimum in electrical and mechanical behavior observed during the graphitization of PAN- and pitch-based carbon fibers, except that the optimum heat treatment temperature for CNFs is lower. Perhaps the lower optimum heat-treatment temperature for CNFs results from the initial purity of the exterior turbostratic layer, which is most influenced by the heat-treatment process.

Acknowledgments

The authors would like to thank the NIST ATP who supported this work under Cooperative Agreement Award No. 70NANB5H1173, *Vapor-Grown Carbon Fiber Composites for Automotive Applications.*

R.H. Blunk of General Motors graciously allowed us to use his apparatus and contributed useful suggestions on the intrinsic conductivity measurements. Thanks also to T. Burchell of Oak Ridge National Laboratory for many helpful suggestions. Thanks to Jane Howe of Oak Ridge National Laboratory for the TEM shown in Figure 5.1. Much of the data and work shown are due to the efforts of C. Kwag, Joana C. Finegan, M.J. Matuszewski, and K.R. Walters.

References

1. G.G. Tibbetts et al., Physical properties of vapor-grown carbon fibers, *Carbon,* 31, 1039, 1993.
2. J.R. Sarasua and J. Pouyet, Dynamic mechanical behavior and interphase adhesion of thermoplastic (PEEK, PES) short fiber composites, *J. Thermoplastic Composite Mater.,* 11, 2, 1998.
3. R.F. Gibson, *Principles of Composite Materials Mechanics,* McGraw-Hill, New York, 1994, p. 21.
4. S. Dong and R. Gauvin, Application of dynamic mechanical analysis for the study of the interfacial region in carbon fiber/epoxy composite materials, *Polymer Composites,* 14, 414, 1993.
5. G.G. Tibbetts et al., A review of the fabrication and properties of vapor-grown carbon nanofiber/polymer composites, *Comp. Sci. Technol.,* 67(7–8), 1709–1718, 2007.
6. G.G. Tibbetts, C.A. Bernardo, D.W. Gorkiewicz, and R.L. Alig, Mechanical properties of vapor-grown carbon fiber composites with thermoplastic matrices, *Carbon,* 32, 569, 1994.
7. G.G. Tibbetts and J.J. McHugh, Mechanical properties of vapor-grown carbon fiber composites with thermoplastic matrices, *J. Mater. Res.,* 14, 2871, 1999.
8. I.C. Finegan, G.G. Tibbetts, D.G. Glasgow, J.M. Ting, and M.L. Lake, Effect of carbon nanofiber-matrix adhesion on polymeric nanocomposite properties. II., *J. Mater. Sci.,* 38, 3485, 2003.
9. I.C. Finegan and G.G. Tibbetts, electrical conductivity of vapor-grown carbon fiber/thermoplastic composites, *J. Mater. Res.,* 16, 1668, 2001.
10. G.G. Tibbetts, I.C. Finegan, and C. Kwag, Mechanical and electrical properties of vapor-grown carbon fiber thermoplastic composites. *Mol. Cryst. Liq. Cryst.,* 387, 129, 2002.
11. Ehrburger, Surface properties of carbon fibers, in *Carbon Fibers, Filaments, and Composites,* J.L. Figueiredo, C.A. Bernardo, R.T.K. Baker, and K.J. Huettinger, Eds., Dordrecht: Kluwer, 1990, p.147–161.
12. G.G. Tibbetts and J.J. McHugh, Mechanical properties of vapor-grown carbon fiber composites with thermoplastic matrices, *J. Mater. Res.,* 14, 2871, 1999.
13. G.G. Tibbetts et al., Physical properties of vapor-grown carbon fibers, *Carbon,* 31, 1039, 1993.
14. K. Lafdi and M. Matzed, SAMPE 2003, *35th Int. Technical Conf.,* Dayton, OH, September 2003.
15. J.Y. Howe et al., Heat treating carbon nanofibers for optimal composite performance, *J. Mater. Res.,* 21, 10, 2006.

6

Compounding Layered Silicate Nanocomposites

Paul Andersen

CONTENTS

6.1 Introduction

Layered silicate-based polymer nanocomposites have demonstrated a significant potential to become the basis for development of the next generation of enhanced performance polymer compounds. Incorporation of only a small loading (4 to 5%) of properly treated, well-dispersed/exfoliated organoclay into the base polymer results in a compound with a substantial improvement in thermal, mechanical, as well as other physical properties over those of the base polymer. Until recently, most development efforts have focused on determining proper surface treatment to make the clay compatible with the base polymer and therefore improve the ease with which it can be dispersed in the polymer.

Most publications still concentrate on the importance of the chemistry used to modify the surface of the clay. They provide a description of resultant product properties but do not include the role of processing or give details of the compounding setup. Therefore, the key challenge facing many new entrants into the field is to determine how to maximize the clay exfoliation. Of course, using clay modified specifically for compatibility with the polymer matrix is extremely important; however, proper design and operation of the compounding system is equally critical. This chapter reviews the design flexibility associated with co-rotating twin-screw extruders and discusses key unit operations necessary to obtain clay that is well dispersed.

Interest in nano-particle-based polymer composites has expanded significantly since the late 1980s when the patent of Okada et al.[1] (assigned to Kabushiki Kaisha Toyota Chou Kenkyusho) for *in situ* polymerization of a Nylon 6/clay nanocomposite with, as stated in claim 1, "high mechanical strength and excellent high-temperature characteristics" was issued.[1] The results presented in the patent show that polymer nancomposites based on layered silicates provide a significant potential for development of a wide range of enhanced performance polymer compounds. As demonstrated by several researchers, a relatively small loading of properly dispersed (well-exfoliated) organoclay provides a substantial improvement in a polymer's properties.[2-6] These include improved thermal properties such as heat distortion temperature (HDT), mechanical properties such as flexural strength and modulus (without significant loss of impact), barrier properties, flame resistance, and abrasion resistance. However, until the early to mid-2000s, there were few commercial materials. Those in the market were mostly based on Nylon 6, and were for niche market applications. The reason for this, at least in part, is that many of the initial composites, such as the Toyota material previously noted, were developed using direct polymerization of a monomer clay mixture. While this method is suitable for certain polymers such as Polyamide 6, the complexity and expense of building a production facility limits entry of many smaller firms into the market. Also, until recently, most development has focused on determining proper surface treatment to make the clay (typically montmorillonite) compatible with the base polymer and therefore improve the ease with which it can be dispersed.

Using properly treated clay (i.e., more compatible with the matrix polymer) significantly reduces the degree of difficulty in compounding nanocomposites. For example, a Polyamide 6 and compatible organoclay can be compounded using a mixing-type single-screw extruder. The result is a nanocomposite where the clay (as determined by XRD and TEM) is partially exfoliated.[7] However, comparison of physical properties between the nanocomposite compounded on the single-screw extruder and one compounded on a twin-screw extruder configured with a very mild mixing zone shows significant differences. The modulus of the single-screw extruder compounded material compared favorably with that of material compounded on the twin-screw extruder. However, while the yield strength of the single-screw processed material was better than the base polymer, it was inferior to the twin-screw compounded material. Finally, it was worse than both the base polymer and the twin-screw compounded material in Izod Impact strength. A similar study determined that a single-screw extruder could partially exfoliate a very compatible organoclay to form a Polyamide 6 based nanocomposite; but when a less compatible organoclay was used, a twin-screw extruder was required to improve dispersion.[8]

While a number of treated clays have been developed for Polyamide 6, finding the proper "chemistry" to make a "fully" compatible organoclay for use with other polymers—such as polyethylene(PE), polypropylene (PP), and PET—has proven more elusive.[8] The result has been that to enhance compatibility, a third component has been utilized. Currently, compatibilizer material such as maleated PP is used with varying degrees of success to enhance the dispersion of the organoclay in PP. For example, when an organoclay modified specifically for use with PP is incorporated into the PP, the flexural modulus of the blend increases somewhat. However, the further addition of the maleated PP compatibilizer provides a significant additional increase in the flex modulus.[9]

As a consequence of this concentration on chemistry, there has been a lag in understanding the mechanics and mechanisms needed to implement commercially viable production methods. This scenario is slowly changing, as evidenced by increasingly more attention being given to the development of a direct compounding process, as well as cooperative development relationships among industrial companies such as that among General Motors (GM), Basell, and Southern Clay Products.[7–16] The result of this specific cooperation has been the introduction of olefin-based nanocomposites by GM on several vehicle lines. These applications include body side panels on mid-sized vans,[17] body side moldings on sedans,[18] and various exterior parts on sport utility vehicles.[19]

Although more development effort is being directed toward production technology, with few exceptions, most publications continue to concentrate on describing resultant product properties and do not give many, if any, details of the processing equipment setup.[7, 8, 11, 15, 16] Therefore, a key challenge facing many companies that want to enter the nanocomposite field is to determine the best process to use and how to maximize clay exfoliation.

Of course, as indicated above, choosing an organoclay specifically modified for compatibility with the polymer matrix will go a long way toward achieving good exfoliation. However, if these and other polymers are to be used effectively in nanocomposites, then for many polymers, such as polypropylene, the clay treatment chemistry must be further improved and the processing/compounding step needs to assume a more prominent role. That is, proper design and operation of the processing system must be better understood. To realize the latter, the key operating parameters that influence the dispersion and incorporation of the clay platelets in the polymer must be determined. Based on some initial work done in this area, it is also important to use processing equipment that is flexible in design and has the capability to combine residence time with dispersive and distributive mixing.[7] One piece of equipment that provides these characteristics is the co-rotating, fully intermeshing, twin-screw extruder (Figure 6.1).

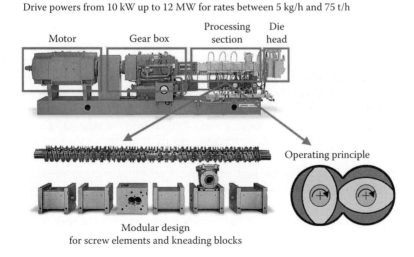

Drive powers from 10 kW up to 12 MW for rates between 5 kg/h and 75 t/h

Motor   Gear box   Processing section   Die head

Operating principle

Modular design
for screw elements and kneading blocks

FIGURE 6.1
Basic layout and main components of a twin-screw extruder.

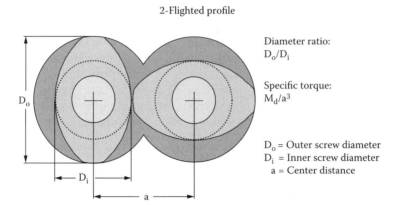

FIGURE 6.2
Characteristic dimensions of co-rotating twin-screws

6.2   Process Equipment: The Co-rotating Twin-Screw Extruder

6.2.1   Design Parameters

The fully intermeshing co-rotating twin-screw extruder has three degrees of design free-dom: (1) geometry, (2) power transmission, and (3) screw rpm.[20] Also, the cross-sectional geometry of the extruder is defined by three dimensions: (1) screw diameter (either the outer diameter (D_o) or the inner (root) diameter (D_i)); (2) D_o/D_i ratio; and (3) center-line distance (a) between the two screw shafts (Figure 6.2). Once two of the above three cri-teria are defined, the other is fixed. For a constant centerline distance, as the D_o/D_i ratio is increased, the extruder D_o increases and the D_i decreases (Figure 6.3). Therefore, the diameter ratio can be used as a comparative measure of the free X-section area among extruders and thus the internal free volume per unit length as well. The larger the D_o/D_i ratio becomes, the greater the internal free volume per unit length. Concomitant with an increase in internal volume is an increase in the amount of material that can be trans-ported by the screws. Therefore, D_o/D_i is a relative measure of the volumetric throughput capacity of the extruder. However, D_o/D_i cannot be increased indefinitely. It is limited by geometry constraints and mechanical properties of the screw shaft and elements.[2]

D_o/D_i has several additional influences on the extruder design and operating conditions. In addition to being a comparative measure of internal free volume, D_o/D_i determines the extruder shear rate constant, as well as the maximum diameter of the screw shaft. As D_o/D_i is increased, the channel depth of the extruder increases and the shear rate constant decreases (Figure 6.4). Additionally, Figure 6.4 illustrates that average shear rate (shear rate constant × rpm) also decreases and consequently energy imparted on the material being processed is also reduced. Finally, as D_o/D_i increases, shaft diameter available for power transmission is reduced (Figure 6.3). This creates a conflicting scenario. An extruder with a larger D_o/D_i has a greater volumetric throughput capacity but geometric constraints limit the diameter of the screw element shaft. This has a negative impact on the power transmission, and thus throughput capacity of the extruder. Therefore, an appropriate balance between required power transmission capability and available free volume

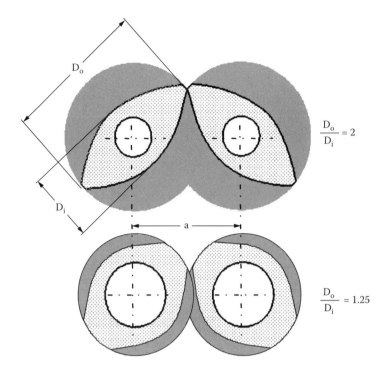

FIGURE 6.3
Relationship between D_o to D_i ratio and free volume.

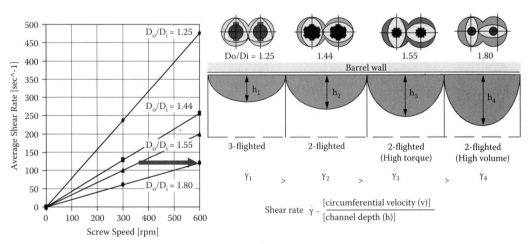

Higher screw speeds compensate for lower average shear rates
of filled channels at the same total shear and temperature stress.

FIGURE 6.4
Effect of D_o/D_i and rpm on average shear rate.

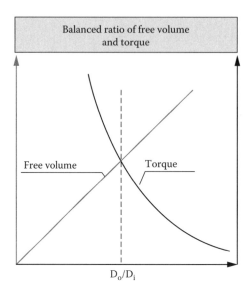

FIGURE 6.5
Free volume and permissible torque as a function D_o/D_i.

(Figure 6.5) must be determined so that neither one is a process-limiting step. The ideal situation exists when a process is simultaneously power limited and volume limited.

Table 6.1 shows the difference in free cross-sectional area, average shear rate, and power/volume ratio expressed as [torque capacity/centerline distance cubed (M/a^3)] of four co-rotating twin-screw extruders with the same centerline distance, but D_o/D_i ratios ranging from 1.25 to 1.80. (Although the extruder with the 1.55 D_o/D_i has thinner shafts, the allowable power transmission goes up as a result of a shaft/element geometry interface improvement. Energy is transferred through involute splines as opposed to one or two or even six keys (Figure 6.6). However, note that due to shaft limitations, the power/volume ratio of the extruder with the 1.8 D_o/D_i ratio is decreased.) While all four extruders could be used for similar processing tasks, the processing length, screw configuration, and operating conditions would need to be different. The resulting rate typically would be less for the early-generation machines with lower power volume factor (M/a^3) and smaller D_o/D_i.

TABLE 6.1

Comparison of Twin-Screw Extruders with the Same Centerline Distance

Machine Size (300 rpm)	OD/ID	Free Cross-sectional Area (cm²)	Average Shear rate (s⁻¹)	Torque/ Centerline³ (M/a³)
ZSK-53	1.25	10.1	225	4.7–5.5
ZSK-57	1.44	16.7	125	4.7–5.5
ZSK-58	1.55	18.2	100	13.6
ZSK-62	1.80	25.1	~60	11.3

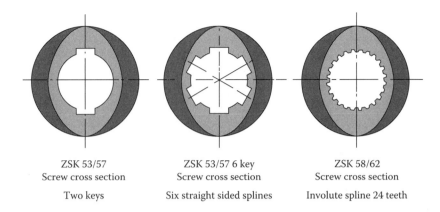

ZSK 53/57
Screw cross section

Two keys

ZSK 53/57 6 key
Screw cross section

Six straight sided splines

ZSK 58/62
Screw cross section

Involute spline 24 teeth

FIGURE 6.6
Shaft/element geometry interface.

However, when power vs. volume is considered, the 1.55 D_o/D_i ratio extruder would be capable of the highest rates for energy-intensive processes such as glass-filled nylon, but the 1.8 D_o/D_i ratio extruder would win out for processing low bulk density material not requiring a significant energy input such as 60% talc-filled PP.

6.2.2 Process Section

The process section of the twin-screw extruder consists of modular components (barrel section, shaft, and screw bushings) that are assembled in the optimum configuration to meet required unit operations. Figure 6.7 shows a typical process section and required unit operation for glass incorporation into a polymer matrix. In this case, they include feeding, bulk transport and preheating, melting, glass feeding, mixing glass and matrix, devolatilization, metering, and discharge pressurization.

The length of the process section depends on the number of unit operations required, as well as the power consumed. For mixing/compounding lower viscosity materials, 60 to 1 L/D (extruder length/screw diameter and longer) process sections have been assembled.

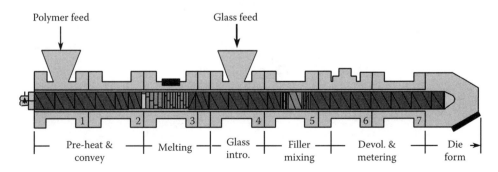

Polymer feed

Glass feed

| 1 | 2 | 3 | 4 | 5 | 6 | 7 |

Pre-heat &
convey

Melting

Glass
intro.

Filler
mixing

Devol. &
metering

Die
form

FIGURE 6.7
Polymer/glass fiber compounding setup.

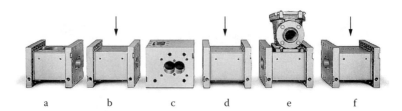

FIGURE 6.8
Building block system of twin-screw extruder barrel sections.

However, for standard compounding tasks, the L/D ranges from 24 to 48, with most process sections ranging between 32 and 40 L/D.

6.2.2.1 Barrels

Most twin-screw extruders have barrel sections to accommodate every process unit operation. These include upstream and downstream feeding of diverse material forms such as polymer pellets, irregular flake, or low bulk density powders; inert fillers (both low and high aspect ratio); or even liquids. Also, vent barrels have been designed that will handle a wide range of vapor loads—from very high levels of solvent or monomer to miscellaneous moisture removal. In either case, the emphasis is on material quality because vent sections are notorious as a source of contamination. Figure 6.8 shows a series of several barrel segment designs. Recently, systems for side venting with a "vent stuffer," have been introduced to accommodate difficult-to-handle material that has a tendency to vent flow (Figure 6.9).

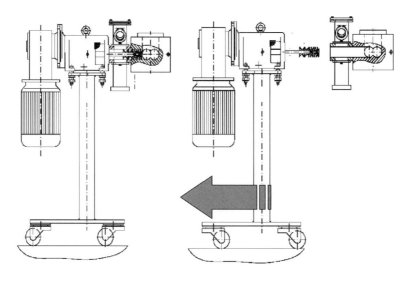

FIGURE 6.9
Side vent system design and setup.

3-flighted profile   2-flighted profile

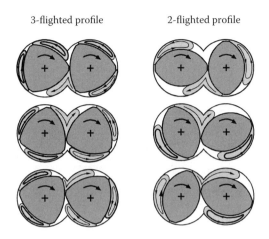

FIGURE 6.10
Cross-section profile and material flow pattern.

6.2.2.2 Elements

Material transport in intermeshing, co-rotating twin-screw extruders generally depends on drag flow. The screws pick up the material as they rotate; and where the two screws meet, a complete transfer of the material from one screw to the other takes place (Figure 6.10). The tip of one screw wipes the flanks and roots of the other screw, resulting in a self-wiping action. As the material is transferred from one screw to the other, the direction of material flow is changed and new material surfaces are created with each screw revolution. This operation provides the mechanism of conveying, mixing, and pressure build-up.

Screw bushings. Standard screw bushings are constructed with pitches ranging from approximately 1/2 D to 2.0 D, where D is the machine diameter (Figure 6.11). High pitch elements (Figure 6.11a) might typically be used in feed or devolatilization areas of the extruder. Narrower pitch elements (Figure 6.11b and d) are used in areas where compaction of material and 100% fill is desired, such as before kneading blocks or seals, or between unit operations (i.e., feeding and vacuum devolatilization). Up to approximately 2.5 D, increasing element pitch results in a decrease in residence time and degree of fill, a more narrow residence time distribution, increased drag flow capacity, but also increased sensitivity to pressure flow. That is, as the pitch of an element increases, drag flow conveys

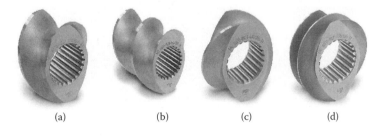

(a)   (b)   (c)   (d)

FIGURE 6.11
Typical twin-screw extruder conveying screw bushings.

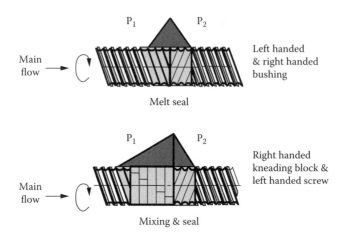

FIGURE 6.12
Working principles of restrictive elements.

material in the down-channel direction at a faster rate. However, if there is a restrictive force placed in the flow path, the higher pitch element is less efficient in building up the pressure necessary to push material past the restriction.

Reverse pitch elements (Figure 6.11c) are used to generate back-pressure (Figure 6.12) and therefore create sections of 100% fill that, for example, can be used to separate unit operations or totally fill a mixing section.

Kneading blocks and special mixing elements. The basic building blocks for mixing in the co-rotating, intermeshing type, twin-screw extruder are kneading blocks and special mixing bushings.

While screw bushings are characterized by pitch (i.e., flight angle), kneading blocks are characterized by individual disc length (width) and stagger angle between successive discs (Figure 6.13). They have the same cross-section as the screw bushings. The offset angle between elements causes an axial displacement flow due to staggered pressure profiles (Figure 6.14).

Kneading blocks introduce both a distributive and dispersive mixing component into the system. The relative intensity of each component depends on disc length as well as

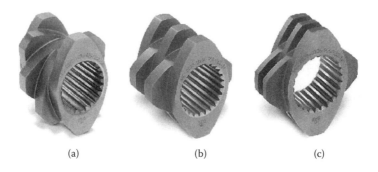

(a)   (b)   (c)

FIGURE 6.13
Typical twin-screw extruder kneading blocks.

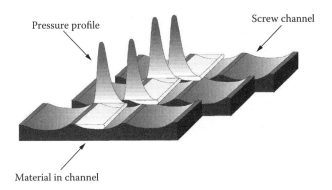

FIGURE 6.14
Typical pressure profile in the kneading discs of a co-rotating twin-screw extruder.

stagger angle. For constant stagger angle (Figure 6.15), increasing the disc width leads to an increase in the dispersive mixing component per unit mixing length, but at the expense of distributing mixing (stream splitting). That is, as a disc gets wider, it can more easily trap particles in the high shear region but there are fewer discs per unit length to distribute them. For a constant disc width (Figure 6.16), the greater the stagger angle the more back mixing and the higher the residence time distribution. Finally, just as there are reverse pitch elements, there are reverse stagger kneading blocks. As with reverse pitch elements, reverse kneading blocks are used to create back-pressure. They are, however, less restrictive than the reverse pitch elements.

Toothed/gear type elements (Figure 6.17) are the most commonly used of the special mixing elements. The number of teeth around the circumference as well as the tooth angle define them. The former contributes to stream splitting for generation of interfacial surface and the latter conveying capacity. The main function of these elements is to provide the maximum amount of distributive mixing (little if any dispersive mixing) with minimal energy input.

Additional special bushings include slotted screw bushings (Figure 6.18a) and conveying screw bushings over-cut with a reverse flight (Figure 6.18b). Standard conveying-type screw bushings are also used in certain circumstances.

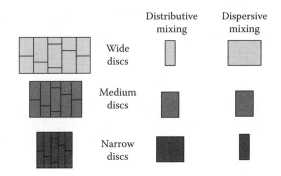

FIGURE 6.15
Forwarding kneading blocks: disc width vs. mixing characteristics.

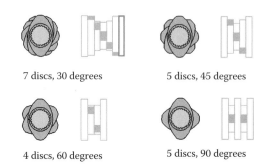

7 discs, 30 degrees   5 discs, 45 degrees

4 discs, 60 degrees   5 discs, 90 degrees

FIGURE 6.16
Kneading block geometry: stagger angles of kneading discs.

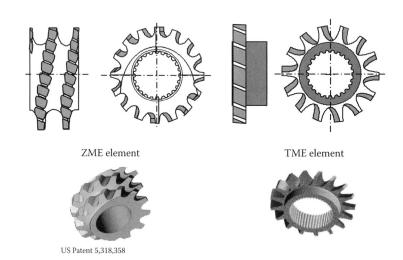

ZME element   TME element

US Patent 5,318,358

FIGURE 6.17
Comparison of "toothed" type mixing elements.

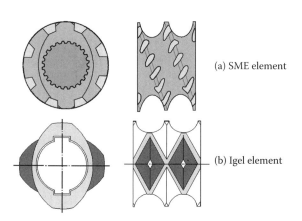

(a) SME element

(b) Igel element

FIGURE 6.18
Other types of distributive mixing elements.

6.3   Compounding Objectives

While nanocomposites contain usually no more than 5% loading of organoclay, they present a unique combination of compounding challenges not faced with other polymer systems, such as fiber- or mineral-filled compounds, impact-modified resins, as well as additive or color concentrates.

For fiber-reinforced compounds, the fibers, whether roving or chopped strand, need to be unbundled, then distributed, and finally wetted out to have maximal bonding with the polymer matrix. For the fibers to be effective in reinforcing the compound, all of these steps must be accomplished with minimal attrition of fiber length. While not a simplistic compounding task, using the correct processing techniques (such as downstream addition of the fiber) and proper mixing element sequencing in the screw design will result in a properly mixed material.

In most mineral-filled compounds, the filler (typically talc, calcium carbonate, or titanium dioxide) must be distributed and then, if necessary, dispersed. Typically, these mineral particles need not be smaller than a micron. Even when these fillers are present at higher loading amounts, or are introduced as sub-micron particles, they are still processed with relative ease as long as the compounding extruder configuration is properly designed.

As the loading of mineral filler in these formulations increases, or when processing materials such as color concentrates based on organic pigments that tend to "cake," in addition to distribution and dispersion of the filler/pigment, the compounder must be vigilant that the filler/pigment is not compacted into agglomerates prior to incorporation into the polymer matrix. If the filler/pigment is present in significant enough volume or, more likely, in such form that it segregates from the polymer feed to form localized pockets of pure filler (such as in powder filler and blends of polymer pellets), the filler can be compressed under significant force in the nip area of the KB (Figure 6.19) to a point that agglomerates are formed. These agglomerates, which later must be broken up to achieve the desired particle size distribution, can be more difficult to disperse than filler aggregate particles in the original feed stream. However, as outlined in the next several paragraphs, good compounding techniques such as proper feed sequencing of ingredients and use of proper screw elements will virtually eliminate agglomerate formation.

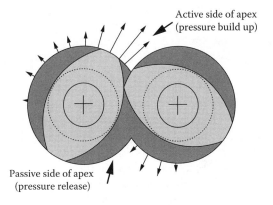

FIGURE 6.19
Schematic of radial pressure profile in the twin-screw extruder.

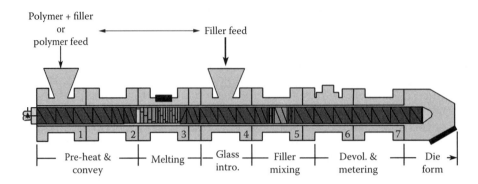

FIGURE 6.20
Typical setup options for filler incorporation.

The twin-screw extruder modular design allows it to be configured to permit specific sequencing of desired unit operations (Figure 6.20). For example, the polymer and filler can be introduced at the same feed location at the beginning of the extruder, or the filler can be metered in further downstream after the polymer has melted. Both of these techniques have potential advantages and also shortcomings. If the two materials are introduced together and uniformly distributed, then as the polymer experiences the pressure and requisite stress needed to induce melting, the filler is also subjected to high dispersive stresses. This may be the first critical step in breaking down filler particles (including organoclay) into smaller units. On the other hand, if as the filler enters the melting section, it is not well distributed with the polymer, then, as shown previously in Figure 6.19, it can be trapped in the apex region as the two kneading discs come together. Without polymer present to absorb and dissipate the energy, the filler can be subjected to extremely high compressive forces. As mentioned previously, these forces may actually cause filler particles to form agglomerates that must subsequently be dispersed. One option to avoid making agglomerates is to use a type of kneading block in the melting section that is designed to minimize the generation of high force in the apex area.[22]

Three-lobe eccentric kneading blocks for two-lobe extruder systems were designed for more uniform energy input on the two-lobe system (Figure 6.21), particularly in the melting zone. As shown in Figure 6.22, the elements are offset so that one tip wipes the barrel wall while the other two have a significantly greater clearance (3 to 5 times) that permits polymer melt to more easily flow over these tips. This sets up a more circumferential rather than down-channel material flow in the element. The result is more energy efficient as well as more homogeneous polymer melting.[23] The material exits the melting zone with few (if any) remaining unmelted particles.

In addition to providing more uniform melting, three-lobe kneading blocks have been shown to have a beneficial impact on reducing agglomerates formed during melt mixing of polymer/filler preblends.[24] That is, they are less prone to compacting pigments, fillers, or additives segregated during the melting process. The three-lobe kneading blocks have a lower pressure peak at the apex than two-lobe kneading blocks (Figure 6.23). The lower apex pressure for eccentric kneading blocks is primarily the result of the increased clearance of two tips, but also three-lobe kneading blocks have a smaller crest angle (Figure 6.24).

Because these elements have a reduced diameter and a minimum crest angle, they do not generate high pressure in the apex. As noted, it is this pressure that can compact non-melting material into agglomerates.

US patent 4,824,256

FIGURE 6.21
Three-lobed kneading elements.

An alternative approach to address the issue of agglomerate formation is to use two-lobe elements with very narrow width discs. The probability of material being subjected to pressure great enough to form agglomerates depends on the KB crest angle and individual disc width. The larger the crest angle and the wider the disc element, the more likely that the element will generate filler agglomerates. The difficulty with this approach is that as the discs get thinner, mechanical strength decreases. One way to circumvent the strength issue is to make the tip of the disc narrow but leave the root thick. Figure 6.25 shows one disc from such an element, while Figure 6.26 illustrates a complete element.[25] Additionally, this element reduces pressure further by having the discs from the element on the opposite screw shaft offset. In this way, as the elements rotate, the two crests are offset as they pass through the apex. These elements are typically used in masterbatch preblends of compactable organic pigments.

An additional alternative is to feed filler downstream of the polymer melting zone. The advantage to this approach is that the molten polymer will act as a buffer to keep the filler from being subjected to high compressive force. On the other hand, because the polymer

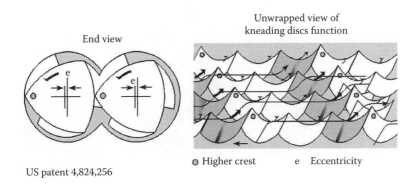

End view

Unwrapped view of kneading discs function

US patent 4,824,256

⊙ Higher crest   e   Eccentricity

FIGURE 6.22
Operating principle of three-lobe KB.

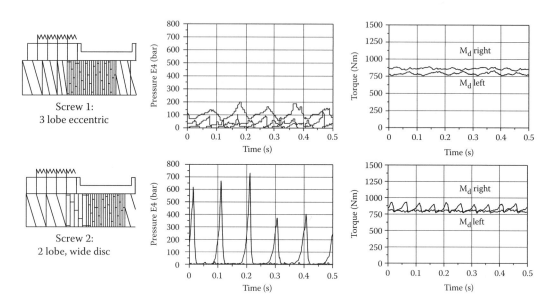

FIGURE 6.23
Comparison of pressure in two-lobe vs. eccentric three-lobe KB.

is already molten, the dispersive stress transmitted to the filler particles will be less than in the melting section.

The feed location of clay into the compounding line as well as the type of clay have a significant impact on physical properties. Heidemeyer et al. found that the particle size of the clay, the meq of the quat, and the feed location all influenced properties.[26] The larger clay particle required greater energy input to break down the particle. Therefore, it benefited

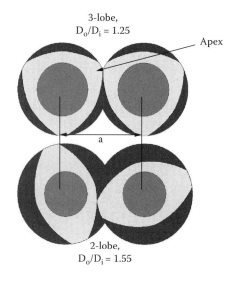

FIGURE 6.24
Crest width comparison three-lobe vs. two-lobe geometry.

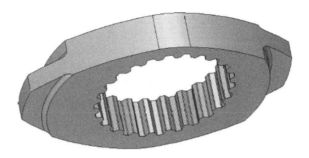

FIGURE 6.25
Low shear reduced tip width kneading disc.

from being added at the first feed barrel along with the polymer. The additional meq of quat for the same particle size had a significant impact on modulus, but the addition of clay in the first barrel still had the most significant impact. When the particle size of the clay is reduced by almost an order of magnitude, the feed location of the clay is virtually inconsequential. The modulus of all the samples were within a few percent of each other.

Nanocomposites are not immune to the potential of agglomeration issues. Although the treated clay used in nanocomposites is not particularly compactable, the processor must be aware that, as indicated in the previous paragraphs, high pressures can be generated by standard kneading blocks in certain orientations. This is especially important for production of nanocomposite masterbatch material.

All this being noted, the fact remains that in compounding nanocomposites, the first challenge is to uniformly disperse clay particles that are approximately 8µ in diameter into possibly up to a million nanometer thick platelets. To exacerbate the problem, clay, although it has been treated, is most likely not fully compatible with the polymer. Therefore, the compounder faces a task similar in difficulty to trying to break rocks into dust and then dissolve the dust in water. The question that must be answered is whether this is best accomplished by feeding the clay upstream with the polymer for better stress transfer, but with a greater risk of particle agglomeration, or feed the clay downstream into the already molten polymer.

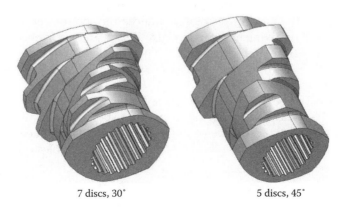

7 discs, 30°   5 discs, 45°

FIGURE 6.26
Low shear kneading blocks (SAM).

In addition to choice of feed location, a critical design decision is choice and sequencing of mixing unit operations. One needs to determine the most effective sequence of steps in transforming a less than fully compatible 8-μ clay particle into up to a million nanometer thick platelets. For example, if pure dispersive mixing is all that is required, then a series of wide disc kneading blocks would be sufficient. However, based on the conclusions of Dennis et al., the amount of shear introduced does not directly correlate to dispersion.[37] They concluded that when using less-compatible organoclay, a combination of residence time and stress was most effective. They propose an exfoliation mechanism whereby the organoclay particles are dispersed, through shear breakup or cleavage along the platelet interface, into thousands of indeterminate-size particles. These particles can no longer be affected by additional shear to break up into even smaller particles. Subsequent dispersion results from polymer penetrating between platelets and forcing either the surface layer to peal off, thus attacking the clay layer by layer, or interplatelet fracture (as water might crack a rock apart through a continuous freeze-thaw cycle).

While the development of clay-based nanocomposite materials still has a long way to go, preliminary experimental findings indicate that the best physical properties are obtained when all ingredients are fed together in barrel 1. However, clay exfoliation does not always agree with the physical property values.[27]

If polymer nanocomposites are going to live up to their potential, then as noted previously, either the clay treatment chemistry must be improved significantly or the compounding step needs to assume a more prominent role. The following section describes a preliminary evaluation to determine the key processing parameters that influence the dispersion and incorporation of the platelets in the polymer.

6.4 Process Variables

A compounding line is set up with a specific sequence of unit operation, as shown previously in Figures 6.7 and 6.20. Additionally, there are process variables that can have a significant impact on the resulting product quality. These include but are not limited to screw configuration, rpm, temperature profile, product through put rate, and recipe. The screw configuration can be configured to influence the intensity of mixing, type of mixing (dispersive vs. distributive), as well as total residence time and residence time distribution. Total residence time and mixing intensity are also significantly influenced by rpm. As rpm is increased, residence time decreases while the unit energy input increases. On the other hand, as the overall throughput rate increases, with all other variables held constant, the overall unit energy input decreases.

These variables have been evaluated in a recent publication where a series of polypropylene-based samples was run using the following experimental outline as a guide[27]. It is important to note that the objective of this work was not to produce completely exfoliated clay, but rather highlight the efficacy of the process variables on intercalation and exfoliation.

Process setup:
Feed locations
Dispersive vs. distributive mixing

Mixing section sequencing

Residence time

Operating Parameters (mixing intensity):

Screw rpm (N) increase at constant Q/N

Screw rpm increase at constant Q

Formulation:

Matrix polymer viscosity

Compatibilization agent

Masterbatch vs. direct compounding

6.5   Process Variable Evaluation

6.5.1   Equipment Setup

To evaluate the impact of process variables, a 25-mm, 10-barrel, 41 to 1 L/D twin-screw extruder with feed locations at barrels 1 and 5, and vacuum at barrel 9 was set up to compound all samples. Three screw configurations were used, two of which are shown in Figure 6.27. Both configurations have downstream feeding options. The melting sections are composed of three-lobe kneading blocks to minimize the potential for any agglomerate formation. The downstream section of screw 1 consists of all tight pitch conveying bushings and distributive type elements. It is designed for increased residence time and distributive mixing in a pressurized environment. The downstream section of screw 2 consists mainly of forwarding kneading blocks that are not backed up by any restrictive elements except at the very end of the zone. Therefore, except at the end of the zone, these elements will be only partially filled and the material will traverse this section relatively

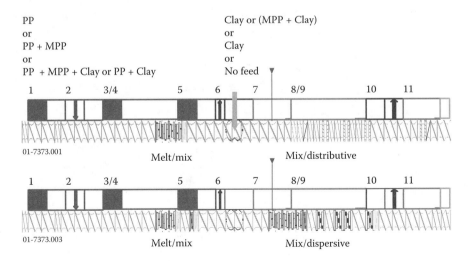

FIGURE 6.27
Screw configurations and feed sequencing.

quickly. The third screw configuration was set up to run masterbatch, masterbatch let-down, and direct compounded product. It also had downstream feed capacity. In total, more than 100 samples were run.

6.5.2 Feedstock

The main polymer feedstock was a 4.8 MFR polypropylene. The clay was an alkyl quaternary ammonium montmorillonite (Cloisite 20A) manufactured by Southern Clay Products. The compatibilizer was a maleated PP (Polybond 3200) manufactured by Crompton Corporation (now Chemtura). The standard formulation for all compounds was 90 parts PP, 5 parts maleated PP (MaPP), and 5 parts clay. If MaPP was omitted from the formulation, then PP was substituted.

6.5.3 Product Testing

All samples were evaluated by x-ray diffraction at a 1°/min scan rate. The area under the curve was used as a general evaluation of clay exfoliation (smaller is better). The d-spacing was used as a relative determination of intercalation (greater is better). Additionally, ash was measured for each sample to confirm formulation accuracy. Select samples were sub-mitted for transmission electron microscopy and physical property evaluation.

6.5.4 Observations

The first point to address is the impact of a compatibilizing agent on the clay dispersion. Figure 6.28 shows a graph of d-spacing and area under the XRD (x-ray diffraction) curve for two sample formulations processed on the dispersive screw configuration over a range of screw rpm values. The data clearly show that the samples with the maleated polypro-pylene compatibilizing agent have a greater increase in d-spacing and a smaller area under the XRD curve. The former indicates improved intercalation, while the latter implies better exfoliation. This figure as well as several that follow also show that as the material is pro-cessed at higher rpm values, the d-spacing of the clay in the sample decreases, indicating less intercalation, but the area under the XRD curve appears to also decrease. This would

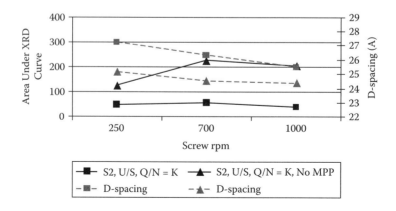

FIGURE 6.28
Effect of maleated PP on clay exfoliation and intercalation.

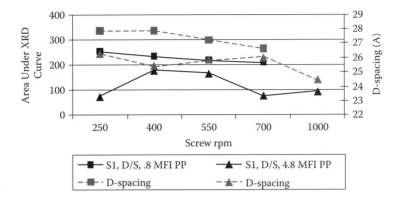

FIGURE 6.29
Effect of PP MW on clay exfoliation and intercalation.

be an indication that more clay had been exfoliated. One possible explanation is that at higher screw rpm, the energy input is such that intercalation does not play as significant a role in promoting exfoliation.

Another interesting and seemingly contradictory observation is that the higher molecular weight (MW) PP shows (as indicated by a greater d-spacing) enhanced intercalation with respect to the lower MW PP (Figure 6.29). This is noteworthy because it is somewhat contrary to the results of Wang et al. who report that lower MW maleated PP improves intercalation.[15] At the same time, the area under the x-ray diffraction peak (Figure 6.29) is significantly larger for the high MW PP than for the lower MW PP. This indicates poorer exfoliation of the platelets in the higher MW sample. One possible explanation for these seemingly contradictory observations might be that the high MW PP can penetrate the clay platelets enough to show separation but not achieve exfoliation. The lower MW PP penetrates and exfoliates. Therefore, these samples exhibit smaller d-spacing as well as area under the x-ray diffraction curve. Additional experiments with low MW PP would help clarify this observation.

Figure 6.30 shows that adding clay together with PP and MaPP in barrel 1 is more effective in the generation of material with better intercalation (d-spacing) and exfoliation (area

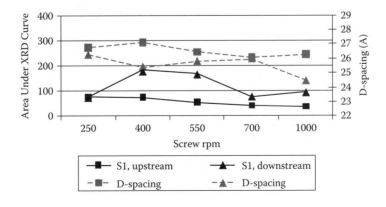

FIGURE 6.30
Clay feed location: effect on exfoliation and intercalation.

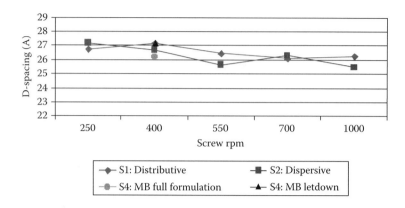

FIGURE 6.31
Screw configuration: effect on intercalation of clay.

under the curve) than is feeding clay downstream. Additionally, but not shown on this chart, adding clay plus MaPP downstream produced virtually the same result as feeding clay by itself downstream. Another point to be noted is that the data for feeding clay downstream are more erratic based on the trending of the data with respect to rpm. The final point to make regarding Figure 6.30 is that, just as with the previous figures, it shows intercalation to decrease slightly with increasing rpm at constant Q/N. However, the area under the curves is a little more inconsistent.

This result tends to support the concept that residence time impacts intercalation. Another observation is that the dispersive, distributive, and masterbatch screw configurations show minimal difference in intercalation efficiency (Figure 6.31).

While analysis of x-ray data may be interesting, it is a compound's physical properties that will define success. Subsequent to the runs on screw 1 and 2, a third configuration was used to compare masterbatch vs. direct compounded products. Figure 6.32

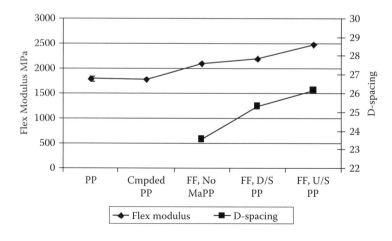

FIGURE 6.32
Direct compounded product: physical property comparison.

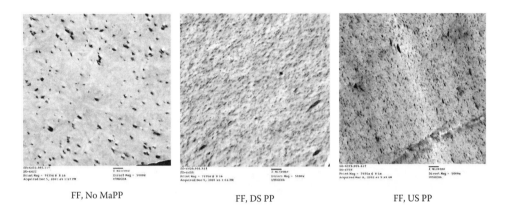

FF, No MaPP   FF, DS PP   FF, US PP

FIGURE 6.33
Direct compounded product: electron micrograph comparison.

shows flexural modulus for neat PP, compounded PP, and direct compounded products. The increase in flex modulus tracks the increase in d-spacing and provides further support for feeding all ingredients into the first barrel. It also shows that while removal of the MaPP compatibilizer results in lower physicals, clay by itself still provides some improvement.

In addition to physical properties, the samples were evaluated by transmission electron microscopy. Figure 6.33 shows TEMs (transmission electron micrographs) of the three clay-containing samples. The material containing no MaPP exhibits considerably larger clay aggregates than do either of the samples containing MaPP. The micrographs further reinforce the positive impact of feeding all ingredients into barrel 1 so that clay is dispersed by mechanical means to the smallest particle size. This exposes more surface area to improve the efficiency of the intercalation and exfoliation process.

Samples were also made by letdown into PP of a clay, MaPP, and PP masterbatch. The final formulation still maintained the 90/5/5 relationship. The ratio of ingredients in the masterbatch ranged from equal parts of each to equal parts of clay and MaPP with 25% PP. The order of addition for masterbatch ingredients was also varied. The masterbatch was first compounded on the twin-screw extruder and then mixed with the appropriate amount of PP for subsequent processing. Figure 6.34 shows that the masterbatch samples have a slightly greater flex modulus than their direct compounded counterpart, although the d-spacing is somewhat erratic. This particular masterbatch was compounded, adding the PP downstream. However, masterbatch efficiency does not seem to be impacted by the order of ingredient addition. The same masterbatch recipe as discussed previously but with PP added upstream together with clay and MaPP produced physical property results in letdown compounds equivalent to those shown in Figure 6.34.

Figure 6.35 shows TEMs comparing the direct compounded material vs. masterbatch (MB) product. It is somewhat more difficult to differentiate between the two materials, but the MB sample appears to have the smaller clay aggregates. Figure 6.36 shows the same two samples at higher magnification where it is somewhat more evident that the MB material is better exfoliated.

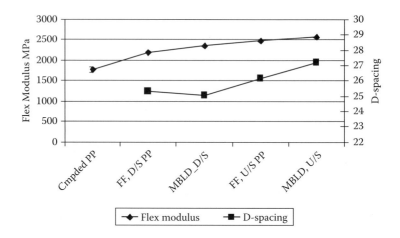

FIGURE 6.34
Physical properties: recipe and process conditions (MB configuration).

Based on the samples shown in the previous examples, it is evident that clay need not be fully exfoliated to improve physical properties, but each step forward adds to the improvement of the material properties.

Finally, one parameter that has not been explicitly evaluated during the previously noted work is material process temperature. The organic component in the clay will degrade at elevated temperature. This is a time at temperature function, which in the framework of compounding systems having a residence time of 10 to 30 seconds means that the organic component will be eliminated at melt processing temperatures above 250°C.

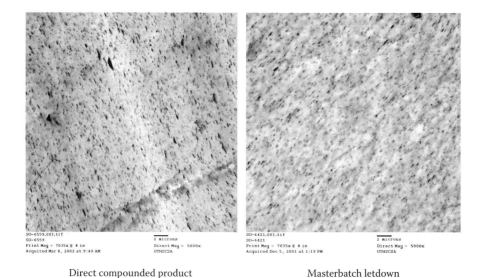

Direct compounded product   Masterbatch letdown

FIGURE 6.35
Direct compounded (PP added upstream) vs. MB: electron micrograph comparison (low magnification).

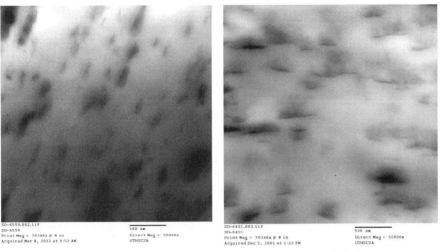

Direct compounded product   Masterbatch letdown

FIGURE 6.36
Direct compounded (PP added upstream) vs. MB: electron micrograph comparison (high magnification).

6.6 Summary

Published studies have shown that successful compounding of nanocomposites depends on the organoclay-polymer compatibility and processing conditions. However, these conditions have not been very well defined. This chapter highlighted some of the key variables that must be determined if a more quantitative understanding of the role that process setup and operating conditions play is to be obtained. The most significant process variable appears to be *feed location*.

References

1. Okada et al., U.S. Patent No. 4,739,007 (1988).
2. Giannelis, E.P., Polymer layered silicate nanocomposites, *Adv. Mater.*, 8, 29, 1996.
3. Lan, T. and Pinnavaia, T.J., Clay-reinforced epoxy nanocomposites, *Chem. Mater.*, 6, 2216, 1994.
4. Shi, H., Lan, T., and Pinnavaia, T.J., Interfacial effects on the reinforcement properties of polymer-organoclay nanocomposites, *Chem. Mater.*, 8, 1584, 1996.
5. Wang, Z. and Pinnavaia, T.J., Nanolayer reinforcement of elastomeric polyurethane, *Chem. Mater.*, 10, 3769, 1998.
6. Yao, K.J., et al., Polymer/layered clay nanocomposites: 2 polyurethane nanocomposites, *Polymer*, 43, 2981, 2001.
7. Cho, J.W. and Paul, D.R. Nylon 6 nanocomposites by melt compounding, *Polymer*, 42, 1083, 2001.
8. Dennis, H. R. et.al. Nanocomposites: the importance of processing, *SPE Antec 2000*, 428, May 7–11, 2000, Orlando, FL.
9. Andersen, P.G., Twin screw extrusion guidelines for compounding nanocomposites, *SPE Antec 2002*, 219, May 58, 2002, San Francisco, CA.

10. O'Neil, C.J., Acquarulo, L.A., and Xu, J., Optimizing nano filler performance in selected nylons, *SPE Antec 2000,* 1522, May 7–11, 2000, Orlando, FL.

11. Qian, G., Cho, J.W., and Lan, T., Preparation and properties of polyolefin nanocomposites, *SPE Polyolefins 2001,* February 25–28, 2001, Houston, TX.

12. Wolf, D., Fuchs, A., Wagenknecht, U., Kretzschmar, B., Jehnichen, D., and Häußler, L., Nanocomposites of polyolefine clay hybrides, *Eurofillers '99,* Lyon, September 6–9, 1999, Proceedings.

13. Wagenknecht, U., Pötschke, P., Kretzschmar, B., and Wolf, D., Formation of polymer-clay nano-composites during the compounding process, a novel method of clay dispersion, *Eurofillers '99,* Lyon, September 6–9, 1999, Proceedings.

14. Dahman, S., Melt compounding nanocomposites: the drive towards commercialization, *Nanocomposites 2000,* Brussels, 6-7 November 2000.

15. Wang, H., Zeng, C., Swoboda, P., and Lee, L.J., Preparation and properties of polypropylene nano-composites, *SPE ANTEC 2001,* 2307, May 6–10, 2001, Dallas, TX.

16. Andersen, Paul G., Processing nanocomposites on a kneader reciprocating single screw compounding system, *SPE ANTEC 2003,* May 4–8, 2003, Nashville, TN.

17. GM Press Release August 28, 2001.

18. Rogers, W.R., Polymer nanocomposite applications in the automotive industry, at *Polymer Nanocomposites 2005,* September 28–30, 2005, Boucherville, Quebec, Canada.

19. Sherman, L.M., Chasing Nanocomposites, Plastics Technology On-line, October, 2004.

20. Andersen et al., Understanding high rate and high speed compounding on co-rotating twin-screw extruders, *SPE ANTEC 1997,* 238, April 27–May 2, 1997, Toronto, Canada.

21. Andersen, P.G., The Werner and Pfleiderer twin-screw co-rotating extruder system, in *Plastics Compounding,* Todd, D.B., Editor, Hanser, Munich, Germany, 1998.

22. Haering et al., U.S. Patent No. 4,824,256 (1989)

23. Yu et al., Comparative melting trials in ZSK extruders, *SPE ANTEC 2001,* 139, May 6–10, 2001, Dallas, TX.

24. Rogers et al., The effect of three-lobe, offset kneading blocks on the dispersion of calcium carbonate in polystyrene resin, *SPE ANTEC 2001,* 129, May 6–10, 2001, Dallas, TX.

25. Haering et al., U.S. Patent No. 6,048,088 (2000).

26. Heidemeyer, P. et al., Nanoskalige pertikel als funktionelle fuellstoffe, *GAK,* 59, 96, 2006.

27. Andersen, P.G., Compounding nanocomposites, *SPE Polyolefins 2005,* February 28–March 2, 2005, Houston, TX.

7

Dispersion of Agglomerated Nanoparticles in Rubber Processing

Kwang-Jea Kim and James L. White

CONTENTS

7.1 Introduction

Small particles are used for reinforcing fillers in elastomers and thermoplastics.[1–7] These include synthetically prepared carbon black, silica and zinc oxide, as well as naturally occurring calcite, clay, and talc. These particles are commercially available in various sizes. Many particles are quite small, on the order of 0.01 μm (or 10 nm) but others are on order of 0.1 μm. The tire industry, in its early years, used primarily zinc oxide as a reinforcing filler. The industry then began using carbon black.[8] Recently, attention has been given to silica as particle reinforcement because it lowers rolling resistance, thus causing lower fuel consumption.[9] Carbon blacks have been traditionally used in tire manufacturing industry because of their reinforcing effect when mixed into elastomers. Today, small amounts of zinc oxide are retained in rubber compounds for their ability to enhance vulcanization/

cross-linking processes involving sulfur. Titanium dioxide has also been used for white tire sidewalls.

The purpose of this chapter is to compare the property of nanoparticles and agglomerate dispersion, processability, extrusion, and mechanical behavior of various nano-sized particles, such as silica, carbon black, zinc oxide, talc, and calcium carbonate, in elastomers.

7.1.1 Nanoparticles

7.1.1.1 Carbon Black[1, 3, 6, 7, 10]

The oldest method of manufacturing carbon black is by burning vegetable oil in a small lamp and collecting the carbon black accumulated on the tile cover. Later, natural gas was used as a source for what is called the "channel black" process, acetylene black process, or oil furnace process. Currently, carbon blacks are produced commercially by the furnace procedure, which is the incomplete combustion of refinery oils.

The structure of carbon black is similar to graphite and consists of large sheets of hexagonal rings formed by carbon atoms separated from each other by a distance of 1.42Å (or 0.142 nm). Carbon exists in two allotropic crystalline forms.[10] In both forms, carbon has a valence of four and has covalent bonds with other carbon atoms. These forms are generally referred to as "diamond" and "graphite."

The graphite form of carbon involves a resonating structure of carbon atoms. Carbon black is a largely amorphous form of carbon whose structure most closely resembles the graphite form as shown in Figure 7.1. It thus has strong covalent bonds. The small primary particles are strongly connected by covalent bonds into an aggregate.

The primary aggregate structure of carbon black is determined indirectly by measuring the amount of a liquid required to fill the voids in a specified mass of carbon black. This is based on the idea that irregular aggregate particles will pack more poorly than spheres and leave more voids for a liquid to fill. Generally, dibutyl phthalate (DBP) is used. Procedures are discussed in ASTM D2414 and D3493.

The arrangements of graphite layers in carbon black are parallel to each other, forming concentric inner layers; this arrangement has been called a "turbostratic" structure. The distance between parallel layers of carbon blacks varies in the range of 3.50 to 3.65 Å. The inner layers of a carbon black aggregate are less ordered than the outer layers. Carbon black exposed to high temperature undergoes a graphitization process. Oxygen present in

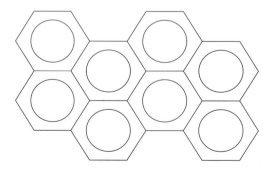

FIGURE 7.1
Graphite form structure of carbon black.

TABLE 7.1

Carbon Blacks and Their Classification

ASTM N-type	Old Designation	ASTM Particle Size (nm)	Example	BET (m²/g)	DBP No D2414 (cm³/100g)
N100 to N199	SAF	11–19	N110	145	113
N200 to N299	ISAF, CF, GPT	20–25	N220	121	114
N300 to N399	HAF	26–30	N330	82	102
N400 to N499	FF, XCF	31–39			
N500 to N599	FEF	40–48	N550	43	121
N600 to N699	HMF, GPF, APF	49–60	N660	36	90
N700 to N799	SRF	61–100	N762	28	64
N800 to N899	FT	101–200			
N990 to N999	MT	201–500	N990	7–12	44

Note: N (normal curing), SAF (super abrasion furnace), ISAF (intermediate super abrasion furnace), HAF (high abrasion furnace), CF (conductive furnace), GPT (general-purpose tread), FF (fine furnace), XCF (extra conductive furnace), FEF (fast extrusion furnace), HMF (high modulus furnace), GPF (general-purpose furnace), APF (all-purpose furnace), SRF (semi-reinforcing furnace), FT (fine thermal), and MT (medium thermal).

the system reacts with the carbon atoms in the center of particles, resulting in the formation of hollow spheres having increased crystallinity. Graphite layers are stacked on each other regularly so that each carbon atom has directly above and below it another carbon atom, which has a three-dimensional order. The distance between carbon atoms in each layer is about 3.4 Å.

ASTM D1765 describes a standard classification that involves a letter (N or S) and three numbers. The letter indicates normal curing (N) or slow curing (S) (which is found in oxidized carbon blacks). The first number following the letter indicates the ultimate particle size range; for example, "1" represents average carbon black size between 11 and 19 nm, "2" represents sizes between 20 and 25 nm, and "3" represents sizes between 26 and 30 nm, whereas "9" represents sizes between 201 and 500 nm. The last two digits are arbitrarily assigned by the ASTM. Prior to the ASTM classification described above, letter names (e.g., SAF, ISAF, FEF, etc.) were applied to carbon blacks, which sought to represent particle size and structure. Major carbon blacks are described in Table 7.1 together with their BET (Brunauer, Emmett, and Teller) surface areas.

7.1.1.2 Silica[4, 5, 10–18]

The structure of silica (SiO_2) involves each silicon atom having four somewhat polar covalent sp^3 bonds with oxygen as shown in Figure 7.2. This resembles the bonds in diamond.

Silicas (SiO_2) exist in crystalline and amorphous forms. Crystalline silica has three different allotropic crystalline forms: (1) hexagonal with trigonal crystal structure, which occurs in the mineral quartz; (2) cubic with tetragonal crystal structure, which occurs in the mineral crystabolite; and (3) a second hexagonal form in the mineral tridymite.

The most widely occurring of silica minerals, quartz, is a hard, colorless substance, well known for rotating the plane of polarized light. Quartz forms hexagonal crystals.[15] When silica minerals are melted ($T_m \sim 1600°C$) and then cooled, they vitrify into a glass, which is called silica glass.

$$\begin{array}{c} \diagup \quad \diagup \quad \diagup \quad \diagup \\ -\text{Si} - \text{O} - \text{Si} - \text{O} - \\ \diagup \quad \diagup \quad \diagup \quad \diagup \\ -\text{O} - \text{Si} - \text{O} - \text{Si} - \\ \diagup \quad \diagup \quad \diagup \quad \diagup \\ -\text{Si} - \text{O} - \text{Si} - \text{O} - \\ \diagup \quad \diagup \quad \diagup \quad \diagup \end{array}$$

FIGURE 7.2
Silica structure.

The commercial silica used in thermosets and elastomers is amorphous. It has the form of aggregates of ultimate particles as shown in Figure 7.3.[14]

Synthetic silicas are manufactured by various methods: fumed, electric arc, fused, gel, and precipitated. All synthetic silicas are amorphous in nature, have a chemical composition of SiO_2, and are colorless, odorless, tasteless, fine-particle, white powders. Silicas are

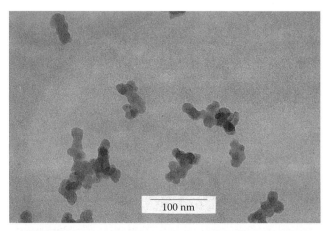

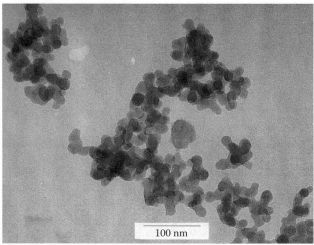

FIGURE 7.3
TEM photomicrographs of silica aggregates at 300,000 magnification (Hi-Sil 190).

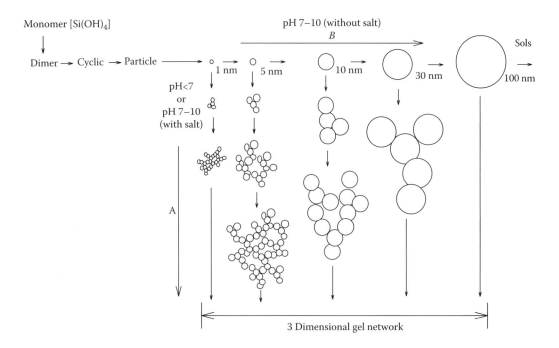

FIGURE 7.4
Synthesis of SiO$_2$ via polymerization of Si(OH)$_4$. (Source: Redrawn from Iler, R.K., *The Chemistry of Silica*, Wiley-Interscience Publication, New York, 1979.)

chemically inert, insoluble powders. Synthetic silicas are considered polymers of monosilicic acid (Si(OH)$_4$). Polymers of silicic acid were classified by Goto and Okura into two types, designated *"A"* and *"B."*[16, 17] To produce *"A,"* α-silicic acid is reacted rapidly in the presence of molybdic acid (α means silicic acid reacting in less than 5 min). The product has a low degree of polymerization, less than four (monomer and dimer species). In *"B,"* β-silicic acid reacts more slowly to produce polymers of higher molecular weight (β means silicic acid reacting in 10 to 30 min.).[12, 13] This gives the ultimate silica particles; reaction rates vary in proportion to the specific surface area.[12] Iler[18] has developed a model of polymerization starting from monomer Si(OH)$_4$ to form silica. This is shown in Figure 7.4. In basic solution (*B*), particles in solution grow in size but decrease in number. In acid solution or in the presence of flocculating salts (*A*), particles aggregate into three-dimensional networks and form gels.[12]

Typical primary particle sizes of synthetic silicas are 0.01 to 0.10 μm and a BET surface area of 50 to 800 m^2/g. The refractive index of synthetic silica is 1.45, which is close to the refractive indices of polymers, making it possible to produce translucent and transparent filled polymer compounds. The surface of synthetic silica contains hydroxyl groups called silanol groups, which easily tend to form hydrogen bonds with water molecules and demonstrate hydrophilic character.

Among the synthetic silicas, fumed silica is the most expensive and also has special industrial applications. Fumed silica is prepared by the hydrolysis of silicon tetrachloride vapor above 1000°C. Fumed silica was first developed in the early 1940s and is mainly used as reinforcing filler in silicone rubber and as a thixotrope in polyester and epoxy resins.[18] Small levels of fumed silica are also used in thermoplastics as a "rheology control agent" and a release agent. The primary particles of fumed silicas are connected by several other primary particles to show a chain-like aggregate structure. This chain-like structure is liable for the well-known thixotropic properties observed with fumed silica. The primary particles of aerogel silica form a large sponge-like mass that is supposedly responsible for the very high internal surface area and microporosity in silica gels. The primary particles of high-structure precipitated silica are connected to form an open, grape-like aggregate of varying size.

Electric arc silicas are produced by reacting quartz with coke at about 2000°C.[5]

Silica gels are manufactured by the reaction of sodium silicate solutions (water glass) with sulfuric acid under acidic conditions. This process dates to the 1940s. Silica gels are used as covering agents in plastics, and they are sold in small amounts to plastic compounders and processors as process aids and adsorbents. Silica gels exhibit a high BET surface area due to high internal porosity in the silica gel structure. One instance of silica internal surface area is selective adsorption, in which the internal portion of the BET surface area is useful in obtaining satisfactory performance outcomes.

Precipitated silicas are made by the reaction of sodium silicate solution and sulfuric acid under alkaline pH conditions. Precipitated silicas are predominantly used as reinforcing fillers in elastomers and, to a smaller amount, as thixotropes and viscosity builders in thermoset resins. Their application in elastomers includes the new green tires pioneered by Michelin. A new generation of high-structure, precipitated silicas is finding application in silicone rubber and in many specialty plastics. Precipitated silica is the lowest priced synthetic silica.

The properties of the synthetic silicas are related to the BET surface area, particle size, and particle shape, as are carbon blacks, but also to silanol group density. Table 7.2 summarizes the characteristics of synthetic silicas. In rubber processing applications, the degree of reinforcement obtained with silicas is related, like carbon black, to the external surface area. The internal surface area is not conducive to reinforcement; it is believed that the internal surface is "inaccessible" to large polymer molecules and thus excluded from reinforcement.

Silica primary particles exist in aggregate form (secondary particle structure), similar to carbon black. The aggregate size ranges from a few to a hundred primary particles. The agglomerate form (ternary structures) is formed when the secondary particle further aggregates during the manufacturing process. Thus, silica particle size structures exist at three levels: (1) primary particles (ultimate particles), (2) aggregates (secondary particles; primary particles fused form), and (3) agglomerate particles (tertiary particles; aggregates

TABLE 7.2

Synthetic Silicas and Their Comparative Properties

Synthetic Silicas	Fumed	Electric Arc	Silica Gel	Precipitated
Surface area (m^2/g)	50–400	150–200	300–1000	60–300
Oil absorption (cc/g)	150–250	80–120	150–250	50–250
Bulk density (g/l)	90–120	120–150	90-160	160–200
5% pH	3.6–4.3	3.5–4.2	4.0–7.5	6.5–7.5
% Moisture	<1.5	<2.0	5.0	6.0
Silanol groups ($1/nm^2$)	2–4	2–3	4–8	8–10
Primary particle size (nm)	7–40	—	—	10–30
Average particle size (µm)	0.8	4–8	4–10	1.5–10

loosely attached form). Agglomerates are broken down to functional aggregates under the application of high shear stress.

Naturally occurring microcrystalline silicas are used in polyester compounds, adhesives, electric resistors, refractory products, polyurethane elastomers, injection molded thermosets, epoxy compounds, silicone rubber, protective coatings, and ceramic glazes. Microcrystalline novaculite tripoli ore (natural silica) is used with poxies, polyurethanes, phenolics, polyamides, thermoset and thermoplastic polyesters, poly-p-phenylene sulfide, silicone rubber, and RTV silicones. Diatomaceous silicas are used for anti-blocking of processing low-density polyethylene film. Fused silicas are used for high-loading compounds and have light weight, low shrinkage, low thermal expansion, excellent thermal shock resistance, and good electric properties. They are used in silicones, epoxies, high-molecular-weight fluorocarbons, and other resins.

7.1.1.3 Zinc Oxide (ZnO)

Zinc oxide is a crystalline, odorless, and white or yellowish white powder. Chinese white or zinc white, zinc oxide (ZnO) is prepared either by oxidization or by burning zinc. It can be produced by pyrometallurgical techniques in which zinc metal in a vapor state reacts with oxygen.[19, 20] A newer method is a vapor synthesis process in which zinc metal is vaporized. The vapor is quickly cooled in the presence of oxygen, causing nucleation and condensation of small particles of zinc oxide.

The crystal structure of zinc oxide is hexagonal and has been investigated by many researchers.[21–23] Zinc oxide crystallizes in two forms. The first form, wurtzite zinc oxide (wZnO), has a hexagonal unit cell with a = 3.250 Å and c = 5.207 Å, as shown in Figure 7.5. The second form is metastable zinc oxide, which has a hexagonal unit cell with a = 4.280 Å, c = 5.207 Å.[21]

The general particle size of zinc oxides usually ranges from 0.1 to 0.4 µm, and its BET area ranges from 10 to 20 m²/g. Particles having an average size of 0.036 µm (36 nm) have been produced with a higher BET area at 15 to 45 m²/g.[24]

Zinc oxide absorbs carbon dioxide from the air, has high UV absorption, and is used as an antiseptic additive. It is insoluble in water and alcohol but soluble in some acids, ammonium carbonate, and alkali hydroxide solutions. Zinc oxide is used as a

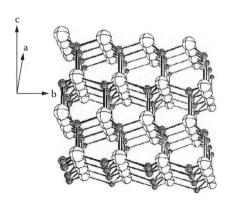

FIGURE 7.5
Crystal structure of wurtzite form of zinc oxide (wZnO) (3D) (small balls represent the O^{2-} atoms and the larger balls represent Zn^{2+} atoms).

filler and accelerator-activator in rubber and plastics. It improves resistance to weathering when used with polypropylene, promotes hardness, is a flame retardant, and increases electrical conductivity in polymers. It has been used in silicones, polyesters, and polyolefins.

Zinc oxide is a photochemically active material and has many applications due to its photochemical properties and chemical reactivity. It forms white zinc sulfides, thus preventing product discoloration. Zinc oxide is a semiconductor.

7.1.1.4 Titanium Dioxide

Titanium dioxide (TiO_2) produces white colors and has a high refractive index (n ~ 2.6) that leads to significant light scattering. It is added to synthetic fibers in the manufacturing process to make them seem white rather than translucent or transparent.

Titanium dioxide exists in different crystalline forms (anatase and rutile), which have a tetragonal crystal structure with unit cell dimensions.[25, 26] The unit cell structure of rutile (octahedrite) is a = b = 4.594 Å, c = 2.958 Å, $\alpha = \beta = \gamma = 90°$. The unit cell structure of anatase (sagenite or reticulated) is a = b = 3.793 Å, c = 9.51 Å, $\alpha = \beta = \gamma = 90°$.

Titanium dioxide is commercially produced by two different processes. In the sulfate process, titanium dioxide is prepared by reacting titanium ores with sulfuric acid. In the chloride process, titanium dioxide is produced by reacting titanium ores with chlorine gas.

Compounds with anatase TiO_2 show an outstanding bluish white color. Rutile types exhibit a creamy white color.[6]

7.1.1.5 Talc

Talc is called soapstone and has a Mohs hardness of 1.0. Talc has the chemical structure $MgO \times Si_4O_9 \times (OH)_2$. Talc particles have flake-like shapes as seen in Figure 7.6.

Talc is classified in the phyllosilicate group. It may be considered a hydrated magnesium silicate. Pauling first reported the structure of talc particles.[27, 28] It is one of a series of lamella silicate minerals involving two-dimensional silicate sandwiched with other minerals.[11, 27] The silicate layers of talc, MgO, are electrically neutral and loosely superimposed on one another to form a crystalline material. The layers slide readily over one another, resulting in easy cleavage and a soapy feeling.

The basic structure of talc is a sheet of "brucite" ($Mg(OH)_2$) in between two silicate layers. The mean particle size of most industrial talc is in the 2 to 20 µ range. Talc is usually in platelet form with an aspect ratio between 10 and 30. Gruner reported the diffraction patterns based on a monoclinic unit cell with a = 5.26 Å, b = 9.10 Å, c = 18.81 Å, $\beta = 100.0°$.[29] Later, Rayner and Brown reexamined the crystal structure of talc and reported a triclinic unit cell with a = 5.293 Å, b = 9.179 Å, c = 9.496 Å, $\alpha = 90.57°$, $\beta = 100.0°$, $\gamma = 90.03°$.[30] Here, the c-axis repeat is one-layer silicate spacing. Other researchers have also reported triclinic unit cells for talc as shown in Figure 7.6b.[31–33]

7.1.2 Processing Equipment

The coming of the pneumatic tire industry, associated with the rise in popularity of the automobile, increased production of rubber compounds and brought about large quantities of fine particles and hazardous vulcanization accelerators. This change made necessary the introduction of internal mixers with isolated mixing chambers to the rubber

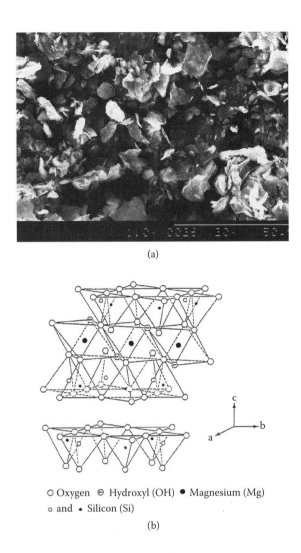

(a)

(b)

○ Oxygen ⊖ Hydroxyl (OH) ● Magnesium (Mg)
o and • Silicon (Si)

FIGURE 7.6
(a) SEM photomicrograph of talc particles at 1000X magnification, and (b) crystal structure of talc.

industry by the second decade of the 20th century. Before the introduction of the internal mixer, the early rubber industry was largely based on mixing with two roll mills, a system originating with Edwin M. Chaffee.[42]

There has been significant research on the mechanical properties of rubber compounds and more recently how visualization of mixing processes for particle dispersion in internal mixers.[37–40]

7.1.3 Dispersion Problems with Small Particles

Carbon black replaced zinc oxide in the 1910s and has been used in the tire manufacturing industry.[35] Subsequently, silica has become the most important reinforcing filler in the tire manufacturing industry, partially substituting carbon black. In recent years, tire industries have focused on the better performances of tires, such as rolling resistance, hysteresis, snow traction, and wet traction. Higher rolling resistance increases gas mileage. Silica-filled tires satisfy the above requirements. Silica particles have polar characteristics while carbon black has nonpolar characteristics. Silica-filled rubber compounds have been known as "green tires" because of the lower rolling resistance, which thus decreases hydrocarbon fuel consumption. However, silica particles are difficult to disperse compared to carbon black particles due to their agglomeration induced by polar or van der Waals forces.

The dispersion of carbon black has been studied by various researchers.[37–50] Rwei et al. have observed the breakup of spherical agglomerates of carbon black suspended in silicone oil and subsequently in thermoplastics. Two distinct breakup mechanisms, denoted as erosion and rupture, were described by these authors. They found that the agglomerate size decreases exponentially in time, indicating that the rate of breakdown is proportional to agglomerate size. They also found that the rate of agglomerate breakdown is "rate constant" and is proportional to the applied shear stress.[46–49] Coran and Donnet presented a second fundamental study of the breakup of carbon black agglomerates during mixing.[44, 45] They found a first-order rate in terms of the rate of decrease in undispersed carbon black with time.

Dispersion of silica particles has been studied by various researchers.[5, 51–57] Bachmann et al. were the first to review the intensive mixing of silica and silicates in 1959.[51] Dannenberg later described the mechanism of the surface chemical interactions of filler-reinforced rubbers for carbon black and silica filled system in 1975.[52] The mechanisms of silica-rubber coupling and rubber properties were discussed by Wagner in 1976.[5]

Wang et al. suggested dispersive components of silica surface energies, which are higher for the fumed silicas and much higher for precipitated silicas.[58] Wang et al. compared silica and carbon black surface energies.[59] The surface energies of dispersive components were very high for carbon blacks. Carbon blacks showed strong interaction with nonpolar or low-polarity polymers. Silica showed very high adsorption values, which represents strong particle-particle interaction in the resulting filler network. Horwatt et al. sought to stimulate the dispersion behavior of agglomerates at high stresses.[60] The simulation showed that the breakup of agglomerates follows a power law relationship between the average cluster size and the level of stress applied, which indicates that the fracture process becomes more abrupt as the clusters become more dense.

7.2   Rubber-Nanoparticle Compounds

7.2.1   Particle Dispersion and Agglomerate Characteristics upon Processing

The characteristics of the primary particles of carbon black, silica, calcite, talc, and zinc oxide are quite different from each other. Talc consists of flakes; calcite and zinc oxide are more isotropic. The roughly spherical primary particles of carbon black and silica were fused into other primary particles through covalent chemical bonds to form "aggregates." This gives carbon black and silica what is called "structure."

Typical scanning electron microscope (SEM) photographs of carbon blacks (CB143, CB29), silicas (S210, S185, S150), calcite, talc, and zinc oxide particle agglomerates in EPDM (ethylene propylene diene monomer) after internal mixer mixing are shown in Figures 7.7a–i.[61, 62] The internal mixer processed silica (S185) particles (Figure 7.7d) are further processed with a two-roll mill and they (S185) further break down as shown in Figure 7.8.[63, 64] The SEM pictures of agglomerate particles were characterized using an image analyzer and their sizes were determined by the "mass" or "z+1" average.[65]

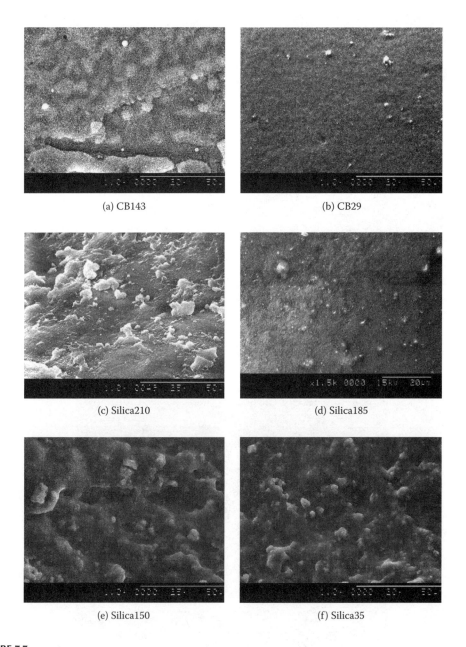

(a) CB143

(b) CB29

(c) Silica210

(d) Silica185

(e) Silica150

(f) Silica35

FIGURE 7.7
SEM photographs of particle agglomerates in EPDM at 20 vol% (a) CB143, (b) CB29, (c) S210, (d) S185 (1500X), (e) S150, (f) S35, (g) Cal20, (h) Tal17, and (i) ZnO9 at 1000X magnification.

(Continued)

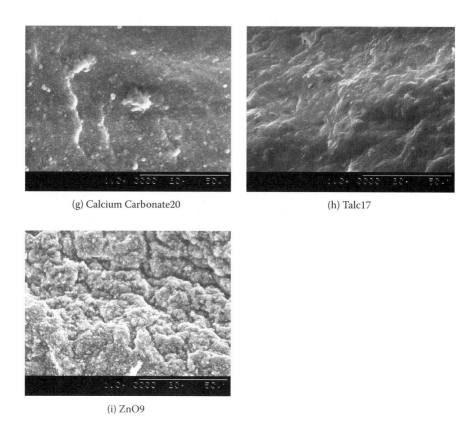

(g) Calcium Carbonate20   (h) Talc17

(i) ZnO9

FIGURE 7.7 *Continued.*

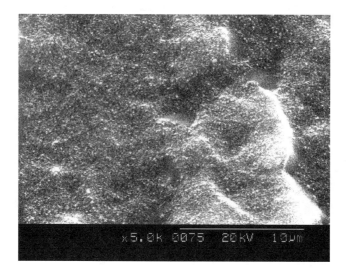

FIGURE 7.8
SEM photographs of S185 processed with internal mixer + two-roll mill at 20 vol% (5000X magnification).

When agglomerated particles are added to an elastomer and processed in an internal mixer, they are dispersed and distributed in the elastomer matrix, regardless of their sizes, concentrations, polarities, and process speeds, as shown in Figures 7.9a–e.[61] Low BET (9–35 m²/g) particle agglomerates (silica [S35], zinc oxide [Zn09], carbon black [CB29], calcite [Cal20], and talc [Tal17]) are decreasing exponentially as mixing time increases (see Figure 7.9a). Highly concentrated particles show lower agglomerate particle size than low concentration compounds due to particle interference over percolation concentration (as shown in Figures 7.9b–d). Increasing the rotor speed reduces agglomerate sizes due to the increased stress on agglomerated particles (see Figure 7.9e).

Correlations between agglomerate sizes in the particles of internal mixer mixing study with BET surface area, as shown in Figure 7.10, They indicate[61]

1. The silica agglomerates are larger than those of the other particles. However, for the same specific BET surface areas, the agglomerate sizes of calcite, talc, and zinc oxide are similar to those of carbon black.

2. The agglomerates become larger as the particles become smaller and the BET surface area is increased.

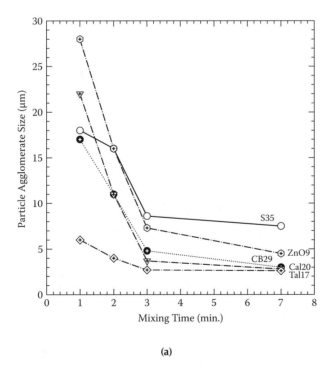

(a)

FIGURE 7.9

(a) Low BET (9–35 m²/g) particles dispersion as a function of mixing time at 100°C, 30 rpm, and 20 vol%; (b) carbon black agglomerate dispersion as a function of mixing time at 100°C, 30 rpm, 10 vol%, and 20 vol%; (c) silica agglomerate dispersion as a function of mixing time at 100°C, 30 rpm, 10 vol%, and 20 vol%; (d) calcite, talc, zinc oxide agglomerate dispersion as a function of mixing time at 100°C, 30 RPM, 10 vol%, and 20 vol%; and (e) carbon black, silica, calcite agglomerate dispersion as a function of mixing time at 30 rpm and 60 rpm (20 vol%).

(Continued)

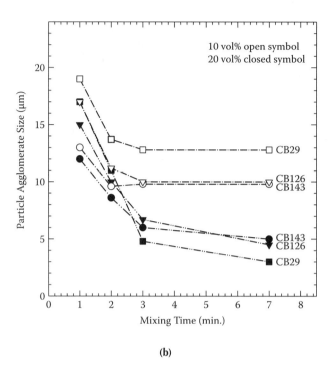

(b)

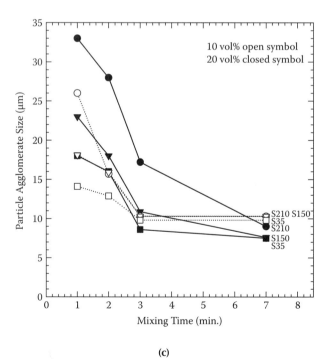

(c)

FIGURE 7.9 *Continued.*

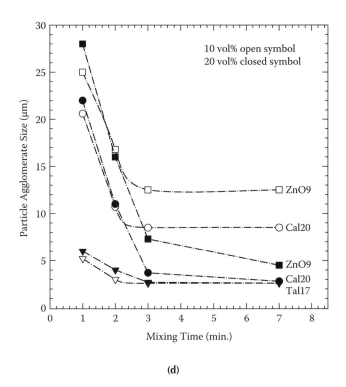

(d)

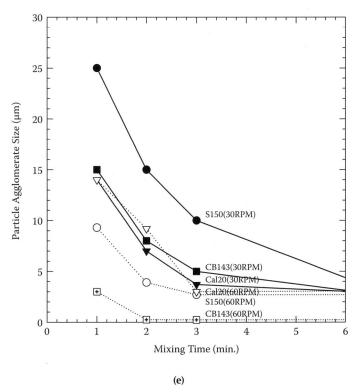

(e)

FIGURE 7.9 *Continued.*

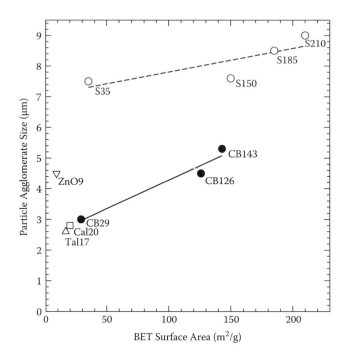

FIGURE 7.10
Particle agglomerate sizes as a function of BET area at 20 vol%, 30 rpm, and 7 min.

The number of primary particles in an agglomerate diameter is plotted as a function of the BET surface area of the particle as presented in Figure 7.11.[61] For large particles, the number of primary particles per agglomerate is roughly the same, and is between about 5 and 20. For small particles, the numbers of primary particles per agglomerate increase up to 170 for silica and 60 for carbon black.

Comparing two-stage mixing (internal mixer+two-roll mill) and one-stage mixing (internal mixer) systems, the silica particle (S185) exhibits a considerable particle agglomerate reduction (0.7 μm) at two-roll mill stage compared to internal mixer processing stage (about 3.3 μm). The agglomerate size of silica (S185) as a function of BET surface area is shown in Figure 7.12.[63] Among them, the two-roll mill processing system exhibited the lowest number of ultimate particles in an agglomerate, as did a carbon black filled system.

7.2.2 Processability

The viscosity of silica-filled systems exhibited the highest values for both low and high BET surface area compared to other particles filled compounds. Presumably, strong silica-EPDM networks were formed by silica. As the primary particle sizes decreased, the viscosities increased due to increased surface area contacting the polymer matrix. This was more significant in the case of silica particles because of their strong polarity.

The shear viscosity of low BET particles (BET surface areas between 9 and 35 m²/g), carbon black (CB29), silica (S35), calcite (Cal20), talc (Tal17), and zinc oxide (ZnO9) filled

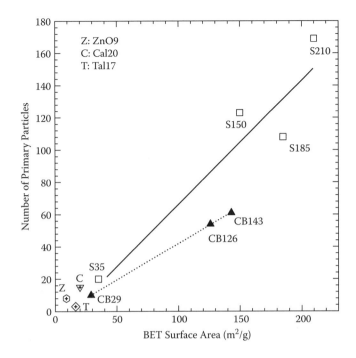

FIGURE 7.11
Number of primary particles in an agglomerate diameter after compounding in an internal mixer as a function of BET area at 30 rpm and 7 min.

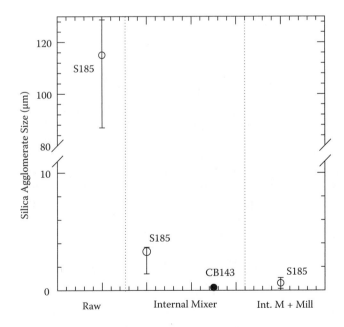

FIGURE 7.12
Silica agglomerate size processed with internal mixer and two-roll mill at 20 vol% and comparison with carbon black.

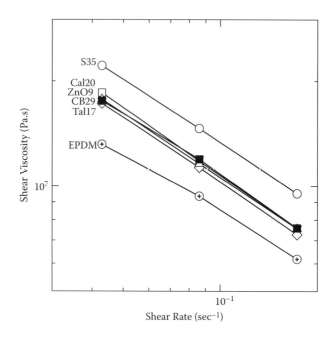

FIGURE 7.13
Shear viscosity of low BET particles (carbon black, silica, calcite, and zinc oxide at 10 vol%) as a function of shear rate.

EPDM compounds, is measured in a pressurized rotational rheometer with a biconical rotor, which was described by Montes et al. (see Figure 7.13).[61, 66] The shear viscosities of all particles overlap each other, with the exception of silica, which exhibits the higher viscosity due to the rough surface character of silica.

The shear viscosities of small-sized silica (S210, S150) and carbon black compounds (CB143, CB126) are higher than those of larger particles (S35, CB29), as shown in Figure 7.14.[61] As the surface area of small particles becomes larger, the contacting area between the filler and polymer chain increases, which results in a higher viscosity. Furthermore, for the same/similar particle sizes (e.g., S210 vs. CB143, S150 vs. CB126, S35 vs. CB29), the viscosities of the silica-filled compounds are much higher than those of carbon black (i.e., S210 > CB143, S150 > CB126, S35 > CB29). This is due to the higher polar characteristics of silica particles compared with nonpolar carbon blacks. When polymer chains are trapped in silica agglomerates, which have strong attraction forces between silica agglomerates, polymer chain mobility decreases. This results in a higher viscosity of silica-filled compounds than nonpolar carbon-black-filled compounds, which show less polar attraction forces between carbon black particles.

7.2.3 Extrusion Characteristics

Both silica and carbon black compounds exhibit a reduction in extrudate swell.

Typical photographs of extrudates of carbon black (CB126 (see Table 7.3)), silica (S210, S35), calcite, talc, and zinc oxide compounds at various loading levels and the same die wall shear rate ($32Q/\pi D^3 = 1000$ s^{-1}) are shown on Figures 7.15a–f61. As the concentration

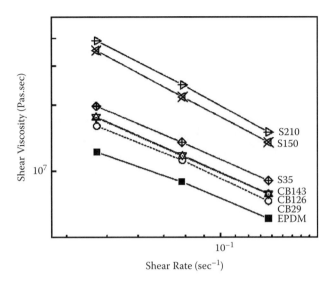

FIGURE 7.14
Shear viscosities of various silicas and carbon blacks at 10 vol% as a function of shear rate.

of small particles into EPDM increases, the quality of extrudate smoothness improves; while, the magnitude of the swell value d/D decreases. The improvements in extrudate quality are greatest for silica and carbon black and the least for the zinc oxide. The extrudate swell ratio d/D is highest for the gum rubber, which is about 1.5 for EPDM (compared

TABLE 7.3

Particles Used in This Study

Material	Supplier; Product Name	BET Surface Area (m²/g) (Size of Equivalent sphere μm (nm))	Code
Silica	PPG;		
	*Silene 732D	35 (0.075)	S35
	*Hi-Sil 233	150 (0.019)	S150
	*Hi-Sil255LD	185 (0.018)	S185
	*Hi-Sil 190	210 (0.017)	S210
Carbon black	Cabot;		
	*Sterling NS-1-N762	29 (0.089)	CB29
	*Vulcan 7H N234	126 (0.021)	CB126
	*Vulcan 9-N110	143 (0.021)	CB143
Calcium carbonate	Specialty Minerals;		
	*Multifex-MM	20 (0.07)	Cal20
Talc	Specialty Minerals;		
	*Ultratalc-609	16.5 (0.8)	Tal17
Zinc oxide	New Jersey Zinc Co.;		
	*KADOX-911	9 (0.12)	ZnO9

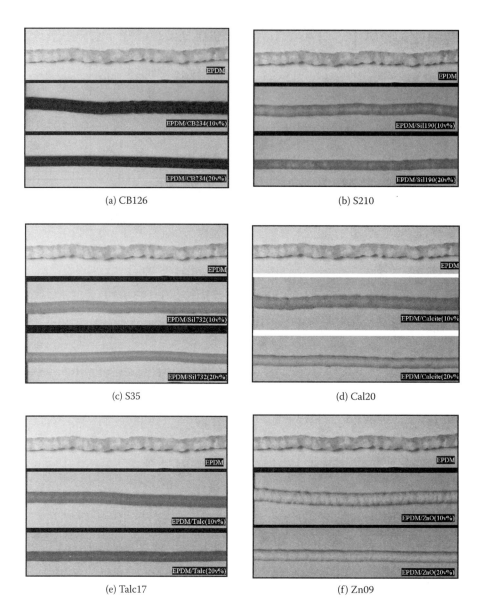

FIGURE 7.15
Capillary extrudates at 1000 (sec⁻¹) (a) CB126, (b) S210, (c) S35, (d) Cal20, (e) Tal17, and (f) ZnO9.

to the accepted Newtonian fluid value of 1.12). The swell ratio value of each compound is summarized in Figure 7.16.[61] The silica compounds show the most rapid decrease in die swell, then the carbon black, and then the least is observed for the calcite and zinc oxide. The carbon-black- and silica-filled swell data indicate that the smaller-particle filled systems show a greater extent of reduced swell more than the larger-particle filled systems.

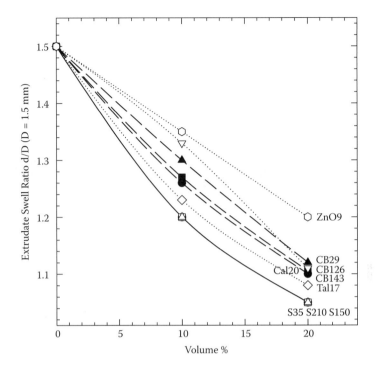

FIGURE 7.16
Extrudate swell ratios of various particles as a function of volume loadings.

7.3 Further Comments on Processing

7.3.1 Intermeshing vs. Banbury Mixer[40]

For particle dispersion study in an internal mixer, various rotors were compared. Koolhiran and White compared traditional nonintermeshing (tangential/separated) double-flighted rotor machines of F.H. Banbury design and intermeshing rotor machines of R.T. Cooke design to disperse silica, carbon black, and talc in rubber (styrene-butadiene rubber).[40, 67–69]

They observed particle dispersion in elastomer matrix using visualization instruments and found that the intermesh design rotors circulated and mixed the material at a higher rate because of the kneading action between the rotors. However, the tangential design rotors showed that the material was stagnant between rotors. They also found that the agglomerates of compounds mixed by intermeshing rotors were more rapidly dispersed than those mixed by tangential rotors.

The compounds processed by the intermeshing mixer exhibited lower viscosity than those produced by the tangential mixer. Viscosity levels of silica-filled compounds were higher than those of carbon-black-filled compounds at the same particle size. As the particle sizes decreased, their compound viscosities increased.

(a)   (b)

FIGURE 7.17
SEM photographs of S185 processed with ultrasound at (a) 10 vol% and (b) 20 vol% (10,000X magnification).

They also showed that high surface area silica was the most difficult to disperse and incorporate.

7.3.2 Ultrasound Extruder Processing

Typical SEM photographs of ultrasound-treated S185 are shown in Figure 7.17. Ultrasound-processed systems showed the smallest agglomerate sizes and the lowest number of ultimate particles in an agglomerate close to carbon-black-filled system, while the untreated silica system exhibited the largest sizes and the highest numbers in an agglomerate. The S185 silica particles were processed in an internal mixer and their agglomerate sizes were significantly reduced with two-roll mill processing from 3.3 to 0.7μm, as shown in Figure 7.16. After two-roll mill processing, the S185 compounds were further processed with ultrasound and the silica agglomerate sizes were further reduced to dimensions on the order of 0.3 to 4 μm.[63, 64]

7.3.3 Particle Effects on Mechanical Properties of Elastomer Vulcanizates

The size and concentration of particles affects the mechanical properties of the elastomer vulcanizates. For small strains of elastic solids, the hydrodynamic theory of Einstein and of Guth and Simha could be applied.[70–73] This theory was compared to experimental data on rubber-carbon black vulcanizates. There seemed only agreement at the lowest concentrations. They showed experimentally that the modulus was higher than the predictions at higher concentrations, and increasingly deviated to higher values at smaller particle sizes as shown in Figure 7.18.[51, 73, 74] The modulus increased as the carbon black particle size decreased in SBR matrix.[51]

Mullins reported experimental studies of the mechanical properties of rubber-carbon black vulcanizates.[74, 75] He studied the engineering stress (F/A_o)-strain curves of elastomer compounds. Unlike the gum elastomers, the compounds showed a softening (modulus reduction) behavior if successively stretched. Most of the softening occurs in the first deformation. After a few stretching cycles, a steady state is reached. Subsequently,

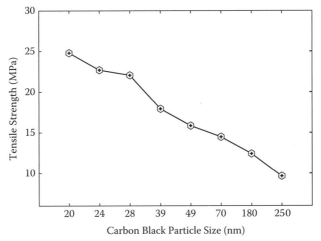

(a) Influence of carbon black particle size (carbon black 50phr in SBR)

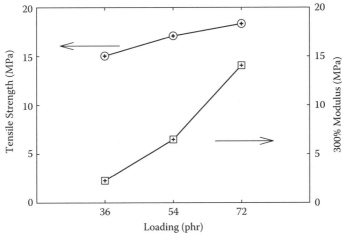

(b) Influence of carbon black (High Aggregate Furnace (HAF), 26–30 nm)
concentration in SBR

FIGURE 7.18
Influence of carbon black size and concentration on tensile properties of rubber vulcanizates. (From Bachmann, J.H., Sellers, J.W., Wagner, M.P., and Wolf, R., Fine Particle Reinforcing Silicas and Silicates in Elastomers, *Rubber Chem. Technol.*, 32, 1286, 1959.)

the samples gradually stiffened on standing. This behavior is often called the "Mullins effect."

Payne found another interesting and related mechanical effect.[76, 79] He observed that the dynamic storage modulus $G'(\omega)$ of rubber-carbon black compounds depends on the strain amplitude of an oscillatory deformation, decreasing with increasing strain level, which is shown in Figure 7.19.

High moduli, high viscosities, and thixotropy were observed for the small carbon-black-filled elastomer compounds. These seem associated with particle-particle interactions.

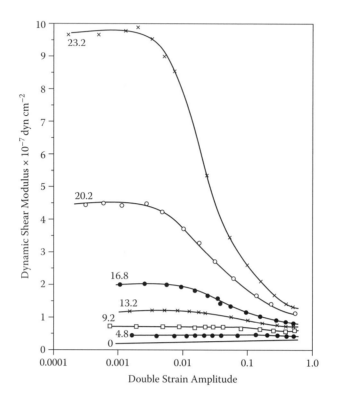

FIGURE 7.19
Influence of strain on the dynamic storage modulus of butyl rubber/carbon black compounds up to 23.2 vol%. (From Payne, A.R., *Rubber Plast. Age*, 42, 963, 1961; Payne, A.R. and Whittaker, R.E., Low Strain Dynamic Properties of Filled Rubbers, *Rubber Chem. Technol.*, 44, 440, 1971.)

This would be seen to be associated with networks of carbon black aggregates, which gradually form under quiescent conditions. These break up during severe deformations but can subsequently reform.

7.4   Conclusions

Conclusions from various types and sizes of particles processed through internal mixers and two-roll mill processing are as follows:

1. Processed through an internal mixer, larger particles (BET surface area less than 35 m^2/g), calcite, carbon black, talc, or zinc oxide, filled rubber compounds (exemplified by EPDM compounds) exhibit similar agglomerate dispersion behavior; however, silica exhibits larger agglomerates.

2. In smaller sized carbon blacks and silica particles (BET surface area greater than 120 m^2/g and less than 210 m^2/g) filled systems, the particle agglomerates are difficult to disperse, and the difficulty increases as the BET surface area is increasing. This effect is much greater for silicas than for carbon blacks. This is due

to the polar characteristics of silica particles compared to the nonpolar carbon blacks.

3. The 20 vol% filled compounds show smaller agglomerate sizes compared to 10 vol% ones. These compounds have higher viscosity and experience higher stresses during mixing.

4. The addition of carbon blacks and silicas into rubber increases the viscosity. However, it reduces extrudate swell and extrudate surface roughness. This was much more pronounced with the smaller particles of silica than those of carbon black.

5. The intermeshing rotors more rapidly dispersed silica, carbon black, and talc agglomerates than the tangential rotors.

6. Ultrasound processing significantly reduced silica agglomerate sizes.

References

1. Donnet, J.B. and Vogt, A., *Carbon Black, Physics, Chemistry and Elastomer Reinforcement*, Dekker, New York, 1976.
2. Stephens, H., *Rubber Technology*, 3rd ed., Morton, M., Ed., Van Nostrand-Reinhold, 1987.
3. Byers, J.T. *Rubber Technology*, 3rd ed., Morton, M. Ed., Van Nostrand-Reinhold, 1987.
4. Wagner, M.P. *Rubber Technology*, 3rd ed., Morton, M., Ed., Van Nostrand-Reinhold, 1987.
5. Wagner, M.P., Reinforcing Silicas and Silicates, *Rubber Chem. Technol.*, 49, 703, 1976.
6. Hofmann, W., *Rubber Technology Handbook*, Hanser, Munich, 1989.
7. White, J.L. and Kim, K.J., *Thermoplastic and Rubber Compounds: Technology and Physical Chemistry*, Hanser, Munich, 2007, chap. 1.
8. Boonstra, B.B. and Medalia, A.I., Effect of Carbon Black Dispersion on the Mechanical Properties of Rubber Vulcanizates, *Rubber Chem. Technol.*, 36, 115, 1963.
9. McNeish, A.A. and Byers, J.T., Low Rolling Resistance Tread Compounds—Some Compounding Solutions, Degussa Corporation, ACS *Rubber Division Meeting*, Anaheim, CA, May 1997.
10. Pauling, L., *General Chemistry*, 3rd ed., Dover, New York, 1988.
11. Moeller, T., *Inorganic Chemistry*, Wiley, New York, 1952.
12. Iler, R.K., *The Chemistry of Silica*, Wiley-Interscience Publication, New York, 1979.
13. Iler, R.K., *Surface and Colloid Science*, E. Matjevic, Ed., Vol. 6, Wiley, New York, 1973, p. 11.
14. Vital, A., Klotz, U., Graule, T., Mueller, R., Kammler, H.K., and Pratsinis, S.E., Synthesis of Spherical, Non-aggregated Silica Nanoparticles, *NATO Advanced Research Workshop: Nanostructured Materials and Coatings for Biomedical and Sensor Applications*, Kiev, Ukraine, August 4–8, 2002, in *NATO Science Series, II. Mathematics, Physics and Chemistry*, 102, 203–210, 2003.
15. Bolton, H. C. and McIntyre, P.N., The Electric Field in Crystals. III. The Refractivity of α-Quartz, *J. Phys. C: Solid State Phys.*, 1, 889, 1968.
16. Goto, K. and Okura, T., *Jap. Anal.*, 4, 175, 1955.
17. Goto, K., *J. Chem.Soc. Japan Pure Chem. Sect.*, 76, 729, 1955.
18. Klopfer, H. (Degussa), German Patent 762,723 (1942).
19. Andrews, K.W., An X-Ray Examination of a Sample of Pure Calcite and of Solid-Solution. Effects in Some Natural Calcites, *Mineral Mag.*, 29, 85, 1950.
20. Graf, D.L., Crystallographic Tables for the Rhombohedral Carbonates, *Am. Mineralogists*, 46, 1283, 1961.
21. Lowenheim, F.A.,. *Faith, Keyes & Clark's Industrial Chemicals*, 4th ed., Moran, M., Ed., Wiley Interscience, New York, 1975, p. 882.

22. Zaborski, M., Slusarski, L., Donnet, J.B., and Papirer, E., Surface Properties of Zinc Oxide and Their Effect on the Reinforcement of Elastomers, *Kautsc. Gummi Kunstst.*, 47(10), 730, 1994.

23. Gordienko, V.P. and Dmitriev, Y.A., Formation of Zinc Carboxylates in a Polyethylene-Zinc Oxide System Subjected to UV Irradiation, *J. Polym. Sci., Ser. B*, 37, 249, 1995.

24. Tanaka, T., Waki, Y., Hamamoto, A., and Nogami, N., The Effect of Surface Treatment on Mechanical Properties of Injection Molded Composites, *SPE ANTEC Tech. Papers*, 43, 3054, 1997.

25. Bennett, R.A., Stone, P., Price, N.J., and Bowker, M., Two (1′2) Reconstructions of TiO_2(110); Surface Rearrangement and Reactivity Studied Using Elevated Temperature STM, *Phys. Rev. Lett.*, 82, 3831, 1999.

26. McCavish, N.D. and Bennett, R.A., Ultra-Thin Film Growth of Titanium Dioxide on W(100), *Surf. Sci.*, 47, 546, 2003.

27. Pauling, L., The Structure of the Micas and Related Minerals, *Proc. Natl. Acad. Sci. U.S.A.*, 16, 123, 1930.

28. Pauling, L., The Structure of the Chlorites, *Proc. Natl. Acad. Sci. U.S.A.*, 16, 578, 1930.

29. Gruner, J.W., *Z. Kristallogr.*, 88, 412, 1934.

30. Rayner, J.H. and Brown, G., Triclinic Form of Talc, *Nature*, 212, 1352, 1966.

31. Ross, M., Smith, W.L., and Ashton, W.H., *Am. Mineralogists*, 53, 751, 1968.

32. Akizuki, M. and Zussman, J., The Unit Cell of Talc, *Mineral. Mag.*, 42, 107, 1978.

33. Perdikatsis B. and Burzlaff, H., *Z. Kristallog.*, 156, 177, 1981.

34. Chaffee, E.M., U.S. Patent No. 16, 1836.

35. Wagner, M. P., *Rubber World*, 164, 45, 1971.

36. Funt, J.M., Dynamic Testing and Reinforcement of Rubber, *Rubber Chem. Technol.*, 61, 842, 1988.

37. Kim, P.S. and White, J.L., Flow Visualization of Intermeshing and Separated Counter-Rotating Rotor Internal Mixer, *Rubber Chem. Technol.*, 67, 880, 1994.

38. Kim, P.S. and White, J.L., Comparison of Black Incorporation Development of Dispersion in Intermeshing and Separated Counter Rotating Rotor Internal Mixers, *Kautsch Gummi Kunstst*, 49, 10, 1996.

39. Cho, J.W., Kim, P.S., White, J.L., and Pomini, L., Flow Visualization in an Internal Mixer using an Adjustable Rotor System—Comparison of Double Flighted and Four Flighted Rotors, *Kautsch. Gummi Kunstst.*, 50, 496, 1997.

40. Koolhiran and White, J.L., Comparison of Intermeshing Rotor and Traditional Rotors of Internal Mixers in Dispersing Silica and Other Fillers, *J. Appl. Polym. Sci.*, 78, 1551, 2000.

41. Cotton, G.R., Mixing of Carbon Black with Rubber. I. Measurement of Dispersion Rate by Changes in Mixing Torque, *Rubber Chem. Technol.*, 57, 118, 1984.

42. Cotton, G.R., Mixing of Carbon Black with Rubber. II. Mechanism of Carbon Black Incorporation, *Rubber Chem. Technol.*, 58, 774, 1985.

43. Lee, S.D., White, J.L., Nakajima, N., and Brzoskowski, R., A Comparative Study of Characterization Materials of Carbon Black Dispersion Methods in Simulation and Emulsion SBR Compounds Prepared at Various Mixing Levels, *Kautsch. Gummi Kunstst.*, 42, 992, 1989.

44. Coran, A.Y. and Donnet, J.B., The Dispersion of Carbon Black in Rubber. I. Rapid Method for Assessing Quality of Dispersion, *Rubber Chem. Technol.*, 65, 973, 1992.

45. Coran, A.Y. and Donnet, J.B., The Dispersion of Carbon Black in Rubber. II. The Kinetics of Dispersion in Natural Rubber, *Rubber Chem. Technol.*, 65, 998, 1992.

46. Rwei, S.P., Manas-Zloczower, I., and Feke, D.L., Analysis of Dispersion of Carbon Black in Polymeric Melts and Its Effect on Compound Properties, *Polym. Eng. Sci.*, 32, 130, 1992.

47. Rwei, S.P., Horwatt, S.W., Manas-Zloczower, I., and Feke, D.L., Observation and Analysis of carbon black agglomerate dispersion in simple shear flows, *Int. Polym. Process.*, 6, 98, 1991.

48. Rwei, S.P. Manas-Zloczower, I., and Feke, D.L., Characterization of agglomerate dispersion by erosion in simple shear flows, *Polym. Eng. Sci.*, 31, 558, 1991.

49. Rwei, S.P. Manas-Zloczower, I., and Feke, D.L., Observation of carbon black agglomerate dispersion in simple shear flows, *Polym. Eng. Sci.*, 30, 701, 1990.

50. Wang, M.J. and Wolff, S., Filler-Elastomer Interactions. VI. Characterization of Carbon Blacks by Inverse Gas Chromatography at Finite Concentration, *Rubber Chem. Technol.*, 65, 890, 1992.
51. Bachmann, J.H., Sellers, J.W. Wagner, M.P., and Wolf, R., Fine Particle Reinforcing Silicas and Silicates in Elastomers, *Rubber Chem. Technol.*, 32, 1286, 1959.
52. Dannenberg, M., The Effects of Surface Chemical Interactions on the Properties of Filler-Reinforced Rubbers, *Rubber Chem. Technol.*, 48, 410, 1975.
53. Boonstra, B.B., Cochrane, H., and E. M. Dannenberg, E.M., Reinforcement of Silicone Rubber by Particulate Silica, *Rubber Chem. Technol.*, 48, 558, 1975.
54. Wolff, S. and J., Donnet, Filler-Elastomer Interactions. IV. The Effect of the Surface Energies of Fillers on Elastomer Reinforcement, *Rubber Chem. Technol.*, 65, 329, 1992.
55. Mandal, S.K. and Basu, D.K., Reactive Compounds for Effective Utilization of Silica, *Rubber Chem. Technol.*, 67, 672, 1994.
56. Roy Choudhury, A., De, P.P., and Roy Choudhury, N., Chemical Interaction between Chlorosulfonated Polyethylene and Silica—Effect of Surface Modifications of Silica, *Rubber Chem. Technol.*, 68, 815, 1995.
57. Ismail, H. and Freakley, P.K., The Effect of Multifunctional Additive on Filler Dispersion in Carbon Black and Silica Filled Natural Rubber Compounds, *Polym.-Plast. Technol. Eng.*, 36, 873, 1997.
58. Wang, M.J., Wolff, S., and Donnet, J., Filler-Elastomer Interactions. I. Silica Surface Energies and Interactions with Model Compounds, *Rubber Chem. Technol.*, 64, 559, 1991.
59. Wang, M.J., Wolff, S., and Donnet, J., Filler-Elastomer Interactions. III. Carbon-Black-Surface Energies and Interactions with Elastomer Analogs, *Rubber Chem. Technol.*, 64, 714, 1991.
60. Horwatt, S.W., Manas-Zloczower, I., and Feke, D.L., Dispersion Behavior of Heterogeneous Agglomerates at Supercritical Stresses, *Chem. Eng. Sci.*, 47, 1849, 1992.
61. Kim, K.J. and White, J.L., Breakdown of Silica Agglomerates and Other Particles during Mixing in an Internal Mixer and Their Processing Character, *J. Ind. Eng. Chem.*, 6(4), 262, 2000.
62. Kim, K.J. and White, J.L., Silica Surface Modification Using Different Aliphatic Chain Length Silane Coupling Agents and Their Effects on Silica Agglomerate Size and Processability, *Composite Interfaces*, 9(6), 541, 2002.
63. Kim, K.J. and White, J.L., Silica Agglomerate Breakdown in Three-Stage Mix including a Continuous Ultrasonic Extruder, *J. Ind. Eng. Chem.*, 6(6), 372, 2000.
64. Isayev, A.I., Hong, C.K., and Kim, K.J., Continuous Mixing and Compounding of Polymer/Filler and Polymer/Polymer Mixtures with the Aid of Ultrasound, *Rubber Chem. Technol.*, 76, 923–947, 2003.
65. White, J L., *Rubber Processing Technology; Materials and Principles*, Hanser, Cincinnati, OH, 1995.
66. Montes, S., White, J.L., and Nakajima, N., Rheological Behavior of Rubber Carbon Black Compounds in Various Shear Histories, *J. Non-Newtonian Fluid Mech.*, 28, 183, 1988.
67. Banbury, F.H., U.S. Patent No. 1,200,700 (1916).
68. Banbury, F.H. ,U.S. Patent No. 1,227,522 (1917).
69. Cooke, R.T. U.S. Patent No. 2,015,618 (1935).
70. Einstein, A., Eine neue Bestimmung der Moleküldimensionen, *Ann. Phys.*, 19, 289, 1906.
71. Einstein, A., Investigations on the Theory of the Brownian Movement, Dover, NY, 1956.
72. Guth, E. and R. Simha, *book Kolloid Z.*, 74, 266,1936.
73. Guth, E., Theory of Filler Reinforcement, *J. Appl. Phys.*, 16, 20, 1945.
74. Mullins, L., *The Chemistry and Physics of Rubberlike Substances*, L. Bateman, Ed., Mac Laren and Sons, London, 1963.
75. Kelly, A., *Strong Solids*, Clarendon Press, Oxford, 1966.
76. Payne, A.R., *Rubber Plast. Age*, 42, 963, 1961.
77. Payne, A.R. and Whittaker, R.E., Low Strain Dynamic Properties of Filled Rubbers, *Rubber Chem. Technol.*, 44, 440, 1971.

8

The Rheology of Polymeric Nanocomposites

Subhendu Bhattacharya, Rahul K. Gupta, and Sati N. Bhattacharya

CONTENTS

8.1   Introduction

The rheological characterization of polymeric materials is important with respect to the processing of polymeric materials. Polymeric materials, unlike other materials, exhibit liquid-solid (viscoelasticity) behavior in the melt phase and this leads to added complications in processing. The morphological evolution and the final properties of the polymeric system to a large extent depend on the processing technique used and also on the melt-phase properties of the polymeric material. The rheology of polymeric materials often provides clear insight into the molecular structure of the polymer under varied conditions and thus helps in controlling the desired final properties of the materials after processing.[1–4]

Most polymers are amorphous and possess less crystallinity. It is well known that the addition of small amounts of fillers, even in amorphous polymers, leads to substantial improvement in overall properties, such as an increase in viscosity, and good thermal and mechanical properties. The concept of inherent free volume in polymeric materials resulted in the development of filled polymer systems. It was believed that if the free volume within the polymeric system could be filled, then there would be improvement in the thermomechanical properties of the polymeric materials. This proposition led to a significant amount of research in the area of filled polymer systems, finally leading to the advent of polymer nanocomposites technology.

A typical polymer composite is a combination of a polymer and filler. Because compounding is a technique that can produce a filled polymeric material and ameliorate the drawbacks of conventional polymers, it has been studied over a long period of time with well-known practical applications. Reinforcing materials such as "short-fiber" are often used for compounding with thermoplastic polymers to improve their mechanical and/or thermal properties. In a thermoplastic polymer such as polyamide (nylon), glass and carbon fibers are used mainly as reinforcing materials. A filler, typically micron-sized, is incorporated into polymeric materials to improve their mechanical properties by producing composites. The polymer matrix and the fillers are bonded to each other by weak intermolecular forces; usually, chemical bonding is rarely involved. If the reinforcing material in the composite could be dispersed on a molecular scale (nanometer level) and interacted with the matrix by chemical bonding, then significant improvements in the thermal, rheological, and mechanical properties of the material or unexpected new properties might be realized. These are the general goals of polymer nanocomposite studies. To achieve this objective, clay minerals (montmorillonite, saponite, hectorite, etc.) have been used as filler materials. A single layer of silicate clay mineral is about 1 nm (nanometer) thick and about 100 nm wide, so it has a significantly large aspect ratio. In comparison, a glass fiber 13 µm (micrometer) in diameter and with a length of 0.3 mm (millimeter) is 4×10^9 times larger in size compared to a typical silicate layer. That is, if the same volumes of glass fiber and silicate were evenly dispersed, there would be a roughly 109-fold excess of silicate layers, with an exponentially higher specific surface area available. Thus, it is expected that the addition of fillers in the nanometer size range would provide the highest amount of improvement in properties.[5–7]

The rheology of polymer micro- and nanocomposites is affected by the level of interaction between the polymer and the filler.[8] For nanocomposites, the level of interaction is a function of the shape, size, concentration, and functionality of the filler. The low dynamic mobility of the filler under the application of an external flow field results in an improvement in matrix viscosity and other viscoelastic and dynamic properties.[9] The intercalation/exfoliation of polymer chains—or rather the adsorption of polymer chains on the filler particles—leads to the formation of different configurations or structures when subjected to a flow field resulting in a different rheological behavior.[10]

8.2 Fundamentals of Rheology

Rheology is basically a science used to determine the flow behavior of complex materials under different flow conditions. Chemists and physists use rheology to characterize the structure of various complex fluids at the macromolecular level. Chemical engineers are more concerned with characterization of the viscosity of these complex fluids. The focus of this chapter is to provide insight into the rheological techniques used to characterize polymeric composites and their subsequent flow behavior. There are three basic techniques normally utilized to determine the rheological behavior of a material: (1) shear, (2) dynamic, and (3) extensional rheology. Shear rheological techniques are used to determine the flow behavior and structure evolution of a material under the influence of a steady-state deformation. Dynamic rheology is a structural analysis tool because it uses a frequency excitation method to determine the molecular-level structure of a polymeric system. Finally, extensional rheology is used to determine the structural changes in a material undergoing stretching in the complete absence of surface effects. Thus, basically the major difference between the three rheological tools is the method of deformation employed to generate the rheological response to investigate the materials properties.

In case of polymer nanocomposites, rheological analysis has been extensively employed to study the structural fruition of the materials under different processing conditions. The advantage of using rheology as a tool in morphological analysis is that it not only provides detailed information on the molecular level changes in structure, but also at the same time provides strong theoretical reasoning to validate the results obtained from other characterization techniques such as SEM, TEM, XRD, FT-IR, etc. The only drawback of rheology is that it is an indirect method as compared to other direct methods of morphological analysis (e.g., TEM, SEM, and XRD). Alternatively, rheology can study the configuration of a system in the melt state, making it an advantageous tool for analysis of multiphase systems, especially those employing filler particles. A discussion of the rheological behavior of polymer nanocomposites is presented in a recent book by Bhattacharya et al. (2007).

8.2.1 Shear and Extensional Rheology for Polymer Nanocomposites

The use of dynamic rheological analysis for nanocomposites has been primarily in the dynamic shear analysis area, the reason being that these methodologies are the simplest methods that can effectively provide structural information. The shear rheological measurements have been effectively used to study the outcome of nanofiller loading and polymer-nanofiller interactions on the shear thinning behavior and dynamic moduli. The dynamic modulii have been used in many instances to examine the pseudo-solid-like behavior of the nanocomposites at long times and the reinforcement of properties caused by the presence of the nanofillers. Linear and nonlinear viscoelastic behavior has been investigated mainly through dynamic measurements. Dynamic properties have also been used to study the formation of a three-dimensional percolated network and the estimation of percolation threshold for filler loading beyond which the network formation is established and filler-filler interactions become significant. Dynamic properties have been further used to differentiate between an intercalated and exfoliated structure and to assess the degree of filler dispersion within the polymer matrix. Moreover, dynamic measurements have been used to establish the mixing sequence for clay dispersion. Extensional rheological tests are a bit complex to initiate and involve more complicated theoretical analysis as well but this drawback is balanced by the fact that extensional rheological methods are the best method to determine the structure of a surface in the presence of a highly variable flow field (or higher shear rates) simply because extensional rheological analysis is normally carried out in the absence of surfaces and hence the structural changes within the material are a sort of an inherent characteristic of the material involved in the study. Thus, extensional rheology can provide first-hand information on polymer nanocomposites. Extensional rheological measurements have been extensively used to study strain hardening behavior in polymer nanocomposites.

Time temperature superposition (TTS) has been attempted for the dynamic shear data to achieve a master curve. The Cox-Merz rule has been applied below a critical filler concentration range. Limited research on normal stress behavior has been carried out with some anomalous behavior with change of filler concentration. Die swell effect has been found to decrease with the addition of nanofillers compared to the matrix polymer alone. Extensional viscosity and melt strength have been studied for the intercalated and exfoliated systems with anomalous behavior reported at high uniaxial extension of the macro-molecules. Rheology has also been used to investigate the polymer melt intercalation kinetics.

8.3   Measurement Techniques

Most steady shear measurements for nanocomposites have been carried out using the rotational parallel plate, and cone and plate geometries. Rotational rheometry is being employed for the measurement of dynamic properties as well. Viscometric flow is assumed to have been generated in the fluid layer for an applied rate of shear. A variety of rheometers are available for the steady shear measurements. Temperature control is critical because most measurements are carried out at or near the melting point of the matrix polymer. Temperature control is also necessary to avoid the effect of viscous heating. Rotational rheometers are suitable for low to medium range shear rate measurements of nanocomposites. Usually, any measurement above a shear rate range of approximately 10 s^{-1} in a rotational plate rheometer is not appropriate because the material tends to extrude from the small gap of the installed plate assembly, providing an inaccurate reading. The measurement is based on the assumption that the angular motion of the rotating plate has persisted for a sufficiently long time to attain a steady-state condition of flow. The measurement is also made under the assumption that there is no slip at the solid-fluid boundary. For higher filler concentrations, occurrence of slip at the plate surface is a distinct possibility.

The occurrence of slip can be tested using parallel plates of different gap settings or cone and plates with varying cone angles. Comparison of data from parallel plates and cone and plates for the same material under identical conditions could also be used to assess the presence or absence of slip.

8.3.1 Dynamic Measurements

Although steady-state techniques are widely used for the measurement of viscous and elastic properties, steady shear methods can alter or destroy the microstructure and morphology of nanocomposites. Dynamic measurement, on the other hand, is a very useful technique for investigating the structure of delicate materials and deals with the state of the material due to unperturbed structure at small deformations. Dynamic measurements yield valuable information regarding the extent and dynamics of structure formed by particles in viscoelastic fluids. When tested in the molten state, a parallel plate, or a cone and plate assembly, is normally used. Most dynamic tests are conducted in the linear viscoelastic range of the material. This is tested by

- Dynamic strain sweep test
- Dynamic time sweep test
- Dynamic frequency sweep test
- Stress relaxation test

Strain sweeps are undertaken at different frequencies (0.1 to 100 rad/s) to determine the linear viscoelastic region of the material. The samples are subjected to a shear stress at a given frequency. As stress increases, the corresponding shear strain also increases accordingly, and the rheological response of the material is recorded. The most sensitive parameter, the storage modulus (G'), is monitored as a function of strain or stress. The range in which G' remains constant gives the linear viscoelastic region for the material at the given temperature and frequency. This test indicates the region in which the deformation is small enough for the modulus to be independent of deformation. A dynamic time sweep test is conducted to establish any variation in measurement in a given condition of temperature and frequency. In this test, the sample is subjected to an oscillatory stress, which lies in the linear viscoelastic region, and the dynamic response is recorded with time at the test temperature. The variation of the dynamic response (G' and G'') is attributed to degradation or changes in the properties of the nanocomposites.

The stress relaxation test is normally used to characterize a ternary polymer nanocomposite containing blends of two different polymers along with a filler material. The test involves the application of a stress for a certain period of time and then the stress is subsequently allowed to decay at different rates. This process provides detailed information about the structure of the interphase between two polymers in the presence of clay particles because the two polymers characteristically have different relaxation spectra but the interphase between the polymers, on the other hand, has a relaxation spectrum governed by the combined effect of the two polymer components along with the influence of the clay particles present at the interphase; hence, a composite relaxation spectrum generated for the sample can be used to study the nature of the interphase and the dispersion state of the clay particles in the composite system.

8.3.2 Extensional Rheological Measurements

The commonly used methods for extensional viscosity measurements include

- Constant stress measurements that involve sample end separation (Cogswell, 1969) or constant gauge length (improvisation of Meissner-type equipment (Meissner, 1972))
- Constant strain rate measurements that involve sample end preparation (Ballman, 1965; Meissner, 1972)
- Continuous drawing of filament

Although there are various techniques for measuring extensional flow properties, the two methods that have been mostly used for polymer nanocomposites are the Meissner-type rheometer and the continuous drawing of a monofilament.

8.3.2.1 *Meissner-Type Extensional Rheometer*

In the original Meissner-type rheometer, a rod-shaped sample floating on an oil bath was uniaxially drawn by a pair of rotary wheels by clamping the rod-shaped sample at either end between the wheels. The main material parameter that is often studied using this extensional apparatus is the transient extensional viscosity or tensile stress growth rate at constant strain rates. This instrument was later modified by Meissner, replacing the oil bath with an air-cushion on which a small rectangular specimen (not cylindrical) of the sample was floated while uniaxially stretched by means of two pairs of rotating metal belts, in order to eliminate slippage of the melt between the rotating wheels used in the earlier instrument. An idealized illustration of a polymer sample undergoing uniaxial extensional flow is shown in Figure 8.1. Consider a rod of initial length L_0 that is stretched to a final length L_f at time t as indicated in Figure 8.1.

Equations 8.1 through 8.4 describe the relationship between various time-dependent parameters during the measurement of uniaxial extensional viscosity at constant strain rate. The total extension at any time is defined as the Hencky strain ε^H (Equation 8.1). When considering isothermal, uniaxial extension using the Meissner-type rheometer, it is important to realize that the strain rate is constant and the change in cross-sectional area of the sample is given by Equation 8.2:

$$\varepsilon(t) = \dot{\varepsilon}t = \ln\frac{L_f}{L_o} = \ln\lambda = \varepsilon^H \tag{8.1}$$

$$A(t) = A_0 \exp(-\dot{\varepsilon}t) \tag{8.2}$$

$$\tau(t) = F(t)/A(t) \tag{8.3}$$

$$\eta_E^+ = \tau(t)/\dot{\varepsilon} \tag{8.4}$$

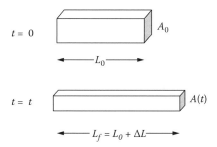

FIGURE 8.1
An idealized illustration of a stretching sample.

where $\varepsilon(t) = \varepsilon^H$ is the Hencky strain; $\dot{\varepsilon}$ is the constant strain rate; L_f and L_0 are the final and initial lengths of the sample, respectively; $\tau(t)$ is the tensile stress; and η_E^+ is the transient extensional viscosity.

The total strain is usually referred to as Hencky strain and is related to sample stretch ratio λ (= L_f/L_0). With the knowledge of the cross-sectional area $A(t)$, true tensile stress $\tau(t)$ can be calculated using Equation 8.3. Extensional viscosity, given by Equation 8.4, is a true material function and should be independent of the measuring technique and any assumptions concerning the constitutive behavior of the material. It is, however, a function of stretch rate and temperature.

8.3.2.2 Drawing of Molten Filament After Extrusion

Continuous drawing experiments are frequently used as a qualitative measure of the extensional rheology because of their similarity to practical processing operations such as fiber spinning. It is important to realize that many polymer processes deal with molten polymers that emerge from dies into stress fields and are then subjected to extensional deformation. The main parameters in these experiments are the extension rate, die dimensions, draw height, take-up speed of the rollers, and die temperature. Figure 8.2 provides a schematic diagram of the process.

Cogswell (1972) noted several advantages of this method for understanding extensional rheology, including

- Large deformations and high rates of strain may be studied.
- Low viscosity systems may be investigated.
- The experiments cater to convenient and rapid measurements over a wide range of conditions due to their dynamic equilibrium state and the variability of stretch rate.
- The study is of a fundamental character (e.g., it measures "melt strength" and draws instabilities like melt fracture and draw resonance).

The main drawbacks of these experiments are

- Nonuniform axial and radial temperature distribution of the drawn fiber due to ambient cooling
- Varying stress and strain rates along the length of the fiber

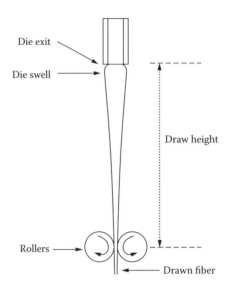

FIGURE 8.2
Schematic representation of the drawing process. The pair of rollers continuously draws the extruded monofilament or fiber at a pre-set velocity or acceleration.

Research conducted with drawing experiments has been extensive. The following will be very relevant for the drawing of nanocomposites:

- Influence of applied stress on crystalline behavior and crystalline morphology
- Heat transfer from the drawn filament
- Instabilities, neck formation, and failures during drawing or pinning operations
- Comparison between extensional viscosity obtained from draw-down experiments and steady uniaxial extensions

8.3 The Rheology of Composites

Work in the area of filled polymers started with macrocomposites and then moved on to microcomposites and finally to nanocomposites.

Macrocomposites are polymer composites formed just by mixing two polymer components or by addition of large solid particles within the polymer matrix. The microcomposites are formed by the addition of micron-sized particles or reinforcing fibers to the polymer matrix, and nanocomposites are formed by the addition of small nanoparticles of various shapes to the polymer matrix. The basic difference in the behavior of all the above-mentioned polymer composites arises due to the difference in the level of interaction between the polymer matrix and the filler or the component used to make the composite. As the size of the filler material becomes smaller, the surface charge density of the particles starts increasing due to an increased surface area and, as a result, the level of

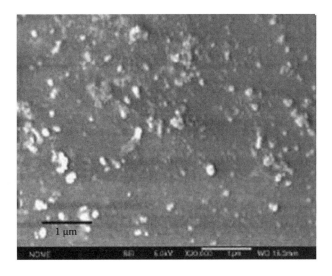

FIGURE 8.3
SEM images of distribution of Ca on surface of PC matrix (adopted from Zhong and Wang (2003)).

interaction between the filler and the matrix material also subsequently increases, which further causes a change in the morphology and rheological behavior of the material.[11, 12] On the other hand, in the case of a polymer blend composite, different structures are formed based on the compatibility and the concentration of the added component, leading to various types of morphology ranging from a suspension type morphology to a co-continuous inter-penetrating network type system.[13]

8.3.1 Rheology of Polymer Nanocomposites

Polymer nanocomposites have been produced either by embedding various types of nanoparticles in the polymer matrix or by the development of nanostructure within the polymer matrix by subsequent addition of a second polymer to the system. The rheological study of various kinds of nanocomposites has revealed some interesting structural parameters and behavior under different operating conditions and processes.

Wang et al. (2005, 2006) have investigated the rheology of polymer nanocomposites formed by the addition of $CaCO_3$ particles into the polymer (polycarbonate; PC) matrix and have found a substantial improvement in the polymer melt viscosity (high melt strength) and a decrease in impact strength.[14] Interestingly, without the existence of interaction of $CaCO_3$ with the polymer matrix, there was a subsequent random dispersion of the $CaCO_3$ particles within the polymer matrix, which further proves that the processing parameters of polymeric composites play an important role in controlling the equilibrium dispersion state of the nanoparticles within the system.[15]

The EDS composition distribution map of Ca element on the surface and the SEM images (Figure 8.3) show a random distribution of $CaCO_3$ particles within the polymer matrix. The SEM images also reveal that there is agglomeration of $CaCO_3$ particles within the polymer matrix at some places. The size of the nanoparticles within the polymer matrix is around 250 to 300 nm. The dispersion state of the particles can be further improved by processing the nanocomposites at higher shear rates while melt-blending.

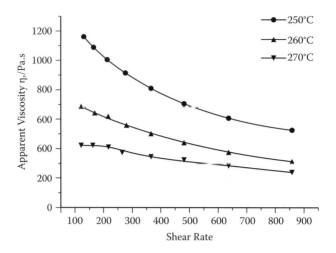

FIGURE 8.4
Steady shear measurements for PC/CaCO3 composite at different temperatures (adopted from Zhong and Wang (2003)).

Figure 8.4 shows the flow curves for pure PC at different temperatures. It can be seen that PC shows weak pseudo-plastic flow behavior. The melt viscosity or the apparent viscosity only decreases slightly with increasing shear rate. The pure PC displays a weak shear thinning behavior but is more sensitive to temperature. Increasing the temperature by 10°C results in a substantial reduction in viscosity and improvement in the rheological response by showing a relatively stable gradient to shear rate.[16] This behavior of pure PC could be attributed to the presence of a lot of acryl groups in its primary chain, resulting in a rigid structure and a very high resultant flow activation energy.

Figure 8.5 shows the effect of the addition of $CaCO_3$ particles on the melt viscosity at 260°C. It can be seen that the addition of just 1 wt% $CaCO_3$ particles reduces the apparent

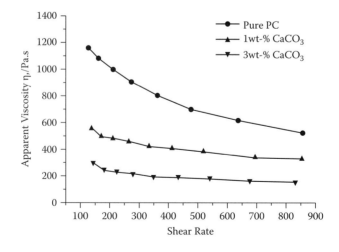

FIGURE 8.5
The variation in viscosity with different proportions of CaCO3 loading (adopted from Zhong and Wang (2003)).

viscosity of $CaCO_3$ substantially. The reduction in apparent viscosity could be attributed to the enhanced chain mobility of PC in the melt in the presence of $CaCO_3$. The development of a stable viscosity reduction gradient further indicates a sort of rheological enhancement caused by addition of $CaCO_3$ nanoparticles to the polymer matrix. The interesting thing to note is that the nanoparticles do not really interact with the polymer matrix and hence do not cause a substantial change in its structure but still result in reduction in viscosity by enhancing the chain dynamics in the melt state.[17]

8.4   Nanocomposites with Carbon Nanotubes

Potschke et al. (2002) have investigated the rheological behavior of a PC/carbon nanotube nanocomposite for a 0.5 to 15 wt% loading of carbon nanotubes.[18] The increase in melt viscosity for carbon nanotubes is higher than that when other nanofillers (including carbon nanofibers) are used as filler materials. This could be attributed to the increased aspect ratio of carbon nanotubes. In general, in most of the research performed thus far for noninteracting particles, it has been found that the change in rheological behavior of the polymer matrix by the addition of nanofillers is highest in the case of additives with the highest aspect ratio.[19] Figure 8.6 shows the variation in complex viscosity of the nanocomposite with the addition of different wt% of carbon nanotubes.

As shown in Figure 8.6, the complex viscosity of pure PC is much lower as compared to that of the master batch containing 15 wt% carbon nanotubes. The master batch shows a strong shear thinning behavior with increase in frequency, whereas the pure PC polymer shows limited frequency dependency. The interesting thing to note is that unlike the previous case, the complex viscosity increases with the addition of nanotubes to the pure polymer. Also, the polymers with 0.5 and 1 wt% nanotubes closely resemble a rheological response like the pure polymer, indicating that with a low wt% of nanotubes there is no significant difference in the rheological response of the material. Also, the samples with 0.5 and 1 wt% show very little dependency on frequency and would probably exhibit a

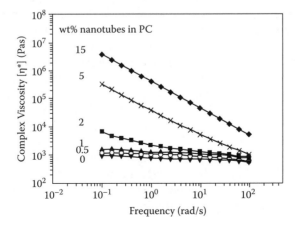

FIGURE 8.6
The variation in complex viscosity with frequency for different wt percent of nanotubes as a filler in case of polycarbonate/carbon nanotubes nanocomposites (adopted from Potschke et al. (2002)).

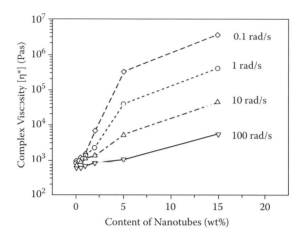

FIGURE 8.7
Variation of complex viscosity with variable amounts of nanotube loadings at different frequencies (adopted from Potschke et al. (2002)).

Newtonian plateau at lower frequencies, as shown by the pure sample as well. All other samples with increasing amounts of nanotube loading do not show a typical terminal region in the complex viscosity. Also interesting is that the samples with 5 wt% and 15 wt% denote a similar dependency on frequency and the viscosity is almost linear within the frequency range investigated at and after 5 wt% addition of nanotubes to the system.

Figure 8.7 shows the variation in complex viscosity with the addition of carbon nanotubes at different frequencies. As observed initially at low frequencies, the increase in viscosity is non-linear with respect to composition. In fact, the rise in viscosity is very low at low frequencies for lower amounts of nanotube loading, specifically until 1 wt%; after that, the rate of increase in viscosity increases with nanotube loading. At higher frequencies, the increase in viscosity is almost linear with the increase in nanotube loading. In fact, for all the samples after 5 wt% of nanotube loading, the change in viscosity with respect to composition is almost at the same rate for all frequencies explored. This behavior could be attributed to similar physical changes taking place within the polymer matrix after addition of 5 wt% nanotubes to the system.

Figure 8.8 shows a plot of elastic modulus (G') against frequency for different amounts of nanotube loadings in polycarbonate. It can be demonstrated that G' increases with increasing amounts of nanotube loading, as well as with an increase in frequency. The loss modulus (G'') also shows a similar kind of behavior but the rate of increase is much lower as compared to that of G'. Both G' and G'' converge at higher frequencies, indicating that the rate of increase is much higher at lower frequencies as compared to higher frequencies. The elasticity of the melt increases with 0.5 and 1 wt% nanotubes loadings but the curves are almost similar in nature with respect to frequency. Above 1 wt% nanotube loading, the slope of the modulus curves changes significantly; but then again, above 5 wt%, the clay loading the behavior becomes independent of frequency. It is known from the literature that an interconnected filler structure develops within a polymer filler system with increasing concentration of filler. This behavior could be attributed to the fact that with increasing filler loading, the filler-to-filler interaction increases, resulting in an interconnected network structure. This structure results in the development of a yield stress in the material, as indicated by the presence of a plateau region at lower frequencies in G' and

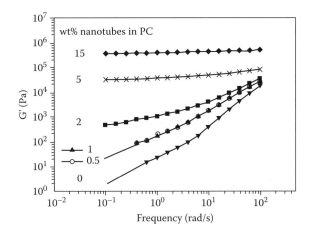

FIGURE 8.8
The variation of storage modulus (G′) with frequency (adopted from Potschke et al. (2002)).

G''. In this case, the effect is more pronounced for G', which shows that the melt elasticity increases more with the addition of filler to the system. As shown in the case in Figure 8.8, G' tends to reach a plateau after 2 wt% carbon naotubes loading, indicating that probably the percolation threshold for this material is reached.

Figure 8.9 shows a modified Cole–Cole plot for PC/nanotubes nanocomposites with different amounts of clay loadings. Cole–Cole plots are a good way to see the structural differences arising from the use of fillers in polymers as compared to the pure system. In this case, it can be seen that for a given value of G'', G' increases with increasing carbon nanotube loading, although the rate of increase decreases with increasing amounts of nanotube loading. At 5 wt% loading, the value of G' is greater than that of G'', indicating that the melt elasticity increases more rapidly after the addition of 5 wt% nanotube.

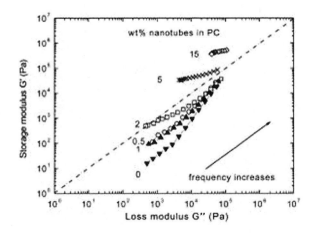

FIGURE 8.9
Comparison of storage and loss modulus at different frequencies with variable amount of clay loading (adopted from Potschke et al. (2002)).

8.5 Polymer-Clay Nanocomposites

Krishnamoorti and Yurekli (2001) investigated the rheological behavior of polymer-clay nanocomposites and reported results that are in close agreement with those for other filled systems.[20] The difference was noticed specifically in the terminal zone behavior and at lower frequencies. In addition, it was noted that for polymer-clay nanocomposites, a small amount of clay loading provides significant improvement in rheological and mechanical response as compared to conventional filled polymeric systems. They used PCL (polycaprolactone) and nylon 6 as base materials in their experiments. The clay used was montmorillonite. The clay particles had the presence of functional groups that readily interacted with the polymer chains. The clay particles also had significantly large surface areas due to the high aspect ratio. As discussed previously, the clay particles are normally 1 nm thick and around 100 nm in length. Sometimes, the lateral dimensions go up to few microns as well.

Figure 8.10 shows a plot of storage modulus with frequency, where the time temperature superposition principle (TTS) is used to calculate the values of G' at lower frequencies. The measurements were made at different temperatures, and then a subsequent shift factor was used to generate the master curves shown in Figure 8.10. At lower silicate loadings, only a horizontal shift factor was needed because at lower silicate loadings there was no particular feature in the curves that needed a vertical shift factor. For higher silicate loadings, the use of a vertical temperature shift factor is required. It is clear from Figure 8.10 that G' increases monotonically with increasing clay loading and with increasing frequency as well, except for the case of PCL2, which has a somewhat lower value than PCL1 at the highest frequency. G'', on the other hand (not shown here), shows nonmonotonic behavior with increasing frequency and clay loading, with the value for PCL2 being higher than PCL5 and PCL10, but the general trend in loss modulus is that it increases with increasing clay loading. At higher frequencies, G' and G'' show a power law dependency with respect to frequency. The nonmonotonic behavior in the samples could be attributed to the change in molecular weight distribution caused by addition of clay particles. Both curves exhibit

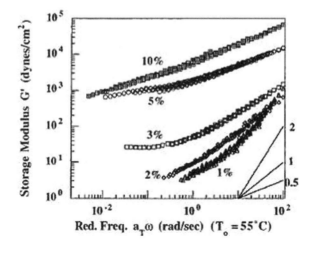

FIGURE 8.10

Time Temperature Superposition (TTS) plots for the variation of storage and loss modulus with frequency for different amounts of silicate particle loading for PCL nanocomposites (adopted from Krisnamoorti et al. (2001)).

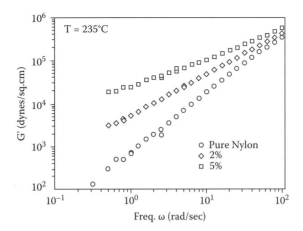

FIGURE 8.11
Storage modulus variation with frequency for nylon 6-clay nanocomposite (adopted from Krisnamoorti et al. (2001)).

a plateau region at low frequencies for small amounts of clay loading, but at higher clay loadings, the plateau region disappears.

Figure 8.11 and Figure 8.12 show the dynamic measurements for pure nylon 6 (Zhong and Wang, 2003) and nylon 6-clay nanocomposite for different clay loadings.[21] For the case of nylon 6-clay nanocomposites, G' and G'' increase monotonically with frequency with increasing amounts of clay loading. The nanocomposites exhibit a nonterminal behavior at small amounts of clay loadings; but with increasing clay loadings, the power law dependency of G' and G'' begins to decrease and then a plateau region appears at 5 wt% clay loading. The values of G' and G'' converge at higher frequencies, as shown in Figure 8.12. The behavior of the end tethered nanocomposites can be contrasted with that of filled polymer systems. In the case of other filled polymer systems, the values of G' and G'' increase

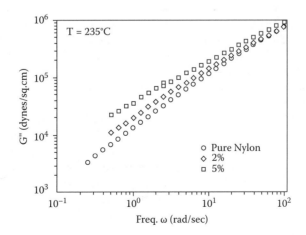

FIGURE 8.12
Loss modulus variation with frequency for nylon 6-clay nanocomposite (adopted from Krisnamoorti et al. (2001)).

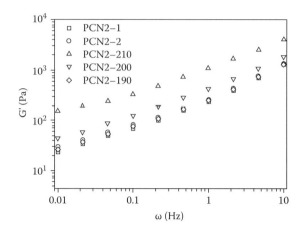

FIGURE 8.13
Variation of G′ with temperature for PBT/clay nanocomposites crystallized isothermally at different temperatures (adopted from Wu et al. (2007)).

monotonically with frequency with no observable change in frequency dependency, and nonterminal behavior or power law behavior is not exhibited by them as compared to the case of clay-based nanocomposites at lower frequencies. Also, the filled polymer systems exhibit yield stress behavior at lower frequencies with increasing filler loadings.

Wu et al. (2007) have studied the rheological behavior of PBT-clay (monmorillonite) nano-composites under the percolation threshold for an isothermally crystallized PBT.[22] The study reveals an interesting result: that the detachment and dispersion of clay is higher in the case of a sample crystallized at high temperature as compared to a normal sample.

Figure 8.13 shows the G′ of samples crystallized isothermally at different temperatures. Polymer nanocomposites PCN-1 and PCN-2 denote samples obtained by crystallization in the first and second cycle. It can be seen that the value of G′ increases with increasing crystallization temperature (T_c). Also, the storage modulus for PCN-2 is approximately an order higher as compared to that of PCN2-1. This result indicates that there is possibly a better dispersion of clay particles at higher crystallization temperatures. The frequency dependency of the terminal zone is decreasing in the curves although there is a slight deviation from the Cox-Merz rule.

Figure 8.14 represents the dynamic rheological measurements made on samples obtained from the second sweep and crystallized at different temperatures. It can be seen that the value of G′ for PCN2-1 is lower than the value of G″, indicating that the viscoelastic properties of the PBT/clay system are still governed by the polymer matrix. But in case of samples crystallized at 210°C, it can be seen that the value of G′ is higher than G″. Also, an equilibrium plateau region is observed at lower frequencies for G′ and G″ and there is an intersection of the two curves at higher frequencies. This solid-like rheological response at lower frequencies could be attributed to the formation of a percolation network at lower frequencies for samples crystallized at higher crystallization temperatures. The reason for the formation of a percolation network at a clay loading below that of the required value of percolation threshold (3%) could be attributed to the introduction of long-range order leading to a percolation network generated by the short-range order introduced during crystallization at higher temperatures within the clay tactoids.

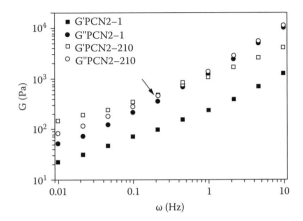

FIGURE 8.14
Dynamic material functions (G) with frequencies isothermally crystallized in the second cooling cycle at different temperatures (adopted from Wu et al. (2007)).

Figures 8.15a and b show the TEM image for a normally crystallized PCN sample. It can be seen in the images that there are several tactoids of clay particles present within the matrix of the polymer. The clay particles are present in the form of aggregates within the matrix. It was demonstrated that for samples crystallized at 210°C, the clay tactoids had delaminated and there were several single layers of clay within the matrix resulting in random distribution. Thus, crystallization at higher temperatures does help in delamination of the clay particles. The presence of clay particles as such does enhance the nucleation rate of crystals within the matrix; and at a higher temperature, the polymer viscosity is also low and hence the clay particles are in an activated state and are free to move. Along with the growth of the crystals, the randomization in the final structure of the crystals ensures delamination of clay particles as well as better dispersion.

Oksman et al.[23] studied the rheological and morphological characteristics of cellulose whisker nanocomposites. PLA was the base polymer used to prepare the nanocomposites. In this work, 5 wt% cellulose nanowhiskers was added to the base polymer and different amounts of anionic surfactant were used to improve the dispersion of the nanowhiskers in the polymer matrix. PLA is a polar molecule due to the presence of the acetate group. The presence of an anionic surfactant creates a repulsive interaction between the nanowhiskers and the base polymer, thus improving the dispersability in the matrix. Interestingly, the dispersion of the nanowhisker in the polymer matrix increases with increasing amounts of nanowhisker loading but the mechanical properties—specifically the tensile strength—decrease after 5 wt% nanowhisker addition to the samples.

Figures 8.16a and b show the DMTA test conducted on samples reinforced with nanowhiskers in the presence and absence of a surfactant. It can be seen from the Figures 8.16c and d that there is not a very significant improvement in storage modulus after addition of nanowhiskers to the sample. There is a significant amount of degradation of the polymer chains in the sample after addition of surfactant to the samples because as it can be seen from this figure that an increased surfactant content shifts the tan delta peak and the cold crystallization region to lower temperatures. This effect could be attributed to improved segmental mobility. The improvement in segmental mobility is due to the presence of smaller chains caused by degradation of the polymer sample. The addition of

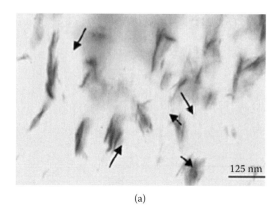

(a)

FIGURE 8.15a
TEM images of a PBT nanocomposite crystallized under normal processing (adopted from Wu et al. (2007)).

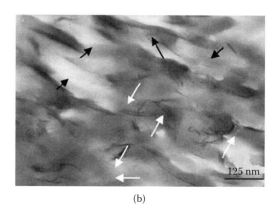

(b)

FIGURE 8.15b
The TEM image of PBT/clay conditions isothermally crystallized at 210°C (adopted from Wu et al. (2007)).

nanowhisker due to its reinforcing effects counteracts chain scission to a certain extent. Here, the presence of anionic surfactant creates a repulsive interaction between the chain and the nanowhiskers producing this kind of a behavior; it would be interesting to see the effects after the addition of a cationic surfactant because in that case there will be an attractive interaction and hence the system might behave like a conventional nanocomposite with a positive deviation in the cold crystallization peaks.

Wang et al. (2003, 2006) investigated the rheological behavior of PP-organophilic montmorillonite nanocomposites.[24] The addition of organically modified MMT leads to an increase in relaxation time and melt viscosity of the polymer. All the composites were further found to exhibit nonterminal behavior in the lower frequency regions. The polymer exhibited a pseudo-solid-like behavior in the presence of the nanofiller particles.

Figure 8.17 and Figure 8.18 show plots of the storage and loss modulus with respect to frequency for differing amounts of filler content. It can be seen that the storage and the loss modulus increase with increasing amounts of filler content. The increase in the storage modulus as compared to the pure polymer hints toward a possible increase in relaxation time as well because now the matrix has a greater tendency to store energy than to disperse it away

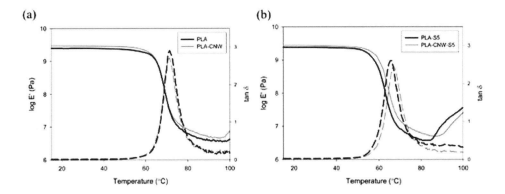

FIGURE 8.16a, b
The variation in elastic modulus with temperature for a PLA/nanowhisker nanocomposite (adopted from Oksman et al. (2007)).

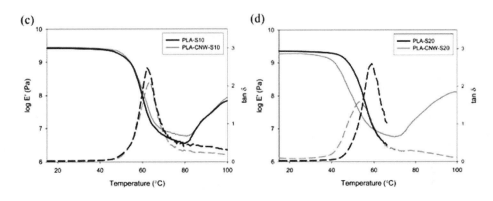

FIGURE 8.16c, d
The variation of elastic modulus with temperature for nanowhisker loading of 10 and 20% (adopted from Oksman et al. (2007)).

as compared to the polymer. The polymer-MMT system also shows a complete absence of nonterminal behavior with increasing clay content. The G' and G'' exhibit power law dependencies with the power law index much smaller than 2 and 1, respectively. The presence of nonterminal behavior could be attributed to the presence of an ordered domain structure in the polymer system under flow due to the presence of interacting particles. A long-range order is further established in the melt state as well. This orderly structure leads to the development of nonterminal behavior by restricting the segmental motion of the polymer chains.

Figure 8.19 shows a plot of the terminal slope of G'' at different temperatures for different amounts of clay loadings. It can be seen from this figure that the terminal slope increases with increasing temperature going more toward the ideal slope regimes. This effect could be attributed to the improved chain mobility at higher temperatures. But it can be also seen that the rate of increase of the terminal slopes is an order of magnitude less than that for the pure polymer and becomes even less as clay content is increased gradually. This behavior provides ample evidence toward increasing relaxation times after addition of clay particles to the polymer system. The formation of an orderly structure even at higher temperatures is also evident by the reduced rate of change of the terminal slopes.

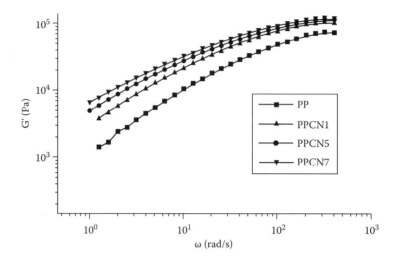

FIGURE 8.17
The variation in storage modulus with frequency for a PP/MMT nanocomposite (adopted from Wang et al. (2006)).

Fornes et al.[25] explored the rheological behavior of EVA-based nanocomposites with montmorillonite (MMT) and flurohectorite and found results similar to the filled polymer melts: for example, increases in melt strength, storage and loss modulus, and complex viscosity. The interesting aspect noticed in this research was the reduction in oxygen permeability of the system with the addition of clay, as indicated by the reduced percentage of oxygen required during thermo-oxidation. Also, for flurohectorite clay exchanged with amino dodeconoic acid, a microcomposite was formed. The formation of a microcomposite due to the use of a different modifier for the clay indicates that the functionality of the clay

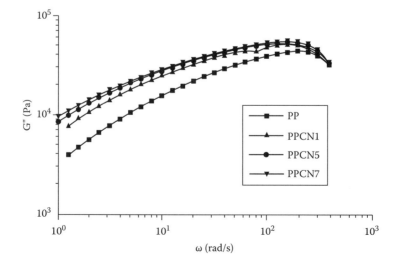

FIGURE 8.18
The variation of loss modulus with frequency for PP/MMT nanocomposite (adopted from Wang et al. (2006)).

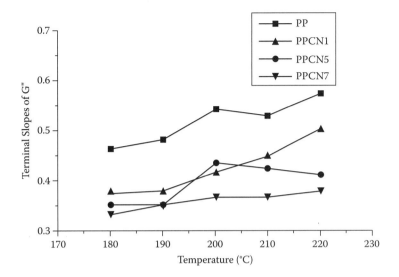

FIGURE 8.19
Variation of terminal slopes for G″ with temperature for PP/MMT nanocomposite (adopted from Wang et al. (2006)).

particles governs the dispersion state of the clay particles with a polymer matrix resulting in the formation of a nano- or a microcomposite.

Figure 8.20 and Figure 8.21 show the melt elasticity of EVA with different amounts of clay content that is modified with different amounts and types of modifier, respectively. It can be seen that the EVA modified with clay-ADA (secondary modifier) shows less improvement in melt elasticity as compared to that modified with clay-ODA. Thus, the functionality of the clay particles does control the final properties of the polymer-clay

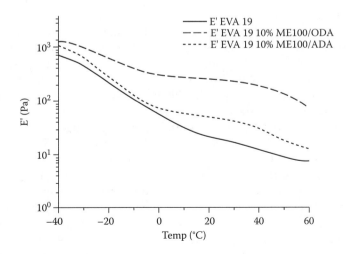

FIGURE 8.20
Variation of melt elasticity with temperature for variable amounts of clay and modifier content for EVA/clay nanocomposites (adopted from Fornes et al. (2001)).

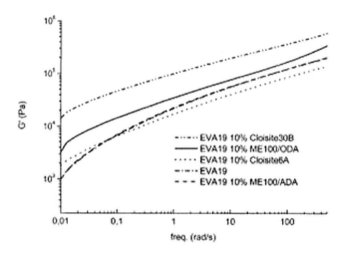

FIGURE 8.21
The variation in storage modulus variable amounts of clay loading and modifier content for EVA clay nanocomposites (adopted from Fornes et al. (2001)).

system. ADA-modified clay showed a strong tendency of aggregation within the polymer matrix, resulting in the formation of a microcomposite, whereas the use of clay modified with ODA resulted in complete delamination of clay particles within the polymer matrix. The same type of behavior was shown in the curves of storage modulus with frequency. It can also be seen from Figure 8.21 that the curves of the nanocomposite (cloisite 30B, cloisite 6A, MeODA) show a constant slope throughout the frequency region examined; on the other hand, the curves for the microcomposite show terminal region behavior at lower frequencies. Terminal region is defined when the storage modulus approaches a limiting value. This region will be inherently linear; that is, G' will show a slope of 2. The absence of terminal region behavior in nanocomposites can be attributed to the confinement of the polymer chains within the clay galleries. Also, the clay particles are well dispersed within the polymer matrix. The functionality of the clay particles governs the interlayer distance of the clay platelets and also controls the dispersion state of clay within the polymer matrix resulting in different rheological responses—because with microcomposites the polymer chains do not cause delamination of clay particles and hence actually intercalate the clay galleries to a limited extent. As a result, the rheological response of the polymer microcomposite is very similar to that of the pure polymer.

Yang et al.[26] prepared polyamide-attapulgite fiber nanocomposite modified with cetyl trimethylammonium bromide (CTAB) and toluene 2,4-di-isiocynate (TDI), which resulted in the formation of an exfoliated morphology. The attapulgite fibers were well dispersed within the polymer matrix and resulted in the formation of a percolation network structure.

Figure 8.22 and Figure 8.23 show a plot of G' and the relaxation spectrum for pure polyamide and polyamide-clay nanocomposite. It can be seen that the pure polyamide samples exhibit homopolymer-like terminal behavior with G' proportional to ω^2. As the attapulgite (type of clay) content increases, the frequency dependency of the polymer samples keeps on decreasing and almost becomes independent of frequency at lower values. Thus, the samples exhibit solid-like behavior at lower frequencies. Also, it can be seen that for samples with attapulgite, G' is higher than G'' (not shown here) at lower frequencies, which is

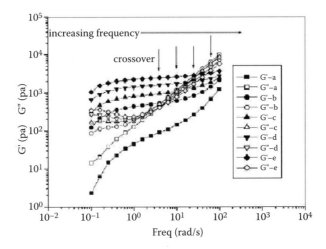

FIGURE 8.22
Dynamic material functions for pure clay polyamide and polyamide clay nanocomposites loading (adopted from Yang et al. (2008)).

not a very common phenomenon at lower frequencies of nanocomposites. This indicates a typical solid-like behavior of the sample and the formation of percolation networks within the polymer matrix. Also, there is a distinct crossover frequency for all the samples when the values of G' become less than G'', resulting in a liquid-like behavior. The liquid-like behavior at higher frequencies could be attributed to the orientation of attapulgite particles in the flow direction, resulting in deviations away from the solid-like behavior and the development of typical liquid-like behavior in these samples. The generation of an

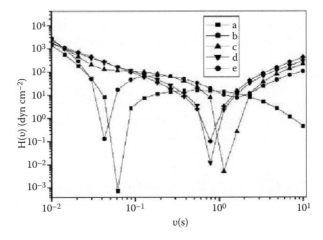

FIGURE 8.23
Relaxation spectrum for polyamide nanocomposite with variable amount of clay (adopted from Yang et al. (2008)).

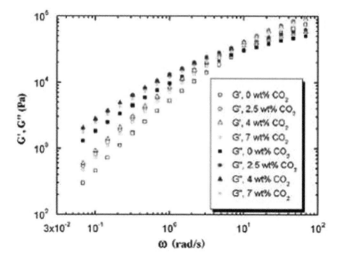

FIGURE 8.24
The variation of storage and loss modulus for PP/clay nanocomposite with variable amounts of CO_2 (adopted from Zhao and Huang (2008)).

additional relaxation spectrum near 1 second (Figure 8.23) as compared to the pure polymer hints toward the formation of percolation networks and longer relaxation times with the addition of clay particles. Thus, the addition of clay particles slowly changes the liquid-like behavior of the polymer into solid-like behavior with increasing content of functionalized modifiers.

8.5.1   Effect of Carbon Dioxide on Nanocomposites

Zhou and Huang[27] studied the rheological properties of PP-clay nanocomposites using super-critical CO_2 while processing the nanocomposites and found the effect of the use of CO_2 on the morphology of the nanocomposites.

Figure 8.24 shows the dynamic rheological behavior of PP-clay (Gu et al., 2004) nanocomposites prepared at 190°C using different weight percents (wt%) of CO_2. The clay concentration in all the above samples was kept constant at 3 wt%. As shown in the Figure 8.24, the values of G' and G'' increase with increasing CO_2 concentration, but interestingly the cross-over frequency for the samples with different concentrations of CO_2 remains the same. The increase in the material functions with increasing CO_2 concentration can be attributed to the reduction in matrix viscosity on addition of CO_2 to the system; as a result, the dynamics of the polymer chain is improved, which results in a better intercalation of the polymer chains into the clay galleries causing an improvement in G' and G'', respectively.

The complex viscosity (Figure 8.25) also increases with increasing CO_2 loading. This could be attributed to the increase in melt strength caused by improved dispersion of the clay particles within the polymer matrix. Also, it can be seen that after using 7 wt% CO_2 in the samples, the complex viscosity and G' and G'' decrease. This effect could be attributed to the fact that the reduction in matrix viscosity on addition of CO_2 follows an exponential behavior and, as a result, with a large amount of reduction in matrix viscosity the clay particles have a greater tendency to aggregate, thus reducing the polymer intercalation and subsequently the material functions and final matrix viscosity as well. Thus, there is

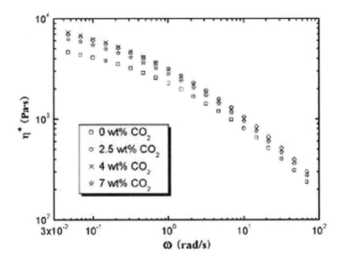

FIGURE 8.25
The variation in complex viscosity with different amounts of clay loading (adopted from Zhao and Huang (2008)).

an optimum amount of super-critical CO_2 to be used with the samples while preparing polymer nanocomposites or filled polymers.

The linear viscoelastic behavior, as characterized by the storage and loss moduli, was previously reported by Krishnamoorti and Giannelis[28] for exfoliated nylon 6 and poly(ε-caprolactone) formed by end tethering of polymer chains by layered silicates. At high frequencies, both G' and G'' showed solid-like (nonterminal) behavior, which was attributed to the tethering of the soft poly (ε-caprolactone) chains to the hard silicate layers. At the low frequencies, G' and G'' showed a frequency-independent plateau, with G' exceeding G'', which does not normally occur in ideal situations. This kind of behavior can be linked to a pseudo-solid-like response due to the incomplete relaxation of the polymers tethered to the silicate layers. Nonterminal flow behavior has also been observed in intercalated poly(styrene-isoprene) di-block co-polymer.[29] At all frequencies, both G' and G'' for the nanocomposites increased monotonically with increasing silicate loading. The viscoelastic behavior at high frequencies was unaffected by the addition of the layered silicate, with the exception of a monotonic increase in the modulus value. Further, at low frequencies, where the unfilled system exhibited liquid-like behavior, both G' and G'' moduli for the nanocomposites showed a diminished frequency dependence.

Galgali et al.[30] (2001) reported the difference in rheological behavior of intercalated polypropylene nanocomposites with and without compatibilizer. The possibility of exfoliation was greatly enhanced by the presence of compatibilizer. These exfoliated silicate layers easily produced percolated networks that strongly resisted shear deformation.

8.5.2 Rheology and Clay Structure within Nanocomposites

The relationship between rheological behavior and the nanostructure of the polymer-layered silicate nanocomposites (PLSN) was investigated by Lim and Park.[31, 32] The nanostructural change of intercalated polystyrene (PS) silicate nanocomposites was monitored by the rheological measurement.[33] It was observed that any change in the interface properties of PS nanocomposites during the intercalation (the annealing at 200°C in a rheometer heating chamber in N_2 atmosphere) was reflected in the storage modulus. For the PS nanocomposites, the storage modulus increased with increasing annealing time of up to a steady value, suggesting that saturated intercalation had occurred. Furthermore, Lim and Park[34] have reported a difference in rheological behavior between intercalated and exfoliated morphology of polymer silicate nanocomposites. The polystyrene nanocomposites with simple intercalated structure exhibited a slight enhancement at low frequency having a distinct plateau-like behavior, while the exfoliated PE-*g*-MA silicate nanocomposites exhibited both a distinct plateau-like behavior at low frequency and enhanced moduli at high frequency, due to strong attractive interaction with the silicate layers. A similar type of rheological response (solid-like behavior, enhanced G' and G'') had been observed in macroscopic filled systems, such as carbon black filled polystyrene.[35] The major difference between these materials and PLSNs is the high loading of fillers in microcomposites (about 25 wt% for carbon black and 40 to 60% for glass) compared to PLSNs loadings of 2 to 5 wt%. The PLSNs show solid-like behavior at such a low loading due to the very high aspect ratio of silicate layers, their enhanced dispersion into the polymeric matrix, and their good interaction between silicate layers and polymer chains.

The important findings reported by various authors are as follows:

- With anisotropic fillers (e.g., layered silicates), the formation of percolated network superstructure occurs at a much lower filler loading.
- Exfoliated systems have shown dramatic increases in linear viscoelastic properties compared to the intercalated systems.

The transition from a liquid-like to solid-like nature of the unfilled and filled polymers can be analyzed from the power-law slopes of G' at low frequencies. This slope characterizes the quiescent nature of these nanocomposites. Ferry[36] noted that for non-cross-linked homopolymers, the power-law linear viscoelastic slopes can be expressed as $G' \propto \omega^2$ and $G'' \propto \omega^1$ (and $\eta^* \propto \omega^0$). G' is used in this analysis because it is very sensitive to changes in the meso-structure of the material. The formation of such structures restricts the mobility of the polymer chains, thus enhancing the ability to store energy. This energy storage capacity is depicted as the solid-like response of G' at low frequencies.

8.5.3 Rheology of Silica Particles and Clay-Based Nanocomposites: Differences and Similarities

To a large extent, the rheological behavior of clay-based nanocomposites and silica-particle-based microcomposites is governed by the kind of interaction forces existing between the particles and the polymers, the functionality of the fillers, and the chain length and the molecular weight distribution of the matrix polymer. Based on the above-mentioned facts, different morphologies are established within the polymer system under different flow conditions. The microcomposites are generally formed using finely divided silicon dioxide as filler, which builds up into micron-size aggregates within the polymer, whereas clay-based nanocomposites are formed by addition of nanometer-size clay particles that finally end up as nanometer-size aggregates within the polymer.

8.5.3.1 Rheology of Silica-Particle-Based Nanocomposites

The silica particles used to make the microcomposites can be amorphous as well as colloidal silica particles. These particles have a very high specific surface area and have a tendency to self-aggregate, leading to the formation of three-dimensional networks in the molten polymer matrix. The nanocomposites made out of silica particles, similar to conventional clay based nanocomposites, also show a nonterminal zone, apparent yield stress, and shear thinning dependency of viscosity. The network structure formation leads to this kind of rheological behavior and in the case of silica particles, the network structure is formed by the adsorption of polymer chains on the surface of the particles. The adsorption, in turn, is a function of the surface functionality of the particles used.[37, 38] The functionality of the polymer chain grafted onto the surface-functionalized silica particles governs the network or the gel type behavior displayed by the polymer-silica particle system. When the molecular weight distribution of the grafted polymer chains is of the same order of magnitude as that required for physical entanglement and the polymer is nonpolar in nature, then a steric repulsion is created by the presence of the polymer in between the interacting particles; this helps in avoiding the formation of a three-dimensional network.[39] On the other hand, when the polymer chains are larger than the chain length required for physical entanglement, then the polymer chains interact with each other as well with silica particles, resulting in the formation of networks and development of a gel-type behavior. This kind of behavior results in the generation of a percolation threshold, after which no further enhancement in properties can be achieved by the addition of fillers to the system.

Figure 8.26 shows the linear viscoelastic behavior of a PS (polystyrene) fumed silica nanocomposite studied by Cassagnau et al.[40] The solid lines represent the dynamic rheological response of pure PS, and the dashed lines represent the rheological response of PS-silica nanocomposites. It can be seen that pure PS shows typical slopes of 2 and 1 for G' and G'', respectively, and also a terminal region at lower frequencies. The silica

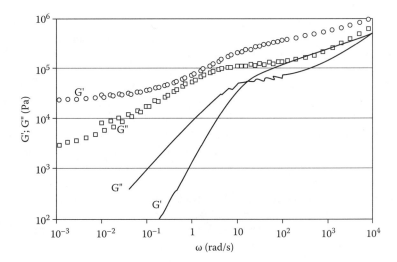

FIGURE 8.26
The variation in storage and loss modulus with frequency for pure PS and PS/fumed silica composites (adopted from Cassagnau et al. (2005)).

composites, on the other hand, show the origin of a secondary plateau region and non-terminal behavior at higher concentrations. The addition of fumed silica particles or clay platelets evokes almost the same kind of rheological response from the polymeric matrix.

Zhang et al.[41] studied the rheological response of a colloidal silica-based nanocomposite. Their aim was to determine how polymer particle and particle-particle interactions influence the viscoelastic properties of nanocomposites. PEO (polyethylene oxide) was used as the base polymer and colloidal silica particles were used as the filler material. The polymer chains were able to adsorb onto the surface of the silica particles due to the high surface energy of the silica particles. A large enough polymer molecule is able to adsorb onto the surface of more than one particle, causing a strong attractive interaction within the system, resulting in a solid-like behavior. Hence, the molecular weight distribution of the polymer chains plays an important role in governing the rheological behavior of polymer-colloidal silica nanocomposites. Now if the silica particles are surface modified, then the adsorption of PEO chains onto the silica particles is limited. As a result, the complex shear modulus shows different results, such as the presence of a terminal region for the system. The interesting fact to observe in this research is the contrast in the reasons behind the development of a solid-like behavior for fumed silica and colloidal silica particles as fillers. In the case of fumed silica, the particle-to-particle interaction leads to strong interaction potential within the system, resulting in a solid-like behavior; whereas in the case of colloidal silica, the polymer-to-particle interaction governs the solid-like behavior in the system and also the final morphology. If the silica particles are surface treated, then the adsorption of the polymer chains onto the surface of the silica particles is limited; as a result, the formation of three-dimensional networks is also restricted by the strength of the surface treatment of the particles. Also, the surface treatment subsequently reduces the effective interaction between the particles. Now the solid-like behavior generation becomes a strong function of the polymer chain length and the molecular weight distribution of the host polymer, because longer chain lengths can lead to more possible particles per chain, which can increase the interaction between the particles and therefore result in the generation of a solid-like behavior.

The rheological behavior of polymer clay-based nanocomposites is very similar in nature to that of other filled polymers. The only difference is that the functionalization of the clay particles and their plate-like structure result in a variation in the rheological behavior of the samples as compared to other conventional fillers. Also, in the presence of functionalized clay particles, polymer molecular weight distribution, interfacial properties, and surface charge density play an important role in controlling the rheological response of the polymer system. The liquid/solid transition normally takes place at lower clay loadings in comparison to the use of fillers.[42, 43] This could be explained by the fact that the predominant mechanism in the formation of a polymer-filled network is one of adsorption of polymer chains onto the filler particle surface but in the case of polymer nanocomposites, it is caused by the intercalation of polymer chains into clay galleries as well as tethering of the polymer chains to the surface of the clay particles, thus spiraling the move toward a network structure at much lower concentrations. The improvement in mechanical properties using clay particles is higher than that for conventional filler particles. This effect could be attributed to the fact that the clay particles themselves are immobile and, unlike conventional fillers, strongly affect the dynamic mobilities of the polymer chains. The reduction in mobility of the chain under an applied stress improves the elasticity of the system and the polymer now behaves more like a strong solid material, causing an improvement in mechanical properties. Also, under applied normal load, the clay particles have a tendency

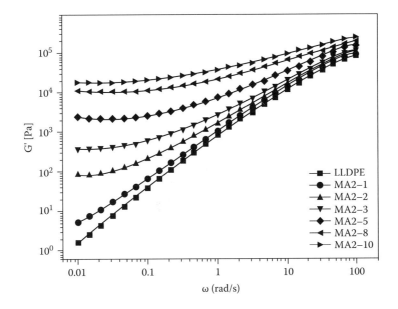

FIGURE 8.27

The variation in storage modulus with Frequency for LDPE/clay nanocomposites for different amounts of clay loading (adopted from Cassagnau et al. (2005)).

to align in the stretch direction, resulting in improved resistance to applied load. The only drawback in using clay compared to conventional nanofillers could be the quicker generation (at lower concentration and loads) of stress absorption and concentration within the polymer matrix causing failure. This behavior of clay particles could be attributed to the formation of an interconnected structure at particular loadings and applied load, which generate a clearly defined path through which the applied stress energy can pass. This further speeds up the process of stress intensification at the surface of clay particles, causing quicker failure.

Figure 8.27 and Figure 8.28 show the storage modulus and complex viscosities for an LDPE-clay nanocomposite for different amounts of clay loadings with PE-MA used as a compatibilizer. It can be seen that a secondary plateau region develops at lower frequencies as the clay concentration increases. This behavior could be attributed to the liquid-to-solid transition with increasing clay concentration at lower frequencies. It could be interpreted that at lower frequencies with increasing clay concentration, a network-like structure is established, leading to a solid-like behavior. The important aspect to note here is that the addition of PE-MA to functionalize LDPE does play an important role in controlling the morphology of the polymer-clay system.

Hiljanen-Vainio et al.[44] studied the structure-property relationships for polyester urethane (PEU)-elastomer/filler composites. The addition of a fiber-like filler improved the stiffness of the nanocomposites. The addition of a plate-like or spherical filler improved the tensile strength. The addition of an elastomer improved the impact resistance of the composites. The difference in improvement in mechanical properties could be attributed to the variable filling mechanism of the different composites. The presence of a fiber-like filler increases the chain density per unit volume of nanocomposite and hence makes it stiffer, whereas the addition of a plate-like or spherical filler results in the adsorption of chains onto the filler surface, resulting in an improvement in tensile properties propelled by a

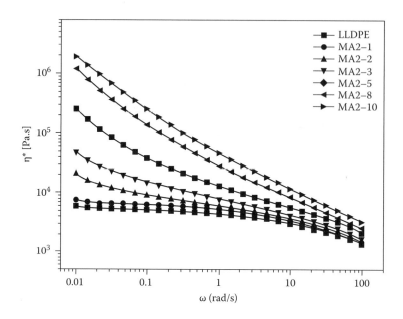

FIGURE 8.28
The variation of complex viscosity with frequency for LDPE/clay nanocomposites for different amounts of clay loading (adopted from Cassagnau et al. (2005)).

decrease in chain mobility. The interesting aspect to note is that the rheological behavior is also affected, for similar reasons, in the presence of filler. The presence of a nonspherical or a fiber-like filler results in chain alignment in the direction of the applied shear. The formation of three-dimensional networks or cross-links does not normally take place in the case of fiber-based composites because the fibers are aligned in the direction of the applied shear but the chain density increases with the addition of the fiber and hence the elasticity of the matrix increases due to the increased resistance to motion of the polymer chains. In the case of fillers that are spherical in nature, network structures are formed at higher filler concentrations, resulting in a solid-like behavior at higher as well as at lower shear rates, which further results in an improvement in the elasticity of the system. Thus, it can be seen that although the rheological behavior is the same under an applied shear, the mechanism is different for the two composites. The effect under an applied extension is interesting. The extension of fiber-based composites results in a strain hardening effect at higher extension rates as compared to that of other filled polymers. This effect could be attributed to the fact that the fiber-reinforced polymer composites are easily aligned in the direction of stretch and hence take a longer time to strain-harden. However, once strain-hardened, the rate of strain-hardening is lower in fiber nanocomposites compared with micro-filled composites because the relaxation times are much higher in the case of filler-based nanocomposites due to increased chain length.

8.5.4 Rheology Based on Preparation Techniques

Leblanc[45] studied the rheological behavior of polymer coconut-based composites using PVC as the base material; the interesting aspect in the study was the quasi disappearance of the linear viscoelastic region in the composites. The preparation technique of the

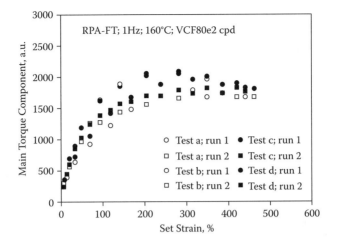

FIGURE 8.29
Variation in rheological response of polymer/coconut based nanocomposites when prepared by dry blending and extrusion (adopted from Leblanc (2006).

nanocomposite had an effect on the linear and the nonlinear rheological response of the the material.

Figure 8.29 and Figure 8.30 show the variation in rheological response of the system based on the type of processing technique used to form the composite. As can be seen from the curves, the graph with dry blending plus extrusion shows a greater amount of scattering for the curves between torque vs. strain rates. Thus, dry blending with internal mixing provided a more homogenous dispersion of the filler in the composite as denoted by the better superposition of data in the second curve. The curves can be used to generate the G' and G'' curves vs. frequency because $\tau(\omega)/\gamma$ denotes the complex modulus G^*. As can be seen,

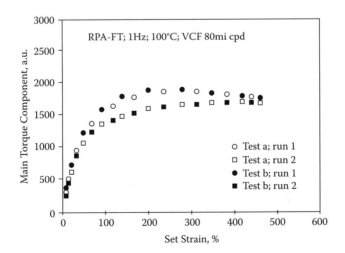

FIGURE 8.30
Variation in rheological response of polymer/coconut nanocomposite when prepared by dry blending and internal mixing (adopted from Leblanc (2006)).

at lower strains, the torque is directly proportional to the strain; but then after some time or at higher strain rates, the data show nonlinear behavior. Thus, a typical nonlinear behavior for filled polymer systems can be observed even in this case at higher frequencies. The data also reveal that there is only a hydrodynamic effect on addition of fillers by the same trend in the curves at lower and higher strain rates. The hydrodynamic force affects the relaxation time slightly and hence only affects the properties slightly. The major lack in improvement in properties can be attributed to the absence of interfacial interactions between the coconut fiber and the polymer matrix. The generated nonlinearity can be attributed to the partial change in chain length by increasing the strain rate. But as such, the absence of permanent interaction results in only a small improvement in the melt and mechanical properties.

Jiang and Kamdem[46] investigated the rheological behavior of PVC-wood fiber composites. The matrix polymer-filler system showed some improvement in mechanical properties. Drastic improvements in mechanical properties can be achieved using a ternary filler like glass or mica with wood fibers within the polymer filler system. The addition of a third particulate filler improved the mechanical properties of the system more than the fiber alone. The reason should be attributed to increased adhesion between the fillers, fiber, and polymer, resulting in improved mechanical properties. The melt elasticity, however, decreased, initiated by the improved shear thinning behavior generated by the presence of the glass beads or the fillers.

Kavacevic et al.[47] studied the rheological properties of PVAc/calcite filled nanocomposites with respect to the particle geometry, particle size distribution, and active surface area. The elongation at break of the composite decreases due to restriction in dynamic mobility of the chain, whereas the strength at break increases due to a decrease in stress concentration points upon addition of fillers.

Figure 8.31 shows the shear stress profile with different amounts of filler loadings. In this figure, the letters A, B, and C denote untreated calcite or commercially available calcium carbonate particles; and AS, BS, and CS denote the treated calcium carbonate used as a filler with polymer as well. This is a measurement of shear stress because the sample is stretched under a known applied load. It was observed that the improvement in mechanical properties was a function of the surface treatment and surface coverage of the particles. It can be seen in Figure 8.31 that the polymer samples with treated particles show better improvement in properties compared to samples with untreated particles. The type of

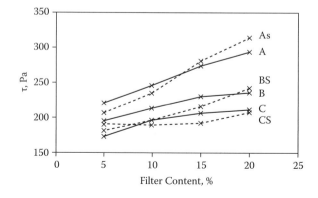

FIGURE 8.31
Variation of shear stress with variable amount of filler content (adopted from Kovacevic et al. (1996)).

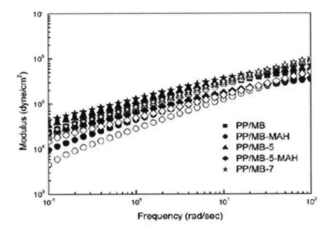

FIGURE 8.32
Variation of dynamic modulus with frequency for PP/starch/silicate ternary nanocomposite (adopted from Kim and Kim (2007)).

surface treatment also determines the extent of improvement possible because it governs the strength of interaction between the polymer chains and the filler.

Kim and Kim[48] studied the rheological properties of PP/starch (master batch) (MB)/silicate nanocomposites. The rheological results reveal an increase in complex viscosity, shear thinning property, and melt viscosity with increasing silicate loading.

Figure 8.32 shows the dynamic modulus and the complex viscosity plots vs. frequency with different amounts of silicate loading as well as after the addition of maleic anhydride (MAH) to the system. The addition of MAH is done to impart functionality to the nonpolar PP molecule. The temperature of measurement used is 200°C at 5% strain. It can be seen in Figure 8.33 that the complex viscosities of the composites without MAH show a monotonic increase with

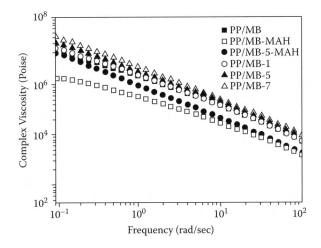

FIGURE 8.33
Variation of complex viscosity with frequency for PP/starch/silicate ternary nanocomposite (adopted from Kim and Kim (2007)).

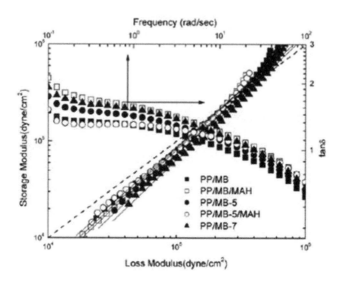

FIGURE 8.34
Storage modulus vs. loss modulus for PP/starch/silicate nanocomposite (adopted from Kim and Kim (2007)).

increasing clay loading. A similar trend is also observed for the case of modulus. The formation of a network due to anisotropy of the silicate particles results in the improvement in flow
properties and melt elasticity. The alignment of the silicate layers and subsequently the polymer chains results in an improved shear thinning behavior as well.

The complex viscosity (Figure 8.33) of the PP/MB/MAH composite increased more than
that of the PP/MB composite at lower frequencies. This effect could be attributed to the
improved functionality of the PP/MB system after addition of MAH. The improvement
in the storage and loss modulus following addition of only MAH is not as significant as
compared to the case with the addition of silicate particles. The addition of MAH does
improve the adhesion between MB and PP but as such does not change the configurational
independence and hence the morphology of the system significantly; as a result, there is
not much change in the storage and loss moduli.

To investigate the effect of silicate loading on the material function of the system, a plot
of storage vs. loss modulus as well as frequency was made. It can be seen in Figure 8.34
that the value of tan δ decreases with increasing clay loading at higher strain rates, causing
a reduction in melt elasticity at higher strain rates. This effect could be attributed to the
flow-induced alignment of the silicate layers. The intersection point of G' vs. G'' also moves
to the right, indicating that cross-over takes place at higher frequencies. This means that
flow-induced alignment at higher frequencies resists the development of solid-like behavior. It can be seen that at lower frequencies the storage modulus is higher than the loss
modulus. This behavior could be attributed to the physical jamming of the silicate layers
with the polymers, which then further resist the deformation resulting in a more solid-like
behavior and hence the storage modulus becomes higher than the loss modulus.

8.5.5 Influence of Filler Shape on Rheology

Knauert et al.[49] studied the rheological influence of variable shapes of fillers used to prepare a polymer nanocomposite.

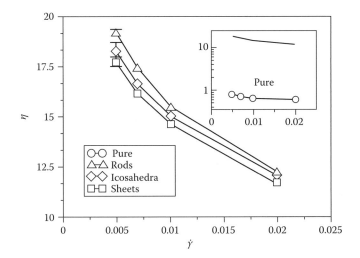

FIGURE 8.35
Shear viscosity variation with shear rates for nanocomposites prepared with different types of fillers (adopted from Knauert et al. (2007)).

Figure 8.35 shows the variation in viscosity with shear rate for different shapes of the filler used to prepare nanocomposites. The use of rod-shaped particles shows the largest complex viscosity at any given shear rate, but at the same time also shows a larger shear thinning behavior. This behavior is self-explanatory because the rod-shaped filler provides the best type of reinforcement by embedding the polymer chains within itself, as compared to other fillers that interact by surface adsorption with the polymer chains,[50] and hence rod-shaped fillers increase the viscosity of the matrix to the highest extent. Furthermore, the ease of alignment of a rod-shaped filler in the flow direction as compared to any other type of filler results in a more pronounced shear thinning behavior. It was shown that tensile strength decreases with an increase in chain length for a pure polymer as compared to the use of a filler where the tensile strength increases with increasing chain length. In contrast, the highest amount of increase in tensile strength can be seen for the case of plate-like nanoparticles, as compared to other shapes, but the rheological influence of rod-shaped particles is better than that for plate-type particles. This behavior arises because plate-type particles provide the highest amount of resistance to an applied normal load and hence the highest tensile strength as well.

8.5.6 Extensional Rheology of Filled Polymer Systems

Extensional viscosity of polymer blends and filled systems has been studied for many years. A review of the studies on uniaxial extensional viscosity of polymer blends was given by Utracki.[51] Most of these studies have dealt with immiscible blends—for example, polymer blends of high-density polyethylene (HDPE) and low-density polyethylene (LDPE), polymer blends of low-density polyethylene (LLDPE and LDPE), polymer blends of polystyrene (PS), and polyethylene (PE), and blends of LDPE/LDPE-*g*-PS/PS.[52–55] In some other studies, immiscible blends, including block copolymers, were used.[36, 37] Knowledge of the extensional rheology for filled polymeric systems is limited due to the general difficulty in measuring the steady extensional viscosity in the presence of solid fillers. Some of the typical examples of fillers used in engineering thermoplastic are carbon black, glass beads, calcium carbonate, and titanium dioxide. Lobe and White,[58] Kobayashi et al.,[59, 60] and Takahashi et al.[61] presented the uniaxial viscosities of LDPE filled with glass beads, glass flakes, talc, and glass fibers, and investigated their effect on the strain hardening property. They found that smaller particles with larger aspect ratios contributed to weaker strain hardening properties. Kotsilkova[62] studied the extensional behavior of PMMA layered silicate nanocomposites and found that unlike microcomposites described by Takahashi et al.,[61] PMMA nanocomposites exhibited strain hardening at high strain rates.

Extensional flow behavior study includes uniaxial extension, biaxial extension, and planar extension. Melt spinning and parison sag in blow molding are examples of uniaxial extension. The film blowing process is an example of biaxial extension, while film casting invokes deformations intermediate between uniaxial and planar.[63] Currently, there is extensive coverage of extensional rheological behavior of polymer melts and equipment. An earlier monograph by Petrie[64] provides detailed theoretical analyses of extensional flows and summarizes all the literature covered during that period. Ziabicki[65] provided a fundamental review of the different forms of fiber spinning operations, their kinematics, and molecular orientations of the polymer. Although extensional rheology has been established as an important aspect of materials processing, such behavior has not been covered in detail with respect to polymer nanocomposites. Studies on melt extensional properties of polymer nanocomposites have been reported previously by Pasanovic-Zujo et al.,[66] Okamoto et al.,[67] and Kotsilkova.[68] These studies measured tensile stress growth or transient extensional viscosities at a constant strain rate. The studies were conducted using a melt-spinning melt technique to investigate fiber structure formation at various take-up velocities.[69] Pavlikova et al.[70] and Zhang et al. (2004) conducted fiber-spinning experiments on polypropylene-clay nanocomposites to study the effect of orientation of the clay and to understand the relationship between the nanocomposite structure and the polypropylene hybrid fiber properties. Okamoto et al.[71] reported on the biaxial flow-induced alignment of silicate layers in polypropylene-clay nanocomposite foams and structure-property relationships.

8.5.6.1 Extensional Rheology of Nanocomposites

Ray and Okamoto[72] conducted extensional rheology studies on PLA-clay nanocomposites. Figure 8.36 shows that for nanocomposite melts, the extended Trouton (ratio of extensional viscosity to shear viscosity) rule is not applicable for extension, compared to the pure polymer melts. Similar results were also observed for PP-OMLS nanocomposites. The results indicate that although there are flow-induced structural changes in the case of extension, they are different compared to the structural changes taking place under shear flow. In the case of PLA-clay samples, the effect of strain-hardening was observed at higher strain rates. This value of extensional viscosity increases with increasing strain rates and shows a similarity with the behavior of the sample under shear measurements.

It could be said that the alignment of clay particles leads to more restricted motion of the polymer chains, subsequently increasing the magnitude of the viscosity.

Pasanovic-Zujo et al.[73] and Gupta et al.[74] also studied the effect of varying the vinyl acetate loading on extension for EVA-clay nanocomposites. Two different EVA-clay samples (EVA18 and EVA28) were prepared with variable amounts of clay loading. In this case, EVA18 and EVA28 refer to 18 and 28 wt% vinyl acetate (VA) content, respectively. It

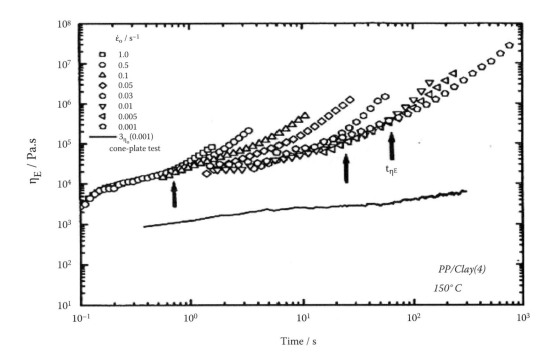

FIGURE 8.36
Time dependence of elongation viscosity for PP/clay (4) (PPCN4) melt at 150°C. The upward arrows indicate up-rising time for different strain rate. The solid line shows three times the shear viscosity, taken at a low shear rate of 0.001 s^{-1} (adopted from Ray and Okamoto (2003)).

was found that neat EVA28 and EVA18 samples showed more strain-hardening behavior compared to their nanocomposites at higher strain rates. However, in the linear region, the neat polymers showed lower extensional viscosity compared to the nanocomposite samples. It has been found that by increasing the vinyl acetate content, a stronger interaction is achieved between the polymer chains and the clay particles. As a result, the extensional viscosity increases up to a Hencky strain rate of 4 for EVA28 nanocomposites, as shown in Figure 8.37. A similar increase in extensional viscosity was observed for EVA18 nanocomposites, but at a lower Hencky strain of up to 2 (Figure 8.37). Surprisingly, at higher strain rates, the EVA samples by themselves show more strain-hardening as compared to the samples with clay. Both these results are observed in the case of uniaxial extensional flow. The decrease in strain-hardening behavior with increasing amounts of clay for EVA is attributed to the stretching of the polymer chains, which in turn leads to the aggregation of clay particles, reducing the overall interaction between the clay particles and the polymer chains, which results in the decrease of strain-hardening behavior.

Xu et al.[75] studied the extensional rheology of carbon nanofiber suspensions using Rheometrics RFX extensional rheometers. They observed the effect of extension rate thinning in the samples and attributed it to the breaking of the network structure present in the system. The break occurs under the application of a constant extension that exceeds the interaction between the fiber and the polymer, resulting in depletion of the network structure. The broken fiber strands align themselves in the flow direction, resulting in a decrease

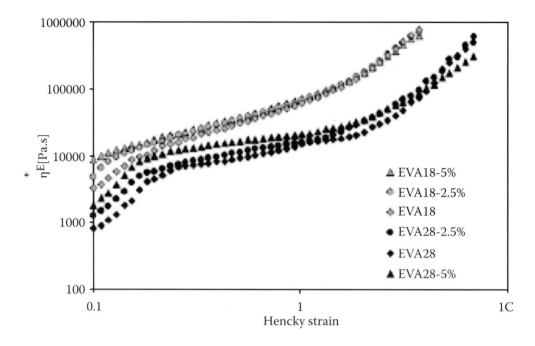

FIGURE 8.37
Extensional viscosity growth as a function of strain for EVA28, EVA18, and their nanocomposites at 130 °C and at a strain rate of 1 s-1 (adopted from Pasanovic-Zujo et al. (2004a)).

of strain-hardening behavior in the sample, although the extensional viscosity of the sample is higher than that for the pure polymer in the linear region at lower values of strain rates.

8.6   The Rheology of Colloidal Suspensions

Colloidal suspensions and polymer nanocomposite melts and solutions have striking similarities in their rheological behavior (shear, dynamic, and extension).

The major cause of variation in the rheological behavior between these two types of complex fluids arises due to the fact that the interaction forces between the dispersed phase and the continuous phase are different in nature for the two systems. For example, in the case of colloidal suspensions, the particle-to-particle repulsion is significant, and only the van der Waals interaction is prevalent in between the particles and the solvent. On the other hand, in the case of polymer nanocomposite melts and solutions, the particle-to-polymer interaction is much stronger due to the combined effect of chemical bonding as well as adsorption, but the particle-to-particle repulsion is weaker. Also, the higher concentration of the dispersed phase plays a significant role in the case of colloidal suspensions while in case of polymer nanocomposites, even a small concentration of clay particles produces a significant change in the rheological response of the material under consideration.

The following factors are the most significant factors that affect the rheological response of colloidal suspensions: (1) surface forces, (2) van der Walls attraction and repulsion forces, (3) electrical double layer repulsion, (4) steric repulsion, and (5) the shape and size of the particles.[76]

8.6.1 Surface Forces

The surface forces and the interaction forces control the interaction between the particles in a given medium for a colloidal suspension. It has been proven that these forces have an influence on the rheology of the system.[77] The interaction between the colloidal particles via surface forces arises due to microscopic or molecular level interaction between the individual molecules. The interaction can be attractive (dissimilar particles) as well as repulsive (similar particles). These forces are generated by Brownian motion, hydrodynamic behavior, as well as by charges created by electrostatic and electrodynamic effects or by volume exclusion.[78] The forces are given by

$$F = -dV/dL$$

where V is the interparticle pair potential and L is the surface-to-surface distance between the two particles. Both the degree and magnitude of the repulsive and attractive forces vary, depending on the mechanism that generates the force. The mechanism can be classified as to bridging, depletion hydrophobic, van der Waals, and electrical double layer (EDL) for attractive forces and as to hydration, EDL, steric, and structural for repulsive forces.

8.6.2 Van der Waals Attraction and Repulsion Forces

These forces are usually generated by secondary interactions between similar and dissimilar particles. The type and strength of the pair potential function between the particles governs the nature of the net interaction between the particles. The forces can be characterized into long-range and short-range repulsive and attractive forces. These forces can generally be characterized mathematically by Lennard's Jones potential or London dispersion forces. In the absence of repulsive forces, the attraction between the particles results in aggregation of the particles and causes the formation of three-dimensional networks due to the close spacing between the particles, which in turn affects their rheological behavior. In case of weak repulsive forces, there is aggregation of particles and the rheological response is different because at very close distances the repulsive forces become important, thereby limiting the approach of the particles toward each other. The presence of the continuous medium between the particles affects the rheology by reducing the resistance to motion under flow, and hence the dynamic rheology changes. The repulsive potential, long or short range, is generally created by the electrical double layer interaction or steric interaction. These interactions can also affect the attraction potential between the particles.

Van der Waals attraction arises when like particles are placed in a medium that has different dielectric properties as compared to the particles. The force arises because of the creation of instantaneous dipoles in the atoms or molecules present within the particles due to a variation in the dielectric property within the system. The attractive potential is given by Equation 8.5 as follows:

$$V_{att} = AR/12D \tag{8.5}$$

In this equation, A is the Hamaker constant, which depends on the dielectric properties of the medium and the particles, and D is the surface-to-surface distance between the particles. If the force is strong enough to overcome repulsion, then it results in coagulation of the particles in a given medium, resulting in sedimentation of the colloidal particles, which in turn changes the rheological behavior of the system by the formation of coagulated or flocculated suspensions. The point to note is that such behavior can only be noticed in particles that are more or less neutral (surface charge-wise).

8.6.3 Electrical Double Layer Formation

An electrical double layer is formed in colloidal suspensions.[78] The thickness of the electrical double layer controls the charge density and the repulsive potential between the particles within the system, which in turn controls the stability of dispersion of the particles within the medium. Certain particles and polymers have the property to ionize in the medium and undergo polarization above a certain pH. The pH below which the particles stay neutral is called the isoelectric point. Above or below the isoelectric point, such behavior is observed. The examples of such surface groups are carboxylate groups (–COOH) on polyacrylic acids, amine groups (–NH$_2$) on polyacrylamide, and hydroxyl groups (–OH) on particle surfaces, or the presence of organic ligands on the surface of the particles. The ionization results in the accumulation of oppositely charged ions near the surface of the particles—for example, the accumulation of chloride ions in case of HCl addition. The accumulation of oppositely charged ions takes place to maintain the overall electro-neutrality of the medium. As the particles approach each other, the layers created by these oppositely charged ions tend to overlap, increasing the counter-ion concentration in between the particles and resulting in the development of a repulsive potential caused by the osmotic pressure exerted by these ions that result in electrical double layer repulsion. The thickness of this layer is measured by the Debye length.[79] When this length is small, the repulsive forces are short range and the net force can be attractive, resulting in a van der Waals attractive potential; but in cases where the distance is large, the repulsive potential is long range, overcoming the attractive forces and hence the interaction is of the van der Waals repulsive type. This repulsion force (Vrep) between particles of radius R is given by

$$Vrep = 2\pi\varepsilon\varepsilon_o R\psi_o 2 \exp(-KD) \tag{8.6}$$

where R is radius of the particle, K is inverse Debye length, ε is the relative permittivity of water, ε_o is the permittivity of space, and ψ represents the surface potential of the particles.

8.6.4 Steric Repulsion

A steric interaction occurs in a system due to the adsorption or grafting of polymers on particle surfaces, or vice versa.[80] The strength of the interaction is a function of the mixing interaction and the volume exclusion term. The nature of the mixing interaction force depends on the nature of the continuous phase or the solvent. If the polymer is miscible in the solvent, then the interaction will be repulsive in nature. On the other hand, immiscibility of the polymer in the solvent will result in attractive interactions.[81]

The volume exclusion interaction is caused due to the fact that polymers adsorbed onto the surface lose additional configurational entropy when approached by another particle. This loss creates a steric repulsion force. These steric forces are normally short-range forces. The adsorbed layer thickness (δ) controls the strength of interaction. The estimation of the potential energy of a system consisting of terminally anchored polymer molecules on a flat plate surface provides a mathematical approach to quantifying this effect.

The potential per unit surface area of the particle (V) is given by De genes (1987) as follows:

$$V = 3\pi PkT/11S^2 \, (\delta^{5/3}/D^{11/3}) \tag{8.7}$$

where kT is thermal energy of the system, S is average distance between the polymer strands, δ is adsorbed layer thickness, and D is the separation distance between the interacting particles.

8.6.5 Interactions in Polymer Nanocomposites

This section provides insight into the various possible interactions in a polymer nanocomposite. There are four possible interactions in a polymer nanocomposite:[83, 84]

1. Polymer-polymer interaction
2. Polymer-solvent interaction
3. Polymer-particle interaction
4. Particle-particle interaction

The above-mentioned interactions generally affect the rheology of a polymer nanocomposite under various rheological conditions. The point to note is that most of these interactions can be characterized by the theory developed for colloidal suspensions—except for the case of polymer-particle interaction and polymer-polymer interaction. The difference in the nature of the above-mentioned interactions in a polymer nanocomposite system as compared to a colloidal system basically leads to a change in the rheological behavior of polymer-clay nanocomposites as compared to colloidal suspensions. The similarity of the majority of the interaction forces also leads to certain similarities between the rheological responses of the two systems. In short, it could be concluded that the nature and trend of the interaction within the two systems lead to similarities and differences in the rheological behavior of the systems.

8.6.5.1 Polymer-Solvent Interactions

The polymer-to-solvent interaction is governed by the free energy of mixing between the components.[85] The free energy of mixing helps in determining the interaction parameter X, which is quantitatively used to predict the level of interaction between the polymer chains and the solvent molecules. The interaction parameter can be calculated by the difference in the cohesive energy density between the polymer and the solvent. The cohesive energy density is, in turn, a function of the intereraction pair potentials between the two components in the system. A solvent in which the intra- and intermolecular interaction potential remains the same is called a theta solvent; and for such a system, the interaction potential is 0.5. For any hindrances in the interaction potential, the interaction parameter is either greater than or less than 0.5. Now, in case of a theta solvent, the polymer occupies an approximately spherical volume (excluded volume) such that the inter- and intramolecular potential stay in balance. The small concentration of polymer chains in the excluded volume ensures that such a condition is achieved because when the concentration is low, each polymer molecule is independent of the other molecule. A large portion of the excluded volume is occupied by solvent molecules trapped within the polymer chains. The excluded volume can be characterized by determining the radius of gyration of the polymer chains. In the case of a good solvent ($x > 0.5$), the polymer chains tend to interact more with solvent molecules. As a result, they swell and the excluded volume increases with increasing solvent molecules within the excluded volume. In the case of a poor solvent, this interaction is further reduced, causing a reduction in the excluded volume. The point here is that the polymer-solvent interaction is difficult to characterize because the interaction in this case is a strong function of the configuration of the chain, which changes with conditions, unlike solid particles where the configuration of the particles stays the same irrespective of the type of interaction. With increasing concentration of the polymer chains, the concentration of polymer chains within the excluded volume increases (x < 0.5), resulting in the entanglement of chains. The entanglement of the chains and the configurational state of the system affect the resistance to flow for a system (entanglement increases resistance and decreases swelling), affecting the rheological behavior of the system.

8.6.5.2 *Polymer-Particle Interactions*

The polymer particle interaction has a significant influence on the rheological response of the system.[86-88] The polymer-to-particle interaction is governed by three important particle characteristics: (1) the concentration of the particles, (2) the shape and size of the particles, and (3) the functionality of the particles. The repulsive–attractive potential between the particles and the polymer-particles is a function of particle concentration.[89] At very low particle concentrations, the particles are well dispersed within the polymer matrix and as a result the interaction potential between the particles is weakly attractive. This attractive potential restricts the polymer chain interaction within the particles and therefore affects the configuration of the polymer chain within the system and hence the rheological response of the polymer. The size and shape of the particles determine the available surface area for interaction with polymer chains and therefore affect the morphology of the polymer chains within the matrix and hence the rheological response. The functionality of the particles governs the dispersion state of the particles within the polymer system. Also, the introduction of functionality within the particles increases the repulsive interaction potential between the particles and, as a result of aggregation of particles, is inhibited until there are higher concentrations of particles within the polymer matrix. In addition, the polymer tends to entangle itself near the particles, which are functionalized. As a result, the polymer-particle system then behaves more like a solid. Thus, the rheology of the polymer is affected overall by the net interaction potential between the particles and the polymer within a polymer-particle system.

8.6.5.3 *Particle-Particle Interactions*

The particle-to-particle interaction also has a significant influence on the rheology of the polymer-particle system. The effect of the particle-particle interaction depends on whether or not the particles are functionalized.[90] When the particles *are not* functionalized, the attractive potential between the particles is higher. As a result, the particles have a greater tendency to form aggregates and the polymer-particle system behaves like a conventional filled polymer system. In the case where the particles *are* functionalized, the polymer-particle interaction improves due to the increase in the repulsive potential between the particles. This further causes a change in the rheological response of the system.

8.6.5.4 *Polymer-Polymer Interaction*

The polymer-to-polymer chain interaction also affects the rheological response of the system. The interaction potential of the chains results in a variation in the apparent free volume of the system, which in turn governs the available volume for the filler particles to fill in. The dynamics of the chain within the matrix is also affected by the chain-to-chain interaction.[91, 92] When the interaction is low, the chains are shorter in length and are freer to move, making them behave more like a fluid; and when the interaction is large, the chains are larger and are not very free to move within the matrix, and hence behave more like a solid. The polymer-to-particle interaction is also to some extent dependent on the chain-to-chain interaction because the chain-to-chain interaction in a way governs the configuration of the pure polymer system and hence the level of interaction between the two.

8.6.6 Difference in Rheological Behavior of Nanocomposites and Colloidal Suspensions

As stated previously, the differences and similarities in the rheological response for the colloidal suspensions and polymer nanocomposite arise from the variations and similarities in the interaction forces within the two systems. The various possible interactions, their determination, and physical significance were discussed above. These principles can be used to study the rheological response of colloidal suspensions and polymer nanocomposites accordingly.[93–95]

Table 8.1 shows the similarities and differences among some of the rheological parameters for colloidal suspensions and polymer nanocomposites. It can be said that although at certain places the rheological behavior is common, morphological evolution and its effect on the structure and properties of the system is significantly different in the two cases.

TABLE 8.1
Summary of rheological behavior of colloidal suspensions and polymer nanocomposites

	Colloidal suspensions (Insoluble particles)			Polymer nanocomposites	
	Hard spheres	Repulsive stabilized particles	Attractive particles	Functionalized filler (FF)	Non-functionalized filler (NFF)
Shear viscosity	Shear thinning	Shear thinning	Shear thinning	Shear thinning	Shear thinning
Yield stress	Absent**	At high volume fractions	At low volume fractions	At low volume fractions as compared to NFF	At relatively high volume fractions compared to FF
Storage modulus	$G'=G'(w)$	$G'=G'(w)$ $G' = $ const.*	$G' = $ const.	$G' = G'(w)$***	$G'= G'(w)$ $G' = $ constant
Loss modulus	$(G''<G')$# Or $(G''>G'')$##	$G'' = G''(w)$ $G'' < G'$ $G'' >G'$*	$G'' =G''(w)$ $G''<<G'$	$G'' = G''(w)$ $G'' = $ constant $G''>G'$**** $G'' <G$!!	$G'' = G''(w)$ $G'' = $ constant

* Depending on the concentration of the suspended particles
** Except at volume fractions exceeding random dense packing
# At lower frequencies if not then there is a presence of yield stress in the material
Normally at higher frequencies
*** The dependency of G' on w depends on the concentration of the filler. Also it can be greater or lesser than G'' depending on the cross over frequency and the concentration and type or rather shape of the filler used
****At lower frequencies or in the presence of a yield stress the material and measurements become difficult, yield stress development takes place normally at higher filler concentrations
!! Normally at higher frequencies depending on the cross over frequency and the type of filler used

8.7   Rheological Modeling of Polymer Nanocomposites

The mathematical modeling of the rheological behavior of polymer nanocomposites has been investigated for cases where the clay loading is less than 10 wt%. In most cases, a homogenous dispersion of clay in the polymer melt matrix is assumed to reduce the mathematical complexity of the overall model. Among a host of parameters affecting the viscometric functions of the polymer nanocomposite, most of the models imply the importance of molecular weight distribution and clay polymer interaction as the most important molecular parameters governing the viscoelastic behavior of polymer nano-composite melts.[96-98] Three different molecular weight regimes are identified with respect to the molecular weight (M), the critical molecular weight (Mc), and the molecular weight distribution of the polymer.

Case 1: M < Mc—the case of unentangled polymer chains, which is solved using the Network model.

Case 2: M = Mc—the case where the molecular weight is near the threshold concentration between an entangled and an unentangled system, which is solved using the FENE Dumbell model.[99]

Case 3: M > Mc—the case of an entangled system, which is solved using the generalized Rouse model for polymers where the effect of entanglement on diffusion rate is determined using the well-known tube model in which Reptation theory is also employed. The generalized Rouse model has been used to determine the anisotropic diffusion rate because it takes into account the bead density in Khun segments by considering the parameter delta, which is wall-to-wall-particle distance in its form.

In short, it could be said that the Network model, the FENE Dumbell model, Tube theory, and the Reptation theory are applied to the case of viscoelastic modeling of polymer nanocomposites; the Cross-Carreau model, Williamson-Carreau model, Herschel-Buckley model are applied to the case of steady shear viscosity of polymer nanocomposites; and the K-BKZ model is applied to the case of extensional viscosity for polymer nanocomposites. The models that are applied to polymer nanocomposites can be summarized as follows:

1. Herschel-Buckley model
2. Williamson-Carreau model
3. Network model
4. FENE Dumbbell model
5. K-BKZ model

8.7.1   Herschel-Buckley Model

The Herschel-Buckley equation was found effective in determining the rheological behavior of fluids such as mud, clay suspensions, oil, and drilling fluids. The Herschel-Buckley equations are given as follows:

$$\tau = \tau_y + k\dot{\gamma}^{\,n} \qquad (8.8)$$

$$\eta = \frac{\tau_Y}{\dot{\gamma}} + k\dot{\gamma}^{n-1} \qquad (8.9)$$

Equation 8.9 was used to calculate the viscosity η of suspensions or the fluid in consideration. Here $.\tau$ is the shear stress, $\tau._y$ is the yield stress, $\dot{\gamma}$ is the shear rate, k is a constant, and n is the flow index and depends on the type of fluid. When $n > 1$, the fluid shows a shear thickening behavior; for $n = 1$, the fluid shows a Bingham plastic type of behavior; and for $n < 1$, the fluid shows a shear thinning behavior, which is what is commonly observed with polymer nanocomposites. The model reduces to a power law model when $.\tau_y$ is zero. It is very well known that the values of n, k, and $.\tau_y$ are functions of the type of fluid.

8.7.2 Williamson-Carreau Model

The Williamson-Carreau equation is given by

$$\eta = \frac{\eta_0}{1 + \left|\lambda\dot{\gamma}\right|^{1-n}} \qquad (8.10)$$

where η_0 is the zero shear viscosity, λ is the characteristic time, $\dot{\gamma}$ is the shear rate, and n is a constant that depends on the type of fluid and the overall structure of the polymer-clay system (such as intercalated or exfoliated).

8.7.3 Viscoelastic Models

8.7.3.1 The Network Model

A limited amount of work was undertaken to develop viscoelastic models for polymer nanocomposites. Viscoelasticity is affected by the interfacial surface area of the nanoparticles and the amount of interaction between particles and the matrix polymer. The model works on the principle that the attachment-detachment kinetics of the grafted chains play an important role in determining the viscoelastic response of the nanocomposite because the above-mentioned parameter effectively governs the relaxation time of the polymer nanocomposite. The relaxation time is significantly affected by the polymer structure, which evolves due to the interaction with the nanoparticle (e.g., straightening of the polymer chains improves the solid-like behavior or the elastic response of material). In the network approach, a strong filler polymer interaction is assumed. It is also assumed that the particles have an effective diameter that is equivalent to the gyration radius of the host polymer chain.

8.7.3.2 The FENE Dumbbell Model

In cases where the polymer molecular weight distribution is very close to the threshold value, it approaches toward entanglement or a regime where the chains are very closely spaced, resulting in entanglement of the entire system. The entanglement of polymers and the presence of nanoparticles make the diffusion of of the polymer chains nonisotropic with respect to the longitudinal and transverse directions. As a result, the polymer particle system shows different relaxation times in the two directions.

The nonisotropy can be explained by the variation in particle concentration along the polymer in both directions. As a result, the frictional coefficients also vary in both directions, resulting in variable relaxation times. The continuum approach is used for the polymer particle system with the condition that the frictional coefficient is considerably increased at the polymer-particle interface. As per the combined form of the Doi Edwards formalism and the Reptation theory, the polymer chain is supposed to be confined by a tube where the tube diameter is equivalent to the entanglement distance. Although the Reptation approach takes into account the hydrodynamic effects of entanglement due to diffusion, it does quantitatively account for the nonisotropy created by the Brownian motion of the polymer chains within the confined tube. The Brownian motion is found to significantly affect the diffusion rate due to variation in the frictional coefficients. As a result, both the above-mentioned factors have a significant role in controlling the relaxation spectrum of the polymer nanoparticle system and, in turn, its rheological behavior. Thus, the modified encapsulated dumbbell model is combined with the tube model to account for the above effects. As in the case of the network model, the generalized bead spring model structure is also employed here. The FENE Dumbbell approach is employed in cases where the polymer nanoparticle system is at concentrations close to entanglement because it takes into account the nonlinear viscoelastic behavior arising in polymer nanoparticle systems very close to the threshold concentration because, at threshold, any given volume of the polymer matrix will consist of a large entangled polymer chain with a small volume fraction of disentangled chains. In fact, at concentrations very close to the threshold, the polymer system tends to oscillate between the above two equilibrium structures because the total energy remains the same.

Very few studies exist in the literature related to the prediction of extensional properties of polymeric nanocomposites. Pasanovic-Zujo et al.[100] applied the K-BKZ model to predict the extensional behavior of nanocomposites. K-BKZ is an extension of the generalized Maxwell model. This model is modified by incorporating a damping function that is representative of relaxation mechanism in extensional flow. Detailed analysis of this model can be found in Pasanovic-Zujo et al.[101] In particular, the model was applied to the case of extension of EVA-based nanocomposites. The model was able to use the material function β, which is also included in the equation for the damping function used in the model. The suitable use of this parameter (β) has a significant effect on predicting the uniaxial extensional viscosity. The parameter β was adjusted to match the experimental data more accurately. The experimental results of extensional viscosity vs. strain rates and extensional viscosity vs. time were generated for EVA18 nanocomposites using RME equipment. As stated previously, the damping function, which is only a function of shear rate, and a relaxation modulus, which is a function of time, can be compared with the experimental viscosity data. The memory function as such was calculated from the experimental data.

Figure 8.38 shows the extensional behavior of EVA18 nanocomposite for 5 wt% of clay at various strain rates. As can be seen from this figure, EVA18 nanocomposites exhibit strong strain hardening behavior at higher strain rates, but the strain hardening behavior reduces with increasing amounts of clay loadings compared to that of pure EVA18. The terminal value of the extensional viscosity (at lower strain rates) continuously increases with increasing values of clay loadings, as compared to pure EVA. This is due to the increase in modulus, which is caused by the solid-like behavior of the material caused by the higher amount of interaction between the polymer and the clay particles. This can be attributed to the fact that at higher strain rates, the polymer chains are stretched and, as a result, there is an aggregation of the clay particles, which in turn reduces the overall interaction between the polymer and the nanoparticles. The K-BKZ model, as shown in Figure 8.38, does fit the

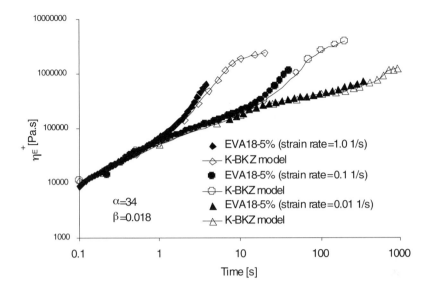

FIGURE 8.38
Variation of extensional viscosity with time at different values of strain rates for EVA18 nanocomposite with 5 wt% clay, showing experimental as well as theoretical values (adopted from Pasanovic-Zujo et al. (2004)).

experimental data accurately. More details on the modeling of polymeric nanocomposites have been discussed in a recent book by Bhattacharya et al. (2007).[102]

8.7.4 Molecular Dynamic Simulation

Molecular dynamic simulation is a technique where first the polymer particle system is placed inside a suitable lattice; the placement of the chains and nanoparticles inside the lattice is governed by the principle of minimum energy. Hence, after placing the chains randomly and assuming various kinds of probability distribution for the free energy of the chain, an average value for the free energy from the iterations is assumed and, accordingly, an equilibrium structure is determined at time zero for the system. Now, various standard equations for chain dynamics and polymer-particle-polymer interaction are applied to the lattice and the structural changes in the system are noted. Thousands of such iterations help in analyzing the behavior of the system under different conditions. Thus, molecular dynamic simulation closely simulates real working conditions and then tries to achieve an optimum scientific explanation for the real-time observed behavior by a trial-and-error procedure in selecting parameters and equations governing the behavior.

8.8 Conclusions

The rheological behavior of various types of nanocomoposites and composites with different types of fillers has been explored. There are many similarities in the rheological behavior of polymer nanocomposites processed with different types of fillers. Although the overall rheological response is the same for different types of nanocomposites, the structural evolution of the system under different rheological conditions is different, and hence other properties of the system are affected in different ways. The variation in the type of interaction forces and mechanism of interaction within a given polymer filler system introduces a variability in the overall rheological response of the system. The system with nonfunctionalized nanoparticles adsorb the polymer chains onto to their surface, whereas functionalized particles are chemically linked to the polymer chains, with secondary surface adsorption. The dynamic mobility of polymer chains is higher in the case of nonfunctionalized particles and hence the rheological response is more like that of the pure polymer. The improvement in mechanical properties by different types of fillers (based on functionalization and shape) shows interesting trends. In the case of a nonfunctionalized filler, the elongation at break is just a bit less than that of the pure polymer, and the tensile modulus and impact strength are improved. The functionalization of the particles produces the same kind of results at lower loadings but the tensile strength at break decreases with fuctionalization due to flow-induced alignment and hence increased resistance to an applied load. Similarly, the rheological response also shows a change for smaller amounts of loading in the case of functionalized particles. The liquid-to-solid-like transition, the absence of a terminal zone for G' and G'', and the development of a secondary plateau region in cases of complex viscosity all take place at lower filler concentrations. The significant difference arises in cases of extensional rheology where polymers with functionalized particles show a more pronounced strain-hardening effect compared to normal filled polymers. Furthermore, there is a percolation threshold development for interacting particles as compared to noninteracting particles. The morphological evolution as observed under SEM and TEM is also different in the case of interacting particles. Thus, it could be concluded that the rheological differences arising in filled polymers and polymer nanocomposites are due to the following parameters: shape, size, functionalization, level of interaction, and concentration.

General Bibliography

Ahmad, S., Bohidar, H.B., Ahmad, S., and Agnihotry, S.A., *Polymer*, 47, 3583–3590, 2006.
Ballman, R.L., *Rheologica Acta*, 4(2), 137–140, 1965.
Bar-Chaput, S. and Carrot, C., *Rheologica Acta*, 45, 339–347, 2006.
Berger, L. and Meissner. J., *Rheologica Acta*, 31, 63–74, 1992.
Bertrand, E., Bibette, J., and Schmitt, V., *Physical Review E*, 66, 60401, 2002.
Bhattacharya, S.N., Gupta, R.K., and Kamal, M.R., *Polymeric Nanocomposites—Theory and Practice*, Carl Hanser Verlag, Munich, 2007.
Boek, E., Coveney, P., Lekkerkerker, H., and van der Schoot, P., *Physical Review E*, 55, 3124–3133, 1997.
Breedveld, V., *Journal of Chemical Physics*, 114, 5923, 2001.
Buxton, G. and Balazs, A., *Journal of Chemical Physics*, 117, 7649, 2002.
Cassagnau, P., *Polymer* 49, 2183, 2008.
Chen, L., Zukoski, C., Ackerson, B., Hanley, H., Straty, G., Barker, J., and Glinka, C., *Physical Review Letters*, 69, 688–691, 1992.
Cogswell, F.N., *Transactions of the Society Rheology*, 16, 383–403, 1972.
Dealy, J.M. and Wissburn, K.F., *Melt Rheology and Its Roles in Plastics Processing*, Chapman and Hall, London, 1990.
Doyle, P., Shaqfeh, E., and Gast, A., *Physical Review Letters*, 78, 1182–1185, 1997.

Du, F., Scogna, R., Zhou, W., Brand, S., Fischer, J., and Winey, K., *Macromolecules*, 37, 9048–9055, 2004.
Ferry, J.D., *Viscoelastic Properties of Polymers*, John Wiley & Sons, New York, 1980.
Fornes, T.D., Yoon, P.J, and Keskkulaand Paul, H.D.R., *Polymer*, 42, 9929–9940, 2001.
Fu, X. and Qutubuddin, S., *Materials Letters*, 42, 12–15, 2000.
Fuchs, M. and Cates, M., *Physical Review Letters*, 89, 248304, 2002.
Galgali G., Ramesh C., and Lele, A., *Macromolecules*, 34, 852–858, 2001.
Gates, T., Odegard, G., Frankland, S., and Clancy, T., *Composites Science and Technology*, 65, 2416–2434, 2005.
Gelfer, M., Burger, C., Chu, B., Hsiao, B., Drozdov, A., Si, M., Rafailovich, M., Sauer, B., and Gilman, J., *Macromolecules*, 38, 3765–3775, 2005.
Giannelis, E., Krishnamoorti, R., and Manias, E., *Advances in Polymer Science*, 138, 107–148, 1998.
Giza, E., Ito, H., Kikutani, T., and Okui, N., Journal of Macromolecular Science, Phys., B, 39(4), 545–559, 2000.
Gu, S., Ren, J., and Wang, Q., *Journal of Applied Polymer Science*, 91, 2427–2434, 2004.
Gupta, R.K., Pasanovic-Zujo, V., and Bhattacharya, S.N., *Journal of Non-Newtonian Fluid Mechanics*, 128(2-3), 116–125, 2005.
Hattori, T., Takigawa, T., and Masuda, T., *Journal of the Society of Rheology Japan*, 20, 141–145, 1992.
Hiljanen-Vainio, M., Heino, M., and Seppala, J., *Polymer*, 39, 865–872, 1998.
Huang, W. and Han, C.D., *Polymer*, 47, 4400–4410, 2006.
Jiang, H. and Kamdem, D., *Journal of Vinyl and Additive Technology*, 10, 59–69, 2004.
Kim, Y. and Kim, J., *Journal of Industrial and Engineering Chemistry*, 13, 1029–1034, 2007.
Kobayashi, M., Takahashi, T., Takimoto, J., and Koyama, K., *Polymer*, 36(20), 3927–3934, 1995.
Kobayashi, M., Takahashi, T., Takimoto, J., and Koyama, K., *Polymer*, 37(16), 3745–3747, 1996.
Kotsilkova, R., *Rheology–Structure Relationship of Polymer/Layered Silicate Hybrids, Mechanics of Time Dependent Materials*, Kluwer Academic: New York, 6, 283–300, 2002.
Kovacevic, V., Lucic, S., Hace, D., and Glasnovic, A., *Polymer Engineering and Science*, 36, 1135, 1996.
Krishnamoorti, R. and Giannelis, E., *Macromolecules*, 30, 4097–4102, 1997.
Krishnamoorti, R., Ren, J., and Silva, A., *Journal of Chemical Physics*, 114, 4968, 2001.
Krishnamoorti, R. and Yurekli, K., *Current Opinion in Colloid & Interface Science*, 6, 464–470, 2001.
LeBaron, P., Wang, Z., and Pinnavaia, T., *Applied Clay Science*, 15, 11–29, 1999.
Leblanc.J , *Journal of Applied Polymer Science*, 101, 3638–3651, 2006.
Lee, M., Hu, X., Li, L., Yue, C., Tam, K., and Cheong, L., *Composites Science and Technology*, 63, 1921–1929, 2003.
Lim, S., Lee, C., Choi, H., and Jhon, M., *Journal of Polymer Science, Part B, Polymer Physics*, 41, 2052–2061, 2003.
Lim, Y.T. and Park, O.O., *Macromolecules Rapid Communications*, 21(5), 231–235, 2000.
Lim, Y.T. and Park, O.O., *Rheologica Acta*, 40(3), 220–229, 2001.
Liu, H. and Brinson, L., *Composites Science and Technology*, 68(6), 1502–1512, 2008.
Lobe, V.M. and. White, J.L., *Polymer Engineering and Science*, 19(9), 619–624, 1979.
Maiti, A., Wescott, J., and Goldbeck-Wood, G., *International Journal of Nanotechnology*, 2, 198–214, 2005.
Marshall, L. and Zukoski IV, C., *Journal of Physical Chemistry*, 94, 1164–1171, 1990.
Mewis, J., Frith, W., Strivens, T., and Russel, W., *AIChE Journal*, 35, 415–422, 1989.
Micic, P., Bhattacharya, S.N., and Field, G., *International Polymer Processing*, 12(2), 110–115, 1997.
Okamoto, M., Nam, P.H., Maiti, P., Kotaka, T., Hasegawa, N., and Usuki, A., *Nano Letters*, 1(6), 295–298, 2001.
Okamoto, M., Nam, P.H., Maiti, P., Kotaka, T., Nakayama, T., Takada, M., Ohshima, M., Usuki, A., Hasegawa, N., and Okamoto, H., *Nano Letters*, 1(9), 503–505, 2001.
Oksman, K, Mathew, A., Bondeson, D., and Kvien, I., *Composites Science and Technology*, 66, 2776–2784, 2006.
Pasanovic-Zujo,.V., Gupta, R.K., and Bhattacharya, S.N., *International Polymer Processing*, 4, 388–394, 2004.
Pavlikova, S., Thomann, R., Reichert, R., Mulhaupt, R., Marcinin, A., and Borsig, E., *Journal of Applied Polymer Science*, 89(3), 604–611, 2003.

Petrie, C.J.S., *Elongational Flows: Aspects of the Behaviour of Model Elasticoviscous Fluids*, Pitman, London, 1979.

Picu, C. and Sarvestani, A., *American Physical Society, APS March Meeting*, March 13–17, 2006, Abstract# R30003.

Pötschke, P., Fornes, T., and Paul, D., *Polymer*, 43, 3247–3255, 2002.

Prasad, R., Pasanovic-Zujo, V., Gupta, R., Cser, F., and Bhattacharya, S., *Polymer Engineering and Science*, 44, 1220–1230, 2004.

Qiu, J. and Feng, H., *Transactions of Nonferrous Metals Society of China*, 16, 444-448, 2006.

Ray, S., *Journal of Industrial and Engineering Chemistry*, 12, 811–842, 2006.

Reichert, P., Hoffmann, B., Bock, T., Thomann, R., Muelhaupt, R., and Friedrich, C., *Macromolecular Rapid Communications*, 22, 519–523, 2001.

Riva, A., Zanetti, M., Braglia, M., Camino, G., and Falqui, L., *Polymer Degradation and Stability*, 77, 299–304, 2002.

Sarvestani, A. and Picu, C., *Polymer*, 45, 7779–7790, 2004.

Sarvestani, A. and Picu, C., *Rheologica Acta*, 45, 132–141, 2005.

Starr, F., *Journal of Chemical Physics*, 119, 1777, 2003.

Takahashi, T., Takimoto, J.I., and Koyama, K., *Polymer Composites*, 20(3), 357–366, 1999.

Takahashi, T., Watanabe, J., Minagawa, K., and Koyama, K., *Polymer*, 35(26), 5722–5728, 1994.

Tanaka, K., Kyama, K., and Watanaba, J., *Se-I Gakkaishi*, 50, 41–46, 1994.

Tang, C.Y., Yue, T.M., Chen, D.Z., and Tsui, C.P., *Materials Letters*, 61, 4618–4621, 2007.

Utracki, L.A., *Polymer Alloys and Blends: Thermodynamics and Rheology*, Hanser, New York, 1989.

Utracki, L.A. and Sammut, P., *Polymer and Engineering Science*, 30(17), 1019–1026, 1990.

Valenza, A., La Mantia, F.P., and Acierno, D., *Journal of Rheology*, 30(6), 1085–1092, 1986.

Wang, Z., Xie, G., Wang, X.., Li G., and Zhang, Z., *Materials Letters*, 60, 1035–1038, 2006.

Wu, D., Zhou. C., and Zhang, M., *Journal of Polymer Science: Part B: Polymer Physics*, 45, 229–238, 2007.

Xu, J., Chatterjee, S., Koelling, K.W., Wang, Y., and Bechtel, S.E., *Rheologica Acta*, 44(6), 537–562, 2005.

Yang, H., Li, B., Wang, K.,. Sun, T.E, Wang, X., Zhang, Q., Fu, Q., Dong, X., and Han, C.C., *European Polymer Journal*, 44, 113–123, 2008.

Zhang, Q., Yang, H., and Fu, Q., *Polymer*, 45, 1913–1922, 2004.

Zhang, X., Yang, M., Zhao, Y., Zhang, S., Dong, X., Liu, X., Wang, D., and Xu, D., *Journal of Applied Polymer Science*, 92(1), 552–558, 2004.

Zhao, Y. and Huang, H.-X., *Polymer Testing*, 27, 129–134, 2008.

Zhong, Y. and Wang, S., *Journal of Rheology*, 47, 483, 2003.

Ziabicki, A., *Fundamentals of Fiber Formation: The Science of Fiber Spinning and Drawing*, John Wiley & Sons, London, 1976.

References

1. Doyle, P., Shaqfeh, E., and Gast, A., *Physical Review Letters*, 78, 1182, 1997.
2. Bar-Chaput, S. and Carrot, C., *Rheologica Acta*, 45, 339, 2006.
3. Krishnamoorti, R. and Yurekli, K., *Current Opinion in Colloid & Interface Science*, 6, 464, 2001.
4. LeBaron, P., Wang, Z., and Pinnavaia, T., *Applied Clay Sci.ence*, 15, 11, 1999.
5. Ahmad, S., Bohidar, H.B., Ahmad, S., and Agnihotry, S.A., *Polymer*, 47, 3583, 2006.
6. Du, F., Scogna, R., Zhou, W., Brand, S., Fischer, J., and Winey, K., *Macromolecules*, 37, 9048, 2004.
7. Bhattacharya, S.N, Gupta, R.K., and Kamal, M.R, *Polymeric Nanocomposites-Theory and Practice*, Carl Hanser Verlag, Munich, 2007, p. 383.
8. Huang, W. and Han, C.D., *Polymer*, 47, 4400, 2006.
9. Krishnamoorti, R., Ren, J., and Silva, A., *Journal of Chemical Physics*, 114, 4968, 2001.

10. Ray, S., *Journal of Industrial and Engineering Chemistry*, 12, 811, 2006.
11. Gates, T., Odegard, G., Frankland, S., and Clancy, T., *Composites Science and Technology*, 65, 2416, 2005.
12. Buxton, G. and Balazs, A., *Journal of Chemical Physics*, 117, 7649, 2002.
13. Ray, S., *Journal of Industrial and Engineering Chemistry*, 12, 811, 2006.
14. Wang, Z., Xie, G., Wang, X., Li, G., and Zhang, Z., *Materials Letters*, 60, 1035, 2006.
15. Prasad, R., Pasanovic-Zujo, V., Gupta, R., Cser, F., and Bhattacharya, S., *Polymer Engineering and Science*, 44, 1220, 2004.
16. Riva, A., Zanetti, M., Braglia, M., Camino, G., and Falqui, L., *Polymer Degradation and Stability*, 77, 299, 2002.
17. Giannelis, E., Krishnamoorti, R., and Manias, E., *Advances in Polymer Science*, 138, 107, 1998.
18. Pötschke, P., Fornes, T., and Paul, D., *Polymer*, 43, 3247, 2002.
19. Knauert, S., Douglas, J., and Starr, F., *Journal of Polymer Science: Part B: Polymer Physics*, 45, 1882, 2007.
20. Krishnamoorti, R. and Yurekli, K., *Current Opinions in Colloid & Interface Science*, 6, 464, 2001.
21. Zhong, Y. and Wang, S., *Journal of Rheology*, 47, 483, 2003.
22. Wu, D., Zhou, C., and Zhang, M., *Journal of Polymer Science: Part B: Polymer Physics*, 45, 229, 2007.
23. Oksman, K., Mathew, A., Bondeson, D., and Kvien, I., *Composites Science and Technology*, 66, 2776, 2006.
24. Wang, Z., Xie, G., Wang, X., Li, G., and Zhang, Z., *Materials Letters*, 60, 1035, 2006.
25. Fornes, T.D., Yoon, P.J., Keskkulaand Paul, H.D.R., *Polymer*, 42, 9929, 2001.
26. Yang, H., Li, B., Wang, K., Sun, T.E, Wang, X., Zhang, Q., Fu, Q., Dong, X., and Han, C.C., *European Polymer Journal*, 44, 113, 2008.
27. Zhao, Y. and Huang, H.-X., *Polymer Testing*, 27, 129, 2008.
28. Krishnamoorti, R. and Giannelis, E., *Macromolecules*, 30, 4097, 1997.
29. Krishnamoorti, R., Ren, J. and Silva, A., *The Journal of Chemical Physics*, 114, 4968, 2001.
30. Galgali, G., Ramesh, C., and Lele, A., *Macromolecules*, 34, 852, 2001.
31. Lim, Y.T. and Park, O.O., *Macromolecules Rapid Communications*, 21(5), 231, 2000.
32. Lim Y.T and Park O.O., *Rheologica Acta*, 40(3), 220, 2001.
33. Lim, Y.T. and Park O.O., *Macromolecular Rapid Communications*, 21(5), 231, 2000.
34. Lim Y.T. and Park O.O., *Rheologica Acta*, 40(3), 220, 2001.
35. Lobe V.M. and White J.L., *Polymer Engineering and Science*, 19(9), 619, 1979.
36. Ferry J.D., *Viscoelastic Properties of Polymers*, John Wiley & Sons, New York, 1980.
37. Liu, H. and Brinson, L., *Composites Science and Technology*, 68 (6), 1502, 2008.
38. Tang, C.Y., Yue, T.M., Chen, D.Z., and Tsui, C.P., *Materials Letters*, 61, 4618, 2007.
39. Picu, C. and Sarvestani, A., *American Physical Society, APS March Meeting*, March 13–17, 2006, Abstract # R30003.
40. Cassagnau P., *Polymer*, 49, 2183, 2008.
41. Zhang, X., Yang, M., Zhao, Y., Zhang, S., Dong, X., Liu, X., Wang, D., and Xu, D, *Journal of Applied Polymer Science*, 92(1), 552, 2004.
42. Lim, S., Lee, C., Choi, H., and Jhon, M., *Journal of Polymer Science, Part B, Polymer Physics*, 41, 2052, 2003.
43. Kim, Y. and Kim, J., *Journal of Industrial and Engineering Chemistry*, 13, 1029, 2007.
44. Hiljanen-Vainio, M., Heino, M., and Seppala, J., *Polymer*, 39, 865, 1998.
45. Leblanc, J., *Journal of Applied Polymer Science*, 101, 3638, 2006.
46. Jiang,. H. and Kamdem, D., *Journal of Vinyl and Additive Technology*, 10, 59, 2004.
47. Kovacevic, V., Lucic, S., Hace, D., and Glasnovic, A., *Polymer Engineering and Science*, 36, 1135, 1996.
48. Kim, Y. and Kim, J., *Journal of Industrial and Engineering Chemistry*, 13, 1029, 2007.
49. Knauert, S., Douglas, J., and Starr, F., *Journal of Polymer Science: Part B: Polymer Physics* , 45, 1882, 2007.
50. Bertrand, E., Bibette, J., and Schmitt, V., *Physical Review E*, 66, 60401, 2002.

51. Utracki, L.A., *Polymer Alloys and Blends: Thermodynamics and Rheology,* Hanser, New York, 1989.
52. Valenza, A., La Mantia, F.P., and Acierno, D., *Journal of Rheology*, 30(6), 1085, 1986.
53. Micic, P., Bhattacharya, S.N., and Field, G., *International Polymer Processing*, 12(2), 110, 1997.
54. Utracki, L.A. and Sammut, P., *Polymer Engineering and Science*, 30(17), 1019, 1990.
55. Takahashi, T., Watanabe, J., Minagawa, K., and Koyama, K, *Polymer*, 35(26), 5722, 1994.
56. Hattori, T., Takigawa, T., and Masuda, T., *Journal of the Society of Rheology of Japan*, 20, 141, 1992.
57. Tanaka, K., Kyama, K., and Watanaba, J., *Se-I Gakkaishi*, 50, 41, 1994.
58. Lobe, V.M. and White, J.L, *Polymer Engineering and Science*, 19(9), 619, 1979.
59. Kobayashi, M., Takahashi, T., Takimoto, J., and Koyama, K., *Polymer*, 36(20), 3927, 1995.
60. Kobayashi, M., Takahashi, T., Takimoto, J., and Koyama, K., *Polymer*, 37(16), 3745, 1996.
61. Takahashi, T., Takimoto, J.I., and Koyama, K., *Polymer Composites*, 20(3), 357, 1999.
62. Kotsilkova, R., *Rheology–Structure Relationship of Polymer/Layered Silicate Hybrids, Mechanics of Time Dependent Materials,* Kluwer Academic, New York, 6, 2002, p. 283–300.
63. Dealy, J.M. and Wissburn, K.F., *Melt Rheology and Its Roles in Plastics Processing,* Chapman and Hall, London, 1990.
64. Petrie, C.J.S., *Elongational Flows: Aspects of the Behaviour of Model Elasticoviscous Fluids,* Pitman, London., 1979.
65. Ziabicki, A., *Fundamentals of Fiber Formation: The Science of Fiber Spinning and Drawing,* John Wiley & Sons, London, 1976.
66. Pasanovic-Zujo, V., Gupta, R.K., and Bhattacharya, S.N., *International Polymer Processing*, 4, 388–394, 2004.
67. Okamoto, M., Nam, P.H., Maiti, P., Kotaka, T., Hasegawa, N., and Usuki, A., *Nano Letters*, 1(6), 295, 2001.
68. Kotsilkova, R., *Rheology–Structure Relationship of Polymer/Layered Silicate Hybrids, Mechanics of Time Dependent Materials*, Kluwer Academic, New York, 6, 2002, p. 283–300.
69. Giza, E., Ito, H., Kikutani, T., and Okui, N., *J. Macromolecular Science -Physics B*, 39(4), 545, 2000.
70. Pavlikova, S., Thomann, R., Reichert, R., Mulhaupt, R., Marcinin, A., and Borsig, E., *Journal of Applied Polymer Science*, 89(3), 604, 2003.
71. Okamoto, M., Nam, P.H., Maiti, P., Kotaka, T., Nakayama, T., Takada, M., Ohshima, M., Usuki, A., Hasegawa, N., and Okamoto, H., *Nano Letters*, 1(9), 503, 2001.
72. Ray, S., *Journal of Industrial and Engineering Chemisty*, 12, 811, 2006.
73. Pasanovic-Zujo, V., Gupta R.K., and Bhattacharya, S.N., *International Polymer Processing*, 4, 388–394, 2004.
74. Gupta, R.K., Pasanovic-Zujo, V. and Bhattacharya, S.N., *Journal of Non-Newtonian Fluid Mechanics*, 128(2-3), 116, 2005.
75. Xu, J., Chatterjee, S., Koelling, K.W., Wang, Y., and Bechtel, S.E., *Rheologica. Acta,,* 44 (6), 537, 2005.
76. Boek, E., Coveney, P., Lekkerkerker, H., and van der Schoot, P., *Physical Review E*, 55, 3124, 1997.
77. Breedveld, V., *Journal of Chemical Physics*, 114, 5923, 2001.
78. Chen, L., Zukoski, C., Ackerson, B., Hanley, H., Straty, G., Barker, J., and Glinka, C., *Physical Review Letters*, 69, 688, 1992.
79. Fuchs, M. and Cates, M., *Physical Review Letters*, 89, 248304, 2002.
80. Marshall, L. and Zukoski IV, C., *Journal of Physical Chemistry*, 94, 1164, 1990.
81. Gates, T., Odegard, G., Frankland, S., and Clancy, T., *Composites Science and Technology*, 65, 2416, 2005.
82. Mewis, J., Frith, W., Strivens, T., and Russel, W., *AIChE Journal*, 35, 415, 1989.
83. Zhao, Y. and Huang, H.-X., *Polymer Testing*, 27, 129, 2008.
84. Starr, F., *The Journal of Chemical Physics*, 119, 1777, 2003.
85. Huang, W. and Han, C.D., *Polymer*, 47, 4400, 2006.

86. Zhao, Y. and Huang, H.-X., *Polymer Testing*, 27, 129, 2008.
87. Tang, C.Y., Yue, T.M., Chen, D.Z., and Tsui, C.P., *Materials Letters*, 61, 4618, 2007.
88. Qiu, J. and Feng, H., *Transactions of Nonferrous Metals Society of China*, 16, 444, 2006.
89. Maiti, A., Wescott, J., and Goldbeck-Wood, G., *International Journal of Nanotechnology*, 2, 198, 2005.
90. Picu, C. and Sarvestani, A., *American Physical Society, APS March Meeting*, March 13–17, 2006, Abstract # R30003.
91. Zhong, Y. and Wang, S., *Journal of Rheology*, 47, 483, 2003.
92. Reichert, P., Hoffmann, B., Bock, T., Thomann, R., Muelhaupt, R., and Friedrich, C., *Macromolecular Rapid Communications*, 22, 519, 2001.
93. Huang, W. and Han, C.D., *Polymer*, 47, 4400, 2006.
94. Gelfer, M. et al., *Macromolecules*, 38, 3765, 2005.
95. Fu, X. and Qutubuddin, S., *Materials Letters*, 42, 12, 2000.
96. Gu, S., Ren, J., and Wang, Q., *Journal of Applied Polymer Science*, 91, 2427, 2004.
97. Picu, C. and Sarvestani, A., *American Physical Society, APS March Meeting*, March 13–17, 2006, Abstract # R30003.
98. Lee, M., Hu, X., Li, L., Yue, C., Tam, K., and Cheong, L., *Composites Science and Technology*, 63, 1921, 2003.
99. Sarvestani, A. and Picu, C., *Rheologica Acta*, 45, 132, 2005.
100. Pasanovic-Zujo,.V., Gupta, R.K., and Bhattacharya, S.N., *International Polymer Processing*, 4, 388, 2004.
101. Pasanovic-Zujo,.V., Gupta, R.K., and Bhattacharya, S.N., *International Polymer Processing*, 4, 388, 2004.
102. Bhattacharya, S.N., Gupta, R.K., and Kamal, M.R., *Polymeric Nanocomposites—Theory and Practice*, Carl Hanser Verlag, Munich, 2007, p. 383.

9

Fundamentals of Carbon-Based Nanocomposites

Enrique V. Barrera, Erica Corral, Meisha Shofner, and Daneesh Simien

CONTENTS

9.1 Introduction

Fascination with carbon-based nanotubes is leading to a range of advanced nanocomposites, but not without the developed understanding of how to manipulate the nanotubes for property development. A better understanding of carbon nanotube based nanocomposites can be derived from a matrix of understanding when studying nanotube insertion into a variety of matrices. The general methods to advance carbon nanotube based nanocomposites involves consideration of nanotube starting conditions, dispersion, interaction with the matrices, and alignment. Generally, where advancement has been achieved, nanotube functionalization has been a key contributor either to advance dispersion or to provide fuller linkage to the matrix. Interesting cases have emerged where advanced properties have been seen for entangled nanotube and polymers, as in the case of elastomers, and for forests of nanotubes, as in the case of macroscale fiber systems with well-grown nanotubes on the fibers. In this chapter, the role of nanotubes in nanocomposites will be considered as to property development. While most groups have considered low-concentration systems, there are still opportunities for more research and advanced properties from high-concentration systems. The fuller use of carbon nanotubes may well rest with more self-assembly and synergistic behavior between nanotubes and other nanospecies.

9.2 Key Methods for Developing Composite Materials When Using Nanotube Nanotechnology

The idea of having high-performance fibers, such as a variety of the recently studied nanofibers, with very high strength that are ultra-lightweight and can be manipulated into a variety of matrices for composite development, is enough to intrigue the composite materials enthusiast to investigate opportunities for new advanced nanotechnology-based materials. This has been the idea generated by the discovery of carbon nanotubes and the various types of nanofibers to emerge onto the engineering scene since 1991. In 1991, Iijima followed up the discovery of fullerenes (1985) with the discovery of nanotubes.[1] These nanotubes were multi-walled carbon nanotubes (MWCNTs). Soon thereafter, single-walled carbon nanotubes (SWCNTs) were discovered and a door was opened to a range of nanofibers being discovered, further developed, and refined. In many cases, these nanofibers were used for composite materials developments. These carbon nanofibers — a listing given as vapor-grown carbon fibers (VGCFs), single-walled nanotubes (SWNTs), multi-walled nanotubes (MWNTs), few-walled nanotubes (FWNTs), and double-walled nanotubes (DWNTs) — make up the general varieties of nanofibers for composite materials use. The reader is guided to such works as that by Ajayan, Barrera, Wagner, and Thostenson for a historical preview of nanocomposite development.[2–5] The opportunities expected to come from nanocomposites are cited in works by Files, Meyyapayan, and Yowell.[6–8] Furthermore, the basic aspects of nanocomposites and how nanofibers can advance their properties have been described in works by Viya, Roy, and Maruyama.[9–11] Numerous other references in this field will be given throughout this chapter that will add to a formal background in nanotechnologically based composite materials for those who intend to move into the field. As background for this chapter, we will specifically cite the works by Tour, Park, Winey, Dai, Zhu, and Richard Liang as important foundational articles that will better prepare the reader for the topics enclosed in this current work.[12–17]

For polymer nanocomposite development, it has been felt that there are at least five basic approaches:

Method 1: Nanotube dispersion into resins by various mixing means.

Method 2: Low-concentration approaches to impact shear loading and, in particular, z-axis property enhancement in laminate systems.

Method 3: New resin formulation for hybrid polymer formation.

Method 4: Polymer approaches such as forming interpenetrating network polymers (INPs).

Method 5: Fuller integration that yields paradigm changes in the ways composites are made.

In Method 1, issues with mixing viscosity and porosity have been key factors. In Method 2, laminate composite systems have seen enhancements in short beam shear and double cantilever beam testing.[18, 19] In the case of Method 3, new curing agents have been produced as hybrid systems.[20] For Method 4, INPs have been initiated by Kim et al. that take advantage of tangled nanotube systems.[21] For Method 5, nanotube dispersion in polymers where deformation processing leads to strengthening much like for various alloys and thermoplastics is yet to be achieved although stress-induced reactions have been achieved. Although there have been advances in each of these approaches, the first composite applications will likely come in electrical property advancements of various types. However, it is expected that these opportunities will lead to fuller use as multifunctional nanocomposites get designed.

In this chapter, the role of nanotubes in nanocomposites is investigated, along with the aspects of their use that have resulted in advanced properties for several composite systems.

9.3 Matrix of Understanding

Much of what has been studied in the literature can be compiled into a so-called "Matrix of Understanding." In the chart shown in Figure 9.1, material systems are identified along with key issues and key approaches. The Matrix of Understanding serves as a guide for connecting various aspects of nanocomposite systems to each other. For example, shown in yellow are outcome systems such as nanotubes in fluids, thermoset nanocomposites, thermoplastic nanocomposites, and fully integrated nanotube composites. These systems constitute the four corners of the matrix and are linked to each other in that each of the three lead to fully integrated nanotube composites as processing is further refined. The dark blue squares include

- Mimicking the polymeric properties of nanotubes
- Mimicking the properties of nanotubes to higher scales
- Creating hybrid polymers
- Providing for viscosity control indicates conceptual and methodological approaches to achieving advanced properties in nanocomposites

The concept of mimicking the polymeric properties of nanotubes has been adopted by numerous research groups; some researchers are creating new pathways for mimicking

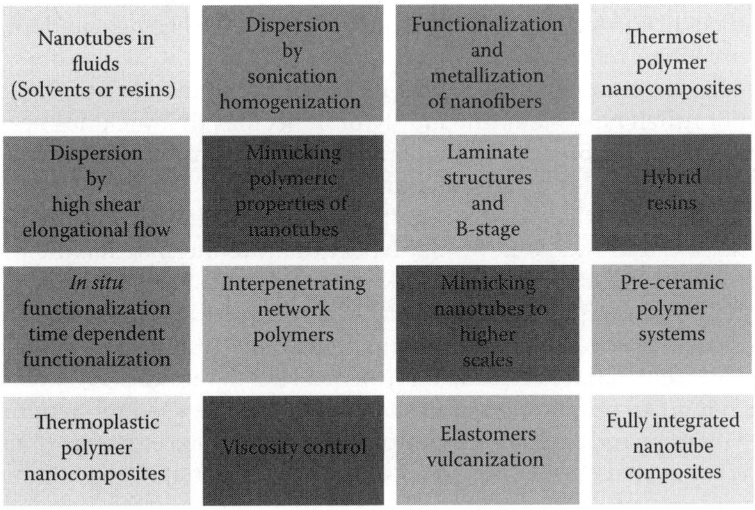

Nanotubes in fluids (Solvents or resins)	Dispersion by sonication homogenization	Functionalization and metallization of nanofibers	Thermoset polymer nanocomposites
Dispersion by high shear elongational flow	Mimicking polymeric properties of nanotubes	Laminate structures and B-stage	Hybrid resins
In situ functionalization time dependent functionalization	Interpenetrating network polymers	Mimicking nanotubes to higher scales	Pre-ceramic polymer systems
Thermoplastic polymer nanocomposites	Viscosity control	Elastomers vulcanization	Fully integrated nanotube composites

FIGURE 9.1
A matrix of understanding for polymer nanocomposites. This matrix serves as a method for growing a fuller understanding of polymer nanocomposites by investigating across varying materials and properties. Along the body diagonals of the matrix are conceptual ideas that foster the general approach to nanocomposite processing. The corners show outcome systems from the various approaches.

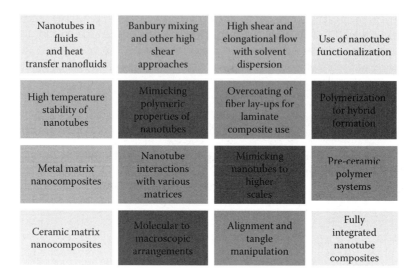

FIGURE 9.2
A matrix of understanding for nanocomposites for nanotubes in a variety of matrices, including polymers, metals, and ceramics. The likeness of nanomaterials across differing matrices is surprising and provides an opportunity for learning new approaches to developing a range of nanomaterials. The opportunity for fully integrated nanotube composites is fostered with each development in the different matrix materials.

the structure of nanotubes to higher scales. It is interesting that most approaches have focused on low-concentration nanotube composites, but it is to be expected because of costs and availability issues. However, greater use will come from these developments as nanotube-polymer interactions and nanotube-nanotube interactions in composites are better understood and implemented. Ideas for higher nanofiber concentration composites are being studied, in particular for mechanical property enhancement and for high electrical conductivity wires and cables.

Just one matrix of understanding is shown in this chapter. However, matrices, such as a function of fullerene system (the family of molecules that includes nanotubes), can be developed to foster a broader understanding when thinking across vastly dissimilar materials and properties. One such matrix is shown in Figure 9.2, but the reader is encouraged to produce other matrices on various levels that would further their own understanding. In Figure 9.2, one would expect that the thermal stability of nanotubes is of key importance. Furthermore, it is likely that processing nanotubes in a polymer might help identify how nanotubes should be processed in a ceramic. This has been the case where shear processing was used in a metal melt, similar to that used for polymers.[22] It is also likely that if nanotubes can be inserted in ceramics for advanced properties, then the nanotube preparation used there could also be used for nanotube insertion in metals.[23, 24] As more researchers are comparing various nano-reinforcements, matrices of understanding such as these could lead to a newer understanding and more effective composite development.

Although the focus of this chapter is how to enable advanced properties in nanocomposites made using nanotubes, it seems important that the range of properties for these composites be identified, in particular for epoxy based composites and for electrical property enhancement. Table 9.1 and Table 9.2 show a listing of properties for various nanocomposites using several nanotube materials. These tables are by no means complete but they do show

TABLE 9.1

Summary of Mechanical Properties of Structural Composites Reinforced with Nanofibers that Show at Least One Property with 30% or More Improvement

Matrix	Fibers	Nanofiber	Wt.%	Property	Control	Value for Composite	% Increase	Ref.
Epoxy		CVD-MWNT	4	Tensile Modulus	0.118 GPa	0.465 GPa	294	C
				Tensile Strength	5 MPa	10 MPa	100	
Epoxy		CVD-MWNT	1	Tensile Modulus	1.2 GPa	2.4 GPa	100	D
				Tensile Strength	30 MPA	41 MPa	37	
Epoxy		MWNT	0.25	Tensile Modulus	1.4 MPa	2.1 MPa	50	E
Epoxy		CVD-MWNT	6	Tensile Modulus	2.75 MPa	4.13 MPa	50	F
Epoxy		NH2-R-DWNT	0.5	Tensile Modulus	2.6 GPa	3.0 GPa	15	G
				Tensile Strength	63.8 MPa	69.13 MPa	8	
				Fracture Toughness	0.65 MPa/m2	0.93 MPa/m2	50	
Epoxy		SWNT-R-NH2	1	Tensile Modulus	2.0 GPa	2.7 GPa	31	H
				Tensile Strength	83 MPa	104 MPa	25	
Epoxy		SWNT-R-NH2	4	Tensile Modulus	2.0 GPa	3.4 GPa	68	H
				Tensile Strength	83 MPa	102 MPa	23	
Epoxy		F-SWNT	1	Tensile Modulus	2.1 GPa	2.6 GPa	30	I
				Tensile Strength	83.2 MPa	95.0 MPa	14	
Epoxy		f-MWNT	1	Tensile Modulus	2.2 GPa	2.3 GPa	4	J
				Tensile Strength	64.67 MPa	64.05 GPa	−1	
				Impact Strength	100 kJ/m2	152 kJ/m2	52	
Epoxy		VGCF	4	Flexual Modulus	3.3 GPa	6.3 GPa	89	K
		VGCF	4	Flexual Modulus	3.3 GPa	4.8 GPa	44	K
		VGCF	4	Flexual Modulus	3.3 GPa	7.9 GPa	137	K
Vinyl Ester	Glass	SWNT	0.015	Inter-laminar Shear Strength	27.2 GPa	39.2 GPa	44	L
Epoxy	SiC	CVD-MWNT	2	Inter-laminar Fracture Toughness	0.95 kJ/m2	4.26 kJ/m2	348	N
				Inter-laminar Shear Sliding Fracture Toughness	91 J/m2	140 J/m2	54	
				Flexural Modulus	23.1 GPa	24.3 GPa	5	
				Flexural Strength	62.1 MPa	150.1 MPa	140	
				Flexural Toughness	5.8 N.mm	30.4 N.mm	424	

the general state of properties in the field. It is clear that properties continue to be advanced via nanotube dispersion, functionalization, and with new ideas for implementation.

Increases as high as 400% have been seen for mechanical behavior, drops in resistivity have occurred over 14 to 15 orders of magnitude, and up to 40 dB of shielding effectiveness has been achieved with nanofiber additions. While multifunctionality is an expected outcome for nanocomposites, no system possesses multifunctional advanced properties to date, although this is expected to occur in the near term.

TABLE 9.2

Percolation Threshold and Resistivity Range of Nanocomposites

Composite	Percolation Threshold (wt%)	Resistivity (Ω/cm)	Ref.
Al/PVC	35	10^5	1
Cu/PE	15	10^{16}	
VGCF/PP	10	10^8	2
VGCF/PP	10	10^1	3
CB/(EVA/HDPE)	4.2	10^5	4
MWNT/PC	2	10^3	5
SWNT/ABS	2.5	10^5	6
SWNT/PA	3	10^6	7
SWNT/PEMA	3	10^6	8
MWNT/Polyimide	7–10	10^4	9
Oxidized MWNT/Epoxy	0.017–0.77 vol%	10^2–10^4	10
MWNT/Epoxy	0.5	10^1	11
MWNT/Epoxy	0.0025	10^4–10^5	12
SWNT/Epoxy	0.1	10^4–10^5	13
MWNT/Epoxy	0.1	10^4–10^5	
DWNT/Epoxy	0.25	10^4–10^5	
DWNT-NH2	0.1–0.3	10^8–10^{10}	
DWNT-NH2	0.3–0.5	10^8–10^{10}	
SWNT/HDPE	4	10^4	14
MWNT/Polyimide	0.15	10^3	15
MWNT/PE	7	10^3	16

9.4   Menu for Processing Nanocomposites

In an early article about nanocomposites, a menu of processing approaches was given and is updated here in Table 9.4.[25] Numerous approaches are at hand for developing nanotube-based composites, and these various approaches have led to advanced composite properties. Generally, combinations of these steps are used based on the starting condition of the nanotubes and the properties requested. For example, to process an epoxy composite with nanotubes for short-beam shear strengthening, the following steps could be used:

1. Start with as-received nanotubes in fluff form that get purified and functionalized for epoxy reaction. More recent approaches include decanting the starting material to remove large agglomerates.

TABLE 9.3

Polymer composites for Electromagnetic Interference Shielding

Composite	Filler Concentration (wt%)	Shielding Effectiveness (dB)	Ref.
MWNT/PMMA	40	27	17
	20	18	
MWNT/PS	7	26	18
VGCF/LCP	15	41	19
MWNT/PUU	5 pph	19	20
CNT/PDMS-PUU	5 pph	26	21
Carbon Fiber/Nylon 6,6	40	67.7	22

TABLE 9.4

A menu of choices for processing nanocomposites based on nanotubes

Starting Nanotubes	Dispersion	Interface	Alignment
Dry as-received	Spreading	Polymer selection	High shear mixing
In solvent as-received	Bench mixing	Tangle reduction	Extrusion
Purified	Homogenization	Dispersion	Elongational flow
Functionalized	Sonication	Alignment	Fiber spinning
Substrate grown	High Shear	Purification	Magnetic fields
Dry aggregate	Stretch drying	Unwrapping	Electric fields
Pearls	Ball milling	Polymerization	Stretch drying
Power	Polymerization	*In situ*	Dip/spin coating
In water	*In situ*	Stabilizating for temperature	Gel/wet spinning
Wrapped	Incipient wetting	Functionalization	Solid freeform fabrication
Masterbatched	Extrusion	Metallization	Tape casting
Bucky tubes	Functionalized	Synergism	Substrate growth
Separated	Defunctionalized	B-stage	Self-assembly
Fluff	Wrapping	Sized	Aligned growth
Bucky paper	Separated	Curing agent advancement	Robocasting
Cut	B-stage	Hybridization	Fused deposition molding
Prepreg	Spraying		Extrusion
Sized	Prepreg		Substrate growth
Metallization	Metallization		
Decanted	Viscosity control		

2. The functionalized nanotubes are entered into a solvent in a well-dispersed condition achieved by sonication. Decanting can be used with each step to foster higher dispersion.

3. The nanotubes are then sprayed onto carbon fibers to produce uniform coverage on carbon fiber plies.

4. Next, these plies are stacked (face-to-face) in the mid-plane.

5. Finally, the processing of the composite occurs by vacuum-assisted resin transfer molding.

Porosity is eliminated as seen in C-scanned samples. Large-scale composites can be processed this way in a variety of thicknesses because the low-concentration nanotubes covered on the carbon fibers do not affect the viscosity of the resin. Alignment steps are not used here but self-assembly through functionalization or through substrate growth could be employed. These are avenues for achieving alignment that are currently being explored.

Future processing methods will combine several of these steps in a single task. This manner of processing will further enable nanocomposite manufacturing on a larger scale. Not included in Table 9.4 are the various approaches to forming composites that are used to produce nanocomposites. There are a range of methods, including vacuum-assisted resin molding transfer, resin transfer molding, vacuum bagging, infiltration methods, etc. For the most part, few are limited by using nanofibers. The only approaches that have some limitation are those where infiltration is required because the viscosity of the resin often increases when nanofibers are added directly into the resin.

Many of the composites seen in Tables 9.1, 9.2, and 9.3 are implemented with the various processing steps seen in Table 9.4. It is important to note that most of the composites employing SWCNTs are using them in the roped form, although some functionalizations lead to reduced rope sizes and may well "unrope" the nanotubes. The initial starting condition of the nanotubes is by far the most important step. It can either lead to property advancement, or it can limit the use of the nanotube in the composite. New investigators should consider this step with great care because composite properties are so dependent on this early condition.

9.5   Nanotubes in the Resin

By far, the easiest processing route is the mixing of nanofibers directly into the resin. In this way, the nanofiber is the sole reinforcement, and a number of opportunities are at hand for taking advantage of their advanced properties. Dry mixing of nanofibers into polymers is not optimal because of handling issues — specifically, the potential for airborne material. This approach does not necessarily fully mix the nanofibers so property advancement is not always seen. The use of solvents and surfactants has been one approach where the nanofibers are dispersed in a fluid to mix with the polymer resin. The potential for retained solvent or surfactant in this approach can cause property-damaging effects unless chemical matching to the system is accomplished. In addition to retained solvents, viscosity tends to increase, and this has limited the ability to go to high concentrations, even greater than a few weight percent. Perhaps one of the best opportunities from using this approach has been seen for elastomers.[26] Small concentrations of purified and functionalized SWNTs have been dispersed in elastomers, showing significant property improvement. Figure 9.3 and Figure 9.4 show the impact that a small amount of

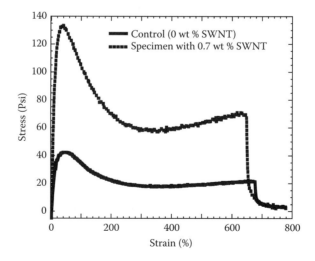

FIGURE 9.3
Enhanced tensile test data for a siloxane rubber with only 0.7 wt% SWNTs added. Note the retained elongation to failure, which does not typically occur when materials are strengthened with fillers.

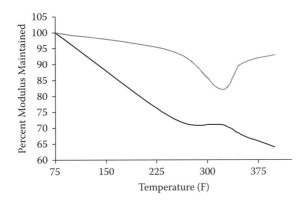

FIGURE 9.4
Rubber (E2M) is seen to hold its strength to higher temperatures while the standard NBR loses its strength as a function of increasing temperature.

nanotubes makes on rubber (siloxane and Standard NBR) when dispersed through a mixing process. Increased strength, retained elongation to failure, and retained strength at high temperature are observed. Retained strength at high temperature has been seen for other polymer systems as a result of nanofiber addition.[27, 28] For elastomers, it is likely that entangled nanotubes are an advantage because the long chains of the elastomer are also highly entangled. Furthermore, the functionalization serves to further cross-link with the elastomer, much like the vulcanized system itself.

9.6 Low-Concentration Nanotube Systems

The percolation threshold for several conducting polymers has been seen to be less than 1% when nanotubes are used.[29] The range varies based on dispersion and segregation level but, ultimately, conducting polymers can be processed from low-concentration nanotube systems.[30] For nanocomposites in general, the electrical properties are expected to be the first to market. For mechanical enhancement, a great deal of focus has been on low-concentration nanotube systems because cost issues are a major concern. One approach that has shown near-term promise is the use of nanotubes for z-axis mechanical property enhancement. For composite panels, nanotubes can be placed in the mid-plane between the plies to improve resistance against shear failure. While many composites are seen to have high strength in the in-plane directions, the out-of-plane strength can be poor because regions between plies do not contain reinforcing fiber. One idea to improve this is to locate nanotubes in the regions between fiber plies so that (i) the matrix region is enhanced or (ii) the resin-fiber interface is enhanced. Recent research by Zhu et al. showed a 45% increase in short-beam shear strength where only 0.1wt% SWNTs were used.[31] Nanotubes, in this case, were sprayed on the fiber plies and provided fiber-matrix interface enhancement. When multiwalled nanotubes were grown on the plies, the double cantilever beam (DCB) strength increased over 250%.[32] Figure 9.5 shows a forest of nanotubes grown on plies that led to high DCB results. When DCB tests were taken

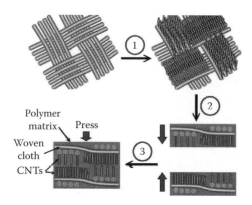

FIGURE 9.5
Nanotubes are grown as forests on fiber plies. The use of nanotubes in well-placed locations leads to enhancements of the mechanical properties. The fracture toughness improved by 254% with nanotubes added to the mid-plane section. (From Ajayan, P.M., Stephan, O., Colliex, C., and Trauth, D., Aligned carbon nanotube arrays formed by cutting a polymer resin—nanotube composite. *Science,* 265 (5176), 1212–1214, 2004.)

of nanotubes sprayed at concentrations of 0.1 wt% in an epoxy/carbon fiber composite, enhancements were at a 40 to 50% increase as well. The next steps in this research area call for manipulating nanotubes to produce well-advanced DCB results where low concentrations of nanotubes are used so that these composites can be moved to applications in the near-term.

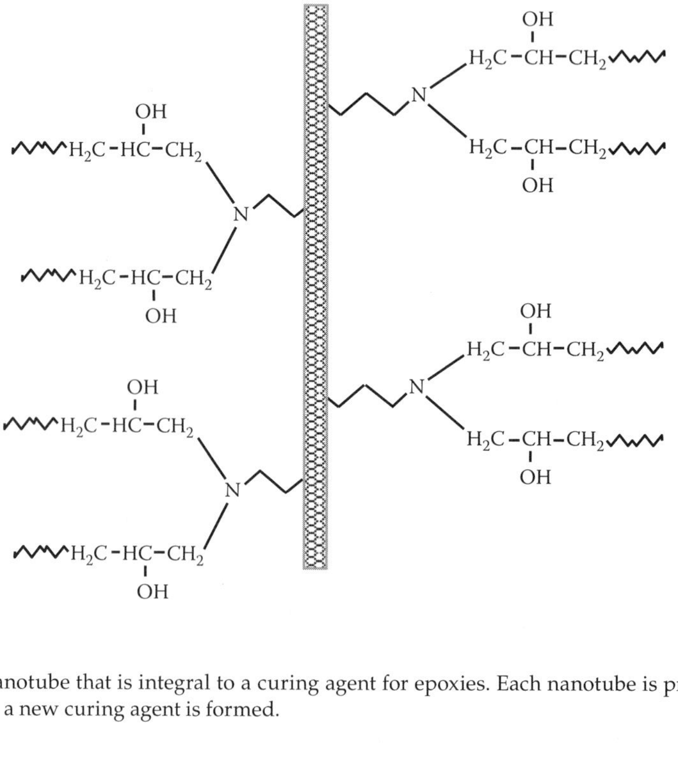

FIGURE 9.6
A schematic of a nanotube that is integral to a curing agent for epoxies. Each nanotube is processed with this formulation so that a new curing agent is formed.

FIGURE 9.7
Side-walled cross-linked nanotubes integrated into the polymer.

9.7 New Polymer Formulation

Because nanotubes are highly polymeric and are molecular structures, it is expected that new polymer formulations will be a direct outcome of nanocomposite development.[33] The idea of new formulations as hybrids involves integrating the nanotubes into the polymer. One aspect of this would be to have the nanocomposite where the nanotube is an integral part of each polymer chain.[34] Figure 9.6 shows a nanotube with functionalization attached that is typical of the curing agent. In this application, the nanotube is part of the formulation. There are numerous approaches where nanotubes can be part of the polymeric formulation, but the length of the polymer linkage may limit the ability of each nanotube to be integrated. The chains may react with themselves or may react with the nanotube, thereby limiting the development of the new formulation and its use in the polymer. Figure 9.7 shows nanotubes side-walled cross-linked within a polymer in a mode that could be more fully integrated and where the system is a hybrid polymer. Integration relies on covalent bonding to provide mechanical strength. The composite should be designed to eliminate secondary bonding such that network bonding, as seen in thermosets, would prevail.

9.8 Interpenetrating Network Polymers

One promising approach to developing nanocomposites is the use of entangled nanotubes. Generally, most forms of nanofibers used for composites tend to start out agglomerated and entangled. Much of the research on nanocomposites has focused on untangling and dispersing these nanofiber systems. Instead, there are opportunities for leaving the nanotubes in the more natural tangled form. Figure 9.8 shows the progression of agglomerated nanotubes to a condition in which the nanotubes are mixed with the polymer and a "chain-mill" condition prevails. In this form, the two polymers (because nanotubes are polymeric) are interlocked with each other in a form that provides strength secondary to the interlocking. Researchers are using this approach more and more, as seen by infiltration approaches with Buckypaper and other forms of nanotubes.[35, 36]

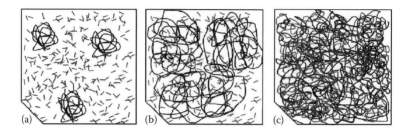

FIGURE 9.8
(a) Agglomerates of nanotubes in a polymer resin; (b) resin infiltrated into the agglomerates before polymerization; and (c) the agglomerates linked and intermingled with the polymer to form an interpenetrating network polymer. (From Kim, J.D., Barrera, E.V., and Armeniades, C.D., Incorporation of Carbon Nanotubes in Epoxy Composites, *35th SAMPE*, Dayton, OH, October 1, 2003.)

9.9   Fully Integrated Nanotube Composites

Fully integrated nanocomposites as shown in Figure 9.8 involve many of the steps shown above and are expected to take advantage of the fuller properties of nanotubes. It is likely that these nanocomposites require higher nanotube loading and more process development to manipulate higher concentration samples. In some forms of these nanocomposites, the use of fiber spinning may be necessary to achieve fuller integration. In this way, the dispersion, matrix interaction, nanotube-to-nanotube interaction, and fuller integration can be optimized. Recent research has moved nanocomposites closer to being fully integrated by having their functional change with increasing shear processing.[37] This idea of strengthening through deformation is similar behavior to that seen for metals when they are strain-hardened.

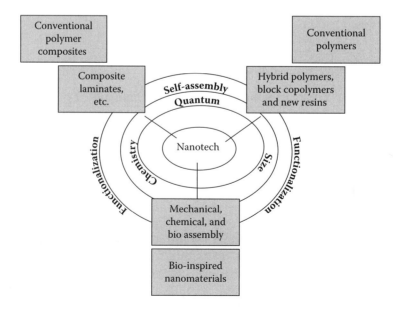

FIGURE 9.9
A design diagram for the development of polymer-based nanomaterials.

9.10 Nanotubes as Platforms for Synergistic Behavior of Materials

In the past 12 years, the study of nanotubes in materials has evolved with several approaches being considered. Initially, like many composites enthusiasts, the focus was on (1) nanotubes for reinforcements. Soon thereafter, the focus was on developing (2) fully integrated nanotube composites to take advantage of the fullest properties of the nanotubes. Since then, (3) the time-dependent nature of nanotubes, (4) the synergistic behavior of nanotubes with other nanoconstituents, and (5) the use of nanotubes as platforms or templates for altering materials have become more recent approaches that have enabled new ideas and research directions. With each of these approaches, a better understanding of the use of nanotubes for composites has been achieved.

As reinforcements, the initial potential for nanotube advancement seemed high. The fiber tensile strength and stiffness were the highest compared to all known fibers; the aspect ratios found were in a useful range; and their light weight suggested that the best reinforcement had finally been found. Still, their high-defect nature for multiwalled nanotubes, the roping of the single-walled nanotubes, and the agglomeration of both decreased the opportunity for new advanced composites to form. Early "benchtop" mixing showed that the viscosity of the polymer mixture increased rapidly when increased concentrations of nanotubes were added. This viscosity increase led to poor mixing conditions. Moreover, the nanotubes tended to stay agglomerated no matter what shear processing condition was used, although some studies showed some promise. Approaches, such as incipient wetting that created a dispersion of nanotubes on or in a polymer powder before mixing, tended to advance dispersion. Other approaches, such as better purification and the use of functionalization, also aided in better dispersion and improved properties. While advances have been made, it is clear that going to larger-sized samples still has processing issues unless higher purification levels and dispersion are achieved, or unless the producer just accepts that the agglomerates will always remain and therefore learns to process with this in mind. Certainly this has been the case for nanotube conducting polymers where "carbon black" routes have been used. In these cases, segregated nanotubes that are interconnected at a network level have yielded electrostatic dissipative (ESD) materials and electromagnetic interference (EMI) materials.

The next idea of fuller integration meant looking past the rule of mixture approaches and basic composite fiber/matrix management toward integrating the nanotubes into the polymer to develop hybrid systems at the nanometer and molecular scale.

9.11 What Are We Missing?

When Tables 9.1 and 9.2 are studied, it is expected that one is asking where the full promise of nanotubes for composites is, because increases are observed yet not at extraordinary levels. It therefore becomes important to ask, "What are we missing in the implementation of nanotubes into composites that has kept us from the higher levels of properties that we would like to expect?" For mechanical property enhancement, a number of articles have given us important insight (see, e.g., references by Barrera, Wagner, Gates, Frankland, and others). These articles either indicate that orders of magnitude advancement should be seen, or that these advances are not seen because incomplete bonding occurs at the nanotube interfaces. The idea of what is complete bonding between the nanotube and the resin is still not well understood. Various levels of functionalization have been produced, and bonding between one in every twenty carbon atoms (1:20) is not uncommon. The question of whether this is enough to constitute complete bonding still remains. When one considers micron-sized reinforcements found in typical composite systems, these reinforcements, although often "sized" with a coating, are not bonded by every available surface atom but differ in that more bonds are present because of their larger size. Although nanotubes have a high surface-to-volume ratio, micron-sized fibers have more surface area compared to nanofibers. Still, the idea is that neither is completely bonded, and therefore some degree of bonding should constitute complete bonding for a rule of mixture condition and this must fully be assessed.[37]

It may not just be the limited bonding condition that prevents the expected advancements. Instead, it may be that many of the nanofibers are not long enough. However, this does not account for enhancements seen when cut nanotubes are used [AF person at].[38] Still, everyone expects that longer nanotubes will lead to better properties if they can be processed into the polymer in typical composite conditions — that is, dispersed and aligned. Yet, the problem may not be length because the aspect ratios of nanotubes tend to be in an acceptable range; that is, hundreds to thousands. Therefore, what could be the cause for not seeing several orders of magnitude advancement? One simplification of a possible solution has to do with load transfer to the matrix. Basic composite loading conditions consider the fiber strength (high for nanotubes), the shear loading (improved when functionalization is used), the misalignment and normal loading (also improved by functionalization), but not often the load as it falls off into the matrix region. The basic understanding of composites suggests that the reinforcements do all the load carrying and the resin is a glue that holds them together. What is typically left out of this understanding is that as the fibers get more and more separated, they are less capable of carrying load and transferring load to the other fibers.

One very interesting aspect of composites is that the load in the matrix between the fibers is more difficult to transfer as the fibers get further and further apart. In fact, it is likely that as the distance increases to dimensions comparable to the fiber diameter, the matrix can no longer carry the load and the matrix fails. This means that for micron-size reinforcements, the distances between fibers should be microns or less, and the distances between nanofibers should be nanometers or less. When nanotubes are agglomerated, the distances between them are larger than nanometer sizes and weak regions of the composite are created. Reinforcement theory also suggests that as the nanofibers get too close, a weakening effect will also occur, as seen when nanotubes are bundled and well entangled. Further investigation of this phenomenon is needed; but if this is the case, high-concentration nanotube composites may well achieve the advanced properties that are expected.

9.12 A Nanocomposite Design Diagram

Figure 9.9 was produced to identify approaches and differences between conventional systems and nanomaterials. The three arms of the diagram illustrate three differing approaches. Other approaches can be added to the diagram when comparing to other conventional materials. One arm notes that conventional composites differ from nano-composites not just by quantum behavior, size, and chemistry effects, but also by the use of self-assembly and functionalization. The more each of these aspects is used, the more the material is considered nanoscale material. For example, a polymer with dispersed nanotubes is still a conventional composite; it just happens to have nanoscale reinforcements added. As functionalization, self-assembly, etc. are brought in, the material becomes more nano-inspired. Another arm shows conventional polymers that in their own right are considered nanomaterials because the molecular chains and structures depend on nanoscale behavior. Still, these are not nanomaterials until integration of the nanospecies involves functionalization, self-assembly, quantum effects, other nanoscopic aspects, and chemistry. Hybrids are therefore the connecting link between traditional polymers and true nanomaterials.

The third arm of the diagram identifies a very different approach and materials system. In this arm, biomolecular materials are not necessarily nanomaterials unless they, too, involve self-assembly, chemistry, etc. The ability to create new materials by bio/nano-combined approaches offers new ways to obtain materials that might not easily be formed by conventional means. In this way, the self-assembly is very important, and in this way the various aspects of fully integrated nanotube composites become critical steps.

9.13 Summary

The purpose of this chapter was to illicit new ideas for continued development of nano-tube- and nanofiber-reinforced nanocomposites. While enhancements have been produced, even higher outcomes are demanded, and thus, the missing steps necessary to achieve exceptional properties are still not available. A range of methods exists for producing nanocomposites and the starting nanotube condition, degree of dispersion, and the matrix interaction are some of the key aspects that must be optimized for property development. It is evident that the reader can learn from vastly differing nanomaterials and that new ideas for developing nanocomposites can easily come from this cross-fertilization.

References

1. Iijima, I., Helical microtubules of graphitic carbon, *Nature*, 354, 56, 1991.
2. Ajayan,, P.M, Stephan, O., Colliex, C., and Trauth, D., Aligned carbon nanotube arrays formed by cutting a polymer resin — nanotube composite, *Science*, 265, 1212, 1994.
3. Barrera, E.V., Key methods for developing single-wall nanotube composites, *J. Minerals, Metals Mater. Soc.*, 52, 38, 2000.
4. Lourie, O. and Wagner, H.D., Evidence of stress transfer and formation of fracture clusters in carbon nanotube-based composites, *Composites Sci. Technol.*, 59, 975, 1999.
5. Thostenson, E.T., Zhifeng, R., and Chou, T.W., Advances in the science and technology of carbon nanotubes and their composites: a review, *Composites Sci. Technol.*, 61, 1899, 2001.
6. Files, B.S. and Mayeaux, B.M., Carbon nanotubes, *Adv. Mater. Proc. (USA)*, 156, 47, 1999.
7. Meyyappan, M., Ed., *Carbon Nanotubes: Science and Applications*, CRC Press, Boca Raton, FL, 2004.

8. Yowell, L., Mayeaux, B., Files, B., and Sullivan, E., Nanotube composites and applications to human spaceflight, IAF abstracts, *34th COSPAR Scientific Assembly, The Second World Space Congress,* 10–19 October, 2002, Houston, TX, p. IAA-12-1-01IAF abstracts.

9. Vieira, R., Pham-Huu, C., Keller, N., and Ledoux, M.J., New carbon nanofiber/graphite felt composite for use as a catalyst support for hydrazine catalytic decomposition, *Chem. Commun.,* 954, 2002.

10. Wei, C., Dai, L., Roy, A., and Benson-Tolle, T., Multifunctional chemical vapor sensors of aligned carbon nanotube and polymer composites, *J. Am. Chem. Soc.,* 128, 1412, 2006.

11. Maruyama, B. and Alam, K., Carbon nanotubes and nanofibers in composite materials, *SAMPE J. (USA).* 38, 59, 2002.

12. Dyke, C.A and Tour, J.M. Functionalized carbon nanotubes in composites, in *Carbon Nanotubes. Properties and Applications,* O'Connell, M.J., Ed., CRC Press, New York, 275, 2006.

13. Park, C. et al., Dispersion of single wall carbon nanotubes by *in situ* polymerization under sonication, *Chem. Phys. Lett.,* 364, 303, 2002.

14. Haggenmueller, R. et al., Aligned single-wall carbon nanotubes in composites by melt processing methods, *Chem. Phys. Lett.,* 330, 19, 2000.

15. Wang, J., Dai, J., and Yarlagadda, T., Carbon nanotube-conducting-polymer composite nanowires, *Langmuir,* 21, 9, 2005.

16. Zhu, J.H. et al., Reinforcing epoxy polymer composites through covalent integration of functionalized nanotubes, *Adv. Functional Mater.,* 14, 643, 2004.

17. Liang, Z. et al., Investigation of molecular interactions between (10, 10) single-walled nanotube and Epon 862 resin/DETDA curing agent molecules, *Mater. Sci. Eng., A,* 365, 228, 2004.

18. Zhu, J.H. et al., Reinforcing epoxy polymer composites through covalent integration of functionalized nanotubes, *Adv. Functional Mater.,* 14, 643, 2004.

18. Ajayan, P., Schadler, L., Giannaris, C., and Rubio, A., Single-walled carbon nanotube±polymer composites: strength and weakness, *Adv. Mater.,* 12, 750, 2000.

19. Zhu, J.H. et al., Reinforcing epoxy polymer composites through covalent integration of functionalized nanotubes, *Adv. Functional Mater.,* 14, 643, 2004.

20. Kim, J.D., Barrera, E.V., and Armeniades, C.D., Incorporation of carbon nanotubes in epoxy composites, *35th SAMPE,* Dayton, Ohio (Oct. 1, 2003).

21. Sitharaman, B. et al., Injectable *in situ* cross-linkable nanocomposites of biodegradable polymers and carbon nanostructures for bone tissue engineering, *J. Biomater. Sci., Polymer Ed.,* 18, 2007.

22. Johannes, L. et al., Survivability of single-walled carbon nanotubes during friction stir processing, *Nanotechnology,* 17, 3081, 2006.

23. Zeng, O. et al., Coating of SWNTs with nickel by electroless plating method, *Mater. Sci. Forum,* 475–479, 1013, 2005.

24. Barrera, E., Key methods for developing single-wall nanotube composites, *JOM: J. Minerals, Metals and Mater. Soc.,* 52, 2000.

25. Tour, T., Bahr, J., and Yang, J., U.S. Patent No. 7,304,103, December 4, 2007.

26. Kozano, K. and Barerra, E.V., Nanofiber-reinforced thermoplastic composites. I. Thermoanalytical and mechanical analyses, *J. Appl. Polymer Sci.,* 79, 125.

27. Zhu, J., et al., Improving the dispersion and integration of single-walled carbon nanotubes in epoxy composites through functionalization, *Nano Lett.,* 3, 1107, 2003.

28. Park, C. et al., Dispersion of single wall carbon nanotubes by *in situ* polymerization under sonication, *Chemical Physics Letters,* 364, 303, 2002.

29. Barrera, E.V., Key methods for developing single-wall nanotube composites, *J. Minerals, Metals Mater. Soc.,* 52, 38, 2000.

30. Zhu, J. et al., Processing a glass fiber reinforced vinyl ester composite with nanotube enhancement of interlaminar shear strength, *Composites Sci. Technol.,* 67, 1509, 2007.

31. Zhu, J. et al., Processing a glass fiber reinforced vinyl ester composite with nanotube enhancement of interlaminar shear strength, *Composites Sci. Technol.,* 67, 1509, 2007.

32. Antonucci, V., Hsiao, K.T., and Advani, S.G., Review of polymer composites with carbon nanotubes, in *Advanced Polymeric Materials: Structure Property Relationships*, Shonaike, G.O., and Advani, S.G., Eds., CRC Press, Boca Raton, FL, 2003.
33. Barrera, E.V., Key methods for developing single-wall nanotube composites, *J. Minerals, Metals Mater. Soc.*, 52, 38, 2000.
34. Liang, Z. et al., Kramer process investigation of carbon nanotube buckypaper/epoxy nanocomposites, *44th AIAA/ASME/ASCE/AHS Structures, Structural Dynamics, and Materials Conference*, 7–10 April, 2003, Norfolk, VA.
35. Xue, Y. et al., Effect of Nanotube Functionalization on Electrical Properties of SWNT Buckypaper Materials, *SAMPE*, October 29–November 1, 2007.
36. McIntosh, D., Khabashesku, V.N., and Barrera, E.V., Nanocomposite fiber systems processed from fluorinated single-walled carbon nanotubes and a polypropylene matrix, *Chem. Mater.*, 18, 4561, 2006.
37. Gates T.S., Odegard G.M., Frankland S.J.V. and Clancy T.C., Computational materials: Multi-scale modeling and simulation of nanostructured materials, *Composites Science and Technology*, 65, 15–16, 2416–2434, 2005.
38. Ziegler, K.J., Gu Z., Shaver J., Chen, J. Flor E.L., Schmidt, D.J., Chan, C, Hauge R.H., and Smalley R.S., "Cutting single-walled carbon nanotubes," Nanotechnology 16) S539–S544, (2005).

10

Polymer Nanocomposites Containing Vapor-Grown Carbon Fibers Aligned by Magnetic or Electric Field Processing

Tatsuhiro Takahashi and Koichiro Yonetake

CONTENTS

10.1 Introduction

Alignment of carbon nanotubes (CNTs) in polymer is of great importance in order to exploit their anisotropic electrical, thermal, mechanical, and optical properties in various industrial applications. Specifically, the alignment of CNTs in polymer by magnetic or electric field processing has received considerable attention, due to the capability of alignment toward various directions, a capability that conventional polymer processing cannot provide. Recently, review articles have been published about CNT/polymer nanocomposites but they cited few articles concerning magnetic or electric processing.[1-3]

Many studies on magnetic or electric field processing of CNTs in solution or polymer matrixes have been published and are divided into the following areas:

1. Material property of CNTs for alignment, such as diamagnetic anisotropic susceptibility[4-11]
2. Alignment of CNTs in solution by magnetic or electric field processing[12-26]
3. Alignment of CNTs in polymer by magnetic or electric field processing[27-42]
4. Orientation of matrix polymer by aligned CNTs and patterning of aligned CNTs[43-46]

The major focus of this chapter concentrates on (1) and (3) above. The effect of magnetic or electric field processing on alignment feature in polymers has not been well understood due to their complexities, that is, always different CNTs in different matrix by different processing. Therefore, the aim of this chapter is to discuss systematically the effect of field types—either magnetic electric DC (direct current) or AC (alternating current)—on fundamental rotational speed, structural development, and volume electric resistivity of composites. To clarify the effects, mainly the same carbon nanofiber, that is, vapor-grown carbon fiber, was used.[47] This nanofiber is graphitized, commercially available from Showa Denko K.K. Japan with an average diameter of 150 nm, and is easy to perform *in situ* observation, large diamagnetic anisotropic susceptibility from graphitization. This chapter also deals with remaining challenges and future directions of magnetic or electric field processing in a polymer matrix.

10.2 Materials and Experimental Equipment

10.2.1 Materials

Commercially available vapor-grown carbon fibers (VGCFs) (trade name of Showa Denko K.K. Japan[®], average diameter = 150 nm) were used as received.[48] The obtained VGCFs were already graphitized at a temperature up to 2800°C, resulting in highly ordered graphite layers and almost no residual Fe catalysts. The diffraction pattern (002) by wide-angle x-ray diffraction (WAXD) analysis suggested 3.39 Å for a d-spacing of (002). The VGCFs, as received, had an aggregated lump structure, the lumps having a diameter of 20 to 50 μm, as shown in Figures 10.1a.[49, 50] By our method, VGCFs were dispersed completely on glass for length analysis by decomposition (at 500°C, only for polycarbonate) of the thin layer, which was prepared by spin-coating a VGCF/polycarbonate solution blend (a well-dispersed suspension) in tetrahydrofuran (THF) (Figure 10.1b).[51, 52] The numbered average length and length distribution of tested VGCFs were evaluated as 6.7 μm and from 1 to 23 μm as measured with the help of computer analysis, respectively (Figure 10.1c).[53, 54] The tested VGCFs were slightly bent or curved, and included VGCFs with partially branched and completely linear structures. Regarding polymer matrixes, polydimethylsiloxanes (Shinetsu Silicone Co. Ltd., Japan, viscosities of 0.097 and 0.97 Pa-s [Pascal-seconds] (25°C)), usually called silicone oil, UV-curable epoxy resin (Arakawa Chemical Industry Co. Ltd. Japan, AQ-9 viscosity of 1.5 Pa-s (25°C)) and low-density polyethylene (LDPE, Tosoh Co. Ltd., Japan, Petrocene 354, MFR = 200, melt viscosity at 150°C = 170 Pa-s) were used. In the case of other materials used for discussion, details were described in figure captions and texts.

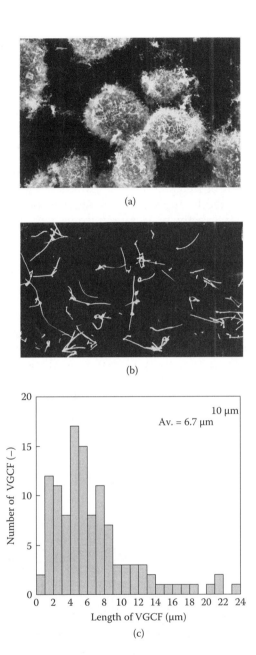

FIGURE 10.1
(a) SEM image of VGCFs as received; (b) SEM image of VGCFs dispersed on glass for length analysis of VGCF®s as received (see preparation method in text); and (c) length distribution of VGCF®s as received. The numbered averaged length was 6.7 µm.

10.2.2 Preparation of Dispersion

A mechanical stirrer was used to mix silicone oil or UV-curable epoxy resin with VGCFs (0.1 to 1 wt%), showing that each VGCF was independently dispersed. The dispersions have excellent stability without precipitation or re-aggregation on the basis of an *in situ* optical microscope observation conducted for several hours. Blends of VGCFs with LDPE were prepared with a batch-type melt mixer (Toyoseiki Seisakusyo Co. Ltd., Japan, Laboplastomil). The electric volume resistivity of composite films was evaluated with a commercially available electric volume resistivity measurement device (Loresta, Hiresta, Diainstrument Co. Ltd., Japan).

10.2.3 *In Situ* Optical Microscope Observation from Two Directions

In situ structural observation was performed along the two directions under a DC electric or magnetic field with our custom-built apparatus.[55] Figures 10.2a and b illustrate the experimental setups used to apply a DC or AC field perpendicular and parallel to the observation direction, respectively. Metal electrodes were used in the setup shown in Figure 10.2a, while transparent indium-tin-oxide (ITO) electrodes were used in the setup shown in Figure 10.2b. In the present study, the DC electric field was fixed at 18 V and supplied by typical DC voltage equipment (PR18-1.2 V, Kenwood Co. Ltd., Japan). The AC electric field was at 20 V (peak to peak, sinusoidal) in the frequency range from 0.1 to 1 MHz with typical AC voltage apparatus (NF Co. Ltd., Japan, multifunction synthesizer). The gap between the electrodes was 125 μm for both setups (Figures 10.2a and b). The gap was carefully adjusted, the polyimide film had a thickness of 125 μm, and 144 V/mm was applied to the dispersion at the initial stage (DC case). The tested concentration was below 1 wt% and the volume electric resistivity (Ω/cm) of the dispersion was similar to that of silicone oil, around 10^{15}. *In situ* observation was performed with two optical microscopes: (1) BX-50, Olympus Co. Ltd., Japan, with a magnification of up to 400X), and (2) Hi-scope HK-2700, Hirox USA, with a magnification of up to 2000X.

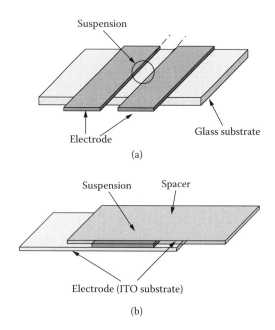

(a)

(b)

FIGURE 10.2
Custom-built apparatus for observing structure of VGCF®s in polymer matrix with optical microscope under electric field (DC or AC). Observation direction is perpendicular (a) and parallel (b) to the electric field direction, respectively. The electrode gap is 125 μm for both setups.

10.3 Theory of Fiber Alignment and Rotation under Magnetic or Electric Field

10.3.1 Alignment under Magnetic Field

Magnetic alignment of diamagnetic fiber was caused by diamagnetic anisotropic susceptibility ($\chi_a = \chi_{//} - \chi_{\perp}$) and was described in several articles.[36-39, 60-63] The alignment of carbon nanotubes (CNTs) along the direction of magnetic field has been attempted utilizing the strong anisotropic diamagnetic property. Magnetically aligned pure CNT thin films were produced using CNT aqueous suspension under 7 to 25T of magnetic field.[64-68] Magnetic alignment of CNTs in polymer has been also carried out, aiming at the enhancement of electric and thermal conductivity in composites.[69-79] Garmestani et al. and Choi et al. prepared CNT (3 wt%)/epoxy composite under 25T, and Kimura et al. made CNTs (1, 2, and 5 wt%)/unsaturated polyester under 10T.[80-82] However, there has been no estimation about necessary alignment time, which should depend on carbon structure, matrix viscosity, and magnetic strength.

A key parameter for magnetic alignment of diamagnetic fiber in polymer is diamagnetic anisotropic susceptibility ($\chi_a = \chi_{//} - \chi_{\perp}$). Kimura et al. proposed a technique called the "suspension method" to determine the diamagnetic anisotropic susceptibility of conventional fibers.[83] The suspension method is more advantageous than other methods because χ_a can be simply determined by observing the rotation of a single fiber through *in situ* observation.[84] Thus, the suspension method was used here for analysis.[85] Figure 10.3a illustrates that only rotation of VGCF occurred under uniform magnetic field.[86]

10.3.2 Alignment under Electric Field

Many reports were published about the alignment of CNTs in solution by DC or AC electric field processing.[87-97] In addition, electric field processing has also been applied to make connected structures by CNTs between comb-like electrodes having several micron distances by dropping a trace amount of CNT suspensions and evaporating solvents.

Figure 10.3b-1,2,3 shows a schematic illustrating the three key forces—that is, torque, Coulombic, and electrophoresis forces—that act on each VGCF due to the electric field.[98] In the presence of the electric field, each VGCF experiences polarization. This polarization leads to a torque force (Figure 10.3b-1) acting on each VGCF. Coulombic attraction (Figure 10.3b-2) was generated among oppositely charged ends of different VGCFs. The electrophoresis force (Figure 10.3b-3) was induced by the presence of charged surfaces. Depending on CNTs and matrices, electrophoresis was sometimes observed. Thus, the torque force, Coulombic attraction, and electrophoresis governed the final formation of the aligned ramified network, that is, the density uniformity between electrodes and the degree of linearity in connected networked structures. The cause of the formation of a ramified network structure, rather than a straight network structure, is that throughout the network structure, more than two fibers having opposite charges became connected to one end of another fiber. Between magnetic and DC electric processing, one can compare orientation relaxation time (i.e., the reciprocal value of rotational speed).

10.3.3 Rotational Analysis under Magnetic or DC Electric Field

By magnifying a part of the photo, including single VGCFs in suspension, suitable VGCFs were selected manually and the angle between fiber axial and magnetic directions was measured by computer analysis. VGCFs having a branched structure were discarded

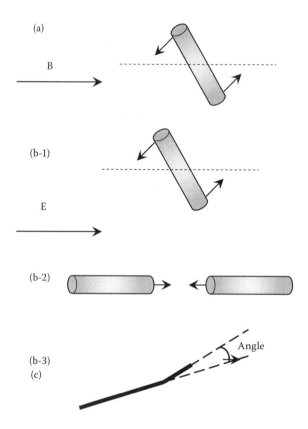

FIGURE 10.3
Schematic illustration of forces to which a VGCF® is subjected to under (a) a magnetic field (rotation force) and (b) an electric field [(b-1) rotation torque, (b-2) Coulomb force, and (b-3) electrophoresis, from top to bottom]. (c) A schematic picture of typical VGCF® structure. VGCF®s having bend angle less than 30° were used for rotational analysis.

manually, because a linear structure was required for analysis idealistically. However, most VGCFs are slightly bent. Therefore, VGCFs having bend angles of less than 30° were utilized for analysis (Figure 10.3c).[99]

In situ structural observation perpendicular to the magnetic or DC electric field direction was carried out (Figures 10.4a and b). A pair of permanent magnets was used to apply a magnetic field (0.23T) (Figure 10.4a) and a DC electric field (113 V/mm) was also applied (Figure 10.4b).

This suspension method is derived from a theoretical equation about the balance of magnetic torque and hydrodynamic torque. The detailed theory of the suspension method was described previously.[100–103] When it was applied to linear fibers, the following equation derived:

$$\tan \theta = \tan \theta_0 \exp\left(-\frac{t}{\tau}\right) \tag{10.1}$$

where θ is the angle between the direction of the magnetic field and the fiber axis at time t, θ_0 is that angle at the initial time, $1/\tau$ is the alignment rate (1/s), and $\underline{\tau}$(s) is the alignment relaxation time described as follows:

$$\tau^{-1} = (F(D)/3\eta)G \tag{10.2}$$

$$G = G_m = \chi_a 1 3^2 / 2\mu_0 \text{ under a magnetic field.} \tag{10.3}$$

(a) Before and after magnetic processing

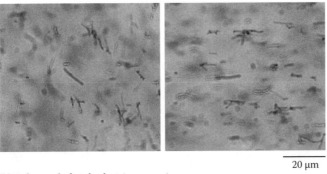

20 μm

(b) Before and after dc electric processing

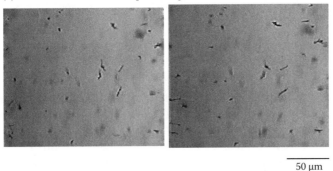

50 μm

(c) Rotational analysis under magnetic (left) and dc electric (right) field

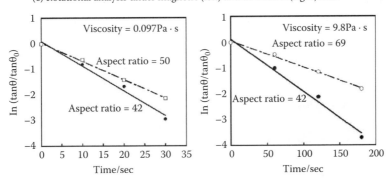

FIGURE 10.4
(a) Optical microscope pictures of VGCF®s (0.01 wt%) dispersed in silicone oil (0.097 Pa-s) before and after (20 s) magnetic processing (0.23T horizontal direction); (b) those in silicone oil (9.8 Pa-s) before and after (60 s) DC electric processing (113 V/mm horizontal direction); and (c) a typical logarithmic plot of the temporal change of the angle θ between the fiber axis and the magnetic (θ_0 is the initial angle) field for VGCF® as a function of time. The slope is equal to $1/\tau$ under magnetic (left) and DC electric (right) field.

$$G = G_E = \varepsilon_0 \varepsilon_a E^2 / 2 \quad \text{under an electric field.} \qquad (10.4)$$

where:

$\mu_0 (= 4\pi \times 10^{-7} \text{ Wb}/(\text{Am}))$ is magnetic permeability of vacuum
$\varepsilon_0 (8.854 \times 10^{-12} \text{C}^2/(\text{N} \cdot \text{m}))$ is electric permeability of vacuum
η (Pa-s) is matrix viscosity

B (Wb/m^2 (=T)) is magnetic flux density

E (V/m) is electric flux density

G (Pa, G_m under magnetic and G_E under electric, respectively) is rotational torque

χ_a (–) is anisotropic diamagnetic susceptibility

ε_a (–) is anisotropic dielectric susceptibility

F(D) (–) is shape factor including aspect ratio of D

The volume of VGCF fiber, which receives rotational torque from magnetic or DC electric field in the present study, is substantially larger than the thermal movement energy of the matrix. It should be noted that τ is not affected by the initial angle.

Then, $\ln(\tan \theta / \tan \theta_0)$ was plotted as a function of t (s) from Equation 10.1, and from the slope $(-1/\tau)$, τ was obtained for magnetic and DC electric fields (Figure 10.4c). The aspect ratio D was calculated using the average diameter of VGCF (150 nm) because it was impossible to measure the fiber diameter, even by optical microscope. The case of error in this method was described in a former report.[104]

10.3.4 Anisotropic Properties of VGCF® from the Suspension Method

Figure 10.5 gives the alignment relaxation time τ as a function of the aspect ratio D. The aspect ratio was calculated based on the measured length with a high-magnification optical microscope under the assumption of an averaged diameter (150 nm). It was possible to measure each VGCF length but not each diameter. With larger aspect ratios, alignment relaxation times became longer. The solid curve in Figure 10.5 is drawn using Equations 10.2 and 10.3 for fitting averaged experimental values using anisotropic magnetic susceptibility χ_a (–) as 3.11×10^{-4}. This value ($\chi_a = 3.11 \times 10^{-4}$) was almost the same as or slightly higher than pitch-based carbon fiber (χ_a as 2.81×10^{-4}) measured in a similar way previously. This is a key material parameter of VGCF under magnetic processing in viscous liquid. When assuming matrix viscosity, magnetic field, and aspect ratio, it is easy to estimate the alignment relaxation time τ, reversely.

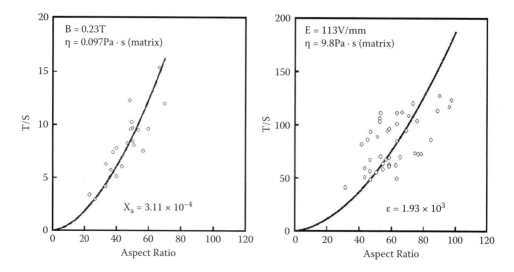

FIGURE 10.5

τ as a function of aspect ratio for VGCF® under magnetic (left) and DC electric (right) field, respectively. Solid curve is drawn by the equations (10.2 and 10.3) for fitting experimental values using χ_a as 3.11×10^{-4} for magnetic and ε_a as 1.93×10^3 for DC electric field, respectively.

In the same way as for DC electric field, ε_a (–) was estimated at 1.93×10^3. This alignment relaxation time is strongly related to the required processing time. It is possible to predict the alignment relaxation time of VGCF under various conditions using Equations 10.2, 10.3, and 10.4. It is noted that the alignment relaxation time τ is widely scattered in an AC electric field using silicone oil or epoxy matrix, and does not depend on the aspect ratio. Therefore, it was impossible to analyze using Equations 10.2 and 10.4. This cause needs to be identified in further research.

10.4 Comparison of VGCF Alignment Rate in Polymer under Magnetic or Electric Field

Figure 10.6a gives the prediction of alignment relaxation time τ as a function of magnetic field and matrix viscosity assuming the averaged length (6 μm, D = 40), and its length distribution of VGCF based on Equations 10.2 and 10.3, which is important to determine the necessary magnetic

(a) Prediction of relaxation orientation time under magnetic field

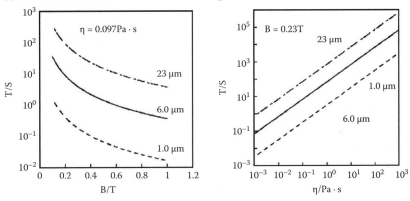

(b) Prediction of relaxation orientation time under dc electric field

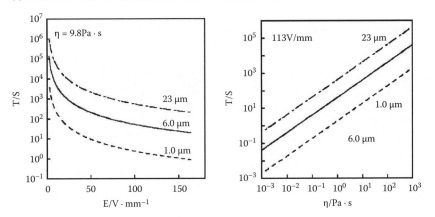

FIGURE 10.6
(a) Prediction of τ using χ_a (3.11×10^{-4}) as a function of magnetic field (in the case of 0.097 Pa-s) and as a function of viscosity (in the case of 0.23T) for VGCF® having 6 μm numbered averaged length and having distribution from 1 to 23 μm. (b) Prediction of τ using ε_a (1.93×10^3) as a function of DC electric field (in the case of 9.8 Pa-s) and as a function of viscosity (in the case of 113 V/mm) for VGCF®.

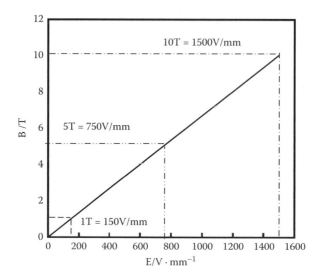

FIGURE 10.7
Comparison of required field strength between magnetic and DC electric fields to obtain the same rotational speed of VGCF®, showing that 1T is equal to 150 V/mm.

processing time. The value of τ decreased with the second order as a function of magnetic field strength, and increased linearly with matrix viscosity. It was confirmed experimentally as well that the value of τ became 10 times larger using 10 times more viscous silicone oil. Figure 10.6b shows the prediction of alignment relaxation time τ as a function of DC electric field and matrix viscosity from Equations 10.2 and 10.4 in the same way.

Generally, it is often asked which field makes multi-walled carbon nanotubes (MWCNTs) align faster. Here, let us compare the alignment rate under magnetic field with that under a DC electric field, focusing only on the rotational mode (Figure 10.3a and b-1). To analyze rotation time accurately (not so fast and not so slow), different silicone oils (0.097 and 0.98 Pa-s) were used in magnetic and DC electric field analyses (see Figures 10.4 and 10.5). However, χ_a and ε_a are not affected by the viscosity of silicone oil. Therefore, it is possible to obtain a comparable field strength to get the same alignment rate in magnetic and DC electric fields by assuming the same matrix viscosity. Figure 10.7 shows a comparison to obtain the same alignment rate, suggesting that 1T is almost comparable to 150 V/mm to rotate VGCF.

10.5 Polymer Nanocomposite Containing VGCFs Aligned by Magnetic Field

Figure 10.8 shows optical micrographs of a VGCF (0.1 wt%)/silicone oil (0.097 Pa-s) suspension under magnetic field (horizontal direction) taken at 0, 20, 40, and 80 s.[105] The VGCF included almost no residual Fe; therefore, the center of gravity of each VGCF hardly moved under the magnetic field. When focusing on shorter VGCFs, it was already aligned along the magnetic direction even after 20 s. After 80 s, all VGCFs were aligned.

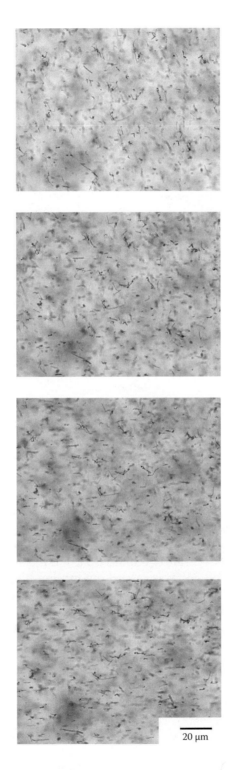

FIGURE 10.8
In situ optical microscope photographs of VGCF®s dispersion in silicone oil (0.097 Pa-s) under 0.23T (horizontal direction) as a function of time: 0, 20, 40, and 80 s from top to bottom.

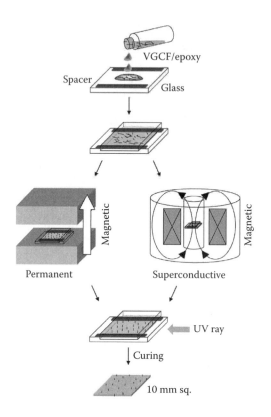

FIGURE 10.9

A schematic procedure about magnetic processing of oriented VGCF®s nanocomposite film (about 10 μm thickness) based on UV-curable epoxy resin using permanent magnets (0.9T) and superconductive magnets (10T).

Alignment of CNTs in polymer after magnetic field processing has been observed.[106–116] However, the improvement in electric conductivity by magnetic processing is only about 1 order of magnitude for 3 wt% CNTs in epoxy (25T) from Choi et al., and for 1 wt% CNTs in unsaturated polyester (10T) from Kimura et al.[117, 118] This slight improvement might be interpreted in terms of the fact that CNTs are not connected to one another by magnetic processing. Therefore, magnetic field processing was carried out for thin films (about 10μm) along the thickness direction, which is almost equivalent to the length of VGCFs (see Figure 10.1).[119–121]

Figure 10.9 shows the experimental procedure of magnetic field processing using VGCFs in UV-curable epoxy resin (1.5 Pa-s) with permanent magnets (1T) and superconductive magnets (10T).[122–124] It was calculated using Equations 10.2 and 10.3 (see Figure 10.6a) that τ for the averaged length of VGCF is about 5 and 0.5 s, respectively. Figure 10.10 shows optical micrographs of cross-sections of VGCFs/epoxy composite film before and after permanent magnet treatment. It was demonstrated that a small amount of VGCFs, longer than 10 μm (see Figure 10.1 for length distribution of VGCFs), went through along the thickness direction even for permanent magnets. The volume electric resistivity along two directions, thickness and surface, was plotted as a function of VGCF before and after processing. Generally, the percolation concentration of VGCFs in polymer is about 10 to 20 wt% for bulk samples. However, only with 0.6 wt% VGCFs by permanent magnet processing exhibited about 5 orders of magnitude improvement in volume electric resistivity, giving an almost transparent appearance with electric conductive property.

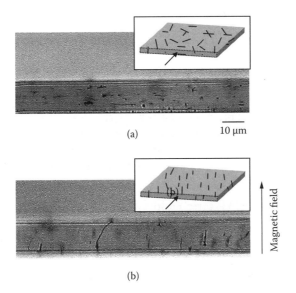

(a)

10 μm

(b)

Magnetic field

FIGURE 10.10
Optical micrographs at the vertical cross-sections of VGCF®s/UV curable epoxy composite films (VGCF®s = 0.05 wt%, thickness = 10 μm): (a) composite film before magnetic treatment, and (b) that treated under magnetic field (1T). Schematic pictures show image of each composite and the arrow is the observation direction.

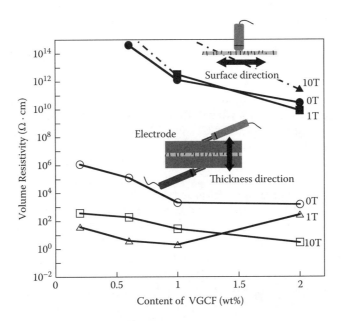

FIGURE 10.11
Volume electric resistivity along surface (●, ▲, ■) and thickness (○, △, □) direction of VGCF®s/UV-curable epoxy composite films, before and after magnetic alignment by using permanent magnets (1T) and superconductive magnet (10T). The measurement limitation of volume electric resistivity was up to 1.0×10^{15} Ω–cm. Dotted lines mean that parts of plots were over the limitation.

10.6 Polymer Nanocomposite Containing VGCFs Aligned by Electric Field

Features of CNT movement in solvents under electric field vary widely, depending on concentration, aggregated structures, electrode distance, shape of electrode, types (DC or AC), evaporation speed of solvent, etc.[125–135] However, this is understandable if one considers three factors (see Figure 10.3b-1,2,3). Polymer containing particles has been investigated as the electrorheological fluid.[136] The formation of a chain-like structure of particles between electrodes by applying electric field is known.[137] Polymers containing carbon black by electric fields have been examined, showing the various networked structure between electrodes.[138, 139]

10.6.1 Processing under DC Electric Field

Martin et al. carried out DC electric processing for MWCNTs/epoxy composite.[140] Figure 10.12 shows an example of epoxy nanocomposites including MWCNTs by DC electric field processing.[141] A fraction of MWCNTs was observed to move toward the anode, with electrophoresis, suggesting the presence of negative surface charges. As soon as these MWCNTs are close enough to the electrode to allow charge transfer, the MWCNTs discharge and absorb onto the anode. The tip of MWCNTs connected to the electrode then become sources of very high field strength and the location for adsorption of further filler particles. As a result, ramified MWCNT network structures extend away from the anode. Prasse also reported electrophoresis of all MWCNTs to anode in an amine curing agent, and it was in good agreement with the former result.

VGCFs are already graphitized in an inert atmosphere and silicone oil has almost no positive or negative charge, compared with an amine curing epoxy system. Figure 10.13 shows a series of structural developments of VGCFs in silicone oil by DC electric field.[142] The VGCFs were rotated and oriented along the direction of the electric field. At the same time, ends of oriented VGCFs, having either a positive or negative charge, became connected with ends of other oriented VGCFs having the opposite charge. It was observed that the dispersed aligned VGCFs were absorbed into an aligned network structure, leading to the formation of an aligned and ramified network (5 min).[143] It is important to note that the concentration of the networks did not vary with distance from either the anode or the cathode. No electrophoresis implies that the VGCFs were not charged or the resistance against their movement was high due to a high matrix viscosity.

The DC electric field processing was applied to VGCFs /low-density polyethylene composite films. Figure 10.14 shows the DC electric processing apparatus and temperature profile to align VGCFs toward the thickness direction in polyethylene film having 125-μm thickness (between ITO electrodes). The volume of electric resistivity of aligned VGCFs/low-density polyethylene composite film was evaluated. Figure 10.15 (top) shows optical micrographs of VGCFs/low-density polyethylene composite film with illustrations. Optical micrographs suggested the orientation of VGCFs even in a thermoplastic viscous polymer.

Figure 10.15 (bottom) gives the resulting volume resistivity of the film along the thickness direction as a function of the VGCF weight fraction. Prior to the application of the DC electric field (18V/125 μm), the volume resistivities of the composites up to 1 wt% of VGCF in low-density polyethylene were beyond the measuring capability (>10^{15} Ω-cm). The higher the concentration, the lower the resistivity, due to the aligned network structure. The randomly dispersed VGCFs in a polymer matrix give a percolation threshold value between 10 and 20 wt%. Starting from 0.1 wt%, the networks form. The saturated resistivity over the percolation threshold concentration was around 10^1. It should be noted that the

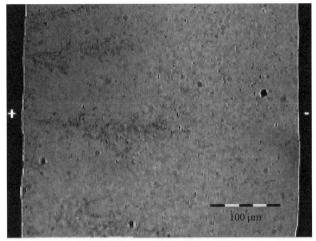

5 min

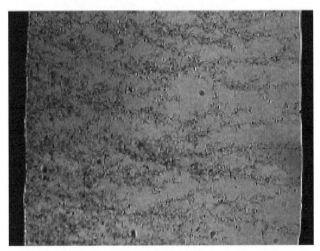

240 min

FIGURE 10.12

Optical micrographs of epoxy composites containing 0.01 wt% multi-walled carbon nanotubes (diameter = 50 nm, length = 43 μm, distribution = 40–46 μm, grown at the University of Cambridge by CVD method) after 5 and 240 min during curing at 80°C in a DC electric field of 100 V/cm. (*Source:* From Martin, C.A., Sandler, J.K.W., Windle, A.H., Schwarz, M.K., Bauhofer, W., Schulte, K., Shaffer, M.S.P., *Polymer,* 2005, 46, 877–886.)

enhancement of electric conductivity does not depend on film thickness under DC electric processing due to connected network structure. However, this was observed only in a thin film under magnetic processing when VGCFs were aligned without connected network.

When DC electric processing was applied to a UV-curable epoxy (1.5 Pa-s) system containing VGCFs, the VGCFs were not aligned even under higher DC voltage. The effect of various matrices having similar viscosities at processing temperature on the aligning property of VGCFs was investigated. Results suggested that DC electric processing was only effective to align and to make network structure in polymer matrices having lower dielectric constants. For example, at 50 Hz:

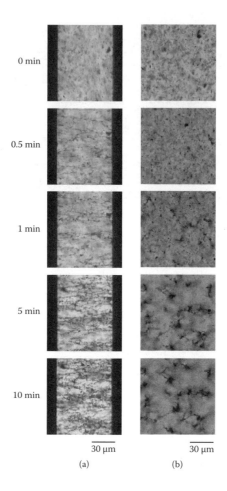

FIGURE 10.13

Structural development of VGCF®s (0.1 wt%) in silicone oil (1000cSt) under DC electric field (18 V) between electrodes (125-μm gap) using dispersed VGCF®s in silicone oil as the initial state as measured with setups (a) and (b) in Figure 10.2. The left side of the figures corresponds to the negative electrode.

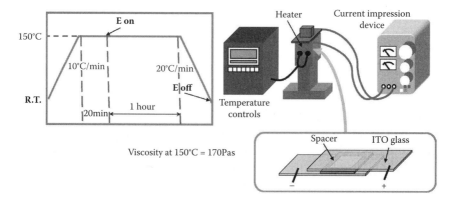

FIGURE 10.14

An illustration for DC electric processing apparatus (18 V, 125 μm between electrodes) and a temperature-control profile for a thermoplastic polymer, low-density polyethylene.

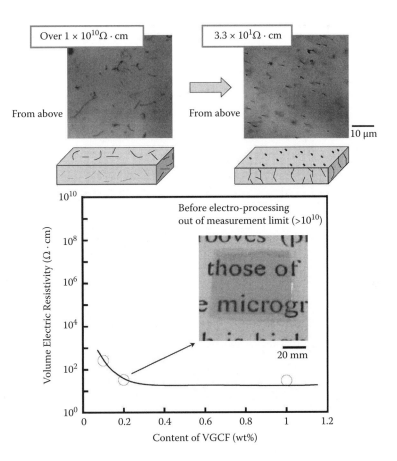

FIGURE 10.15
Optical micrographs of VGCF®s/low-density polyethylene composite film (VGCF®s = 0.2 wt%, thickness = 125 μm, from above) before and after electric processing (18 V) and their volume electric resistivity along thickness direction. Volume electric resistivity of aligned VGCF®s/low-density polyethylene composite film (thickness = 125 μm) by electric processing (18 V) as a function of VGCF® concentration with an inserted photo of an example.

Epoxy = 8.2 (not aligned)

Ethylene vinyl acetate copolymer = 3.1 (aligned)

Silicone oil = 2.8 (aligned)

Low-density polyethylene = 2.7 (aligned)

Thus, these tendencies give the motivation to investigate the aligning property under AC electric field to obtain insight into the mechanism.

10.6.2 Processing under AC Electric Field

Martin et al. also performed AC electric processing for MWCNTs/epoxy composites.[144] Figure 10.16 shows optical micrographs of epoxy nanocomposites containing 0.01 wt% MWCNTs during curing at 80°C in an AC field (100 V/cm, 1 kHz) (see Figure 10.12 by DC electric field).[145] In AC fields, more uniform aligned network structure was achieved between electrodes, suggesting that the electrophoresis movement was eliminated by changing the DC to an AC field.

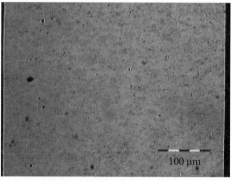

10 min

240 min

FIGURE 10.16

Optical micrographs of epoxy composites containing 0.01 wt% multi-walled carbon nanotubes (diameter = 50 nm, length = 43 μm, distribution = 40–46 μm, grown at the University of Cambridge by CVD method) after 10 and 240 min during curing at 80°C in an AC electric field (100 V/cm, peak value, 1 kHz). (*Source:* From Martin, C.A., Sandler, J.K.W., Windle, A.H., Schwarz, M.K., Bauhofer, W., Schulte, K., Shaffer, M.S.P., *Polymer,* 2005, 46, 877–886.)

Prasse et al. reported electric anisotropy of carbon nanofiber/epoxy composites.[146] Aligned carbon nanofiber/epoxy was prepared by applying a sine wave electric field of 100 V/cm at a frequency of 50 Hz during the curing of the composites. Figure 10.17 shows the resulting electric resistance of the cured composites as a function of the CNF weight fraction in the directions parallel and perpendicular to the field.[147] Starting from 0.75%, the network is formed with a different intensity in both directions. For a fiber content of 1% and above, one can estimate a maximal anisotropy of the resistance of about 10.

Park et al. made aligned single-walled carbon nanotube (SWCNT) urethane polymer composites using AC electric field.[148] Figure 10.5 shows electric conductivities of SWCNT urethane composites prepared at various frequencies. With higher frequencies, higher electric conductivities were observed, suggesting the possibility of controlling the conductivity by AC frequency.[149] However, no structural photographs from *in situ* observation were given. Therefore, it is valuable to perform *in situ* observation to clarify the effect of frequency.

The effects of matrix (i.e., dielectric property) and electric field type (i.e., DC or AC frequency) on the structural development of VGCFs were investigated as shown in Figure 10.19.

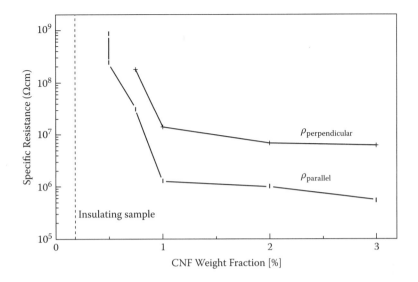

FIGURE 10.17

Specific resistance as a function of the carbon nanofiber[51] (PYROGRAF III (Applied Sci., Inc.) PR-17AG) weight fraction in the cured epoxy composites for the directions parallel and perpendicular to the electric agglomeration field. The dotted line denotes an insulating sample prepared with a CNF weight fraction of 0.2%. The sample was prepared by a sine wave electric field of 100 V/cm at a frequency of 50 Hz during the curing of the composites. (*Source:* From Prasse, T., Cavaille, J.Y., and Bauhofer, W., *Comp. Sci. Technol.*, 2003, 63, 1835–1841.)

Figure 10.19a shows the effect of AC frequency on the structural development of VGCFs after 10 min in silicone oil (dielectric constant = 2.8 at 50 Hz) using two types of apparatus in Figure 10.2. Figure 10.19b shows that after 30 min in UV-curable epoxy (dielectric constant = 8.2 at 50 Hz). Figure 19a suggested that network structures were formed under all conditions; but in terms of the linearity of connected structures, higher frequency gave less branched network structures. On the other hand, no alignment occurred under DC field when using UV-curable epoxy resins, while linearly connected network structure was formed with higher AC frequencies.

To investigate the cause of more linearly connected network structure at higher frequency, rotational speed (Figure 10.3b-1) was analyzed under AC conditions. It was found that the alignment rate was not a function of aspect ratios in AC conditions and was faster with higher frequencies. Careful observation suggested that the branched network derives from the structural development that both rotation and connection occurred at the same time due to the very slow rotational rate, while linearly connected structure derives from the phenomenon that connection occurred after rotation due to a very quick rotational rate. The cause of the very fast alignment rate of VGCFs in matrix with high dielectric constant needs to be identified further, and remains subject to further investigation. Structure development in Figure 10.19b supports the tendencies in Figure 10.18.

Figure 10.20 shows an example of aligned VGCFs (0.5 wt%)/UV-curable epoxy composite after AC electric processing (20 V [peak to peak], 1 MHz, thickness = 25 μm). The film shows the volume electric resistivity along thickness direction of a 10 Ω-cm with transparent appearance.

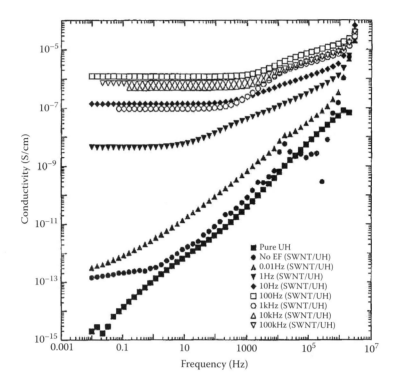

FIGURE 10.18

Conductivities of urethane (UH) and single-walled carbon nanotube/urethane (UH) composite cured with electric field (200 V (peak to peak); 10 min) using various frequencies. Note the alignment of single walled carbon nanotubes along the applied electric field. (*Source:* From Park, C., Wilkinson, J., Banda, S., Ounaies, Z., Wise, K. E., Sauti, G., Lillehei, P. T., and Harrison, J.S., *J. Polym. Sci. Part B Polym. Phys.*, 2006, 44, 1751–1762.)

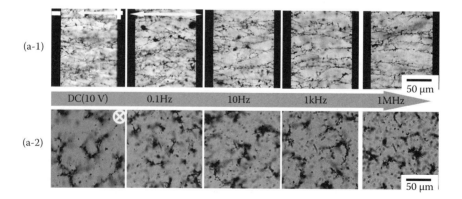

FIGURE 10.19

(a) The effect of various electric processing conditions (DC = 10 V, AC = 20 V (peak to peak)) on aligned structure of VGCF®s (0.1 wt%) in silicone oil (0.97 Pa-s) after 10 min, measured with setups in Figure 10.2; and (b) the effect of various electric processing conditions (DC = 10 V, AC = 20 V (peak to peak)) on aligned structure of VGCF®s (0.1 wt%) in UV-curable epoxy resin (1.5 Pa-s) after 30 min, measured with setups in Figure 10.2. No alignment occurred under DC conditions when using UV-curable epoxy resin, while more connected network structure was formed with higher AC frequencies.

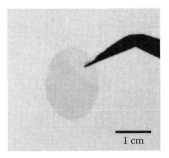

FIGURE 10.20
A photograph of aligned VGCF®s (0.5 wt%)/UV-curable epoxy composite after AC electric processing (20 V (peak to peak), 1 MHz, thickness = 25 μm). The volume electric resistivity of this film along the thickness direction shows 10 Ω-cm.

10.7 Alignment of Matrix Polymer by Aligned VGCF

It is quite interesting to align not only the MWCNT, but also the matrix polymer in nanocomposites. It has been known that conventional glass fiber induces crystallization in some semi-crystalline polymers; however, crystal direction has been random. It was found from x-ray diffraction and optical microscopy that the surface of MWCNTs having a high graphitization degree induces crystallization of polycarbonate along the MWCNT axial direction, although the crystallization speed of polycarbonate is extremely slow, and it is categorized as amorphous polymer. Using this technique, it becomes possible to align all polycarbonate crystals toward one direction using VGCFs aligned by magnetic field.[150, 151] Orientation control of liquid crystal polymers by aligned CNTs has also been reported.[152]

10.8 Patterning of Aligned VGCF

Using a magnetic modulator in a magnetic field, which consists of stacked thin layers (about 100 μm each) of aluminum and iron, it is possible to control the magnetic flux density. With this technique, diamagnetic materials having anisotropic susceptibility, such as VGCFs, can be concentrated in a certain region with aligned structure.[153] The patterning of aligned VGCFs in polymer matrix is of potential interest for future functional devices.

10.9 Conclusion

Recent progress about the effects of field types—either magnetic, or electric (AC/DC)—on alignment rate, structural development, and volume electric resistivity has been systematically overviewed. Figure 10.21 summarizes a schematic picture about aligning VGCFs in polymer matrices described here. Model (b) is from magnetic processing. Models (c)

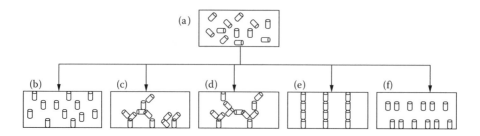

FIGURE 10.21
Illustrated models of aligned VGCF® in polymer by magnetic and electric fields, as described here.

and (d) are from DC electric field; model (e) is from AC electric field; and model (f) is from magnetic field with a magnetic modulator. Magnetic or electric processing offers great capability regarding the alignment of VGCF in various directions with several patterns, something that conventional polymer processing cannot provide.

Acknowledgments

The authors acknowledge financial support from the CLUSTER of the Ministry of Education, Culture, Sports, Science and Technology, Japan, and Grants-in-Aid (No. 17560603) from the Japan Society for the Promotion of Science.

References

1. Coleman, J.N., Khan, U., Blau, W.J., and Gunko, Y.K., Small but strong: a review of the mechanical properties of carbon nanotube-polymer composites. *Carbon*, 44, 1624, 2006.
2. Moniruzzaman, M. and Winey, K.I., Polymer nanocmoposites containing carbon nanotubes. *Macromolecules*, 2006, 39, 5194–5205.
3. Xie, X.L., Mai, Y.W., and Zhou, X.P., Dispersion and alignment of carbon nanotubes in polymer matrix : a review. *Mater. Sci. Eng. R,*, 2005, 49, 89–112.
4. Fujiwara, M., Oki, E., Hamada, M., Tanimoto, Y., Mukouda, I., and Shimomura, Y., Magnetic orientation and magnetic properties of a single carbon nanotube. *J. Phys. Chem.*, 2001, 105, 4383–4386.
5. Kimura, T., Study on the effect of magnetic fields on polymeric materials and its application. *Polym. J.*, 2003, 35, 823–843.
6. Kimura, T., Yamato, M., Koshimizu, W., Koike, M., and Kawai, T., Magnetic orientation of polymer fibers in suspension. *Langmuir*, 2000, 16, 858–861.
7. Ramirez, A.P., Haddon, R.C., Zhou, O., Fleming, R.M., Zhang, J., McClure, S.M., and Smalley, R.E., Magnetic susceptibility of molecular carbon: nanotubes and fullerite. *Science*, 1994, 265, 84–86.
8. Yamato, M., Application of magnetic fields to processing of polymeric materials. *Kobunshi Ronbunshu*, 2004, 61, 433–441.
9. Yamato, M., Aoki, H., Kimura, T., Yamamoto, I., Ishikawa, F., Yamaguchi, M., and Tobita, M., Determination of anisotropic diamagnetic susceptibility of polymeric fibers suspended in liquid. *Jpn. J. Appl.Phys.*, 2001, 40, 2237–2240.

10. Zaric, S., Ostojic, G.N., Kono, J., Shaver, J., Moore, V.C., Hauge, R.H., Smalley, R.E., and Wei, Z., Estimation of magnetic susceptibility anisotropy of carbon nanotubes using magnetophotoluminescence. *Nano Lett.*, 2004, 4, 2219–2221.
11. Ajiki, H. and Ando, T., Magnetic properties of carbon nanotubes. *J. Phys. Soc. Jpn.*, 1993, 62, 2470–2480.
12. Casavant, M.J., Walters, D.A., Schmidt, J.J., and Smalley, R.E., Neat macroscopic membranes of aligned carbon nanotubes. *J. Appl. Phys.*, 2003, 93, 2153–2156.
13. Fischer, J.E., Zhou, W., Vavro, J., Llaguno, M.C., Guthy, C., Haggenmuller, R., Casavant, M.J., Walters, D.E., and Smalley, R.E., Magnetically aligned single wall carbon nanotube films: preferred orientation and anisotropic transport properties. *J. Appl. Phys.*, 2003, 93, 2157–2163.
14. Kordas, K., Mustonen, T., Toth, G., Vaehaekangas, J., Uusimaeki, A., Jantunen, H., Gupta, A., Rao, K.V., Vajtai, R., and Ajayan, P.M., Magnetic-field induced efficient alignment of carbon nanotubes in aqueous solutions. *Chem. Mater.*, 2007, 19, 787–791.
15. Smith, B.W., Benes, Z., Luzzi, D.E., Fischer, J.E., Walters, D.A., Casavant, M.J., Schmidt, J., and Smalley, R.E., Structural anisotropy of magnetically aligned single wall carbon nanotube films. *Appl. Phys. Lett.*, 2000, 77, 663–665.
16. Yamamoto, K., Akita, S., and Nakayama, Y., Orientation and purification of carbon nanotubes using AC electrophoresis. *J. Phys. D: Appl. Phys.*, 1998, 31, L34–L36.
17. Chen, X.Q., Saito, T., Yamada, H., and Matsushige, K., Aligning single-wall carbon nanotubes with an alternating-current electric field. *Appl. Phys. Lett.*, 2001, 78, 3714–3716.
18. Chen, Z., Yang, Y., Chen, F., Qing, Q., Wu, Z., and Liu, Z., Controllable interconnection of single-walled carbon nanotubes under AC electric field. *J. Phys. Chem. B Lett.*, 2005, 109, 11420–11423.
19. Dimaki, M. and Boggild, P., Frequency dependence of the structure and electrical behaviour of carbon nanotube networks assembled by dielectrophoresis. *Nanotechnology*, 2005, 16, 759–763.
20. EL-Hami, K. and Matsushige, K., Alignment of different lengths of carbon nanotubes using low applied electric field. *IEICE Trans. Electron*, 2004, E87–C, 2116–2118.
21. Kumar, M.S., Kim, T.H., Lee, S.H., Song, S.M., Yang, J.W., Nahm, K.S., and Suh, E.K., Influence of electric field type on the assembly of single walled carbon nanotubes. *Chem. Phys. Lett.*, 2004, 383, 235–239.
22. Liu, X., Spencer, J.L., Kaiser, A.B., and Arnold, W.M., Electic-field oriented carbon nanotubes in different dielectric solvents. *Curr. Appl. Phys.*, 2004, 4, 125–128.
23. Walters, D.A., Casavant, M.J., Qin, X.C., Huffman, C.B., Boul, P.J., Ericson, L.M., Haroz, E.H., O'Connell, M.J., Smith, K., Colbert, D.T., and Smalley, R.E., In-plane-aligned membranes of carbon nanotubes. *Chem. Phys. Lett.*, 2001, 338, 14–20.
24. Yamamoto, K., Akita, S., and Nakayama, Y., Orientation of carbon nanotubes using electrophoresis. *Jpn. J. Appl. Phys.*, 1996, 35, L917–L918.
25. Abe, Y., Tomuro, R., and Sano, M., Highly efficient direct current electrodeposition of single-walled carbon nanotubes in anhydrous solvents. *Adv. Mater.*, 2005, 17, 2192–2194.
26. Kamat, P.V., Thomas, K.G., Barazzouk, S., Girishkumar, G., Vinodgopal, K., and Meisel, D., Self-assembled liner bundles of single wall carbon nanotubes and their alignment and deposition as a film in a DC field. *J. Am. Chem. Soc.*, 2004, 126, 10757–10762.
27. Choi, E.S., Brooks, J.S., Eaton, D.L., Al-Haik, M.S., Hussaini, M.Y., Garmestani, H., Li, D., and Dahmen, K., Enhancement of thermal and electrical properties of carbon nanotube polymer composites by magnetic field processing. *J. Appl. Phys.*, 2003, 94, 6034–6039.
28. Garmestani, H., Al-Haik, M.S., Dahmen, K., Tannenbaum, R., Li, D., Sablin, S.S., and Hussaini, M.Y., Polymer-mediated alignment of carbon nanotubes under high magnetic fields. *Adv. Mater.*, 2003, 15, 1918–1921.
29. Kimura, T., Ago, H., Tobita, M., Ohshima, S., Kyotani, M., and Yumura, M., Polymer composites of carbon nanotubes aligned by a magnetic field. *Adv. Mater.*, 2002, 14, 1380–1383.
30. Takahashi, T., Higuchi, A., Awano, H., Yoketake, K., and Kikuchi, T., Oriented crystallization of polycarbonate by vapor grown carbon fiber and its application. *Polym. J.*, 2005, 37, 887–893.
31. Takahashi, T., Suzuki, K., Awano, H., and Yonetake, K., Alignment of vapor-grown carbon fibers in polymer under magnetic field. *Chem. Phys. Lett.*, 2007, 436, 378–382.

32. Takahashi, T., Yonetake, K., Koyama, K., and Kikuchi, T., Polycarbonate crystallization by vapor-grown carbon fiber with and without magnetic field. *Macromol. Rapid Commun.*, 2003, 24, 763–767.

33. Ookubo, T., Awano, H., Takahashi, T., Yonetake, K., and Oishi, Y., Oriented structural development of vapor-grown carbon fiber/polymer composite film by magnetic field and electric conductive properties. *Kobunshi Ronbunshu*, 64, 727–734, 2007.

34. Ookubo, T., Takahashi, T., Awano, H., Yonetake, K., and Oishi, Y., Magnetic processing of polymer composite films including vapor-grown carbon fibers. *TANSO*, 2006, 223, 169–175.

35. Ookubo, T., Takahashi, T., Oh, Y.-J., Awano, H., Yonetake, K., and Oishi, Y., Magnetic processing of polymer composite films including carbon fibers. *TANSO* 2005, 217, 104–110.

36. Tobita, M., Development of new composite material in a high magnetic field. *J. Rubb. Soc. Jpn.*, 2003, 76, 18–23.

37. Yonetake, K. and Takahashi, T., New material design for liquid crystals and composites by magneto-processing. *Sci. Tech. Adv. Mater.*, 2006, 7, 332–336.

38. Martin, C.A., Sandler, J.K.W., Windle, A.H., Schwarz, M.K., Bauhofer, W., Schulte, K., and Shaffer, M.S.P., Electric field-induced aligned multi-wall carbon nanotube networks in epoxy composites. *Polymer*, 2005, 46, 877–886.

39. Park, C., Wilkinson, J., Banda, S., Ounaies, Z., Wise, K. E., Sauti, G., Lillehei, P. T., and Harrison, J.S., Aligned single-wall carbon nanotube polymer composites using an electric field. *J. Polym. Sci. Part B Polym. Phys.*, 2006, 44, 1751–1762.

40. Prasse, T., Cavaille, J.Y., and Bauhofer, W., Electric anisotropy of carbon nanofibre/epoxy resin composites due to electric field induced alignment. *Comp. Sci. Technol.*, 2003, 63, 1835–1841.

41. Takahashi, T., Murayama, T., Higuchi, A., Awano, H., and Yonetake, K., Aligning vapor-grown carbon fibers in polydimethylsiloxane using DC electric of magnetic field. *Carbon*, 2006, 1180–1187.

42. Tanaka, K., Fujita, Y., Kubono, A., and Akiyama, R., Electrically developed morphology of carbon nanoparticles in suspensions monitored by *in situ* optical observation under sinusoidal electric field. *Colloid Polym. Sci.*, 2006, 284, 562–567.

43. Takahashi, T., Higuchi, A., Awano, H., Yoketake, K., and Kikuchi, T., Oriented crystallization of polycarbonate by vapor grown carbon fiber and its application. *Polym. J.*, 2005, 37, 887–893.

44. Takahashi, T., Yonetake, K., Koyama, K., and Kikuchi, T., Polycarbonate crystallization by vapor-grown carbon fiber with and without magnetic field. *Macromol. Rapid Commun.*, 2003, 24, 763–767.

45. Mrozek, R.A. and Taton, T.A., Alignment of liquid-crystalline polymers by field-oriented carbon nanotube directors. *Chem. Mater.*, 2005, 17, 3384–3388.

46. Piao, G., Kimura, F., Takahashi, T., Moritani, Y., Awano, H., Nimori, S., Tsuda, K., Yonetake, K., and Kimura, T., Alignment and micropatterning of carbon nanotubes in polymer composites using modulated magnetic field. *Polym. J.*, 2007, 39, 589–592.

47. Endo, M., Kim, Y. A., Hayashi, T., Nishimura, K., Matusita, T., Miyashita, K., and Dresselhaus, M.S., Vapor-grown carbon fibers (VGCFs): Basic properties and their battery applications. *Carbon*, 2001, 39, 1287–1297.

48. Endo, M., Kim, Y.A., Hayashi, T., Nishimura, K., Matusita, T., Miyashita, K., and Dresselhaus, M.S., Vapor-grown carbon fibers (VGCFs): basic properties and their battery applications. *Carbon*, 2001, 39, 1287–1297.

49. Sato, E., Takahashi, T., and Koyama, K., Comparison of the length of vapor grown carbon fiber before and after mixing process. *Kobunshi Ronbunshu*, 2004, 61, 144–148.

50. Sato, E., Takahashi, T., Natsume, T., and Koyama, K., Development of dispersion and length-evaluation methods of vapor-grown carbon fiber. *TANSO*, 2003, 209, 159–164.

51. Sato, E., Takahashi, T., and Koyama, K., Comparison of the length of vapor grown carbon fiber before and after mixing process. *Kobunshi Ronbunshu*, 2004, 61, 144–148.

52. Sato, E., Takahashi, T., Natsume, T., and Koyama, K., Development of dispersion and length-evaluation methods of vapor-grown carbon fiber. *TANSO*, 2003, 209, 159–164.

53. Sato, E., Takahashi, T., and Koyama, K., Comparison of the length of vapor grown carbon fiber before and after mixing process. *Kobunshi Ronbunshu*, 2004, 61, 144–148.

54. Sato, E., Takahashi, T., Natsume, T., and Koyama, K., Development of dispersion and length-evaluation methods of vapor-grown carbon fiber. *TANSO*, 2003, 209, 159–164.

55. Takahashi, T., Murayama, T., Higuchi, A., Awano, H., and Yonetake, K., Aligning vapor-grown carbon fibers in polydimethylsiloxane using DC electric of magnetic field. *Carbon*, 2006, 1180–1187.

56. Fujiwara, M., Oki, E., Hamada, M., Tanimoto, Y., Mukouda, I., and Shimomura, Y., Magnetic orientation and magnetic properties of a single carbon nanotube. *J. Phys. Chem.*, 2001, 105, 4383–4386.

57. Kimura, T., Study on the effect of magnetic fields on polymeric materials and its application. *Polym. J.*, 2003, 35, 823–843.

58. Kimura, T., Yamato, M., Koshimizu, W., Koike, M., Kawai, T., Magnetic orientation of polymer fibers in suspension. *Langmuir*, 2000, 16, 858–861.

59. Yamato, M., Application of magnetic fields to processing of polymeric materials. *Kobunshi Ronbunshu*, 2004, 61, 433–441.

60. Yamato, M., Aoki, H., Kimura, T., Yamamoto, I., Ishikawa, F., Yamaguchi, M., and Tobita, M., Determination of anisotropic diamagnetic susceptibility of polymeric fibers suspended in liquid. *Jpn. J. Appl.Phys.*, 2001, 40, 2237–2240.

61. Zaric, S., Ostojic, G.N., Kono, J., Shaver, J., Moore, V.C., Hauge, R.H., Smalley, R.E., and Wei, Z., Estimation of magnetic susceptibility anisotropy of carbon nanotubes using magnetophotoluminescence. *Nano Lett.*, 2004, 4, 2219–2221.

62. Ajiki, H. and Ando, T., Magnetic properties of carbon nanotubes. *J. Phys. Soc. Jpn.*, 1993, 62, 2470–2480.

63. Tobita, M., Development of new composite material in a high magnetic field. *J. Rubb. Soc. Jpn.*, 2003, 76, 18–23.

64. Casavant, M.J., Walters, D.A., Schmidt, J.J., and Smalley, R.E., Neat macroscopic membranes of aligned carbon nanotubes. *J. Appl. Phys.*, 2003, 93, 2153–2156.

65. Fischer, J.E., Zhou, W., Vavro, J., Llaguno, M.C., Guthy, C., Haggenmuller, R., Casavant, M.J., Walters, D.E., and Smalley, R.E., Magnetically aligned single wall carbon nanotube films: preferred orientation and anisotropic transport properties. *J. Appl. Phys.*, 2003, 93, 2157–2163.

66. Kordas, K., Mustonen, T., Toth, G., Vaehaekangas, J., Uusimaeki, A., Jantunen, H., Gupta, A., Rao, K.V., Vajtai, R., and Ajayan, P.M., Magnetic-field induced efficient alignment of carbon nanotubes in aqueous solutions. *Chem. Mater.*, 2007, 19, 787–791.

67. Smith, B.W., Benes, Z., Luzzi, D. E., Fischer, J. E., Walters, D. A., Casavant, M. J., Schmidt, J., and Smalley, R.E., Structural anisotropy of magnetically aligned single wall carbon nanotube films. *Appl. Phys. Lett.*, 2000, 77, 663–665.

68. Yamamoto, K., Akita, S., and Nakayama, Y., Orientation and purification of carbon nanotubes using AC electrophoresis. *J. Phys. D: Appl. Phys.*, 1998, 31, L34–L36.

69. Choi, E.S., Brooks, J.S., Eaton, D.L., Al-Haik, M.S., Hussaini, M.Y., Garmestani, H., Li, D., and Dahmen, K., Enhancement of thermal and electrical properties of carbon nanotube polymer composites by magnetic field processing. *J. Appl. Phys.*, 2003, 94, 6034–6039.

70. Garmestani, H., Al-Haik, M.S., Dahmen, K., Tannenbaum, R., Li, D., Sablin, S.S., and Hussaini, M.Y., Polymer-mediated alignment of carbon nanotubes under high magnetic fields. *Adv. Mater.*, 2003, 15, 1918–1921.

71. Kimura, T., Ago, H., Tobita, M., Ohshima, S., Kyotani, M., and Yumura, M., Polymer composites of carbon nanotubes aligned by a magnetic field. *Adv. Mater.*, 2002, 14, 1380–1383.

72. Takahashi, T., Higuchi, A., Awano, H., Yoketake, K., and Kikuchi, T., Oriented crystallization of polycarbonate by vapor grown carbon fiber and its application. *Polym. J.*, 2005, 37, 887–893.

73. Takahashi, T., Suzuki, K., Awano, H., and Yonetake, K., Alignment of vapor-grown carbon fibers in polymer under magnetic field. *Chem. Phys. Lett.*, 2007, 436, 378–382.

74. Takahashi, T., Yonetake, K., Koyama, K., and Kikuchi, T., Polycarbonate crystallization by vapor-grown carbon fiber with and without magnetic field. *Macromol. Rapid Commun.*, 2003, 24, 763–767.

75. Ookubo, T., Awano, H., Takahashi, T., Yonetake, K., and Oishi, Y., Various oriented structural development of vapor-grown carbon fiber/polymer composite film by magnetic field and electric conductive properties. *Kobunshi Ronbunshu*, 2007, 64, 727–734.

76. Ookubo, T., Takahashi, T., Awano, H., Yonetake, K., and Oishi, Y., Magnetic processing of polymer composite films including vapor-grown carbon fibers. *TANSO*, 2006, 223, 169–175.

77. Ookubo, T., Takahashi, T., Oh, Y.-J., Awano, H., Yonetake, K., and Oishi, Y., Magnetic processing of polymer composite films including carbon fibers. *TANSO*, 2005, 217, 104–110.

78. Tobita, M., Development of new composite material in a high magnetic field. *J. Rubb. Soc. Jpn.*, 2003, 76, 18–23.

79. Yonetake, K. and Takahashi, T., New material design for liquid crystals and composites by magneto-processing. *Sci. Tech. Adv. Mater.*, 2006, 7, 332–336.

80. Garmestani, H., Al-Haik, M.S., Dahmen, K., Tannenbaum, R., Li, D., Sablin, S.S., and Hussaini, M.Y., Polymer-mediated alignment of carbon nanotubes under high magnetic fields. *Adv. Mater.*, 2003, 15, 1918–1921.

81. Choi, E.S., Brooks, J.S., Eaton, D.L., Al-Haik, M.S., Hussaini, M.Y., Garmestani, H., Li, D., and Dahmen, K., Enhancement of thermal and electrical properties of carbon nanotube polymer composites by magnetic field processing. *J. Appl. Phys.*, 2003, 94, 6034–6039.

82. Kimura, T., Ago, H., Tobita, M., Ohshima, S., Kyotani, M., and Yumura, M., Polymer composites of carbon nanotubes aligned by a magnetic field. *Adv. Mater.*, 2002, 14, 1380–1383.

83. Kimura, T., Yamato, M., Koshimizu, W., Koike, M., and Kawai, T., Magnetic orientation of polymer fibers in suspension. *Langmuir*, 2000, 16, 858–861.

84. Takahashi, T., Suzuki, K., Awano, H., and Yonetake, K., Alignment of vapor-grown carbon fibers in polymer under magnetic field. *Chem. Phys. Lett.*, 2007, 436, 378–382.

85. Takahashi, T., Suzuki, K., Awano, H., and Yonetake, K., Alignment of vapor-grown carbon fibers in polymer under magnetic field. *Chem. Phys. Lett.*, 2007, 436, 378–382.

86. Takahashi, T., Murayama, T., Higuchi, A., Awano, H., and Yonetake, K., Aligning vapor-grown carbon fibers in polydimethylsiloxane using DC electric of magnetic field. *Carbon*, 2006, 1180–1187.

87. Yamamoto, K., Akita, S., and Nakayama, Y., Orientation and purification of carbon nanotubes using AC electrophoresis. *J. Phys. D: Appl. Phys.*, 1998, 31, L34–L36.

88. Chen, X. Q., Saito, T., Yamada, H., and Matsushige, K., Aligning single-wall carbon nanotubes with an alternating-current electric field. *Appl. Phys. Lett.*, 2001, 78, 3714–3716.

89. Chen, Z., Yang, Y., Chen, F., Qing, Q., Wu, Z., and Liu, Z., Controllable interconnection of single-walled carbon nanotubes under AC electric field. *J. Phys. Chem. B Lett.*, 2005, 109, 11420–11423.

90. Dimaki, M. and Boggild, P., Frequency dependence of the structure and electrical behaviour of carbon nanotube networks assembled by dielectrophoresis. *Nanotechnology*, 2005, 16, 759–763.

91. El-Hami, K. and Matsushige, K., Alignment of different lengths of carbon nanotubes using low applied electric field. *IEICE Trans. Electron*, 2004, E87–C, 2116–2118.

92. Kumar, M.S., Kim, T.H., Lee, S.H., Song, S.M., Yang, J.W., Nahm, K.S., and Suh, E.K., Influence of electric field type on the assembly of single walled carbon nanotubes. *Chem. Phys. Lett.*, 2004, 383, 235–239.

93. Liu, X., Spencer, J.L., Kaiser, A.B., and Arnold, W.M., Electic-field oriented carbon nanotubes in different dielectric solvents. *Curr. Appl. Phys.*, 2004, 4, 125–128.

94. Walters, D.A., Casavant, M.J., Qin, X.C., Huffman, C.B., Boul, P.J., Ericson, L.M., Haroz, E.H., O'Connell, M.J., Smith, K., Colbert, D.T., and Smalley, R.E., In-plane-aligned membranes of carbon nanotubes. *Chem. Phys. Lett.*, 2001, 338, 14–20.

95. Yamamoto, K., Akita, S., and Nakayama, Y., Orientation of carbon nanotubes using electrophoresis. *Jpn. J. Appl. Phys.*, 1996, 35, L917–L918.

96. Abe, Y., Tomuro, R., and Sano, M., Highly efficient direct current electrodeposition of single-walled carbon nanotubes in anhydrous solvents. *Adv. Mater.*, 2005, 17, 2192–2194.

97. Kamat, P.V., Thomas, K.G., Barazzouk, S., Girishkumar, G., Vinodgopal, K., and Meisel, D., Self-assembled liner bundles of single wall carbon nanotubes and their alignment and deposition as a film in a DC field. *J. Am. Chem. Soc.*, 2004, 126, 10757–10762.

98. Takahashi, T., Murayama, T., Higuchi, A., Awano, H., and Yonetake, K., Aligning vapor-grown carbon fibers in polydimethylsiloxane using DC electric of magnetic field. *Carbon*, 2006, 1180–1187.

99. Takahashi, T., Suzuki, K., Awano, H., and Yonetake, K., Alignment of vapor-grown carbon fibers in polymer under magnetic field. *Chem. Phys. Lett.*, 2007, 436, 378–382.

100. Kimura, T., Study on the effect of magnetic fields on polymeric materials and its application. *Polym. J.*, 2003, 35, 823–843.

101. Kimura, T., Yamato, M., Koshimizu, W., Koike, M., and Kawai, T., Magnetic orientation of polymer fibers in suspension. *Langmuir*, 2000, 16, 858–861.

102. Yamato, M., Application of magnetic fields to processing of polymeric materials. *Kobunshi Ronbunshu*, 2004, 61, 433–441.

103. Yamato, M., Aoki, H., Kimura, T., Yamamoto, I., Ishikawa, F., Yamaguchi, M., and Tobita, M., Determination of anisotropic diamagnetic susceptibility of polymeric fibers suspended in liquid. *Jpn. J. Appl.Phys.*, 2001, 40, 2237–2240.

104. Takahashi, T., Suzuki, K., Awano, H., and Yonetake, K., Alignment of vapor-grown carbon fibers in polymer under magnetic field. *Chem. Phys. Lett.*, 2007, 436, 378–382.

105. Takahashi, T., Suzuki, K., Awano, H., and Yonetake, K., Alignment of vapor-grown carbon fibers in polymer under magnetic field. *Chem. Phys. Lett.*, 2007, 436, 378–382.

106. Choi, E.S., Brooks, J.S., Eaton, D.L., Al-Haik, M.S., Hussaini, M.Y., Garmestani, H., Li, D., and Dahmen, K., Enhancement of thermal and electrical properties of carbon nanotube polymer composites by magnetic field processing. *J. Appl. Phys.*, 2003, 94, 6034–6039.

107. Garmestani, H., Al-Haik, M.S., Dahmen, K., Tannenbaum, R., Li, D., Sablin, S.S., and Hussaini, M.Y., Polymer-mediated alignment of carbon nanotubes under high magnetic fields. *Adv. Mater.*, 2003, 15, 1918–1921.

108. Kimura, T., Ago, H., Tobita, M., Ohshima, S., Kyotani, M., and Yumura, M., Polymer composites of carbon nanotubes aligned by a magnetic field. *Adv. Mater.* 2002, 14, 1380–1383.

109. Takahashi, T., Higuchi, A., Awano, H., Yoketake, K., and Kikuchi, T., Oriented crystallization of polycarbonate by vapor grown carbon fiber and its application. *Polym. J.*, 2005, 37, 887–893.

110. Takahashi, T., Suzuki, K., Awano, H., and Yonetake, K., Alignment of vapor-grown carbon fibers in polymer under magnetic field. *Chem. Phys. Lett.*, 2007, 436, 378–382.

111. Takahashi, T., Yonetake, K., Koyama, K., and Kikuchi, T., Polycarbonate crystallization by vapor-grown carbon fiber with and without magnetic field. *Macromol. Rapid Commun.*, 2003, 24, 763–767.

112. Ookubo, T., Awano, H., Takahashi, T., Yonetake, K., and Oishi, Y., Various oriented structural development of vapor-grown carbon fiber/polymer composite film by magnetic field and electric conductive properties. *Kobunshi Ronbunshu*, 2007, 64, 727–734.

113. Ookubo, T., Takahashi, T., Awano, H., Yonetake, K., and Oishi, Y., Magnetic processing of polymer composite films including vapor-grown carbon fibers. *TANSO*, 2006, 223, 169–175.

114. Ookubo, T., Takahashi, T., Oh, Y.-J., Awano, H., Yonetake, K., and Oishi, Y., Magnetic processing of polymer composite films including carbon fibers. *TANSO*, 2005, 217, 104–110.

115. Tobita, M., Development of new composite material in a high magnetic field. *J. Rubb. Soc. Jpn.*, 2003, 76, 18–23.

116. Yonetake, K. and Takahashi, T., New material design for liquid crystals and composites by magneto-processing. *Sci. Tech. Adv. Mater.*, 2006, 7, 332–336.

117. Choi, E.S., Brooks, J.S., Eaton, D.L., Al-Haik, M.S., Hussaini, M.Y., Garmestani, H., Li, D., and Dahmen, K., Enhancement of thermal and electrical properties of carbon nanotube polymer composites by magnetic field processing. *J. Appl. Phys.*, 2003, 94, 6034–6039.

118. Kimura, T., Ago, H., Tobita, M., Ohshima, S., Kyotani, M., and Yumura, M., Polymer composites of carbon nanotubes aligned by a magnetic field. *Adv. Mater.*, 2002, 14, 1380–1383.

119. Ookubo, T., Awano, H., Takahashi, T., Yonetake, K., and Oishi, Y., Various oriented structural development of vapor-grown carbon fiber/polymer composite film by magnetic field and electric conductive properties. *Kobunshi Ronbunshu*, 2007, 64, 727–734.

120. Ookubo, T., Takahashi, T., Awano, H., Yonetake, K., and Oishi, Y., Magnetic processing of polymer composite films including vapor-grown carbon fibers. *TANSO*, 2006, 223, 169–175.

121. Ookubo, T., Takahashi, T., Oh, Y.-J., Awano, H., Yonetake, K., and Oishi, Y., Magnetic processing of polymer composite films including carbon fibers. *TANSO*, 2005, 217, 104–110.

122. Ookubo, T., Awano, H., Takahashi, T., Yonetake, K., and Oishi, Y., Various oriented structural development of vapor-grown carbon fiber/polymer composite film by magnetic field and electric conductive properties. *Kobunshi Ronbunshu,* 2007, 64, 727–734.

123. Ookubo, T., Takahashi, T., Awano, H., Yonetake, K., and Oishi, Y., Magnetic processing of polymer composite films including vapor-grown carbon fibers. *TANSO,* 2006, 223, 169–175.

124. Ookubo, T., Takahashi, T., Oh, Y.-J., Awano, H., Yonetake, K., and Oishi, Y., Magnetic processing of polymer composite films including carbon fibers. *TANSO,* 2005, 217, 104–110.

125. Yamamoto, K., Akita, S., and Nakayama, Y., Orientation and purification of carbon nanotubes using AC electrophoresis. *J. Phys. D: Appl. Phys.,* 1998, 31, L34–L36.

126. Chen, X. Q., Saito, T., Yamada, H., and Matsushige, K., Aligning single-wall carbon nanotubes with an alternating-current electric field. *Appl. Phys. Lett.,* 2001, 78, 3714–3716.

127. Chen, Z., Yang, Y., Chen, F., Qing, Q., Wu, Z., and Liu, Z., Controllable interconnection of single-walled carbon nanotubes under AC electric field. *J. Phys. Chem. B Lett.,* 2005, 109, 11420–11423.

128. Dimaki, M. and Boggild, P., Frequency dependence of the structure and electrical behaviour of carbon nanotube networks assembled by dielectrophoresis. *Nanotechnology,* 2005, 16, 759–763.

129. EL-Hami, K. and Matsushige, K., Alignment of different lengths of carbon nanotubes using low applied electric field. *IEICE Trans. Electron,* 2004, E87–C, 2116–2118.

130. Kumar, M.S., Kim, T.H., Lee, S.H., Song, S.M., Yang, J.W., Nahm, K.S., and Suh, E.K., Influence of electric field type on the assembly of single walled carbon nanotubes. *Chem. Phys. Lett.,* 2004, 383, 235–239.

131. Walters, D.A., Casavant, M.J., Qin, X.C., Huffman, C.B., Boul, P.J., Ericson, L.M., Haroz, E.H., O'Connell, M.J., Smith, K., Colbert, D.T., and Smalley, R.E., In-plane-aligned membranes of carbon nanotubes. *Chem. Phys. Lett.,* 2001, 338, 14–20.

132. Liu, X., Spencer, J.L., Kaiser, A.B., and Arnold, W.M., Electic-field oriented carbon nanotubes in different dielectric solvents. *Curr. Appl. Phys.,* 2004, 4, 125–128.

133. Yamamoto, K., Akita, S., and Nakayama, Y., Orientation of carbon nanotubes using electrophoresis. *Jpn. J. Appl. Phys.,* 1996, 35, L917–L918.

134. Abe, Y., Tomuro, R., and Sano, M., Highly efficient direct current electrodeposition of single-walled carbon nanotubes in anhydrous solvents. *Adv. Mater.,* 2005, 17, 2192–2194.

135. Kamat, P.V., Thomas, K. G., Barazzouk, S., Girishkumar, G., Vinodgopal, K., and Meisel, D., Self-assembled liner bundles of single wall carbon nanotubes and their alignment and deposition as a film in a DC field. *J. Am. Chem. Soc.,* 2004, 126, 10757–10762.

136. Klingenberg, D.J., Dierking, D., and Zukoski, C.F., Stress-transfer mechanisms in electrorheological suspensions. *J. Chem. Soc. Faraday Trans.,* 1991, 87, 425–430.

137. Klingenberg, D.J., Dierking, D., and Zukoski, C.F., Stress-transfer mechanisms in electrorheological suspensions. *J. Chem. Soc. Faraday Trans.,* 1991, 87, 425–430.

138. Prasse, T., Flandin, L., Schulte, K., and Bauhofer, W., *In situ* observation of electric field induced agglomeration of carbon black in epoxy resin. *Appl. Phys. Lett.,* 1998, 72, 2903–2905.

139. Schwarz, M.K., Bauhofer, W., and Schulte, K., Alternating electric field induced agglomeration of carbon black filled resins. *Polymer,* 2002, 43, 3079–3082.

140. Martin, C.A., Sandler, J.K.W., Windle, A.H., Schwarz, M.K., Bauhofer, W., Schulte, K., and Shaffer, M.S.P., Electric field-induced aligned multi-wall carbon nanotube networks in epoxy composites. *Polymer,* 2005, 46, 877–886.

141. Martin, C.A., Sandler, J.K.W., Windle, A.H., Schwarz, M.K., Bauhofer, W., Schulte, K., and Shaffer, M.S.P., Electric field-induced aligned multi-wall carbon nanotube networks in epoxy composites. *Polymer,* 2005, 46, 877–886.

142. Takahashi, T., Murayama, T., Higuchi, A., Awano, H., and Yonetake, K., Aligning vapor-grown carbon fibers in polydimethylsiloxane using DC electric of magnetic field. *Carbon,* 2006, 1180–1187.

143. Takahashi, T., Murayama, T., Higuchi, A., Awano, H., and Yonetake, K., Aligning vapor-grown carbon fibers in polydimethylsiloxane using DC electric of magnetic field. *Carbon,* 2006, 1180–1187.

144. Martin, C.A., Sandler, J.K.W., Windle, A.H., Schwarz, M.K., Bauhofer, W., Schulte, K., and Shaffer, M.S.P., Electric field-induced aligned multi-wall carbon nanotube networks in epoxy composites. *Polymer,* 2005, 46, 877–886.

145. Martin, C.A., Sandler, J.K.W., Windle, A.H., Schwarz, M.K., Bauhofer, W., Schulte, K., and Shaffer, M.S.P., Electric field-induced aligned multi-wall carbon nanotube networks in epoxy composites. *Polymer*, 2005, 46, 877–886.

146. Prasse, T., Cavaille, J.Y., and Bauhofer, W., Electric anisotropy of carbon nanofibre/epoxy resin composites due to electric field induced alignment. *Comp. Sci. Technol.*, 2003, 63, 1835–1841.

147. Prasse, T., Cavaille, J.Y., and Bauhofer, W., Electric anisotropy of carbon nanofibre/epoxy resin composites due to electric field induced alignment. *Comp. Sci. Technol*, 2003, 63, 1835–1841.

148. Park, C., Wilkinson, J., Banda, S., Ounaies, Z., Wise, K.E., Sauti, G., Lillehei, P.T., and Harrison, J.S., Aligned single-wall carbon nanotube polymer composites using an electric field. *J. Polym. Sci. Part B Polym. Phys.*, 2006, 44, 1751–1762.

149. Park, C., Wilkinson, J., Banda, S., Ounaies, Z., Wise, K.E., Sauti, G., Lillehei, P.T., and Harrison, J.S., Aligned single-wall carbon nanotube polymer composites using an electric field. *J. Polym. Sci. Part B Polym. Phys.*, 2006, 44, 1751–1762.

150. Takahashi, T., Higuchi, A., Awano, H., Yoketake, K., and Kikuchi, T., Oriented crystallization of polycarbonate by vapor grown carbon fiber and its application. *Polym. J.*, 2005, 37, 887–893.

151. Takahashi, T., Yonetake, K., Koyama, K., and Kikuchi, T., Polycarbonate crystallization by vapor-grown carbon fiber with and without magnetic field. *Macromol. Rapid Commun.*, 2003, 24, 763–767.

152. Mrozek, R.A. and Taton, T.A., Alignment of liquid-crystalline polymers by field-oriented carbon nanotube directors. *Chem. Mater.*, 2005, 17, 3384–3388.

153. Piao, G., Kimura, F., Takahashi, T., Moritani, Y., Awano, H., Nimori, S., Tsuda, K., Yonetake, K., and Kimura, T., Alignment and micropatterning of carbon nanotubes in polymer composites using modulated magnetic field. *Polym. J.*, 2007, 39, 589–592.

11

Nanocomposites of Liquid Crystalline Polymers Dispersed in Polyester Matrices

Chang H. Song and Avraam I. Isayev

CONTENTS

11.1 Introduction

A number of studies have been done on the nanoparticles and nanotubes in polymer matrices in recent years to find a synergy effect in polymer industries. However, there were few studies of nanofibrillar structured LCP (liquid crystalline polymer) in polymer matrices in the past. The blending technology of thermotropic liquid crystalline polymers (TLCPs) with thermoplastics has evolved to generate the advanced material for high-performance applications and easy processibility. In the past decade, many researchers studied these materials, so-called self-reinforced composites, due to their high potential for *in situ* reinforcement (for a review, see Reference 1). LCPs can also act as a processing aid for thermoplastic processing due to the low melt viscosity of the LCP.

The choice of thermoplastic–LCP pair and the processing conditions should be carefully made to obtain mechanical properties comparable to glass-fiber-reinforced thermoplastics. Extensive studies have been done on polyester/LCP blends (for a review, see References 2 through 4). In the case of PET/PET–HBA-based LCP blends, the LCP phase was well distributed and fibrillated to nano-sized structure, leading to improved mechanical properties [2, 4] (PET = polyethylene terephthalate; HBA = *p*-hydroxy benzoic acid). Because the LCP and polyester phases are immiscible, finely distributed LCP domains can be formed in the melt blending of LCP and polyester using single-screw extrusion with mixing devices. LCP phase domains can be deformed into ellipsoidal or fibrillar shapes by subjecting the melt stream to the elongational flow as it is extruded from a die. A strong reinforcement resulted from the nanofibrillar structure of the LCP phase with a large aspect ratio and uniform distribution of LCP fibrils caused by highly oriented LCP molecules in the flow direction. This reinforcement strongly depends on the processing conditions, LCP composition, extension ratio, viscosity ratio of blend components, interfacial adhesion between the components, and rheological characteristics of the matrix [2–4].

For the injection molding of the self-reinforced composite [5, 6], a new technology has been developed to achieve a sustained microfibrillar structure of the LCP phase in the thermoplastic matrix with improved mechanical properties. The key factor was to carry out the injection molding at the melt temperature below the LCP melting temperature after the formation of microfibrillar LCP structure during extrusion. In this case, the degradation of fibrillar structure in the second step of the operation can be minimized and the benefit of fibrillar structure of the LCP can be retained.

Many studies have been made on the PET/LCP self-reinforced composite in which thermotropic LCP is a copolyester of HNA/HBA [7–13] (HNA = hydroxy naphthoic acid). Ko et al. [7] investigated the structure-property relationship in the extruded cast films prepared from blends of PET/LCP. They were concerned with the effect of process variables such as the extruder screw speed and gear pump speed on the final structure of the blends. Heino and Seppälä [8] investigated the dependence of structure and property of PET/LCP fiber strands on the LCP content and draw ratio. They utilized a twin-screw extrusion for mixing the two components. Then fiber strands of the blends were prepared using a single-screw extruder with a take-up device. Skin-core morphology in all blend fibers was found and structure-property relationships of the blends were investigated. Li et al. [9] observed the change from discontinuous fibrils when the composition was 35 and 60% LCP to continuous fibrils when the composition was 85 and 96% LCP in the fibers spun from blends of PET/LCP. They found that the analytical models for short aligned fibers of Nielsen [14] and Kelly and Tyson [15] were applicable when the LCP fibrils were discontinuous, while the modulus and strength of fibers with continuous LCP fibrils were described by the rule of mixtures. Mithal and Tayebi [10] used two grades of PET with intrinsic viscosities of 0.64 and 0.95 in the study of fiber spinning of PET/LCP. Blends of the higher intrinsic viscosity (high molecular weight) PET and LCP resulted in higher melt strength and viscosity in the processing temperature range of the LCP, while low intrinsic viscosity PET required a low spinning temperature and it led to semi-melted or unmelted LCP phase and poor mechanical properties. Heat treatment of the blend fibers increased the tensile properties of the LCP-rich compositions. Kyotani et al. [11] found that the tensile modulus and strength increase with increasing LCP content and extension ratio in the blends of more than 10% LCP content in the PET/LCP extruded strands. Using x-ray diffraction measurements and scanning electron microscopy (SEM), they demonstrated that the extruded strands of the blends consist of a crystalline and oriented LCP phase and an amorphous and unoriented PET phase. Bonis and Adur [12] studied the compatibilization of the PET/LCP blends and reported that an appropriate compatibilizer can significantly improve the mechanical properties at 10% of the LCP content. The lowering of interfacial tension between the two components and improving the molecular penetration through the interface generate better adhesion and better stress transfer from the continuous phase to the dispersed phase, and resulted in improved mechanical properties of the blend. Silverstein et al. [13] found the macroscopic skin-core morphology in injection molded bars of PET/LCP blends. This structure can be subdivided into ordered and disordered sublayers between tens and hundreds of microns thick that resulted in a hierarchical structure. The LCP orientation in the skin region reflects the elongational flow, while the core reflects the shear flow in the mold. The formation of a highly oriented and highly connected hierarchical structure governs the mechanical properties of the LCP blends.

Several researchers studied PET/LCP self-reinforced composite in which LCP was a copolyester of PET/HBA [16–20]. Yoshikai et al. [16] found the nucleating effect of LCP in PET matrix. They reported that the sonic modulus of the PET/LCP strands subjected to secondary stretching (in an oven at 80°C) increases with increasing LCP content and stretching ratio. The orientation-induced crystallization of PET was accelerated by the presence of LCP. They proposed that an increase in sonic modulus after post-treatment was mainly due to the molecular chain orientation of the PET matrix caused by the accelerated crystallization in the secondary stretching step and not from the additional deformation of the LCP domains. Mehta and Deopura [17] investigated the PET/LCP spun fibers and found that PET is compatible with LCP, and LCP hindered the crystallization of PET, which is the contrast study of Yoshikai et al. [16]. The modulus increased by about 50% after two-stage drawing in the case of 90/10 PET/LCP drawn fiber at the draw ratio of 6 in comparison with that of PET fiber. Sukhadia et al. [18] found that the rods generated more thin reinforcing fibrils than the films in the extrusion of the PET/LCP films and rods in which three LCPs were used: 20/80 PET/HBA (LCP80), 40/60 PET/HBA (LCP60), and a blend of 20/80 and 40/60 PET/HBA by 50 wt% (LCP60-80). The extensional flow fields in the converging section of capillary dies were more effective in the formation of the LCP phase into fibrils. In the film processing, long LCP fibrils of the LCP60-80 phase in PET matrix were not observed due to the high length-to-thickness ratio of the coat-hanger die and partial miscibility of the LCP and PET. Brostow et al. [19] explained the transmission of the mechanical energy through the PET/LCP blend using their island model. They represented the spheres observed in SEM figures as islands and showed that liquid-crystalline sequences of the LCP chains dominated inside the sphere while flexible sequences, either from LCP or PET chains, dominated outside. With their model, when an external mechanical force is applied to a thermoplastic/LCP blend, the lines of force avoid the islands. Zhuang et al. [20] found that the tensile strength and elastic modulus increase and elongation decreases with increasing LCP composition in extruded and melt-spun PET/LCP blends.

PBT/LCP blends have been studied by several researchers [21–30] (PBT = polybutylene terephthalate). Kimura and Porter [21] found a partial miscibility between PBT and LCP based on 40/60 PET/HBA. That is due to the compatibility of the terephthalate-rich phase with PBT. Kiss [22] reported a "gnarled" appearance of LCP phase (Vectra A) in the PBT matrix by SEM observation following melt blending. Ajji and Gignac [23] found an increase in the complex viscosity in the blends of PBT and LCP based on 40/60 PET/HBA copolyester in a batch mixer with an LCP addition up to 20%. It was due to the miscibility between LCP and PBT. At 30% LCP, viscosity decreased due to a coarse morphology. Heino and Seppälä [24] found more improvement in mechanical properties with the extruded blends than the injection-molded blends of PBT- and HBA/HNA-based LCP. It was concluded that the elongational forces are more effective for the deformation of the LCP phase than shear forces. Lee et al. [25] studied the reactive compatibilization of PBT- and 40/60 PET/HBA-based LCP blend with a catalyst. They stated that dibutyltindilaurate catalyst generates a block copolymer at the interface between components in the early stage of the blending, and more random copolymers were generated in the late stage due to transesterification. In the final stage, a homogeneous phase with a single glass transition temperature (T_g) of the blend was found. Pracella et al. [26, 27] studied the blends of PBT/LCP in which LCP is a smectic polyester, poly(decamethylene-4,4'-terephthaloyldioxy-dibenzoate) and they found a single T_g, which indicates a miscibility between components. Seo [28] found good adhesion at the interface of PBT and Vectra A LCP components using a reactive elastomer, SA-g-EPDM, as a compatibilizer, which formed a graft between components. Beery and co-workers [29] studied the flow behavior of PBT/LCP (30/70 HBA/HNA) blends with a capillary rheometer. The blend viscosities were much lower than those of the LCP at low shear rates but approached those of the LCP at high shear rates. Paci et al. [30] found a miscibility in the solution-prepared blends of PBT and a thermotropic homopolyester, poly(biphenyl-4,4'-ylene sebacate) in the thermally isotropic state with transesterification occurring at high temperature.

There have been several studies on the blends of PEN (polyethylene naphthalate) and LCPs. Kim and Jang [31, 32] studied the blends of PEN/Vectra A 950 and found a minimum viscosity of blends at 10% LCP. At high shear rates and concentrations of LCP over 50% LCP, phase inversion and fibrous blend morphology were observed. With increasing LCP content, the storage modulus improved. Kim et al. [33] found an abrupt increase in blend viscosity above 40% LCP content in the ternary blends of PET/PEN/LCP (20/80 PET/HBA). At LCP concentrations greater than 30%, a fibrillar structure of the LCP phase was obtained, resulting in an increase in the tensile strength and modulus.

The morphologies of nanofibrillated LCP phases are primarily controlled by the melt processing conditions. The degree of deformation, the size, and the shape of the LCP phase dictate the properties of polyester/LCP blends [1–4]. However, there has been little work in the literature to establish quantitative predictions of the LCP phase dimension.

Taylor [34, 35] established a deformation theory of a drop in a Newtonian fluid. He calculated the velocity and pressure fields inside and outside the droplet by solving Stoke's equation for creeping motion [34]. He expressed the deformability D of the droplet at low strain conditions as [35]

$$D = \frac{(L-B)}{(L+B)} \tag{11.1}$$

where L is the length and B is the width of the deformed ellipsoid. He derived that the deformation of a droplet and its shape depend on two dimensionless parameters: (1) the capillary number, Ca, and (2) the viscosity ratio, p. The capillary number is the ratio of the hydrodynamic stress to the interfacial stress. For the steady shear flow of a Newtonian fluid, the capillary number is expressed as

$$Ca = \frac{\text{hydrodynamic stress}}{\text{interfacial stress}} = \frac{\eta_m \dot{\gamma} \; R}{\sigma} \tag{11.2}$$

where η_m is the matrix viscosity, $\dot{\gamma}$ is the shear rate, R is the droplet radius, and σ is the interfacial tension. The viscosity ratio, p, which is the ratio of the viscosities of the dispersed phase to that of the matrix expressed as

$$p = \frac{\eta_d}{\eta_m} \tag{11.3}$$

where η_d is the viscosity of the dispersed phase.

If the interfacial tension effect dominates over the viscous effect (i.e., $Ca \ll 1$), the deformation D and the orientation angle α of the droplet are expressed as

$$D = Ca\left(\frac{19p+16}{16p+16}\right), \ \alpha = \frac{\pi}{4} \tag{11.4}$$

Within the entire range of p from zero to infinity, a value of $(19p+16)/(16p+16)$ in Equation 11.4 varies from 1.0 to 1.187; therefore, D is nearly equal to Ca.

If the interfacial tension forces are equal to viscous forces (i.e., $Ca = 1$, $p \gg 1$ [35]),

$$D = \frac{5}{4}p, \ \alpha = \frac{\pi}{2} \tag{11.5}$$

Taylor found that the experimental results of the droplet deformation were well matched with his theory at low deformation rates in both uniform shear and plane hyperbolic flow. But for the case of the interfacial tension effect and the viscous effect being comparable, Taylor's theory was not applicable.

The first-order theory for the deformation of a droplet within a full range of viscosity ratios in the general time-dependent shearing flow field was developed by Cox [36]. At a steady shear flow condition, the droplet deformation is represented as

$$D = \frac{5(19p+16)}{4(p+1)\sqrt{\left(\frac{20}{Ca}\right)^2 + (19p)^2}} \ \alpha = \frac{\pi}{4} + 0.5\tan^{-1}\left(\frac{19pCa}{20}\right) \tag{11.6}$$

There is a critical capillary number, Ca_c, beyond which the droplet can no longer sustain further deformation, and it breaks up into a number of smaller droplets. In steady-state shear flow with $p = 1$, the shape of a droplet becomes unstable for a critical capillary number, Ca_c, of the order of unity [34]. The value of Ca_c depends strongly on the viscosity ratio p and the type of flow (simple shear or extension). Utracki and Shi [37] presented the empirical relationships between the critical capillary number and viscosity ratio.

Below the critical capillary number, the drop reaches a steady state and only slightly deformed equilibrium shape in which the deformation is determined by the order of Ca. When the value of Ca is slightly higher than that of Ca_c, the breakup into smaller droplets can be predicted because of a growing disturbance at the interface. When $Ca \gg Ca_c$, the interfacial stress becomes negligibly small compared to the hydrodynamic stress, and the drop will deform affinely with the applied macroscopic deformation [38].

For time-dependent flows, the flow field in a conical channel for Newtonian fluids was calculated by Chin and Han [39] and Van der Reijden-Stolk and Sara [40] using the Cox theory [36]. Chin and Han [39], Elmendorp and Maalcke [41], and Milliken and Leal [42] considered the viscoelastic droplets or matrices.

The above-mentioned Taylor and Cox theories considered only small deformations. Later, Taylor [43] considered large deformations. He used slender body mechanics to develop a mathematical analysis with small capillary numbers and small viscosity ratios.

Huneault et al. [44] found that the drop deformation and breakup depend on the reduced capillary number, Ca^*, which is the following:

$$Ca^* = \frac{Ca}{Ca_c} \tag{11.7}$$

Depending on the value of Ca^* in the shear and elongation, the droplets will experience either deformation or breakage based on the following criteria:

1. If $Ca^* < 0.1$, there is no deformation of droplets.
2. If $0.1 < Ca^* < 1$, there is no breakup.
3. If $1 < Ca^* < 4$, there is a droplet deformation, but they break conditionally.
4. If $Ca^* > 4$, droplets deform affinely with the rest of the matrix and form long stable filaments.

In the present study, four polyesters (PET, PBT, PEN, and copolyester) were melt blended with LCP based on PET/HBA and extruded by sheet extrusion and fiber spinning processes, and the mechanical properties and morphologies of products obtained at various processing conditions were compared. The optimum processing conditions and the polyester matrix selection leading to the nanofibrillation of the LCP phase in polyester/LCP blends were identified. The mechanical properties of films, fibers, and injection molded bars were investigated and compared. A polyethylene-based compatibilizer was utilized to enhance the interfacial adhesion between the components in the injection molding of polyester/LCP blends. The development of polyester/LCP blend morphology in fiber spinning was studied. The quantitative predictions of the LCP phase dimensions under isothermal and non-isothermal conditions were performed based on the above criteria. The calculated results were compared with the experimental data.

11.2 Experimental

11.2.1 Materials

The studies were performed on four polyester matrices with LCP blends: (1) PET (Eastapak (2) PET Polyester 7352/Eastman Chemical Company) with I.V. of 0.74 dl/g and T_m of 253°C, (20 PBT (Ultradur KR 4036-Q692/BASF AG) with I.V. of 1.24 dl/g and T_m of 225°C, (3) PEN (VFR-40046/Shell Chemical Company) with I.V. of 0.64 dl/g and T_m of 275°C, and (4) copolyester (Eastpak EBM PET Copolyester 13339/Eastman Chemical Company) with I.V. of 1.05 dl/g and T_m of 235°C. The thermotropic LCP used as a reinforcing phase was 18/82 PET/HBA copolyester (Rodrun LC-5000/Unitika Ltd.) with a random chain structure [45] and T_m of 278°C. Chemical structures of polyesters and LCP are shown in Table 11.1 and Table 11.2, respectively. In addition, polyethylene/glycidyl methacrylate (LDPE/GMA) 92/8 (Lotader AX 8840/Elf Atochem North America) was used as a compatibilizer for the blends. Its melting point, T_m, was 110°C. The chemical structure of the compatibilizer is shown in Table 11.2.

TABLE 11.1

The Chemical Structures of Thermoplastic Polyester Matrices

PET

$$-\left[-O\text{-}CH_2\text{-}CH_2\text{-}O\text{-}\overset{O}{\overset{\|}{C}}\text{-}\langle\bigcirc\rangle\text{-}\overset{O}{\overset{\|}{C}}-\right]-$$

I.V. = 0.74 dl/g
T_m = 253°C
Density = 1.400 g/cm^3

PBT

$$-\left[-O\text{-}(CH_2)_4\text{-}O\text{-}\overset{O}{\overset{\|}{C}}\text{-}\langle\bigcirc\rangle\text{-}\overset{O}{\overset{\|}{C}}-\right]-$$

I.V. = 1.24 dl/g
T_m = 255°C
Density = 1.398 g/cm^3

PEN

$$-\left[-O\text{-}CH_2\text{-}CH_2\text{-}O\text{-}\overset{O}{\overset{\|}{C}}\text{-}\langle\bigcirc\bigcirc\rangle\text{-}\overset{O}{\overset{\|}{C}}-\right]-$$

I.V. = 0.64 dl/g
T_m = 275°C
Density = 1.407 g/cm^3

Copolyester

$$-\left[-O\text{-}CH_2\text{-}\langle\bigcirc\rangle\text{-}CH_2\text{-}O\text{-}\overset{O}{\overset{\|}{C}}\text{-}\langle\bigcirc\rangle\text{-}\overset{O}{\overset{\|}{C}}-\right]-$$

I.V. = 1.05 dl/g
T_m = 235°C
Density = 1.400 g/cm^3

TABLE 11.2

The Chemical Structures of LCP and Compatibilizer

Rodrun LC-5000

PRT/HBA
Poly(ethylene terephthalate)/*p*-hydroxy benzoic acid
(18/82)

$$-\left[-O\text{-}CH_2\text{-}CH_2\text{-}O\text{-}\overset{O}{\overset{\|}{C}}\langle\bigcirc\rangle\text{-}\overset{O}{\overset{\|}{C}}-\right]_{18}\left[-O\text{-}\langle\bigcirc\rangle\text{-}\overset{O}{\overset{\|}{C}}-\right]_{82}-$$

T_m = 278°C
Density = 1.380 g/cm^3

Lotader AX 8840

LDPE/GMA
polyethylene/glycidyl methacrylate
(92/8)

$$-\left[-CH_2\text{-}CH_2-\right]_{92}\left[-CH_2\text{-}\underset{\substack{|\\C=O\\|\\O\\|\\CH_2\text{-}CH\text{-}CH_2\\\diagdown O \diagup}}{\overset{\overset{CH_3}{|}}{CH}}-\right]_{8}-$$

T_m = 110°C

11.2.2 Preparation of Polyester/LCP Film and Fiber

The materials were dried in a vacuum oven at 120°C for 24 hours. Then they were physically mixed to make four polyester/LCP compositions by weight: 90/10, 80/20, 70/30, and 60/40. Pure polyester films and fibers were also prepared for comparison purposes. The melt blending was performed using a 1-inch single-screw extruder (Killion Inc.), followed by a six-element static mixer (Koch) at a melt temperature of 285°C in the screw zones, except for a melt temperature of 290°C for PEN/LCP in the screw zones. For the sheet extrusion, a coat-hanger die with a die gap of 1 or 2 mm was attached to the static mixer. A chill-roll with water circulation was used as a take-up and cooling device. It was positioned 5 mm below the die. Extension ratios were calculated from the flow rate in the dies and the take-up speed. The processing conditions for film extrusion with two die gaps are shown in Table 11.3. Depending on the extension ratio and die gap, the thickness of prepared films ranged from 0.02 to 0.23 mm.

For the fiber spinning, a 90° angle fitting die of 10-mm diameter was attached to the static mixer to prevent flow fluctuation due to gravity, and the processing temperature was the

TABLE 11.3

Processing Conditions in Making PET/LCP Films using a Coat-Hanger Die with Gaps of 1 and 2 mm

Temperatures (°C) in Various Zones:						
Zone 1	Zone 2	Zone 3	Zone 4	Zone 5	Static Mixer	Coat-hanger Die
260	285	285	285	285	285	260

	Pressure (psi)		Take-up	Extension Ratio		Shear Rate (sec^{-1})	
PET/LCP	1 mm	2 mm	Speed (FPM)	1 mm	2 mm	1 mm	2 mm
100/0	320–340	260–280	50	—	11.3	38.0	9.9
			100	12.5	24.3		
			150	—	33.7		
			200	21.6	42.2		
			300	35.0	—		
90/10	180–200	160–180	50	—	11.7	36.7	9.5
			100	13.1	25.3		
			150	—	35.0		
			200	22.8	43.9		
			300	36.9	—		
80/20	160–180	110–130	50	—	12.4	34.1	9.0
			100	14.1	26.8		
			150	—	37.2		
			200	24.5	46.5		
			300	39.8	—		
70/30	90–110	80–100	50	—	13.6	32.8	8.2
			100	14.7	29.4		
			150	—	40.8		
			200	25.5	51.0		
			300	41.3	—		
60/40	50–60	50–60	50	—	13.8	32.4	8.1
			100	14.9	29.8		
			150	—	41.4		
			200	25.8	51.8		
			300	41.9	—		

TABLE 11.4

Processing Conditions in Making PET/LCP Fibers using 90° Angle Fitting Die with a Diameter of 10 mm at Screw Speeds of 40 and 70 rpm

Temperatures (°C) in Various Zones:						
Zone 1	Zone 2	Zone 3	Zone 4	Zone 5	Static Mixer	Coat-hanger Die
260	285	285	285	285	285	240

	Pressure (psi)		Take-up Speed (rpm)	Extension Ratio		Shear Rate (sec⁻¹)	
PET/LCP	40 rpm	70 rpm		40 rpm	70 rpm	40 rpm	70 rpm
100/0	100–120	230–250	81.5	57.4	—	5.9	13.4
			108.2	143.4	85.7		
			134.2	190.3	120.8		
			147.4	195.6	124.8		
90/10	40–60	100–120	81.5	140.1	57.8		
			108.2	257.7	101.0	4.8	8.7
			134.2	273.2	108.5		
			147.4	360.1	120.0		
80/20	40–50	90–110	81.5	129.4	73.2	4.3	7.8
			108.2	175.4	105.0		
			134.2	200.6	115.1		
			147.4	278.7	139.1		
70/30	30–50	60–80	81.5	173.1	59.5	3.9	7.0
			108.2	244.9	92.5		
			134.2	279.6	105.6		
			147.4	312.2	130.9		
60/40	10–30	50–60	81.5	178.3	—	2.9	—
			108.2	210.0	—		
			134.2	346.8	—		
			147.4	430.4	—		

same as in the film processing for the purpose of comparison. The processing conditions for the fiber spinning are given in Table 11.4. The extrudate from the die was cooled in a water bath and drawn by a take-up device. Extension ratios were calculated as a ratio of die cross-sectional area to that of fibers. Diameters of prepared fibers ranged from 0.48 to 1.32 mm.

In preliminary experiments, various die temperatures were investigated to find the acceptable die temperature range and the optimum die temperature in the 70/30 polyester/LCP films and fibers, as shown in Table 11.5. The definition of acceptable die temperature

TABLE 11.5

Comparison of Acceptable Die Temperature and Optimum Die Temperature in 70/30 Polyester/LCP Films and Fibers

	Acceptable Die Temperature (°C)		Optimum Die Temperature (°C)	
Matrices	Films	Fibers	Films	Fibers
PET	250–280	230–255	260	240
PBT	250–275	255–285	265	270
PEN	255–285	230–285	260	265
Copolyester	235–285	225–285	265	240

range was limited to the temperatures at which films or fibers can be obtained without a dimensional fluctuation or breakup during take-up. Due to the low melt viscosity of blends at high temperatures, the dimensional fluctuation usually occurs during take-up, while breakup occurs due to the high melt viscosity of blends at low temperatures. The optimum die temperature was decided when the temperature generated the best mechanical properties of films or fibers. In fiber spinning, a more complex change in the viscosity ratio was discovered in comparison with that in films due to the high extension ratio applied and less uniform temperature distribution in the melt stream during spinning. In fiber spinning, the PET and the copolyester showed an optimum die temperature of 240°C, while the PEN and the PBT showed an optimum temperature of 265°C and 270°C, respectively. The discovered optimum die temperatures were applied for film casting and fiber spinning of blends of various LCP concentrations.

11.2.3 Rheological Measurements

The viscosity-shear rate relationship for polyester and LCP blends at various temperatures was measured by means of an Instron Capillary Rheometer (Model 3211). Three capillary dies with a diameter of 1.0668 mm and L/D ratios of 15.7, 30, and 44.5 were used. The diameter of the barrel was 9.525 mm. Five plunger speeds (0.1, 0.3, 1.0, 3.0, 10.0 in./min) were applied. The Bagley end correction was applied for the calculation of viscosities at various temperatures.

11.2.4 Injection Molding

The polyester/LCP fibers were cut into the pellets with pelletizers and dried at room temperature for 24 hours and then in a vacuum oven at 100°C for 4 hours to remove residual moisture, and then injection molded using a Boy 15S injection molding machine. The clamp force was 24 tons and the shot size was $5 \times 10^{-5}\,\text{m}^3$. The melt temperatures for injection molding were 240°C, 220°C, 260°C, and 240°C for PET, PBT, PEN, and copolyester, respectively, and the mold temperature was 30°C. Injection speed setting was 100%. The prepared dumbbell-shaped injection molding bars have the dimensions of $0.0635 \times 0.0031 \times 0.0015$ m.

11.2.5 Tensile Testing

Dumbbell-shaped specimens (ASTM D638 type IV) were cut from the extruded films. The fiber strands were attached to paper tabs. The tensile strength and modulus of films, fibers, and injection molded bar samples were examined using an Instron mechanical testing machine (Model 4204) at a crosshead speed of 5 mm/min at room temperature. The diameters of fiber strands were measured each time because the diameters were varied with processing conditions. The gauge length for fiber strands was fixed at 24.5 mm. For the films and the injection-molded bars, gauge lengths were 40 and 9.5 mm, respectively. For each sample, at least five specimens were tested.

11.2.6 Morphology

Using a scanning electron microscope (Hitachi S-2150), morphological studies on the fractured surface of polyester/LCP blends were performed using the SEM (Hitachi S-2150).

The samples were placed in liquid nitrogen and fractured in the direction perpendicular to flow. The fractured samples were glued on the aluminum sample holders. A coating device was used to coat the fractured surface with a gold-palladium alloy. The SEM photomicrographs of injection-molded bars were taken in the skin region about 20 to 50 μm from the surfaces with the magnification of 2000X. An image analyzer (Leica Q500MC) was used to calculate the average droplet dimensions.

11.2.7 Compatibilizer

A compatibilizer was incorporated during the injection molding of polyester/LCP blends by tumble mixing before molding. Then 1% of the compatibilizer was mixed with 80/20 and 60/40 polyester/LCP blends. Also, 5% of the compatibilizer was mixed with the 60/40 polyester/LCP blends to get the comparison data. The melt temperature in moldings of PET, copolyester, and PBT matrices was 285°C. For the PEN matrix, the melt temperature was 290°C.

11.3   Results and Discussion

11.3.1   Mechanical Properties of PET/LCP Films

The tensile strength and Young's modulus in the machine direction as a function of the extension ratio for 70/30 PET/LCP film obtained at various die temperatures are shown in Figures 11.1a and b. For the film casting, the coat-hanger die gap was 1 mm and the screw speed was maintained at 70 rpm, while the barrel temperature of the single-screw extruder was kept at 285°C. Both the tensile strength and Young's modulus increased with increasing extension ratio at various die temperatures. At any extension ratio, the maximum tensile properties were achieved at the die temperature of 260°C. The tensile properties were plotted as a function of die temperature for 70/30 PET/LCP films obtained at extension ratios of 14, 24, and 38 in Figure 11.2a and b. The tensile properties were calculated at these three extension ratios by the interpolation of data in Figure 11.1. It is obvious that the highest tensile properties of 70/30 PET/LCP film were obtained at the die temperature of 260°C, regardless of extension ratio. With an extension ratio of 38, the tensile strength of 154 MPa and modulus of 9 GPa were achieved at that die temperature. The tensile strength decreased to 140 MPa at 265°C and there was no change at higher temperatures. The tensile strength dropped significantly at temperatures below 260°C. The maximum Young's modulus values were obtained at 260°C, regardless of the extension ratio. The Young's modulus dropped by 1 GPa from its maximum value at temperatures below and above 260°C. The die temperature selection is an important factor in obtaining the maximum tensile properties of PET/LCP films, which is evidenced by the data depicted in Figure 11.1 and Figure 11.2.

The shape and aspect ratio of formed LCP fibrils in the PET matrix are very sensitive to the viscosity ratio of PET and LCP melts at the die outlet. In general, a lower die temperature is desired to obtain a high viscosity of matrix polymer at the die exit. However, the deformation of the LCP phase is not easily obtained at low die

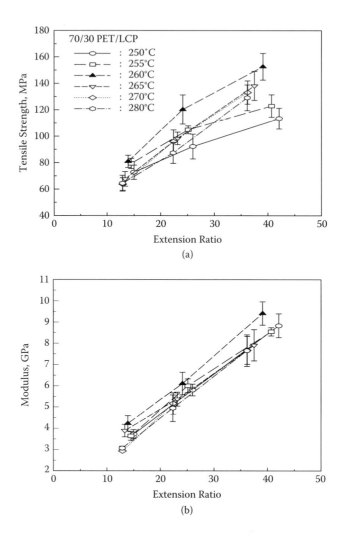

FIGURE 11.1
Tensile strength (a) and modulus (b) in the machine direction vs. extension ratio for 70/30 PET/LCP films at various coat-hanger die temperatures. Die gap is 1 mm.

temperatures because the LCP has a high melting point. On the other hand, a lower viscosity of LCP melt than that of thermoplastic melt is desirable to get the highly elongated LCP phase [1]. Therefore, the search for an optimum die temperature that promotes fibrillation is highly recommended to obtain high mechanical properties of PET/LCP composites.

The importance of the viscosity ratio of blend components in the formation of fibrillar structures in thermoplastic/LCP blends has been stressed by many researchers [1, 8, 45–53]. They suggested that the fibrillation of the LCP phase mostly occurred when the viscosity ratio of LCP to thermoplastic matrix is less than unity. If the viscosity of the dispersed phase (LCP) is relatively small compared to that of the matrix phase (PET), the LCP domains can be easily deformed into continuous long fibers. If the viscosity ratio is greater than unity, the fibrillation of the LCP phase cannot be observed

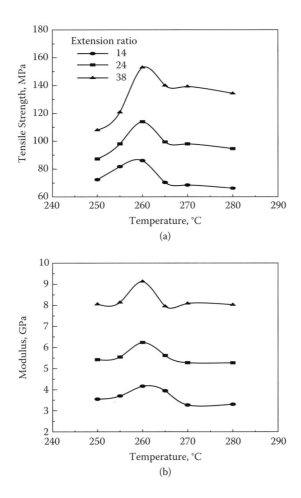

FIGURE 11.2
Tensile strength (a) and modulus (b) in the machine direction vs. coat-hanger die temperature for 70/30 PET/ LCP films at various extension ratios. Die gap is 1 mm.

[54, 55]. The melt viscosity dependence on the shear rate for pure LCP and PET at three temperatures was measured by means of a capillary rheometer. The results are shown in Figure 11.3. The dependence of LCP melt viscosity on temperature is stronger than that of PET melt viscosity in the entire range of shear rates. The crossover of the flow curves of LCP and PET at 260°C occurred at the shear rate 100/s and fibrillation of LCP can be expected above this shear rate. The LCP melt becomes less viscous than the PET melt at higher shear rates, and it creates a possibility of fibrillation with a large aspect ratio of LCP fibrils.

The tensile strength and Young's modulus of PET/LCP films in the machine direction vs. the extension ratio were plotted at various LCP concentrations in Figures 11.4a and b. The die gap was set at 1 mm and the temperature of die was maintained at 260°C. Both the tensile strength and Young's modulus increased with increasing extension ratio and LCP composition. Many researchers studied the effect of the extension ratio on thermoplastic/LCP composites. Dutta et al. [56] and Lin and Yee [57] found

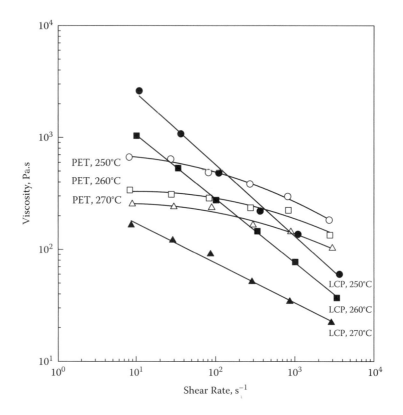

FIGURE 11.3
Viscosity vs. shear rate curves for pure PET and LCP at different temperatures.

that the aspect ratio and molecular chain orientation of the LCP phase increase with increasing extension ratio. The degree of LCP droplet deformation highly depends on the viscosity ratio of the components. The particle size is an important factor for the deformation as well. It is easier to get the LCP phase deformation if the particle size is large, and therefore a larger aspect ratio of fibrils can be obtained. For a large deformation of LCP domains, a sufficient LCP concentration and an appropriate shear rate should be applied.

The effect of LCP concentration on the mechanical properties of thermoplastic/LCP composites has been studied by many researchers [1, 19, 20, 57–59]. They agreed that high LCP content is more favorable for large deformation of LCP domains and high mechanical properties. In the present study of PET/LCP films, 30 % LCP composite generated somewhat higher tensile strength than that of 40 % LCP composite at the highest extension ratio of 42. This is due to the higher fibrillation and more uniform distribution of LCP fibrils in 30 % LCP composite. Therefore, it can be concluded that a higher LCP concentration is necessary but not sufficient for the better reinforcement in PET/LCP films at high extension ratios. Presumably, there exists a critical LCP concentration and extension ratio that can generate the highest tensile strength. The SEM photomicrographs in Figures 11.5a and b illustrate this phenomenon. Only a few micron diameters of LCP

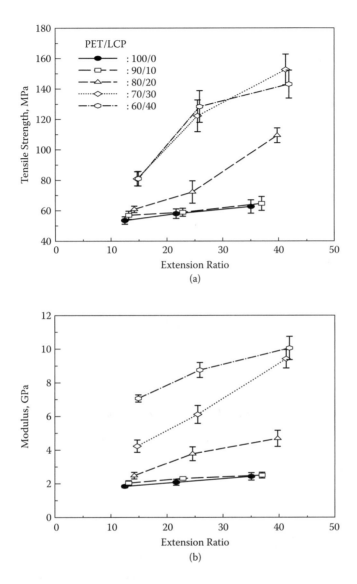

FIGURE 11.4
Tensile strength (a) and modulus (b) in the machine direction vs. extension ratio for PET/LCP films at various LCP concentrations. Die gap is 1 mm.

fibrils in the 70/30 PET/LCP films were observed. They were long and uniformly distributed. The 60/40 PET/LCP film revealed the smaller aspect ratio and somewhat irregular distribution of the LCP phase than the 70/30 PET/LCP film. In the 60/40 PET/LCP film, only a partial amount of LCP droplets were deformed into fibrillar structure and some droplets of ellipsoidal shape were also found in the remainder of the LCP phase. However, the maximum Young's modulus of films was obtained at the 60/40 PET/LCP composition, as shown in Figure 11.5b. Therefore, a high LCP content produces a better stiffness of the PET/LCP films.

(a) 70/30 PET/LCP, Extension Ratio = 42

(b) 60/40 PET/LCP, Extension Ratio = 42

FIGURE 11.5
SEM photomicrographs of 70/30 (a) and 60/40 (b) PET/LCP films at the extension ratio of 42. Die gap is 1 mm.

The tensile properties of PET/LCP films in the transverse direction as a function of extension ratio at various LCP contents are shown in Figures 11.6a and b. The properties decreased slightly at all LCP compositions when the extension ratio increased. At higher LCP concentrations, the larger decrease in tensile properties was observed, which is in agreement with the earlier observation for other thermoplastic/LCP films [58]. This is due to the poor interfacial adhesion between the PET matrix and the LCP fibrillar domain. With increasing LCP content, the interfacial area of the LCP phase increases and it can reduce the tensile properties in the transverse direction. As shown in the photomicrographs of Figures 11.5a and b, the LCP based on PET/HBA is incompatible with the PET matrix even though LCP contains PET moieties in the main chain randomly. The interface between the two components has insufficient adhesion characteristics to keep the transverse tensile strength and modulus the same as those for pure PET material. The transverse properties of thermoplastic/LCP composites can be improved if the interface is enhanced using a coupling agent or other third material [60, 61]. Chin et al. [60] used epoxy resin as a coupling agent for the blending of PET and HBA/HNA-based LCP. Lee and DiBenedetto [61] used a second LCP, which is based on HBA/HNA, as a compatibilizer for the blend of PET

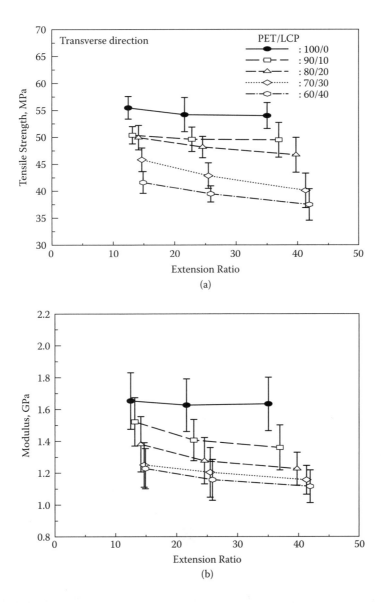

FIGURE 11.6

Tensile strength (a) and modulus (b) in the transverse direction vs. extension ratio for PET/LCP films at various LCP concentrations. Die gap is 1 mm.

and a wholly aromatic copolyester (K161) and found improved adhesion and dispersion over the binary blends of PET/K161.

The variation in mechanical properties is plotted in Figures 11.7a and b as a function of LCP composition in both the machine and transverse directions at three extension ratios. A significant increase in the tensile strength and modulus in the machine direction with increasing LCP concentration is observed, accompanied by some decay in the properties in the transverse direction. In addition, the effect of the extension ratio on properties is larger in the machine direction than that in the transverse direction.

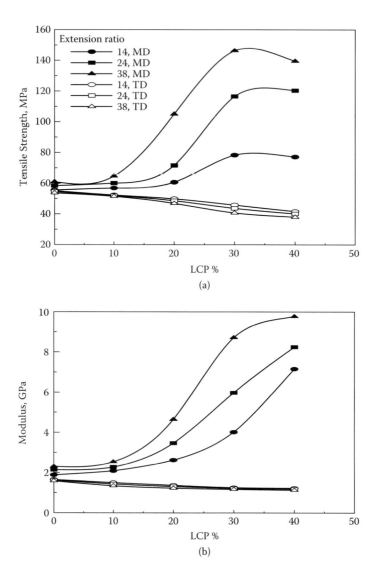

FIGURE 11.7
Tensile strength (a) and modulus (b) vs. LCP concentration for PET/LCP films in the machine (MD) and transverse (TD) directions at various extension ratios. Die gap is 1 mm.

Table 11.3 provides the processing temperature and calculated shear rates from the coathanger die with a gap of 1 mm for the various compositions of LCP in the PET/LCP film. The apparent shear rates ranged from 32 to 38/s. The effect of shear rate in the coat-hanger die was studied by changing the die gap to 2 mm. The processing conditions are listed in Table 11.3. The larger die gap generated the lesser shear rates in the coat-hanger die (8 to 10/s), and less pressure developed in the die section of the single-screw extruder. As shown in Figure 11.3, the melt viscosity of LCP increases with decreasing shear rate in the range from 100 to 10/s, while PET viscosity remains almost constant. Therefore, at low shear rates, the deformation and orientation of LCP molecules are more difficult to accomplish. In Figures 11.8a and b, the tensile strength and modulus of PET/LCP films obtained at low shear rates occurring in

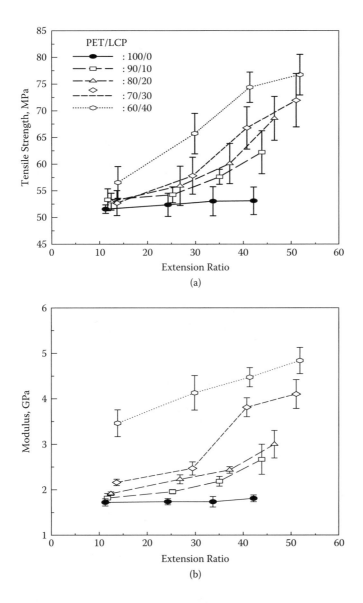

FIGURE 11.8
Tensile strength (a) and modulus (b) in the machine direction vs. extension ratio for PET/LCP films at various LCP concentrations. Die gap is 2 mm.

the coathanger of the larger die gap are presented. The LCP domains at 40% LCP composite seemed to be well deformed and revealed higher mechanical properties than 30% LCP composite, but the maximum values of tensile strength and modulus were significantly lower in comparison with the properties of films (Figures 11.4a and b) obtained at a shear rate of about 30/s in the coat-hanger die with a die gap of 1 mm. The maximum tensile strength of 80 MPa and the maximum modulus of 5 GPa were achieved at 60/40 PET/LCP at a die gap of 2 mm. These values were only half of the maximum values at a die gap of 1 mm. Therefore, it can be concluded that the shear rate in the coat-hanger die is very important for the mechanical properties of the PET/LCP film. For a better fibrillation of the LCP phase and improved

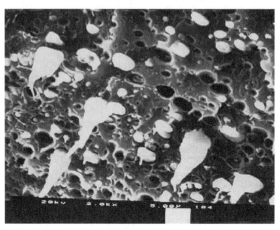

(a) 70/30 PET/LCP, Extension Ratio = 40

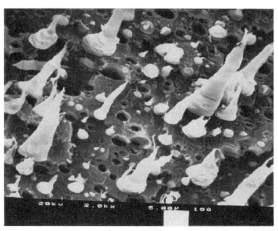

(b) 60/40 PET/LCP, Extension Ratio = 40

FIGURE 11.9
SEM photomicrographs of 70/30 (a) and 60/40 (b) PET/LCP films at the extension ratio of 40. Die gap is 2 mm.

mechanical properties, higher shear rate is essential. The SEM photomicrographs, shown in Figure 11.9, illustrate the morphology of PET/LCP films obtained at a die gap of 2 mm. More fibrillation of the LCP phase was observed in the 60/40 PET/LCP composition than in the 70/30 PET/LCP, but the degree of fibrillation is less uniform and the large aspect ratio of the LCP domains was not observed in comparison with the photomicrographs shown in Figure 11.5 for films obtained at the die gap of 1 mm.

11.3.2 Mechanical Properties of PET/LCP Fibers

The fibers of PET/LCP blends of various LCP compositions were obtained by stretching the melt strand of blends exiting the die of circular cross-section. In fiber spinning, a higher extension ratio was obtained than that in film casting. The extension ratio was defined as the ratio of the cross-sectional area at the die exit to that of the fiber. The tensile strength and Young's modulus of fibers obtained at the screw speeds of 70 rpm and 40 rpm were plotted as a function

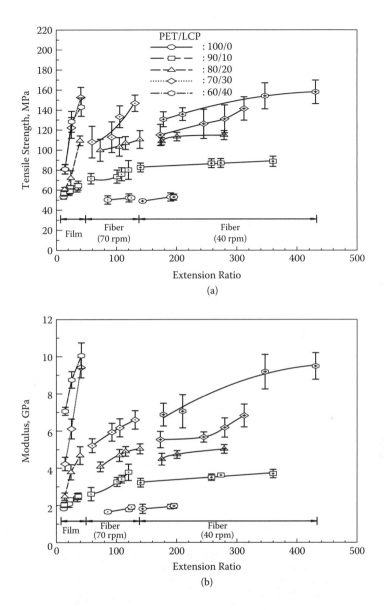

FIGURE 11.10
Tensile strength (a) and modulus (b) in the machine direction vs. extension ratio for PET/LCP films and fibers at various LCP concentrations.

of the extension ratio and are shown in Figures 11.10a and b. The properties of PET/LCP films were also plotted in the same graph for comparison purposes. It is seen that the curves for films and fibers were obtained in different ranges of the extension ratios due to the variation in die geometry and, in the case of fibers, due to screw rpm. In all cases, both the tensile strength and Young's modulus increase with increasing extension ratio and LCP content. The rate of increase is seen to be highest for films in comparison with that of fibers. The effect of LCP content was almost the same at three different conditions. The increase in screw rpm generates a higher shear rate in the die exit, and therefore provides a higher molecular orientation of LCP domains. Ko et al. [7] found an increase in the modulus of the PET/LCP films with increasing screw speed, even at low compositions of LCP based on HBA/HNA.

The highest properties of fibers were obtained for the 60/40 PET/LCP blend at a screw speed of 40 rpm. Such values could not be obtained at a screw speed of 70 rpm. The high screw speed requires the high take-up speed to obtain a large extension ratio, but the 60/40 PET/LCP fibers were easily broken during spinning at such high take-up speeds. Therefore, low screw speed is desirable to obtain large extension ratios in the case of blends with high LCP content. The maximum tensile strength of films and fibers ranged from 150 to 160 MPa but the modulus showed a higher value in the case of films (10 GPa) than in the case of fibers (9 GPa).

The SEM photomicrographs of the 70/30 PET/LCP and 60/40 PET/LCP fibers obtained at a screw speed of 40 rpm and high extension ratios are shown in Figures 11.11a and b, respectively. The finely and uniformly distributed LCP fibrils in the fibers of 60/40 PET/LCP blends were obtained even though a low shear rate in the die was applied. In particular, the diameter

(a) 70/30 PET/LCP, Extension Ratio = 320

(b) 60/40 PET/LCP, Extension Ratio = 440

FIGURE 11.11
SEM photomicrographs of 70/30 (a) and 60/40 (b) PET/LCP fibers obtained at high extension ratios and a screw speed of 40 rpm.

of these fibrils was in the nano-scale range, indicating the possibility of manufacturing nano-composites of PET/LCP blends. This is due to the very high extension ratio applied, and the lack of molecular orientation due to low shear rate can be easily compensated by the high extension ratio. Heino and Seppälä [8] studied the PET/LCP fiber strands and concluded that the effect of draw ratio and LCP content on the improvement of mechanical properties is more significant at higher LCP content (20 to 30 wt%). Farris et al. [62] found a 40% modulus increase of 80/20 PET/LCP fibers compared to PET fibers after their post-treatment (hot or cold drawing). Yoshikai et al. [16] reported an increase in sonic modulus after secondary stretching of blends of PET and LCP based on PET/HBA. The increase in modulus is caused by acceleration of orientation-induced crystallization of PET. Their second step (stretching in an oven) provided a molecular chain orientation of the PET matrix, but the deformation of LCP domains could not be performed due to low oven temperatures.

The SEM photomicrographs of 70/30 PET/LCP fibers obtained at two extension ratios and at a screw speed of 70 rpm are shown in Figures 11.12a and b. In this case, the fibrillar

(a) 70/30 PET/LCP, Extension Ratio = 60

(b) 70/30 PET/LCP, Extension Ratio = 140

FIGURE 11.12
SEM photomicrographs of 70/30 PET/LCP fibers obtained at an extension ratio of 60 (a) and 140 (b) and a screw speed of 70 rpm.

structure of LCP domains in PET matrix is seen only at a higher extension ratio of 140. At a low extension ratio of 60, the LCP phase was seen to be in the form of spheres.

11.3.3 Injection Molding of PET/LCP Composites

By chopping the PET/LCP fibers, pellets were made for the injection molding. Injection molding was performed at the melt temperature below the melting point of the LCP phase to keep the original LCP fibrillar structure intact and avoid the destruction LCP fibrils in the injection molding step [5]. As shown in Figures 11.13a and b, the tensile strengths of

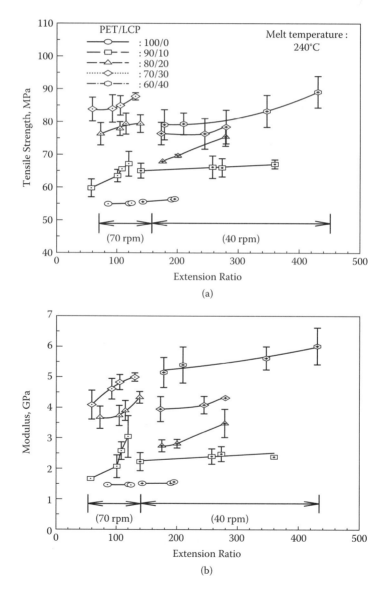

FIGURE 11.13
Tensile strength (a) and modulus (b) in the machine direction vs. extension ratio for injection molded PET/LCP bars at various LCP concentrations.

moldings at the highest extension ratio and LCP content are found to decrease by approximately half compared with the values obtained on fibers before injection molding. The Young's modulus data also revealed a reduction of about 40% after injection molding at the highest extension ratio and LCP content. The observed reduction is due to the breakup and relaxation of the molecular chain orientation of LCP fibrils caused by very high shear rate in injection molding. Due to viscous heating at high shear rates, a higher melt temperature than the setting temperature is induced and it may also lead to degradation of the initial LCP fibrous structure shown previously in Figures 11.11a and b. A clear indication of this effect can be inferred from a comparison of morphologies before injection molding in Figures 11.11a and b and after injection molding in Figures 11.14a and b. The SEM photomicrographs of 60/40 PET/LCP injection molded bars in Figures 11.14a and b clearly illustrate the breakup of LCP fibrous domains and their subsequent agglomeration

(a) 90/10 PET/LCP Injection Molded Bar(extension ratio = 360, screw speed = 40 rpm)

(b) 60/40 PET/LCP Injection Molded Bar (extension ratio = 430, screw speed = 40 rpm)

FIGURE 11.14
SEM photomicrographs of injection molded bars of 90/10 (a) and 60/40 (b) PET/LCP content.

in injection molding samples. On the other hand, the same effect of LCP content and extension ratio on mechanical properties of moldings was found as in the case of fibers and films.

Moldings prepared from the fibers obtained at an extension ratio of 420 showed the highest properties. The tensile strength and modulus of the moldings were 90 MPa and 6 GPa, respectively. McLeod and Baird [6] achieved a tensile strength of 79 MPa and modulus of 6.7 GPa for the 70/30 PET/LCP injection molded bars processed under the LCP melting temperature in which LCP is an amorphous thermotropic LCP (HX1000) made by Du Pont. They suggested that mechanical properties in the transverse direction for the injection-molded PET/LCP composite are enhanced by controlling the barrel temperature, which is lower than the LCP melting point.

11.3.4 Films and Fibers of Polyester/LCP Composites

Using a single-screw extruder, the mixture of 70/30 copolyester/LCP was melt-blended at a melt temperature of 285°C in the screw zones. For film casting, the coat-hanger die temperature was varied from 235°C to 285°C. No breakup or dimensional fluctuation of the film during the casting was observed. In an earlier study of 70/30 PET/LCP film [2], the die temperature variation ranged from 250°C to 280°C. The wider die temperature window in casting the copolyester/LCP film than that of the PET/LCP film was applicable due to the lower melting temperature of copolyester (235°C) than that of PET (253°C). At a die temperature of 265°C, the highest tensile strength and Young's modulus of the 70/30 copolyester/LCP film were obtained, while in the case of 70/30 copolyester/LCP fiber the highest tensile properties were achieved at a die temperature of 240°C and extension ratios of 90, 110, and 130. This fact illustrates that the variation in die geometry requires a different die temperature setting for achieving the high reinforcement.

The tensile strength and Young's modulus of the copolyester/LCP films and fibers in the machine direction are plotted in Figures 11.15a and b as a function of extension ratio at various LCP concentrations. At same LCP concentrations, the tensile strength of fibers was higher than that of films (Figure 11.15a). During fiber spinning, a higher extension ratio was applied, causing a higher deformation of the LCP domain. Dutta et al. [56] studied PC/LCP films and fibers in which the LCP was Vectra RD500 and found that fiber spinning is more effective for the alignment of the LCP molecules into fibrils as demonstrated by a higher-order parameter and a higher average aspect ratio of LCP phase. The modulus of the copolyester/LCP fibers increased up to a factor of 2 in comparison with the copolyester/LCP films at the highest extension ratios utilized. In addition, with increasing LCP concentration, both the tensile strength and modulus of copolyester/LCP films in the machine and transverse direction, respectively, increased and decreased, as shown in Figures 11.16a and b. The tensile strength and modulus as high as 200 MPa and 11.5 GPa were, respectively, achieved for the 60/40 copolyester/LCP fiber at an extension ratio of 250. The skin-core morphologies of the 60/40 copolyester/LCP fiber at an extension ratio of 250 were compared in Figure 11.17. In the skin region, more tiny and fibrillated LCP domains appeared than in the core region. Evidently, these two areas experienced a different thermal history during cooling. The lower diameter and aspect ratios of LCP domains were observed in the SEM photomicrographs of 60/40 copolyester/LCP fiber than that in the PET/LCP fibers [2]. This is due to a higher viscosity of the copolyester matrix, which is attributed to the cyclohexane ring in the flexible chain, than that of the PET, as shown by viscosity data depicted in Figure 11.18.

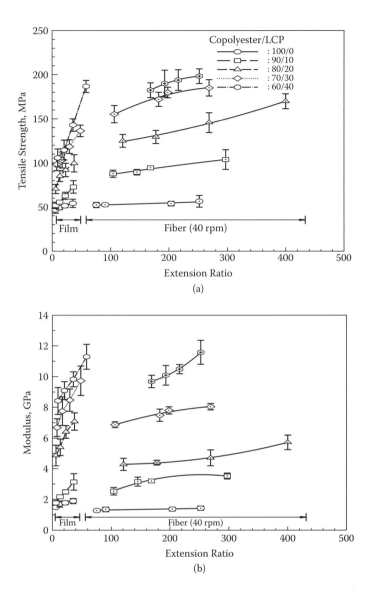

FIGURE 11.15
Tensile strength (a) and modulus (b) in the machine direction vs. extension ratio for copolyester/LCP films and fibers at various LCP concentrations.

In the 70/30 PBT/LCP blend, the highest tensile properties were obtained at die temperatures of 265°C for films and 270°C for fibers. Baird and Ramanathan [63] stressed the importance of the temperature dependence of the viscosity on the fiber spinning of thermoplastic/LCP blends. They demonstrated that the formation of a fibrillar LCP structure is strongly dependent on this parameter, rather than on the viscosity ratio at the extrusion temperature. However, it should be noted that the viscosity of the LCP phase should be lower than that of the matrix in order to generate LCP fibrils in a thermoplastic

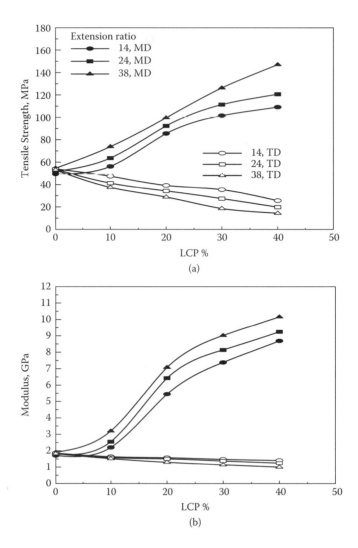

FIGURE 11.16
Tensile strength (a) and modulus (b) vs. LCP concentration for copolyester/LCP films in the machine (MD) and transverse (TD) direction at various extension ratios.

matrix [1]. In addition, it seems that the temperature dependence of viscosity of two components should be comparable, leading to similar conditions during solidification of the two phases.

Figures 11.19a and b demonstrate that the tensile strength and Young's modulus of PBT/LCP films and fibers in the machine direction increase with an increase in the extension ratio. In addition, Figures 11.20a and b show that tensile properties in the machine and transverse directions increase and decrease, respectively, with increasing LCP content. In the case of fibers, tensile properties increased significantly when the LCP concentration increased up to 30%, but increased only slightly at 40% LCP. The partial miscibility between the PBT and LCP components may play an important role in this composition [21, 23, 26, 27, 30]. Due to this effect, the fibrillation of the LCP domain might be hindered.

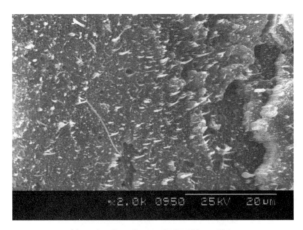

(a) 60/40 Copolyester/LCP Fiber, Skin

(b) 60/40 Copolyester/LCP Fiber, Core

FIGURE 11.17
SEM photomicrographs of skin (a) and core (b) for 60/40 copolyester/LCP fiber at an extension ratio of 250.

As shown in Figures 11.21a and b, it is obvious that the small extension of the 30/70 PBT/LCP melt stream provides highly elongated LCP particles in the PBT matrix during film casting. However, with further extension, more fibrillation or breakup of the LCP domain will occur.

The die temperature dependence on the mechanical properties of 70/30 PEN/LCP blends was studied; it was found that the optimum die temperatures were 260°C for films and 265°C for fibers. As shown in Figures 11.22a and b and Figures 11.23a and b, the tensile strength and Young's modulus in the machine and transverse directions increased and decreased, respectively, with increasing LCP concentration and extension ratio. Higher values were obtained in fibers than in films. Jang and Kim [32] also found that tensile strength and flexural modulus improved with increasing LCP content in the blends of PEN and Vectra A 950. The shape of the LCP domains changed from ellipsoidal particles to fibrillar structures with increasing LCP content. However, the morphologies of the

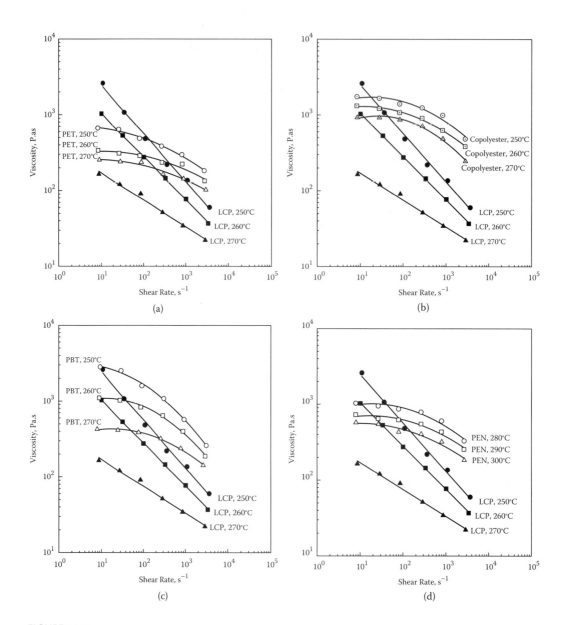

FIGURE 11.18
Viscosity vs. shear rate curves for (a) pure PET and LCP, (b) copolyester, (c) PBT, and (d) PEN at three different temperatures.

PEN/LCP films, shown in Figures 11.24a and b, demonstrated that LCP domain structure changed from ellipsoidal to a ribbon-like structure with increasing LCP content. In this case, a coalescence of the LCP particles seems to be dominant instead of breakup. These effects lead to the formation of large particles, causing the formation of a ribbon-like structure without breakup due to the larger viscosity of the PEN in comparison with the LCP. The viscosity of the LCP decreased significantly with an increase in temperature

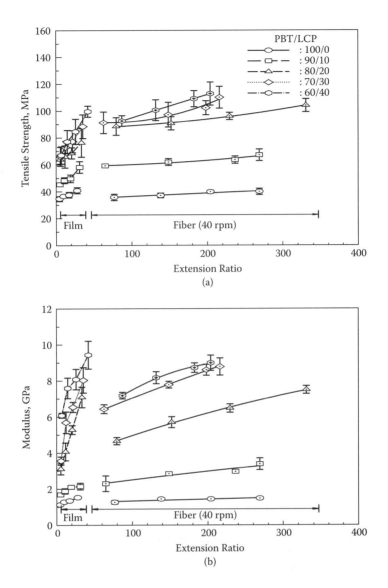

FIGURE 11.19
Tensile strength (a) and modulus (b) in the machine direction vs. extension ratio for PBT/LCP films and fibers at various LCP concentrations.

from 250°C to 270°C, especially at low shear rates (10 to –100/s), as shown in Figure 11.18d. Presumably, the viscosity of the PEN is much higher than the viscosity of the LCP in the shear rate region from 10 to 3000/s at 280°C.

In the 90/10 PEN/LCP film, there was a variation in LCP particle shapes across the film thickness. This is due to a stronger variation of cooling rates between skin and core; and also the deformation mechanisms of LCP phase in the core and skin layers were affected in a different way. More deformed LCP particles were achieved near the chill-roll surface.

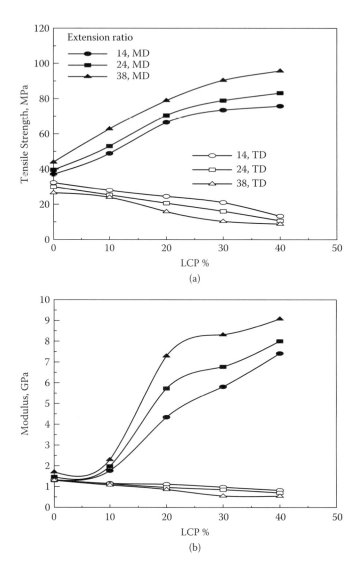

FIGURE 11.20
Tensile strength (a) and modulus (b) vs. LCP concentration for PBT/LCP films in the machine (MD) and transverse (TD) directions at various extension ratios.

11.3.5 Injection Molding of Polyester/LCP Composites

The injection molding of the polyester/LCP blends was performed at melt temperatures below the melting point of the LCP phase to retain intact the LCP fibrils present in the polyester/LCP fibers. The highest values of the tensile strength and modulus of the polyester/LCP blends achieved before and after injection molding are compared in Figures 11.25a and b. It is seen that the properties decreased in all polyester/LCP matrices after injection molding of the polyester/LCP fibers. The latter was due to a change in the LCP fibrillar shape during injection molding. It is concluded that even at injection temperatures below the LCP melting point, sustaining the LCP fibrils preexisting in the polyester/LCP

(a) 70/30 PBT/LCP Film, Extension Ratio = 1

(b) 70/30 PBT/LCP Film, Extension Ratio = 5.9

FIGURE 11.21
SEM photomicrographs of 70/30 PBT/LCP films at extension ratios of 1 (a) and 5.9 (b). Die gap is 1 mm.

fibers seems to be very difficult. This is apparently due to the viscous heating caused by the high injection speed applied during cavity filling. Among four composites, the PEN/LCP showed the highest tensile strength, while the copolyester/LCP showed the highest modulus after injection molding. The SEM photomicrographs of the skin and core of the 60/40 copolyester/LCP injection molded bars made from fibers obtained at an extension ratio of 250 are shown in Figures 11.26a and b. In the core section, coalescence of the LCP domains was observed, while slightly extended LCP domains appeared in the skin region.

11.3.6 Compatibilization of Polyester/LCP Blends

In thermoplastic/LCP blends, a good interfacial adhesion between two components can enhance the mechanical properties of the blends. The effect of incorporating a copolymer,

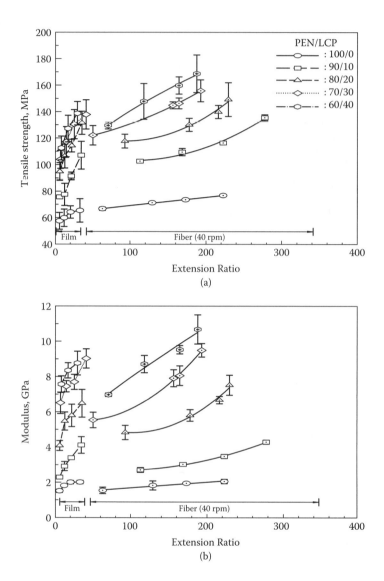

FIGURE 11.22
Tensile strength (a) and modulus (b) in the machine direction vs. extension ratio for PEN/LCP films and fibers at various LCP concentrations.

the polyethylene-glycidyl methacrylate, in polyester/LCP blends was studied to investigate the change in interfacial adhesion. The copolymer was used in the injection molding of polyester/LCP blends. In the film casting and fiber spinning of the polyester/LCP blends, it was noted that the addition of the copolymer up to 5% resulted in a large dimensional fluctuation of the films and fibers due to a large decrease in the viscosities of the blends. Therefore, only injection molding of the polyester/LCP blends containing the copolymer up to 5% was performed.

The tensile strength and modulus as a function of the LCP concentration for injection molded bars of the PET/LCP/compatibilizer blends are listed in Table 11.6. Both the tensile

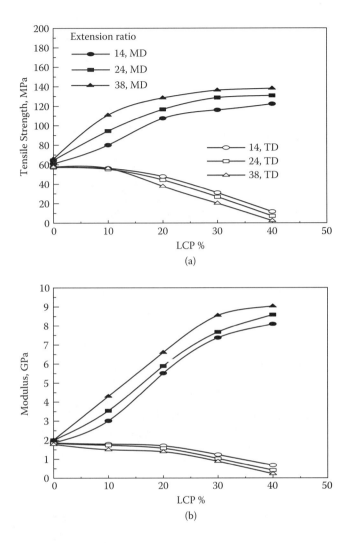

FIGURE 11.23
Tensile strength (a) and modulus (b) vs. LCP concentration for PEN/LCP films in the machine (MD) and transverse (TD) directions at various extension ratios.

strength and Young's modulus increased slightly at 20% and 40% of the LCP blends with 1% compatibilizer addition. However, the properties deteriorated when compatibilizer content increased to 5%. This can be attributed to the negative effect that compatibilizers have on the fibrillation of the LCP phase at this large compatibilizer concentration. As shown in Figures 11.27a and b, the SEM photomicrographs demonstrate the decrease in domain size of the LCP phase with 1% of compatibilizer in the 80/20 PET/LCP injection-molded blends. This reduction in domain size was also found in the 60/40 PET/LCP blends. The large addition (5%) of compatibilizer to the 60/40 PET/LCP blend led to poorer fibrillation of the LCP phase, as shown in Figures 11.28a and b.

In the thermoplastic/LCP blends, the ideal role of a compatibilizer is the creation of finer phase domains, a reduction in coalescence, an increase in interfacial adhesion, and

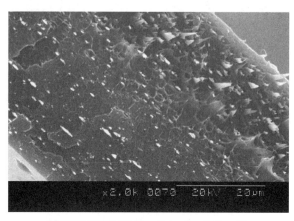

(a) 90/10 PEN/LCP Film, Extension Ratio = 23

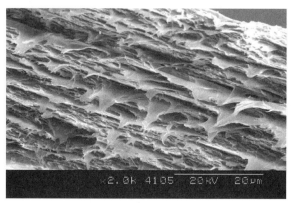

(b) 70/30 PEN/LCP Film, Extension Ratio = 23

FIGURE 11.24
SEM photomicrographs of 90/10 (a) and 70/30 (b) PEN/LCP films at an extension ratio of 23. Die gap is 1 mm.

improved mechanical properties. However, it also affects the particle size and the fibrillation of the LCP domains. Holsti-Miettinen et al. [64] obtained lower values of tensile strength and modulus with the addition of 5% compatibilizer (ethylene/ethyl acrylate/ glycidyl methacrylate, Lotader AX 8860) to PP/Vectra A 950 blends. They also found a decrease in particle size and less fibrillation in the LCP phase. However, enhanced attachment of the LCP domains to the matrix was observed in the compatibilized blends. Heino and Seppälä [65] found improved impact strength of the PP/Vectra A 950 blends at relatively small concentrations of compatibilizer, which was an ethylene-based reactive terpolymer. However, they also observed a reduction in the strength and stiffness in the PP/ LCP blends at a higher concentration of compatibilizer; this was due to the softening effect of the compatibilizer on the material. A negative effect of compatibilizer on the mechanical properties of the blends in the copolyester (Table 11.6), PBT, and PEN (Table 11.7) with LCP blends was found. This property reduction is related to less fibrillation of the LCP phase. It is well illustrated by the SEM photomicrographs of copolyester/LCP/compatibilizer (Figures 11.29a and b), PBT/LCP/compatibilizer (Figures 11.30a and b), and PEN/LCP/

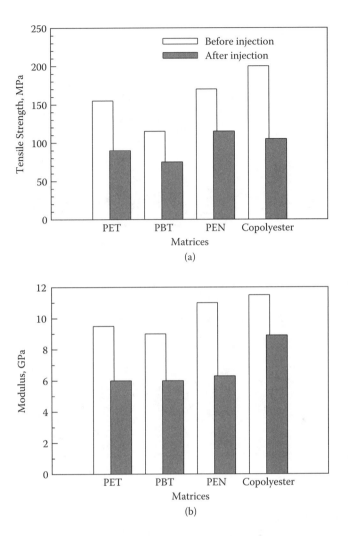

FIGURE 11.25
Highest tensile strength (a) and modulus (b) achieved in polyester/LCP blends before and after injection molding.

compatibilizer (Figures 11.31a and b). Therefore, it is evident that the fibrillation of the LCP phase was hindered by the presence of the compatibilizer in the polyester/LCP blends.

11.3.7 Simulation of Fiber Spinning of Polyester/LCP Blends

Kase and Matsuo [66] studied fiber spinning under the steady-state conditions. They established a model defining the variation of temperature and radius along the spin line as follows:

$$\frac{dT}{dz} = -\frac{2\sqrt{\pi A}\, h(T - T_a)}{WC_p} \tag{11.8}$$

(a) 60/40 Copolyester/LCP Injection Molded Bar, Skin

(b) 60/40 Copolyester/LCP Injection Molded Bar, Core

FIGURE 11.26

SEM photomicrographs of skin (a) and core (b) for 60/40 copolyester/LCP injection-molded bars obtained from fibers at an extension ratio of 250.

TABLE 11.6

Tensile Strength and Modulus in the Machine Direction for Injection Molded Bars of PET/LCP/Compatibilizer and Copolyester/LCP/Compatibilizer Blends

Loading		PET		Copolyester	
LCP (wt%)	Compatibilizer (wt%)	Tensile Strength (MPa)	Modulus (GPa)	Tensile Strength (MPa)	Modulus (Gpa)
0	0	51.0	1.73	53.5	1.60
20	0	77.8	2.69	90.1	2.59
	1	85.4	3.02	87.1	2.58
40	0	110.9	3.71	149.7	3.94
	1	117.4	4.01	127.4	3.67
	5	102.5	3.18	104.1	3.51
100	0	149.1	5.82	149.1	5.82

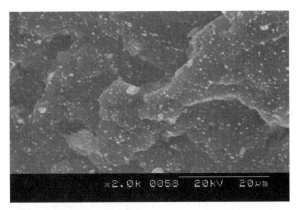

(a) 80/20/1 PET/LCP/Compatabilizer, Core

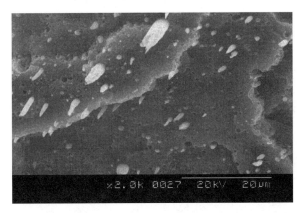

(b) 80/20/0 PET/LCP/Compatabilizer, Core

FIGURE 11.27
SEM photomicrographs of 80/20/1 (a) and 80/20/0 (b) PET/LCP/compatibilizer injection-molded bars.

and

$$\frac{dA}{dz} = -\frac{F\rho}{W\eta_E} A \qquad (11.9)$$

where T is the filament temperature, T_a is the air temperature, z is the spin-line distance from the spinneret, A is the filament cross-sectional area, h is the heat transfer coefficient, W is the melt throughput, C_p is the specific heat, F is the spin-line tension, ρ is the density, and η_E is the elongational viscosity.

From the Nusselt-Reynolds (Nu-Re) relation for air flowing parallel to a cylinder, the heat transfer coefficient, h, is calculated, which is approximated by

$$Nu = 0.42 \ Re^{0.334} \qquad (11.10)$$

(a) 60/40/1 PET/LCP/Compatabilizer, Skin

(b) 60/40/5 PET/LCP/Compatabilizer, Skin

FIGURE 11.28

SEM photomicrographs of 60/40/1 (a) and 60/40/5 (b) PET/LCP/compatibilizer injection-molded bars.

TABLE 11.7

Tensile Strength and Modulus in the Machine Direction for Injection Molded Bars of PBT/LCP/ Compatibilizer and PEN/LCP/Compatibilizer Blends

Loading		PBT		PEN	
LCP (wt%)	Compatibilizer (wt%)	Tensile Strength (MPa)	Modulus (GPa)	Tensile Strength (MPa)	Modulus (GPa)
0	0	47.4	1.55	75.1	1.63
20	0	62.5	2.34	106.1	2.79
	1	72.0	2.46	94.3	2.56
40	0	99.3	3.57	136.2	3.84
	1	92.3	3.49	120.3	3.69
	5	81.2	3.37	106.5	3.33
100	0	149.1	5.82	149.1	5.82

(a) 60/40/1 Copolyester/LCP/Compatabilizer, Core

(b) 60/40/5 Copolyester/LCP/Compatabilizer, Core

FIGURE 11.29
SEM photomicrographs of 60/40/1 (a) and 60/40/5 (b) copolyester/LCP/compatibilizer injection-molded bars.

with Reynolds, Re, and Nusselt, Nu, numbers defined as

$$Re = \frac{2\upsilon R}{\nu_F} \tag{11.11}$$

$$Nu = \frac{2hR}{k_F} \tag{11.12}$$

where υ is the filament velocity, R is the radius of the filament, ν_F is the kinematic viscosity of ambient air, and K_F is the heat conductivity of ambient air. By converting the Equation 11.10 with the definitions of Nu and Re and multiplying the right-hand side of the same equation by the correction factor $(1 + K)$, the heat transfer coefficient, h, is

(a) 60/40/1 PBT/LCP/Compatabilizer, Core

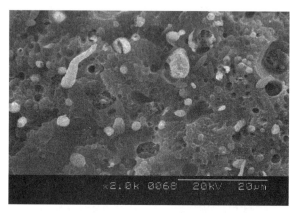

(b) 60/40/5 PBT/LCP/Compatabilizer, Core

FIGURE 11.30
SEM photomicrographs of 60/40/1 (a) and 60/40/5 (b) PBT/LCP/compatibilizer injection-molded bars.

defined as

$$h = 0.21\, k_F \sqrt{\frac{\pi}{A}} \left(\frac{2W}{\sqrt{\pi A}\ \rho v_F} \right)^{0.334} \left(1 + K\right) \tag{11.13}$$

The introduction of the correction factor, $(1 + K)$, in Equation 11.13 is due to the effect of the transverse air velocity on heat transfer. The correction factor determines the cooling rate along the spin line. In calculations, the value of the correction factor, $(1 + K)$, is determined to match the distance between the die and water bath.

Using the Runge-Kutta method [67], the first-order ordinary differential Equations 11.8 and 11.9 are solved numerically. These equations are an initial value problem. The initial temperature, T_0, and radius, R_0, are known, but the initial value of the spin-line tension, F_0, is unknown. With an initially guessed value of F_0, the Newton-Raphson iterative method is used in the calculation. The calculation is continued until the calculated

(a) 60/40/1 PEN/LCP/Compatabilizer, Core

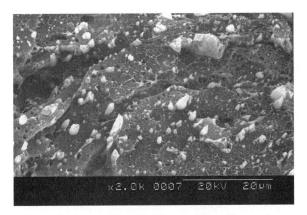

(b) 60/40/5 PEN/LCP/Compatabilizer, Core

FIGURE 11.31
SEM photomicrographs of 60/40/1 (a) and 60/40/5 (b) PEN/LCP/compatibilizer injection-molded bars.

radius of fibers closely approximates the radius calculated from the melt throughput and draw-down ratio. The value of the initial spin-line tension is adjusted until these radii match.

To get the temperature and radius profile along the spin line, the numerical solution is obtained at each point of the axial increment interval, Δz. When the filament radius profile is obtained, the velocity profile at each point on the spin line is calculated from the mass conversion equation, $W = \rho V_z A_z$. The strain rate is calculated as

$$\varepsilon'_z = \frac{V_{z+1} - V_z}{\Delta z} = \frac{dV_z}{dz} \tag{11.14}$$

The initial values and other parameters used in melt spinning simulation are listed in Table 11.8.

TABLE 11.8

Parameters Used in Melt Spinning Simulation

| | Value | | |
Data	60/40 PET/LCP	60/40 PBT/LCP	Unit
Air temperature, T_a	25	25	°C
Glass transition temperature, T_g	78	55	°C
Die temperature, T_d	240	270	°C
Specific heat, C_p	1965	1350	J/kg-K
Initial spin-line tension, F_i	1.0×10^{-1}	1.0×10^{-1}	N
Density, ρ	1400	1389	kg/m³
Melt flow rate, W	0.400×10^{-3}	0.887×10^{-3}	kg/s
Interfacial tension between polyester and LCP, σ	0.175×10^{-3}	0.0597×10^{-3}	N/m
Die diameter, d	0.01	0.01	m
Heat conductivity of air at 25°C, k_F	26.14×10^{-3}	26.14×10^{-3}	W/m-K
Kinematic viscosity of air at 25°C, v_F	15.71×10^{-6}	15.71×10^{-6}	m²/s
Initial LCP droplet diameter, D_0	3.96×10^{-6}	13.52×10^{-6}	m
Viscosity ratio, p	0.01–100	0.01–100	—
Correction factor, (1+K)	100.5	100.5	—

11.3.8 Simulation of LCP Droplet Deformation

Based on Taylor's [34, 35] theory in which a Newtonian fluid was considered, the LCP droplet deformation in a polyester matrix under the melt spinning conditions is simulated. From the SEM photomicrographs and the investigation of average LCP droplet size by image analysis, the initial size of the LCP droplet at the die exit is measured and used in the calculation of droplet deformation.

The reduced capillary number, Ca^* (Equation 11.7), governs the deformation of a droplet. The capillary number, which depends on the strain rate, elongational viscosity, and droplet size, is calculated at each time step. The critical capillary number, Ca_c, depends on the viscosity ratio, p. From Grace's empirical equation [68], a value of Ca_c in the elongational flow is calculated as

$$\log\left(\frac{Ca_c}{2}\right) = -0.64853 - 0.02442(\log p) + 0.02221\left(\log p\right)^2 - \frac{0.00056}{(\log p - 0.00645)} \quad (11.15)$$

A value of Ca is compared with that of Ca_c to determine if the deformation or breakup occurs [44].

According to Taylor's suggestion (Equation 11.4), the deformation is calculated in the region $1.0 < Ca^* < 4.0$. The kinetics of the breakup depend on the viscosity ratio. The dimensionless breakup time is calculated as [37]

$$t_b^* = \frac{t_b \gamma'}{Ca} = 84p^{0.345}\left(\frac{Ca}{Ca_c}\right)^{-0.559} \quad (11.16)$$

where t_b is the time for breakup and γ' is the strain rate. If there is not enough time provided for the breakup before solidification, no breakup occurs.

For $Ca^* > 4$, as indicated by Huneault et al. [44], the affine deformation occurs and the deformation is calculated by the exponential function for uniaxial deformation. In the case of $Ca^* < 0.1$, no droplet deformation takes place. Therefore, calculations of droplet deformation were not performed.

The calculation of the deformation is iterated until either the breakup or solidification of the LCP fibrils occurs. In the case of breakup based on Utracki and Shi's breakup criterion (Equation 11.16) [37], a new dimension of LCP fibril is calculated. At solidification, the calculation is terminated and the final dimension of the LCP fibrils is plotted as a function of spin-line distance. Using this procedure of calculation, the sizes of the LCP droplets under various processing conditions in fiber spinning can be obtained.

The simulation is performed based on the following assumptions:

1. No coalescence of LCP droplets occurs.

2. Shear-induced orientation and crystallization are negligible.

3. The radial temperature profile in melt spinning is negligible.

4. If breakup occurs, new spherical droplets of equal diameter are generated.

The algorithm of the simulation is shown in Figure 11.32.

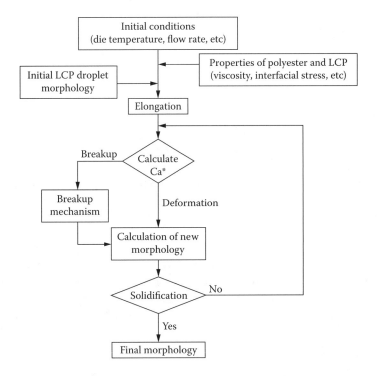

FIGURE 11.32
The algorithm for calculation of LCP droplet deformation under uniaxial elongational flow.

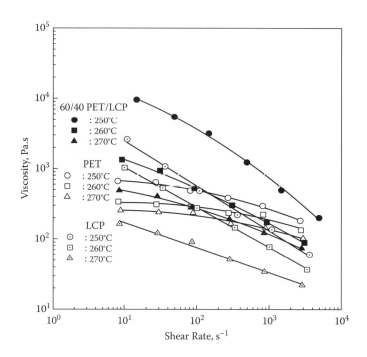

FIGURE 11.33
Viscosity vs. shear rate curves for PET, LCP, and 60/40 PET/LCP at three different temperatures.

11.3.9   60/40 PET/LCP Fibers

The flow curves of the PET, LCP, and 60/40 PET/LCP at various temperatures are depicted in Figure 11.33. The zero shear rate viscosity of the blend was obtained by extrapolating the flow curves at 250°C, 260°C, and 270°C. The temperature dependence of the zero shear rate viscosities, η_0, was obtained as

$$\eta_0 = 1.5810 \times 10^{-31} e^{\frac{42047.4}{T}} \qquad (11.17)$$

The elongational viscosity, η_E, was assumed to obey the Trouton rule:

$$\eta_E = 3\eta_0 \qquad (11.18)$$

The viscosities of the 60/40 PET/LCP blend at 250°C, 260°C, and 270°C is higher than those of the PET at shear rates lower than 100/s. This is due to the high melting temperature of the LCP (278°C). Below the melting temperature of the LCP, the LCP particles behave like a type of filler, which increases the viscosity of blends.

Figure 11.34 shows the SEM photomicrographs of extrudates and fibers prepared from the 60/40 PET/LCP blends. The LCP domains in the extrudates are formed in the shape of spherical droplets. They deformed into long fibrils in the fibers obtained at an extension ratio of 440. They are clearly seen after solvent extraction. The PET matrix was dissolved in boiling phenol for 24 hours and the phenol was washed out with methanol. The distribution of LCP domain diameters in extrudates and fibers was determined by an image

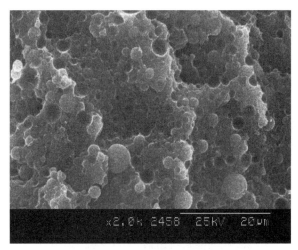

(a) 60/40 PET/LCP Fiber, Extension Ratio = 1

(b) 60/40 PET/LCP Fiber, Extension Ratio = 440

FIGURE 11.34
SEM photomicrographs of 60/40 PET/LCP extrudate (a) and fiber (b).

analyzer. The average diameter of the LCP droplets in extrudates was 3.96 μm. In fibers at an extension ratio of 440, the average diameter of the LCP fibrils was 330 nm. Therefore, it is evident that nano-sized LCP fibrils can be obtained at a high extension ratio during fiber spinning of PET/LCP blends.

The relationship between the interfacial tension and interfacial thickness between two components was established by Wu [69], and the interfacial tension between PET and LCP was calculated from the interfacial thickness of the two components:

$$\gamma_{12} = 55a_I^{-0.86} \tag{11.19}$$

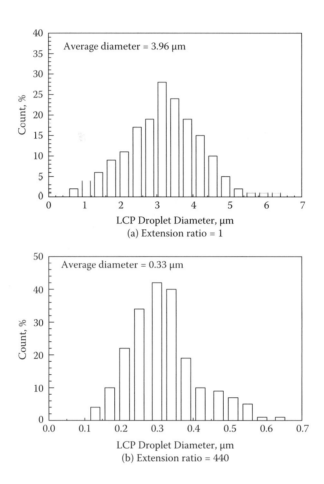

FIGURE 11.35
Diameter of LCP phase in the fiber spinning of 60/40 PET/LCP at the extension ratios of 1 (a) and 440 (b).

where the interfacial tension, γ_{12}, is in dyne/cm and the interfacial thickness, a_I, is in $\overset{o}{A}$. The interfacial thickness observed by the SEM photomicrograph was 800 $\overset{o}{A}$, leading to $\gamma_{12} = 0.175 \times 10^{-3}$ N/m.

The calculated LCP phase diameter along the spin line is shown in Figure 11.36. Under isothermal spinning, a monotonic decrease in the diameter was observed (Figure 11.36a); while under non-isothermal spinning, a fast decrease in the diameter was observed at a very short distance along the spin line. In the latter case due to freezing, the distance where the deformation occurs ranged from 0 to 0.5 cm. The final LCP fibril diameter was 220 nm at a take-up velocity of 1.2 m/s in both isothermal and non-isothermal fiber spinning. This value was close to the experimental diameter of 330 nm.

The deformation of the LCP domain based on Utracki's criterion was calculated [44] using the assumptions that the LCP droplet has a cylindrical shape with its volume being conserved in both isothermal and non-isothermal cases. In all cases, the reduced capillary number, Ca^*, was larger than 4.0. Therefore, the diameter, D, and length, L, of

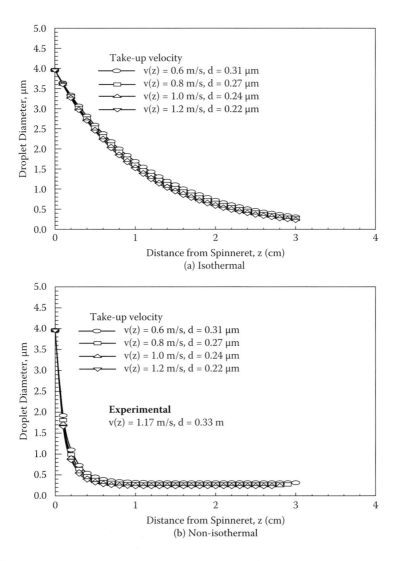

FIGURE 11.36
Variation of LCP phase diameter along the spin line in the fiber spinning of 60/40 PET/LCP in isothermal (a) and non-isothermal (b) simulation.

the LCP fibril after calculation of the deformation was based on the affine deformation theory:

$$D = D_0 \exp\left(-\frac{\varepsilon}{2}\right) \tag{11.20}$$

$$L = L_0 \exp(\varepsilon) \tag{11.21}$$

where D_0 and L_0 are the original droplet diameter and length, respectively, and ε is the strain, which is calculated from the change in fiber diameter. No influence of the viscosity ratio on the LCP droplet deformation was found when the various viscosity ratios between LCP and PET components (approximately 0.01 to 100) were applied in the simulation. It is evident that LCP droplet deformation in the PET matrix does not follow Taylor's deformation theory, but follows the affine deformation theory. There was no LCP breakup observed in the simulation. In both the isothermal and non-isothermal calculations, the final length of the LCP fibril was 1306 μm at a take-up velocity of 1.2 m/s. It was not possible to measure the actual length of the fibrils because their ends were not found in the SEM photomicrographs. The simulation of LCP fibrillar deformation in the 60/40 PET/LCP fiber spinning indicated a maximum aspect ratio $L/D = 6000$ at a take-up velocity of 1.2 m/s.

11.3.10 60/40 PBT/LCP Fiber

The flow curves of 60/40 PBT/LCP and neat PBT at different temperatures are compared in Figure 11.37. Similar to pure LCP (Figure 11.33), the blend of 60/40 PBT/LCP exhibits power-law behavior. The variation in viscosity of the blend with temperature change was less than that in neat PBT. This observation is opposite to that of the PET/LCP case where its viscosity varied with temperature more than that of pure PET. Some researchers found partial miscibility in the blending of PBT/LCP [21, 23, 26, 27, 30]. This might be a reason for the power-law behavior of viscosity and lower variation of the blend viscosity with temperature change. The temperature dependence of a zero shear rate viscosity of the blend is:

$$\eta_0 = 1.1077 \times 10^{-7} e^{\frac{13186.9}{T}} \tag{11.22}$$

The zero shear rate viscosity was taken at the shear rate of 6.5/s .

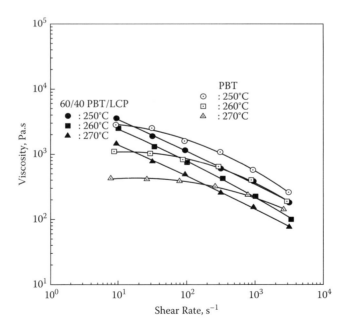

FIGURE 11.37
Viscosity vs. shear rate curves for PBT and 60/40 PET/LCP at different temperatures.

(a) 60/40 PBT/LCP Fiber, Extension Ratio = 1

(b) 60/40 PBT/LCP Fiber, Extension Ratio = 204

FIGURE 11.38
SEM photomicrographs of 60/40 PTB/LCP fiber (a) before and (b) after elongation.

The shapes of the LCP domains in the extrudates and fibers of the 60/40 PBT/LCP blend are compared in Figure 11.38. To obtain the SEM photomicrographs, the PBT matrix was extracted fully by dissolution in 1/3 trifluoroacetic acid/methylene chloride solvent. A much larger initial size of the spherical LCP droplets was found in 60/40 PBT/LCP blends than that in 60/40 PET/LCP blends. The LCP domains were deformed into long fibrils at an extension ratio of 204. The average diameter of the LCP droplets in the PBT matrix in extrudates was 13.52 μm. In fibers, it was reduced to 980 nm.

The interfacial thickness measured from SEM photomicrographs was 2800$\overset{o}{A}$. Based on this value and Wu's relation [69], the calculated interfacial tension was 0.0597×10^{-3} N/m. The obtained interfacial tension in PBT/LCP is lower than that in the PET/LCP blend. This result demonstrates a better adhesion in the PBT/LCP blend than in the PET/LCP blend. This is due to the partial miscibility between PBT and LCP based on PET/HBA.

The variation in the calculated LCP domains along the spin line at various take-up speeds under the isothermal and non-isothermal conditions is compared in Figure 11.39. Under non-isothermal conditions, the largest decrease of diameter was observed at the distance close to the spinneret range from 0 to 0.5 cm. A further decrease was stabilized after 0.6 cm in the non-isothermal spinning. The calculated diameter of the LCP fibrils obtained at the take-up velocity of 2.0 m/s was 860 nm. This LCP fibril diameter was well matched to the experimental value of 980 nm. Under isothermal conditions, the decrease in diameter was more gradual than that under non-isothermal conditions.

At a take-up velocity of 2.0 m/s, a maximum calculated fibril length of 3300 μm was obtained. According to the simulation, there was no breakup of LCP fibrils during melt-spinning of the 60/40 PBT/LCP blend. The deformation of the LCP domain was based on the affine deformation theory given by Equations 11.20 and 11.21. At a take-up velocity of 2.0 m/s, a maximum aspect ratio of 3900 was obtained. A lower aspect ratio of the LCP fibrils was obtained in the PBT/LCP blend than in the PET/LCP blend. Apparently, the partial miscibility between PBT and LCP phases in the PBT/LCP blend hinders fibrillation. This partial miscibility led to less viscosity change with temperature variation. Therefore, a smaller strain was experienced by LCP in the simulation, leading to a lower aspect ratio of the LCP fibrils.

11.4 Conclusions

The effect of the extrusion die temperature on the mechanical properties of PET/LCP films was observed. The best properties were obtained at any extension ratio and a die temperature of 260°C. At an extension ratio of 38, the tensile strength of 154 MPa and a modulus of 9 GPa were obtained. At a low die temperature, the viscosity of the matrix polymer is high. However, at a high die temperature, the large deformation of the LCP phase can be obtained. Therefore, the search for an optimum die temperature is important for achieving a better fibrillation of the LCP domain in PET matrix and consequently high mechanical properties in the PET/LCP films. The formation of microfibrillar structure of LCP domains also depends on the shear rate that governs the viscosity ratio of PET and LCP melts. The shear rate is a dominant factor in introducing more deformation of the LCP domains. At a high shear rate, high LCP content, and high extension ratio, a high aspect ratio and high molecular chain orientation of the LCP phase are obtained. However, in contrast to our expectation, the highest tensile strength of 150 MPa was found in the 70/30 PET/LCP film instead of the 60/40 PET/LCP films. In the 60/40 PET/LCP film, the fibrillar formation of the LCP phase is somewhat hindered. Possibly, there is an optimum particle size of LCP providing the maximum deformation.

With increasing LCP content and extension ratio, properties of PET/LCP films in the transverse direction are reduced. Larger decreases in tensile properties were observed with increasing LCP content than with increasing extension ratio due to the immiscibility and the poor interfacial adhesion between PET and LCP in blends, although the LCP phase has PET structure in the main chain. The improved mutual compatibility between components when using a coupling agent can enhance the mechanical properties. However, the

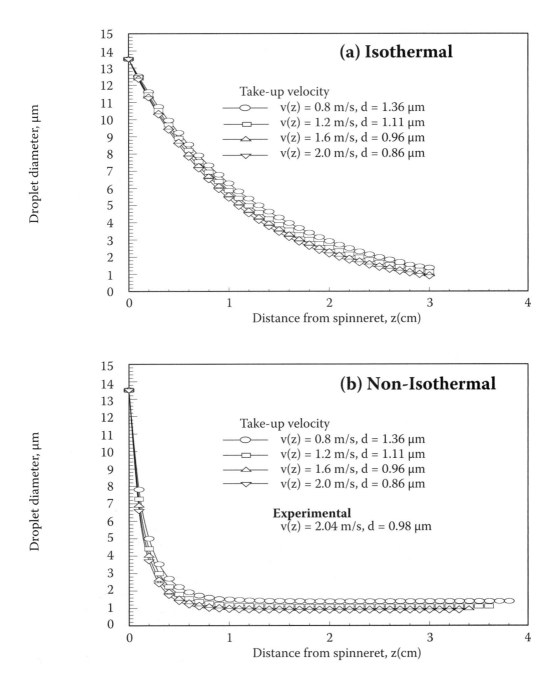

FIGURE 11.39
Variation of LCP phase diameter along the spin line in the fiber spinning of 60/40 PBT/LCP in isothermal (a) and non-isothermal (b) simulation.

change in transverse tensile properties with extension ratio and LCP content is small in comparison with that of properties in the flow direction.

By controlling the output rate in a coat-hanger die and by changing the die gap during film extrusion, the effect of shear rate on the mechanical properties was examined. With the larger die gap, shear rates were reduced and this led to lower mechanical properties of films. Therefore, high shear rate is required for better dispersion and higher orientation of the LCP phase in the PET matrix.

The effects of LCP content and extension ratio in manufacturing of PET/LCP fibers were similar to those in films, although at a higher extension ratio. The higher shear rates were achieved with increasing screw speed, but the breakup of spin line was observed during melt spinning of blends at the high LCP content. A high extension ratio (about 440) was obtained only at low screw speeds. The morphology of 60/40 PET/LCP fibers exhibited the finely and uniformly distributed LCP fibrils in the PET matrix leading to nanocomposite structure. The mechanical properties of these fibers were comparable with those of films. Therefore, the effect of the extension ratio is more important than that of the shear rate in the PET/LCP films and fibers.

In comparison with fiber properties, mechanical properties of the injection molded bars prepared from chopped PET/LCP fibers were reduced. This was due to the high shear rate applied in injection molding that led to a possibility of the destruction of the preexisting LCP fibrils even though the processing temperature is below the LCP melting point. Also, viscous heating due to the high shear rate in injection molding can be the reason for the breakup of LCP fibrils and their subsequent agglomeration.

Variation in the polyester matrices yielded a change in LCP domain deformation, size, and distribution. Therefore, the degree of enhancement in the mechanical properties varied with the type of polyester matrix. The copolyester was the most effective in getting finely distributed LCP domains exhibiting the highest tensile strength and modulus. PET/LCP blends generated the most uniform distribution and largest aspect ratio of the LCP domains. Nanofibrils of LCP in the PET/LCP fibers were obtained at highest extension ratios. In PBT/LCP blends, the fibrillation of the LCP domains was hindered. This led to less improvement in the mechanical properties of the 60/40 PBT/LCP fibers due to a partial miscibility between the PBT and the LCP component. In PEN/LCP blends, a ribbon-like structure of the LCP phase was observed due to the larger viscosity of the PEN in comparison with the LCP. After the injection molding of chopped fibers, tensile properties of moldings decreased for all polyester/LCP blends in comparison with fibers. But the degree of reduction in the tensile strength was found to depend on the matrix, with the smallest reduction observed in the PEN matrix. The addition of a 1% ethylene-based compatibilizer to the PET/LCP blends introduced little increase in both the tensile strength and modulus. However, 5% compatibilizer yielded a reduction in mechanical properties of the 60/40 PET/LCP blends. In other matrices, the mechanical properties decreased with the addition of the compatibilizer, regardless of the concentration of both the LCP and the compatibilizer.

The LCP droplet deformation in the polyester/LCP blends was calculated based on Utracki's criterion. No breakup of the LCP phase appeared at all take-up speeds. Accordingly, the calculation was carried out based on the affine deformation theory. The calculated LCP fibril diameter was well matched to the observation by SEM and the image analyzer. Because of the low viscosity of the LCP, favorable conditions for deformation in the PET matrix in a molten state were obtained. Taylor's deformation theory, which considers the small deformation, was not applicable for LCP deformation, but the affine deformation theory revealed a well-matched LCP deformation in the PET matrix. A lower aspect ratio of LCP fibrils in the 60/40 PBT/LCP fiber was obtained as compared to that in the 60/40 PET/LCP fiber. This was caused by the partial miscibility of the PBT and LCP phase. The obtained diameter of LCP fibrils was 330 nm in 60/40 PET/LCP fibers and 980 nm in 60/40 PBT/LCP fibers.

Acknowledgments

The authors wish to thank the following companies for their generosity in providing materials: Unitika Ltd. (LCP), Shell Chemical Company (PEN), BASF AG (PBT), Eastman Chemical Company (copolyester), and Elf Atochem North America, Inc. (compatibilizer).

References

1. A.I. Isayev, in *Liquid-Crystalline Polymer Systems : Technological Advances*, A.I. Isayev, T. Kyu and S.Z.D. Cheng, Eds., *ACS Symposium Series*, Washington, D.C., 632, 1, 1996.
2. C.H. Song, and A.I. Isayev, *J. Polym. Eng.*, 18, 417, 1999; *SPE ANTEC*, 44, 1637, 1998.
3. C.H. Song and A.I. Isayev, *J. Polym. Eng.*, 20, 427, 2000: *SPE ANTEC*, 45, 2840, 1999.
4. C.H. Song and A.I. Isayev, *Polymer*, 42, 2611, 2001: *SPE ANTEC*, 46, 2523, 1999.
5. A.I. Isayev, U.S. Patent No. 5,260,380, November 9, 1993.
6. M.A. McLeod and D.G. Baird, *SPE ANTEC*, 41, 1420, 1995.
7. C.U. Ko, G.L. Wilkes, and C.P. Wong, *J. Appl. Polym. Sci.*, 37, 3063, 1989.
8. M.T. Heino and J.V. Seppälä, *J. Appl. Polym. Sci.*, 44, 2185, 1992.
9. J.X. Li, M.S. Silverstein, A. Hiltner, and E. Baer, *J. Appl. Polym. Sci.*, 44, 1531, 1992.
10. A.K. Mithal and A. Tayebi, *Polym. Eng. Sci.*, 31, 1533, 1991.
11. M. Kyotani, A. Kaito, and K. Nakayama, *Polymer*, 33, 4756, 1992.
12. L.J. Bonis, and A.M. Adur, *SPE ANTEC*, 41, 3156, 1995.
13. M.S. Silverstein, A. Hiltner, and E. Baer, *J. Appl. Polym. Sci.*, 43, 157, 1991.
14. L.E. Nielsen, *Mechanical Properties of Polymers and Composites*, Vol. 2, Marcel Dekker, New York, Chapter 7 and 8, 1981.
15. D. Hull, *An Introduction to Composite Materials*, Cambridge University Press, 1981, chap 9.
16. K. Yoshikai, K. Nakayama, and M. Kyotani, *J. Appl. Polym. Sci.*, 62, 1331, 1996.
17. S. Mehta, and B.L. Deopura, *Polym. Eng. Sci.*, 33, 931, 1993.
18. A.M. Sukhadia, D. Done, and D.G. Baird, *Polym. Eng. Sci.*, 30, 519, 1990.
19. W. Brostow, T.S. Dziemianowcz, J. Romanski, and W. Werber, *Polym. Eng. Sci.*, 28, 785, 1988.
20. P. Zhuang, T. Kyu, and J. White, *SPE ANTEC*, 34, 1237, 1988.
21. M. Kimura and R.S. Porter, *J. Polym. Sci., Polym. Phys. Ed.*, 22, 1697, 1984.
22. G. Kiss, *Polym. Eng. Sci.*, 27, 410, 1987.
23. A. Ajji and P.A. Gignac, *Polym. Eng. Sci.*, 32, 903, 1992.
24. M.T. Heino and J.V. Seppälä, *Polym. Bull.*, 30, 353, 1993.
25. J.Y. Lee, J. Jang, S.M. Hong, S.S. Hwang, Y. Seo, and K.U. Kim, *Int. Polym. Processing*, 12, 19, 1997.
26. M. Pracella, E. Chiellini, G. Galli, and D. Dainelli, *Mol. Cryst. Liq. Cryst.*, 153, 525, 1987.
27. M. Pracella, D. Dainelli, G. Galli, and E. Chiellini, *Makromol. Chem.*, 187, 2387, 1986.
28. Y. Seo, *J. Appl. Polym. Sci.*, 64, 359, 1997.
29. D. Beery, S. Kenig, and A. Siegmann, *Polym. Eng. Sci.*, 31, 451, 1991.
30. M. Paci, C. Barone, and P. Magagnini, *J. Polym. Sci., Polym. Phys.*, 25, 1595, 1987.
31. B.S. Kim and S.H. Jang, *Polym. Eng. Sci.*, 35, 1421, 1995.

32. S.H. Jang and B. S. Kim, *Polym. Eng. Sci.*, 35, 538, 1995.
33. S.H. Kim, S.W. Kang, J.K. Park, and Y.H. Park, *J. Appl. Polym. Sci.*, 70, 1065, 1998.
34. G.I. Taylor, *Proc. Roy. Soc. London*, A138, 41, 1932.
35. G.I. Taylor, *Proc. Roy. Soc. London*, A146, 501, 1934.
36. R.G. Cox, *J. Fluid Mech.*, 37, 601, 1969.
37. L.A. Utracki and Z.H. Shi, *Polym. Eng. Sci.*, 32, 1824, 1992.
38. H.E.H. Meijer and J.M.H. Jansen, *in Mixing and Compounding-Theory and Practical Progress*, I. Manas-Zloczower and Z. Tadmor Eds., Hanser, Munich, 1994.
39. H.B. Chin and C.D. Han, *J. Rheol.*, 23, 557, 1979.
40. C. Van der Reijden-Stolk and A. Sara, *Polym. Eng. Sci.*, 26, 1229, 1986.
41. J.J. Elmendorp and R.J. Maalcke, *Polym. Eng. Sci.*, 25, 1041, 1985.
42. W.J. Milliken and L.G. Leal, *J. Non-Newt. Fluid Mech.*, 40, 355, 1991.
43. G.I. Taylor, *Proc. Int. Cong. Appl. Mech., 11th*, Munich, Springer, 1964, p. 790.
44. M.A. Huneault, Z.H. Shi, and L.A. Utracki, *Polym. Eng. Sci.*, 35, 115, 1995.
45. S. Hayase, P. Driscoll, and T. Masuda, *Polym. Eng. Sci.*, 33, 108, 1993.
46. A.I. Isayev and M.J. Modic, *Polym. Compos.*, 8, 158, 1987; *SPE ANTEC*, 32, 573, 1986.
47. K.G. Blizard and D.G. Baird, *Polym. Eng. Sci.*, 27, 653, 1987; *SPE ANTEC*, 32, 311, 1986.
48. A.I. Isayev, in *Advanced Composites. III Expanding Technology, ASM*, 1987, p.259.
49. A. Mehta and A.I. Isayev, *Polym. Eng. Sci.*, 31, 963, 1991.
50. J.V. Seppälä, M.T. Heino, and C. Kapanen, *J. Appl. Polym. Sci.*, 44, 1051, 1992.
51. R. Viswanathan and A.I. Isayev, *J. Appl. Polym. Sci.*, 55, 1117, 1995.
52. D. Beery, S. Kenig, and A. Siegmann, *Polym. Eng. Sci.*, 31, 459, 1991.
53. R. Ramanathan, K.G. Blizard, and D.G. Baird, *SPE ANTEC*, 36, 1399, 1987.
54. A.I. Isayev and P.R. Subramanian, *Polym. Eng. Sci.*, 32, 85, 1992.
55. P.R. Subramanian and A.I. Isayev, *Polymer*, 32, 1961, 1991.
56. D. Dutta, R.A. Weiss, and K. Kristal, *Polym. Compos.*, 13, 394, 1992.
57. Q. Lin and A.F. Yee, *Polymer*, 16, 3463, 1994.
58. A.I. Isayev and R. Viswanathan, *Polymer*, 36, 1585, 1995.
59. D. Dutta, R.A. Weiss, and K. Kristal, *Polym. Eng. Sci.*, 33, 838, 1993.
60. H.-C. Chin, K.-C. Chiou, and F.-C. Chang, *J. Appl. Polym. Sci.*, 60, 2503, 1996.
61. W.C. Lee and A.T. DiBenedetto, *Polymer*, 34, 684, 1993).
62. S. Joslin, W. Jackson, and R. Farris, *J. Appl. Polym. Sci.*, 54, 289, 1994.
63. D.G. Baird and R. Ramanathan, in *Contempory Topics in Polymer Science*, Plenum Press, Vol. 6, p. 73, 1990.
64. R.M. Holsti-Miettinen, M.T. Heino, and J.V. Seppälä, *J. Appl. Polym. Sci.*, 57, 573, 1995.
65. M.T. Heino and J.V. Seppälä, *J. Appl. Polym. Sci.*, 48, 1677, 1993.
66. S. Kase and T. Matsuo, *J. Polym. Sci. A.*, 3, 2541, 1965.
67. W.H. Press, *Numerical Recipes in Fortran*, Cambridge University Press, New York, 1992.
68. H.P. Grace, *Chem. Eng. Commun.*, 14, 225, 1982.
69. S. Wu, *Polymer Interface and Adhesion*, Marcel Dekker, New York, , 1982, p. 121.

12

Transmission Electron Microscopy and Related Techniques in the Structure and Morphological Characterization of Polymer Nanocomposites

Vinod K. Berry

CONTENTS

12.1 Introduction

Due to the tremendous growth in research and in the development of nanotechnology, secondary to its applications in various fields of biological and materials science, the need for proper characterization tools has become very important. There is an exponential growth of nanotechnology applications in all fields of science. The scientists and engineers involved in these endeavors are relying more and more on equipment that helps them understand the behavior of materials at the nano level and helps them relate the structure property relationships of those materials. Both in industry and academia, scientists and engineers are trying to get their hands on the most sophisticated equipment, which costs upward of thousands of dollars, to get some meaningful answers. Transmission electron microscopy (TEM) is one tool that enables them to see the structure of materials at almost an atomic level.

12.2 Microscope, Wavelength, and Resolution

The microscope is a device or an instrument that allows us to see and resolve the details that the human eye is not able to see or resolve. An unaided human eye can resolve approximately 0.1 millimeter (mm) between two points (point resolution) or two parallel lines (line resolution). The distance is called the "resolution" or the "resolving power" of the human eye. Thus, resolving power is the ability to see two points of an object separately in an image. If one wants to observe details finer than 0.1mm apart, one uses the help of a microscope. A magnifying glass is the simplest form of microscope. The metric units used in microscopy are fractions of a meter (m). Thus, the units are micrometer (μm; 1 μm = 10^{-6} m) and nanometer (nm; 1 nm = 10^{-9} m). Instruments in this class use visible light or some other kind of radiation as the source of illumination. The human eye is sensitive to radiation in the visible range of the electromagnetic spectrum. This radiation falls in the range of 300 to 700nm; 300 nm is the violet end and 700 nm is the red end of the visible spectrum. Thus, 500 nm is the wavelength (λ) in the middle of the visible region, normally used as the wavelength to define the visible light. Electrons, due to their dual nature (i.e., particle and wave), have been one of the most significant sources of illumination. The wavelength of an electron wave is much shorter than the wavelength of visible light.

The optimum resolution in an optical microscope and in a transmission electron microscope (TEM) is defined by the Rayleigh criteria, which state that the smallest point-to-point distance δ that can be resolved is given by

$$\delta = \frac{0.61\lambda}{\alpha}$$

where:

λ = wavelength of radiation
α = numerical aperture

If we consider α the numerical aperture to be equal to one for simplicity, δ, the resolution is equal to 0.61λ or approximately one half of the wavelength of radiation. For optical microscopes, this is equal to one half of 500 nm, which is approximately 250 nm. Thus, the resolution of a good optical microscope is approximately 250 nm. Using the same criteria, we can approximate the resolution in a transmission electron microscope to

$$\delta = \frac{0.61\lambda}{\alpha}$$

where λ is the wavelength of electrons.

According to deBroglie's famous equation, the wavelength of electrons is related to their energy E. Thus, ignoring the relativistic effects, it can be shown approximately that

$$\lambda = \frac{1.22}{\sqrt{E}}$$

where E = electron volts (in eV) and λ = wavelength (in nm).
So, for electrons with 100-keV energy, the value of λ = 0.0037 nm or 0.004 nm. Thus, a transmission electron microscope operated at 100 kV accelerating voltage should be able to provide a resolution of one half of 0.004 nm, which is 0.002 nm. But in real life, one is not able to approach that resolution. This is due to the constraints placed by the design of the electromagnetic lenses used in present-day electron microscopes. The instruments using electrons as the illumination source are called electron microscopes. It is possible to observe objects almost to the atomic level using a transmission electron microscope (TEM). The best resolution attainable in a modern TEM is approaching 0.1 nm. In addition to the constraints due to aberration effects in the electromagnetic lenses that affect resolution, there are some other factors that help in improving the resolution in a TEM. Table 12.1 shows the relationship between the accelerating voltage and the wavelength of electrons.

TABLE 12.1

Relationship between Accelerating Voltage and Wavelength of Electrons.

kV	Relativistic Wavelength (nm)	Mass (γ)	Velocity 10^8 m/s
100	0.00370	1.196	1.644
300	0.00197	1.587	2.330
1000	0.00087	2.957	2.823

Note: γ is the fractional increase in mass of an electron rest mass m_0; m/s = meters per second. Velocity of light in vacuum (c) is 2.98×108 m/s.

The speed of light in vacuum is 2.998×10^8 m/s. So, at 1000 kV (1 million volt accelerating voltage), the speed of electrons approaches the speed of light. An increase in accelerating voltage increases the speed of the electrons and lowers the wavelength, thereby improving the revolution. In addition to lowering the wavelength, the higher energy electrons, due to their higher velocity, have more penetration in the matter. The disadvantage of having a TEM with very high accelerating voltages is that the cost factor becomes prohibitively high. The slight advantage gained in resolution and penetration depth in a sample does not always justify the cost factor. However, instruments with 200- to 300-kV accelerating voltages have wide applications in materials science and are in widespread use. It is beyond the scope of this chapter to deal with the mathematical interpretation and explanations of the concepts just discussed.

12.3 Electron–Matter Interactions

The term "electron microscope" is generally used for the conventional transmission electron microscope (CTEM), or simply transmission electron microscope (TEM). In the family of electron microscopes, this type of instrument was the first one that was commercially available around the year 1940. There has been a rapid development of electron microscopes in which electrons are used as the source of radiation or illumination to view the sample material. The fact that electrons are a kind of ionizing radiation and when ionizing radiation is used to view a sample material, it has a multitude of interactions with the sample is a familiar fact. Here, what happens when an energetic beam of electrons interacts with matter (specimen) will be examined. This is shown in Figure 12.1.

It is shown that as a result of this interaction, the sample transmits, reflects, absorbs, and emits electrons. These interactions result in reflected signals such as secondary electrons,

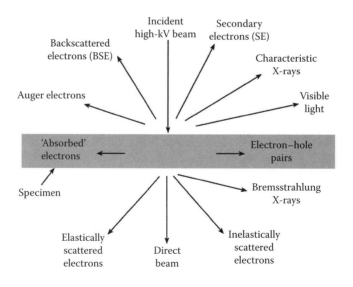

FIGURE 12.1
Interaction of high-energy electron beam encountering a solid thin sample. Most of these signals can be detected in a TEM equipped with the proper detectors. (*Source:* From *Transmission Electron Microscopy*, D.B. Williams and B. Carter, Springer, 1996. With permission from the authors and Springer Science and Business Media.)

characteristic x-rays, cathodoluminescence (visible light), etc., and transmitted signals such as elastically and inelastically scattered electrons. All or many of these signals can be harnessed simultaneously in an instrument, such as a TEM, by employing various detectors above or below the specimen. There are custom-made instruments that have been designed to take advantage of some of these signals. The characteristic x-rays are analyzed, specific to the area of origin, to obtain elemental information; the emitted backscattered electrons, secondary electrons (reflected electrons), and Auger electrons can be analyzed for surface composition and topography, and sample fluorescence. The transmitted electrons that lose some energy (inelastically scattered electrons) can be analyzed for elemental information and for elemental bonding information, in particular for low atomic number elements. The technique is called electron energy loss spectroscopy (EELS). The transmitted diffracted (scattered) electrons can be used to determine the internal structure, crystallography, and composition (thin specimens). To obtain some or all of these signals, one needs to use not only the right sample (electron transparent or electron opaque), but also the right energy (accelerating voltage) electron beam.

Electron probe analyzers (EPMAs) and secondary electron microscopes represent the class of instruments called scanning electron microscopes (SEMs). These instruments essentially use a primary high-energy electron beam to interact with a thick sample and detect the reflected electrons such as secondary electrons, the backscattered electrons, and characteristic x-rays to obtain surface topographical and elemental information.

When an energetic electron beam hits a thin sample, signals are generated below as well as above the sample surface. These signals are harnessed and the class of instruments is known as transmission electron microscopes or scanning transmission electron microscopes (STEMs). When a STEM-type instrument is equipped with a variety of detectors to harness many signals generated above and below a specimen, the instrument is called an analytical electron microscope (AEM). Each class of these instruments is discussed later in a little more detail.

12.4   The Transmission Electron Microscope

It became known that electrons possess a dual characteristic (i.e., particle and wave nature); and when these particles travel fast enough (i.e., approaching the speed of light), they have a characteristic wavelength many orders of magnitude smaller than the wavelength of light. Also, it was known that these particles carry a negative charge and are deflected by electric fields. In 1926, German physicist Busch showed that axial magnetic fields refract electron beams similar to the refraction of light beams by glass lenses. This principle was later used as the basis of deflecting and focusing an electron beam in a transmission electron microscope. Figure 12.2 illustrates the analogy between an optical microscope and an electron microscope. As one can see, the two systems are quite analogous. The arrangement and the ray paths in the two systems are quite similar. There are slight deviations in the basic column design in electron optical instrumentation, depending on the specific analytical applications.

12.4.1   Electron Sources

The basic transmission electron microscope consists of a source of electrons, usually called an electron gun. The gun design can vary, depending on the electron source used. The electron source could be a heated tungsten filament, a heated lanthanum hexaboride

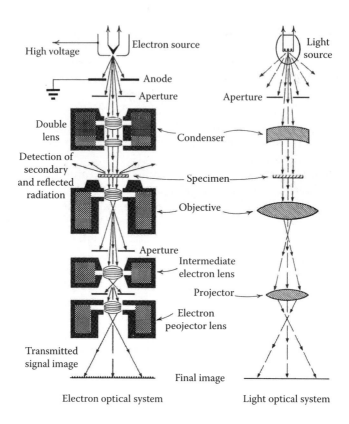

FIGURE 12.2
Schematic comparison of optical and electron microscope systems. (*Source:* From *Electron Optical Applications in Materials Science,* Lawrence E. Murr, McGraw-Hill, 1970. With permission from McGraw-Hill.)

(LaB_6) crystal, or field emission (FE) source; the latter is an extremely sharp tungsten tip. The main objective in using different sources and their design differences is the attainment of maximum electron emission.

The electron emission (or brightness) of the gun used depends on the source. The heated tungsten filament is the most common electron source used in present-day electron microscopes. However, LaB_6 has an electron brightness ten times higher than the tungsten source. The brightness of a cold field emission source is 100 times higher than the LaB_6 source. The entire electron microscope column has high vacuum so that the electrons can travel through the column. The need for a better vacuum in the design of a gun with an LaB_6 source (10^{-7} – 10^{-8} Torr) is imperative, and it is much more so in a gun design with an FE source ($\sim 10^{-10}$ Torr). Instruments with an LaB_6 source are becoming more common although they cost a little more. The cost of gun design with an FE source is significantly more. Thus, electron microscopes with FE sources are mostly used in high-end research labs due to the high initial cost of the instrument and much higher maintenance costs.

12.4.1.1 Thermionic Sources

The electron source, a tungsten filament, or an LaB_6 filament is coupled with beam-forming electrodes. Figure 12.3 shows the top portion of a common form of gun design in such a

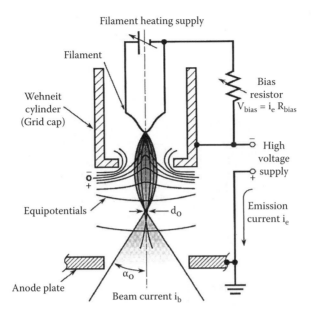

FIGURE 12.3
Schematic diagram of the conventional self-biased thermionic (triode) electron gun. (*Source:* From *Scanning Electron Microscopy and X-Ray Microanalysis*, J.I. Goldstein et al., Eds., Plenum, 1992. With permission from the editors and Springer Science and Business Media.)

system. The electron source in such a system is a V-shaped filament made of fine tungsten wire spot-welded to wire leads attached or mounted on a ceramic disc.

The LaB_6 filament is also mounted in a somewhat similar manner (see Figure 12.4). This design is such that it is easy to replace the filament when the filament burns out after a certain number of hours in use. A DC (direct current) is applied to the filament to provide heat so that it emits electrons, a process called thermionic emission. To facilitate electrons leaving the source (tungsten or LaB_6), their kinetic energy needs to be increased by raising the temperature of the source. This thermionic emission process is applied to separated or detached electrons from the surface. A given amount of work is done against the attractive forces inherent in the source. This attractive force value, called the "work function" (ω), is characteristic of the source material. The emission current density, N, at the cathode surface is defined as follows:

$$N = AT^2 \exp(-\omega/kT) \qquad \text{Richardson's equation}$$

where N is the emission current density or the number of electrons emitted per unit area of the metal surface, T is the absolute temperature, A is Richardson's constant, k is the Boltzmann constant, and ω is the work function (expressed in electron-volts [eV] of energy) and varies from 1.8 eV for B and Sr oxides to 4.6 eV for Mo.

The emission current density N strongly depends on the temperature T and the work function ω. The lower the value of the work function (ω), the more intense the supply of electrons at any given source temperature. Thus, one uses materials that can withstand heating by a few electron volts, do not vaporize, and have very low ω (work function). Tungsten fits into the former category and LaB_6 fits into the latter category.

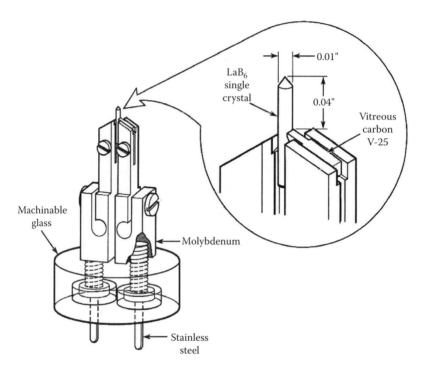

FIGURE 12.4
Schematic diagram of a directly heated LaB_6 emitter that directly replaces tungsten filament assembly. (*Source:* From *Scanning Electron Microscopy and X-Ray Microanalysis*, J.I. Goldstein et al., Eds., Plenum, 1992. With permission from the editor and Springer Science and Business Media.)

12.4.1.2 Field Emission Source

The other category of electron source in electron microscopes comprises field emission gun (FEG). Electrons can also be removed from the metal surface by applying an electric potential between the metal cathode and the anode, resulting in field intensity on the metal surface, $E_f \geq 10^6$ V/mm or 10^9 V/m. When one applies an electric field of this magnitude, electrons escape from the cold cathode by escaping through the surface potential barrier by tunneling. The FE electron sources are now more commonly used in electron microscopes due to lower costs in recent years. There is a need to use much lower vacuum settings (~4 orders of magnitude lower vacuum setting than in a conventional *W* thermionic source). Due to the need for a very high electrostatic field, $E_f \sim 10^9$ V/m, one needs to use an extremely fine tip. The great advantage is the extremely high brightness available with an FE source (almost 10^3 times higher brightness than with an LaB_6 source) and a very long filament life. The table in Figure 12.6 compares some of the important characteristics of three types of electron sources.

As one can see from Figure 12.5 of a field emission source, the field emission tip is an extremely fine tungsten tip, approximately 100 nm in diameter, so that a strong electric field can be concentrated around it.

The field emission tip is the cathode, and there are two anodes with respect to the cathode. The first anode is kept at 3 to 5 kV relative to the cathode. This value of the electric field at the tip is called the extraction voltage, which is pretty high (approximately around 10^9 V/m or 10^6 V/mm). This intense electric field around the tip helps in extracting electrons by easing them

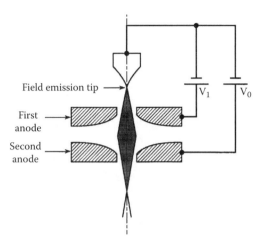

FIGURE 12.5
Schematic diagram of the Butler triode field emission source. V_1 is the extraction voltage, typically a few kilovolts, and V_0 is the accelerating voltage. (*Source:* From *Scanning Electron Microscopy and X-Ray Microanalysis*, J.I. Goldstein et al., Eds., Plenum, 1992. With permission from the editors and Springer Science and Business Media.)

to tunnel out of the fine tungsten tip without the need for any thermal energy to ease them out of the tungsten ω barrier. The second anode, which is at several thousand kilovolts, called the accelerating voltage, helps electrons travel through the column. Because no thermal excitation is provided to the cathode, the electron emission obtained in this manner is called cold field emission. There is no evaporation of tungsten, thus giving the tip a very long life, sometimes several years, before breaking down. One of the important requirements of a field emission electron microscope is a very high vacuum (approximately 10^{-10} Torr or 1.33×10^{-7} Pa).

12.4.1.3 Schottky Emission

This also operates in a manner similar to the cold field emission source except some thermal energy is provided to the tip to reduce the work function barrier of the tip. The tip is usually coated with ZrO to lower the work function of tungsten from 4.5 to approximately 2.8 eV and heated to approximately 1800K to ease the electron emission. The Schottky FE source also requires very high vacuum, as with a cold FE source.

	Units	Tungsten	Lab$_6$	Field emission
Work function ω	eV	4.5	2.4	4.5
Operating temperature	K	2700	1700	300
Current density	A/m^2	5×10^4	10^6	10^{10}
Brightness	A/m^{2}sr	10^9	5×10^{10}	10^{13}
Emission current stability	%/hr	<1	<1	5
Vacuum	Pascals	10^{-2}	10^{-4}	10^{-8}
Lifetime	hr	100	500	>1000

FIGURE 12.6
Characteristics of three principal sources operating at 100-kV accelerating voltage.

12.4.2 Acceleration of Electrons: Accelerating Voltage

After the electron emission from the electron source, the next step is to channel the electrons into the illumination system of the TEM. The gun assembly has a different design for a thermionic source (W or LaB_6) and a field emission source, as mentioned earlier. The cathode (the filament) is covered by a grid cap (see Figure 12.3) with a hole in it, called a Wehnelt cylinder. Below the Wehnelt cylinder is an anode, which is also a metal plate with a hole. Electrons are accelerated by applying a very high electric field parallel to the optic axis. The field is a result of a potential difference V_0 applied between the cathode and the anode, called the accelerating voltage. The anode is kept at earth potential or positive with respect to the cathode and the Wehnelt assembly. Figure 12.3 shows the gun assembly and various fields in a self-biased gun. The electron emission, or the emission current (i_e), is the value assigned to the electrons that leave the grid cap or the Wehnelt cylinder. Only a very small portion of these electrons is capable of escaping through the anode plate. The beam current i_b is the value assigned to electrons that leave the anode and travel through the column and enter the specimen. In the gun configuration just described, it is important that the anode is positive with respect to the cathode. Thus, the anode and the rest of the microscope column are maintained at ground potential and the cathode (the electron source) at a negative potential. A highly stabilized accelerating voltage V_0 is maintained by a high-voltage generator.

The transmission electron microscopes now available have accelerating voltages from 60 kV to 1000 kV (1 MV). Although the instruments available for biological applications have accelerating voltages generally in the range of 60 to 100 kV, the TEMs with materials science applications have accelerating voltages generally in the range from 80 to 200, 300, and even 400 kV. The cost of the basic instrument goes up with higher accelerating voltages. The instruments with 1000 kV or more accelerating voltage are very special research instruments with enormous cost and occupy a huge space. The high-voltage instruments (400 kV and up) have the advantage that the beam electrons have smaller wavelength. This results in slightly better resolution. Also, the higher-voltage instruments, due to higher beam energy, have higher penetration in a sample. Thus, it is very helpful in cases of denser inorganic materials and thicker samples. However, due to relativistic effects, the penetration or the penetrating power does not increase in a linear fashion. In soft polymer samples and biological materials, there is a problem of radiation damage to the sample at higher accelerating voltages. This limits the use of accelerating voltage for such samples to between 80 and 120 kV.

12.4.3 Lens Systems

The function of the lenses in an electron microscope is similar to that in an optical microscope. The major difference is that electron lenses are electromagnetic lenses, versus glass lenses in optical microscopes. The lenses in each system play a very important role. The lenses not only focus the beam on a specimen, but also magnify the image. The focal length of a lens in TEM is variable, unlike in an optical microscope. The focal length is varied by changing the magnetic field (by altering the current in the lenses). Figure 12.7 shows a typical electromagnetic lens used in a modern-day electron microscope. The lens consists of a cylindrically symmetrical core of soft magnetic material with a hole drilled through it. This is called a pole piece. The pole piece is surrounded by coils. In a cross-sectional view, the bore and the gap between the pole piece is visible. There is a jacket around the pole piece; chilled water is circulated to keep the lens cool. Cooling is needed because high current is passed through the coils to generate a magnetic field. The strength of the magnetic field is changed by varying the current flowing through the coils. The magnetic field is strongest closer to the walls of the pole piece and weakest along the axis.

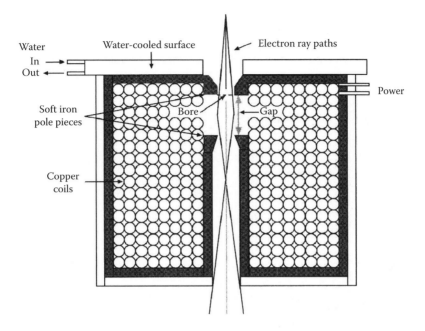

FIGURE 12.7
Schematic diagram of an electromagnetic lens. (*Source:* From *Transmission Electron Microscopy*, D.B. Williams and C.B. Carter, Springer, 1996. With permission from the authors and Springer Science and Business Media.)

12.4.3.1 Condenser Lens

The first condenser lens C_1 is a strong lens with a very small focal length, Figure 12.12. The lens current is varied to change the diameter of the illumination spot called the spot size. The second condenser lens C_2 is relatively weak and works in conjunction with C_1. C_2 allows the fixed spot size obtained by lens C_1 to continuously vary the illumination spot diameter. The second condenser lens C_2 contains a variable condenser aperture (two or three fine holes in a strip). The required aperture is selected by moving the strip forward or backward. By selecting a proper aperture, the diameter D of the electron spot source can be changed in order to control the convergence semi angle of illumination α. Figure 12.8 shows three conditions: focused, under-focused, and over-focused illumination. The C_2 lens current is decreased so that an image of the electron source is found below or after the specimen. In the case of over-focused illumination, the current in C_2 is increased so that the source image occurs above or before the specimen plane. Figure 12.8 thus illustrates the current density profile at the specimen plane that is directly seen as a variation in image intensity on the TEM screen.

12.4.4 The Specimen Stage

The TEM specimens are normally very thin, approximately 50 to 100 nm in thickness, depending on the density of the material. Usually, polymer or polymer composite samples can be viewed slightly thicker due to their lower density compared to ceramic or metal samples. The specimens are supported on a 3-mm copper grid (grids of other materials can be used to support samples). The grids provide thermal and electrical conduction to

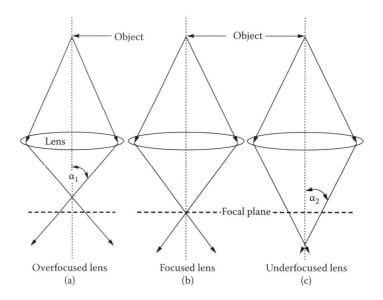

FIGURE 12.8
Ray diagram showing focused, under-focused, and over-focused settings of a lens. (*Source: From Transmission Electron Microscopy*, D.B. Williams and C.B. Carter, Eds., Springer, 1996. With permission from the authors and Springer Science and Business Media.)

the specimen stage, thus minimizing problems of overheating and charging. The design of the stage is such that it allows for specimen insertion into the vacuum without introducing air. This is done by inserting the specimen through an airlock. The specimen, with its holder, is pre-evacuated in the airlock before insertion in the main column.

The specimen stage is available in two basic designs: (1) a top entry stage and (2) a side entry stage. The side entry stage is fairly common with most TEMs. The sample, once inserted in a side entry stage, can be moved in the x and y directions horizontally and in a vertical z direction. Specimen tilt (rotation of the sample along a fixed axis) can be done. Specimen cooling and heating of the sample can also be done, although in the case of polymer or polymer composite samples, heating is rarely needed. Side entry holders are widely used for the majority of imaging.

The top entry stage uses a special conical holder with a collar. That is, when inserted inside, the microscope column sits in a conical well of the specimen stage. This can be moved in the x and y directions by a precision gear mechanism. The advantage of the top entry stage is that it is less prone to picking up vibrations from the TEM environment. However, it is more difficult to provide tilting, heating, or cooling of the specimen. It is also not very practical to do EDS (energy dispersive x-ray) analysis of the samples. This design is primarily used for very high resolution, high magnification work.

12.4.5 TEM Imaging System

The next part is the imaging of the specimen. The final imaging can be done on the viewing screen or on a monitor using a digital camera system. The quality of the final image depends on the imaging lenses, especially the first imaging lens called the objective lens. The other two lenses that follow the objective lens are the intermediate lens and the projector lens; the latter two lenses only magnify the image formed by the objective lens.

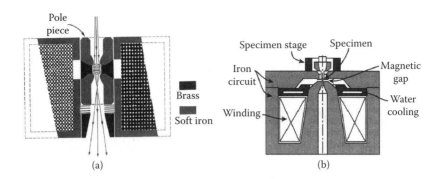

FIGURE 12.9
(a) Objective lens design and (b) a vertical entry objective lens design. (*Source:* From *Electron Optical Applications in Materials Science*, L.E. Murr, McGraw-Hill, 1970; and *Transmission Electron Microscopy*, D.B. Williams and C.B. Carter, Springer, 1996. With kind permission from the authors and McGraw-Hill.)

12.4.5.1 Objective Lens

The objective lens plays one of the most important roles in imaging and resolution in a transmission electron microscope. This is the first image (and diffraction pattern) forming lens. The first image is formed by an objective lens, magnified 50X to 100X, and is enlarged by the subsequent lenses in the electron optical column. The objective lens is a strong lens and has a small focal length. It differs from the condenser lens design described before and intermediate and projector lens combination that follows the objective lens, mostly in terms of geometry. The short focal length of the lens concentrates magnetic field strength in a narrow area on the axis of the pole piece in the column (Figure 12.9a). The geometry or the design allows placing the sample close to the center of the lens due to the strong magnetic field near the center. A conventional objective lens (Figure 12.9a) and a top entry objective lens (Figure 12.9b) are shown. The small focal length of around 2 mm helps in optimizing the image resolution by providing a small coefficient of spherical and chromatic aberration (discussed later). In less dense materials such as polymers or biological materials where extremely high resolution is not needed, slightly higher focal length objectives can be used. The higher focal length objectives also give higher contrast in such materials. It is important to obtain higher contrast from low-density materials.

12.4.5.2 Objective Aperture and Contrast

An aperture is a disc with a hole; the metal surrounding the hole is called a diaphragm. Apertures at various stages in a column play an important role. Apertures are used to allow certain electrons to pass through and exclude others by stopping them by limiting the collection angle. This is shown in Figure 12.10. The diagram illustrates how an aperture disc absorbs or stops the electrons that scatter through larger angles. The electrons that spread less than semi-angle β subtended by the aperture happen to pass through the lens. The stopped or the absorbed electrons do not contribute to the final image. By reducing or limiting the size of the opening (the hole), one can selectively control or stop all the scattered electrons. Thus, the areas of the specimen that scatter electrons strongly will appear as dark areas in the final image. This is called scattering contrast or diffraction contrast. Polymers or biological materials made of low atomic number elements scatter less and thus show lower contrast. There are ways to increase the contrast while viewing

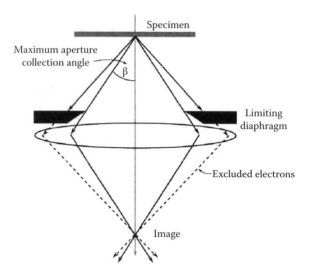

FIGURE 12.10
How the diaphragm of an aperture restricts and allows only electrons contained in a cone less than semi-angle β subtended by the aperture at the specimen are allowed to pass through the lens. Those electrons that are scattered greater than β are stopped. (*Source:* From *Transmission Electron Microscopy,* D.B. Williams and C.B. Carter, Springer, 1996. With permission from the authors and Springer Science and Business Media.)

such materials, for example, by selective staining (introducing high atomic number material selectively). This is discussed later in "Sample Preparation." The materials with high atomic number elements, such as metals, ceramics, etc., have high scattering so they do not need any contrast enhancement.

The role of the objective aperture is not only to produce contrast in the image by blocking the scattered electrons, but also to limit image blurring arising from spherical and chromatic aberrations, thereby affecting the overall resolution. By limiting or decreasing the aperture size, it might be thought that one can improve the resolution. However, the objective aperture gives rise to diffraction effects that become worse with a decrease in the size of the aperture diameter. So, one must strike a balance between the aperture size and the best resolution obtainable.

12.4.5.3 Intermediate Lens

The role of the intermediate lens, situated between the objective lens and the projector lens, is to change the final magnification by changing the focal length of this lens in fine steps. Also, this lens is used to obtain an electron diffraction pattern on the TEM viewing screen. This is done by reducing the current in the intermediate lens so that this lens produces, at the projector image plane, a diffraction pattern rather than an image. This is an important tool (obtaining a diffraction pattern) used in the characterization of crystalline samples for structural determinations.

12.4.5.4 Projector Lens

The projector lens, the final lens in a TEM, produces a final magnified image or a diffraction pattern covering the entire imaging screen. In most cases, this lens is used as a strong lens with a very small focal length to obtain very high final magnification. The lens strength can be varied to obtain images of relatively low magnification (~1000X or less) on the final viewing screen. The final image magnification on the TEM viewing screen is the product of the magnification factors of the objective, intermediate, and projector lenses.

12.4.5.5 Lens Aberrations and Resolution

The lens fields in an electron optical system are imperfectly formed due to the mechanical flaws in the lens design and pole piece construction. Also, the mutual repulsion of electrons at very constricted points, such as lens apertures along the optical axis in an electron optical column, gives rise to image distortions and contributes to the loss of contrast and sharpness. This results in the loss of resolution. There are several lens aberration effects that are primarily responsible for the loss in resolution. They are classified as spherical aberrations, chromatic aberrations, and field effect aberrations. We consider here spherical aberration, chromatic aberration, and astigmatism.

12.4.5.5.1 Spherical Aberration

The electrons originating at different points in the object focus at slightly different points along the optic axis. The further off the axis (peripheral) the electron is, the more strongly it is bent back toward the axis, because it experiences a stronger field. The effect is a characteristic lack of sharpness in the image on the optic axis. Spherical aberration is illustrated in Figure 12.11a. The electrons travel vertically from top to bottom along the Z axis. The Z axis is along the ray path PQ. A point object in a specimen is imaged not as a point image Q but as a halo spread along QQ′. The electrons traveling along path PB, being closer to the pole piece, experience more fields and thus come into focus before the image plane. Whereas, electrons traveling along PA experience less intense field and come into focus at or closer to the image plane. This results in an enlarged image of point P in the Gaussian plane of diameter 2QQ′. Because the electron lenses work only as convergent lenses, it is difficult to minimize spherical aberrations using a convex/concave lens combination as in light optics. The aberration can be minimized using a smaller lens aperture to reduce the electron probe diameter. Spherical aberration is a far more important lens aberration because this limits the resolution of an electron optical system in a transmission electron microscope. The objective lens is the image-forming lens in a TEM, and spherical aberration of this lens is

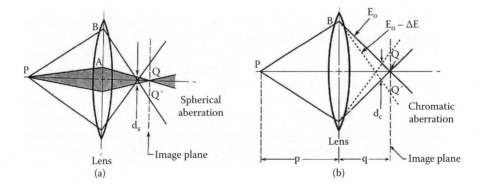

FIGURE 12.11
A schematic diagram showing the (a) spherical aberration effect and (b) chromatic aberration effect. (*Source:* From *Scanning Electron Microscopy and X-Ray Microanalysis*, J.I. Goldstein et al., Plenum, 1990. With permission from the author and Springer Science and Business Media.)

the most important factor in imaging. Spherical aberration depends on changes in the focal length with accelerating voltage and strongly depends on aperture angle α. The correction of spherical aberration largely depends on the design of the lenses, with special field distributions for allowing smaller aperture angles and shorter focal lengths.

12.4.5.5.2 Chromatic Aberration

The term "chromatic" is analogous to "color" in light optics and relates to the wavelength. In electron optics, it also relates to the wavelength and hence the energy of electrons. Chromatic aberration in electron optics is due to velocity (energy and therefore the wavelength of electrons) differences in electrons in the electron beam. The electron lens focuses fast electrons (short wavelength) at a point further from the lens than slow electrons (long wavelength), thus forming a disc image instead of a point image. Figure 12.11b shows that if two electrons in the ray PB have different wavelengths (energies); (*viz.* E_0 and $E_0 - \Delta E$); the two will focus at different points in the image plane. The image of point P, instead of focusing at a point, focuses at different points Q and Q′ in the image plane. Due to this process, the image of point P gets enlarged to 2QQ′. This results in degrading the image resolution. This effect is aggravated for thicker samples and low accelerating voltages. However, the fluctuations in the lens coil supply can be corrected to a great extent by highly stable lens supply. Also, it is advisable to use (1) thinner samples to reduce beam energy losses resulting from inelastic scattering, and (2) higher accelerating voltages.

12.4.5.5.3 Astigmatism

Astigmatism occurs when the focal length in one plane direction differs from the normal direction in the same plane. This is due to the nonuniform magnetic field that arises from the noncircular pole piece bore. This nonuniformity arises from the lack of good machining of the pole piece hole, or if the aperture is not precisely centered on the central axis. This can also happen if the aperture hole is not perfectly clean or gets contaminated during long use. The contamination around the edges of the aperture charges and deflects the beam. This problem is especially troublesome in the condenser and objective lens systems. Astigmatism causes an elliptical distortion in the image plane. It is usually corrected by inserting a stigmater in the appropriate lens system to compensate the field for the nonhomogeneity.

12.4.6 TEM Operation

The illumination system transfers electrons from the electron source to the specimen, either by a broad beam called a parallel beam or a spot mode. In the traditional TEM imaging mode, the current in the two condenser lenses is adjusted to illuminate the specimen with a parallel beam of electrons as shown in Figure 12.12a, or a spot mode Figure 12.12b.

To form an image in a TEM, the electron beam hits the specimen, which is very thin, usually 50 to 80 nm in thickness, in the case of a polymer or a polymer nanocomposite sample. The sample, usually a thin section, is supported on a thin metal mesh called a grid. For a suspension or a powder sample, the grid is supported by a specimen support film that is extremely thin and electron transparent. These support films are generally made of a plastic or carbon, or a plastic film reinforced with a thin layer of carbon.

12.4.6.1 Image-Forming Process

In both light and electron microscopes, there are four fundamental physical processes that take part in the formation of the image: (1) absorption, (2) interference, (3) diffraction, and

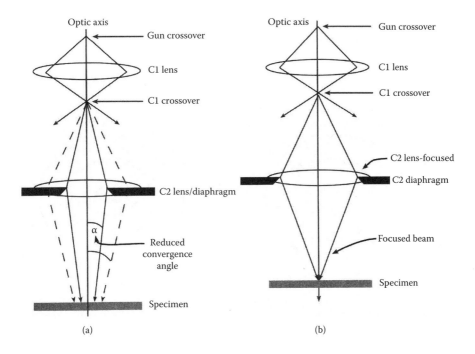

FIGURE 12.12
(a) A parallel beam striking the sample, and (b) a focused beam (spot mode) hitting the sample. (*Source:* From *Transmission Electron Microscopy*, D.B. Williams and C.B. Carter, Springer, 1996. With permission from the authors and Springer Science and Business Media.)

(4) scattering. In general, all four together play some role in image formation. Absorption gives rise to amplitude contrast, which is the variation in intensity to which the eye is quite sensitive. The absorption of photons within different regions of the specimen is the most important factor in the formation of a conventional light microscope image. Interference and diffraction play a very limited role. Also, scattering plays a very limited role in the formation of a light microscope image. In the case of transmission electron microscopy, most of the electrons are transmitted through the specimen. The transmitted electrons go through a complex scattering process. Thus, one needs to understand the scattering process—not only to understand the formation of an image, but also to interpret the image formed.

When a high-energy electron beam strikes a thin specimen, the atoms of the specimen will interact with the striking high-energy electrons. One can understand this interaction by considering one single target atom (specimen) represented in a simplistic way as a positively charged heavy nucleus surrounded by orbiting electrons. The majority of the incident electrons will pass straight through a very thin specimen (Figure 12.13a). Interaction between the incoming fast electrons having little mass and an atomic nucleus (with heavy mass), and passing close to a nucleus (specimen atom), will be deflected through a large angle and suffer almost no energy loss but there will be considerable deviation in trajectory. This type of interaction is called "elastic scattering" (Figure 12.13c). The third kind of interaction is where the fast incoming electrons interact with the slow orbiting electrons of the specimen. In this process, the high-energy electrons and the orbiting electrons will share their energies. The high-energy electrons not only lose some of their energy, but also suffer a change in direction (Figure 12.13b). This type of interaction is called "inelastic scattering." Because there is a larger abundance of electrons in a specimen than the number of

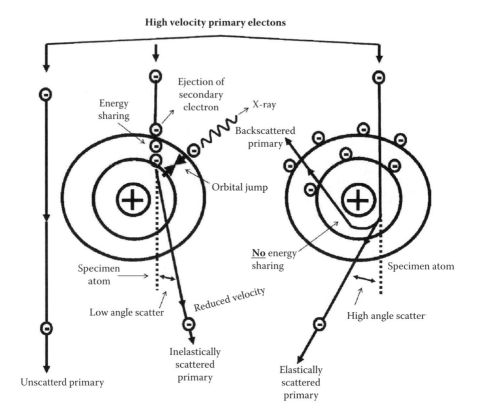

FIGURE 12.13
The high energy electrons' interaction with the atoms of the specimen. (*Source:* Adapted from *Practical Electron Microscopy for Biologists*, G.A. Meek, John Wiley & Sons, New York, 1977.)

nuclei, inelastic scattering is the most important phenomenon in the formation of an image in transmission electron microscopy.

12.4.6.2 Mass Thickness

The probability of scattering of incident electrons by specimen atoms increases with an increase in the number of atoms in the specimen. The amount of scattering of electrons is a strong function of specimen thickness—the greater the thickness, the more atoms in the sample. Also, the amount of scattering is a function of Z, the atomic number of atoms in the specimen. Thus, atoms with high Z have higher density and a higher propensity to scatter than the low Z atoms. So, the higher the specimen density, the greater probability it has to scatter. Therefore, the total scattering is not only proportional to the thickness, but is also proportional to the product of density and thickness. This is called mass-thickness contrast.

12.4.6.3 Absorption

If the mass thickness is high, the beam electrons interacting with the specimen will be completely absorbed. That is, the impinging electrons will go through a sequence of inelastic collisions until their velocity is sufficiently reduced to be captured by the specimen.

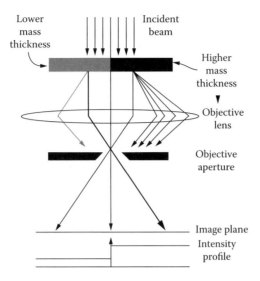

FIGURE 12.14
Shows the mechanism of mass-thickness contrast in a bright field image. For a given thickness of a sample higher Z areas of the samples (darker) will scatter more electrons away from axis than the areas with lower Z. (*Source:* From *Transmission Electron Microscopy,* D.B. Williams and C.B. Carter, Springer, 1996. With permission from the authors and Springer Science and Business Media.)

Absorption in thin samples does not contribute to contrast in the image. Total absorption involves considerable energy transfer to the specimen, which in turn will become hot. This can damage the specimen and is especially true in the case of polymeric materials.

12.4.6.4 Contrast Formation and Objective Aperture

Various phenomena that can result in contrast formation in the final image have been discussed. The process of image formation by a high-energy electron beam is primarily due to the inelastic scattering of electrons, or in other words, due to the deviation of the electron path of the primary electrons after they exit the specimen. Thus, in portions of the specimen where mass thickness is high (materials such as metals), there will be more deviation or change in the path of the primary electrons. The portions of the specimen that have lower mass thickness are traditionally the materials consisting of low Z atoms. There will be less deviation in the path of the primary electrons in the case of biological and polymeric materials. This is illustrated in Figure 12.14.

The objective aperture plays a significant role in the formation of the final image, which is well resolved and has good contrast. The objective aperture controls or stops electrons that are scattered through an angle greater than the acceptance angle of the objective lens (see Figure 12.13). The imaging electrons pass through the hole of the aperture, and the scattered electrons strike the periphery of the aperture and are conducted to the ground. The aperture is a thin disc with a hole or a fine strip with multiple holes of various sizes, and is made of nonmagnetic materials such as molybdenum or platinum. The aperture is placed close to the optical center and the back focal plane of the objective lens. The choice of various sizes of the objective aperture allows the operator to pick the right size (20–100 μm). For low Z materials, such as polymers or biological materials, there is less scattering, and thus smaller apertures are used to enhance contrast.

To view samples of polymers and polymer composites, it is essential to determine the best operating conditions. Not only is it important to obtain maximum contrast, but also it is important to obtain the highest possible resolution. However, these two requirements are mutually incompatible. If the accelerating potential is increased, the speed of electrons in the beam increases. The implication is that the electrons will pass through the specimen without interacting with the atoms and also substantially decrease the ratio of the inelastically scattered electrons to the elastically scattered electrons. Because the inelastic electrons play the dominant role in contrast formation, increasing the accelerating potential reduces contrast in polymeric materials. On the other hand, increased potential helps in viewing thicker samples because of the higher penetrating power at the cost of contrast. Thus, in case of polymeric materials, higher accelerating voltage (100–120 kV) will give higher resolution but lower contrast. On the other hand, lower accelerating voltage (60–80 kV) will increase contrast with a slight decrease in resolution. It is for the operator to judiciously choose the operating conditions to obtain the desired results. Experience has shown that a good accelerating voltage for imaging polymer or polymer composite samples can be varied between 80 and 120 kV. The optimum section thickness can be between 40 and 80 nm. The best resolution, approximately 0.2 to 0.3 nm, at 120-kV accelerating voltage can be attained. Beam damage of polymers and polymeric materials can occur at accelerating voltages equal to or greater than 120kV.

In transmission electron microscopy studies of polymeric materials, one must be very careful in choosing the right operating conditions other than the accelerating voltage. It is very difficult to observe beam-sensitive materials such as polymers at bright levels of illumination. The beam damage to the material can occur. For example, semi-crystalline polymers, such as polyethylene, lose their crystallinity if viewed under very bright illumination. One needs to use a low beam intensity by using low beam current. With a low beam current, the image on the screen becomes faint; using longer exposures can compensate for this. Usually, the area on the imaging screen during the focusing process gets damaged (beam damage). The trick is to use extremely low beam intensity. Many of the newer instruments have a feature called "low dose imaging." This feature allows one to focus on a remote area of the image while the rest of the image is blanked. Once the image is focused on the remote area, the area of interest is illuminated and the image is quickly captured.

12.4.7 Image Capture

The final image can be captured either on a photographic film or digitally using a digital image capture system. The photographic film loaded on film cassettes is kept in a light-tight box immediately below the final viewing screen. The photographic film box usually holds 50 films at a time. The camera system automatically determines the correct exposure for an image and records it on a film just by pressing a button. The exposed film is processed in a dark room and prints are made photographically. Alternatively, the image is captured digitally. The image sensor (camera) is located either above or below the imaging screen. Due to the very large depth of focus of an electron image, the image is in focus at any point in the column. The only difference will be the magnification at a given point in the column. The final magnification will be slightly low if the sensor is above the imaging screen. The imaging screen is calibrated for the actual magnification. The digital camera sensors are also calibrated (above or below the imaging screen). The digitally captured images are stored in a computer and can be recalled later for viewing and printing.

12.5 Sample Preparation

In transmission electron microscopy, sample preparation of materials such as metals, ceramics, other inorganic materials, polymers, and biological samples plays a very important role. In the case of polymeric materials, it requires extreme caution due to the use of toxic chemicals as staining agents. Although these chemicals are used in extremely small quantities, extreme care and proper use of safety procedures are recommended.

Sample preparation procedures and routines for polymer microscopy are exhaustive and broad. It is too large a topic to be discussed in great detail here. The methods range from very simple to very complex and time consuming. Samples, usually thin sections, ready to be viewed in a TEM are mounted on a fine thin mesh called a grid. The grids come in a standard size (3 mm in diameter) and are available in various mesh sizes. The mesh size determines the size of the opening. Mesh openings act as the open area of the sample that can be viewed. The specimen, a thin section, is placed directly on a grid or sometimes on a grid with a very thin electron transparent plastic support film, usually strengthened by a few nanometers (5–10 nm) of carbon coating. There are many different ways to prepare polymeric samples for viewing in a TEM. Only one specimen preparation procedure, called thin sectioning by ultra-microtome, is discussed here.

12.5.1 Ultra-Microtomy

An ultra-microtome is a device used to prepare very thin sections of soft polymeric materials, biological materials, and some inorganic materials. The thin sections are electron transparent and can vary in thickness from approximately 30 nm to 100 nm or so. The thin sections are cut using a freshly prepared glass knife or a diamond knife. The cutting of precise, even thin sections of polymeric material is one of the most difficult procedures in the entire sample preparation procedure sequence. It requires high skill, great patience, and great manual dexterity. The basic principles of ultra-microtomy are shown in Figure 12.15.

In the up-and-down movement of the arm, the sample is sliced by the fixed knife and the thin sections float on a liquid, usually water, in the trough of the knife. After each cut, the arm moves forward by a specified amount, the section thickness, which is pre-set. A thin section is cut in each movement of the arm. A large number of sections form a ribbon. The ribbon of sections is picked up on a grid.

This technique can also be used for nanoparticles or nanofibers when there is a need to study the cross-sectional structure of these materials. The sample is embedded in an epoxy material (one must make sure that the epoxy does not chemically react with the material to be studied) and then thin sections are prepared from a block face of the embedded material.

12.5.2 Cryo-Sectioning

Cryo-ultra-microtoming is the procedure of ultra-microtoming a sample at low temperature (below room temperature). This procedure is essential for polymers or polymer composites where T_g (the glass transition temperature) of the polymer is below room temperature. The T_g of a polymer is the characteristic temperature at which the glassy polymer changes to more flexible and less brittle. Below T_g, the amorphous regions are not flexible and are usually hard and brittle—this is called glassy. The microtoming of such samples is done 10 to 20°C below the T_g of the material in an effort to avoid smearing the sample during the

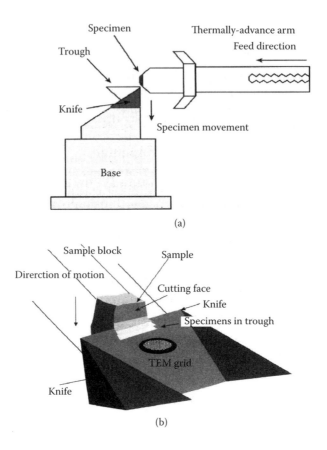

FIGURE 12.15
The basic parts of an ultra-microtome: (A) the sample is held in a chuck at the end of a moving arm that moves in an up and down movement against a knife. (B) The thin sections float off onto water surface, and are picked up on a grid. (*Source:* From *Transmission Electron Microscopy*, D.B. Williams and C.B. Carter, Springer, 1996. With permission from the authors and Springer Science and Business Media.)

cutting process. The T_g of polypropylene is approximately $-120°C$; thus, microtoming of this polymer or its composite is done at approximately $-140°C$.

12.5.3 Staining

Polymeric samples and biological materials are usually made up of elements such as carbon, hydrogen, oxygen, nitrogen, etc. These are low Z elements and therefore have low electron scattering ability. To image such materials, their scattering ability must be enhanced to obtain contrast in the image.

Staining is the most common technique used in polymers to enhance contrast. Staining essentially involves the addition of electron-dense atoms (high Z elements). Staining increases the density of the specimen material by either chemical reaction or physical incorporation by adsorption. However, contrast enhancement of polymers for transmission electron microscopy is quite a comprehensive subject and here we deal with essentially two stains that are commonly used. One reacts chemically with the unsaturated part in a polymer by binding across the double bond. The other works by adsorption. The adsorption by various polymers largely varies; thus, the latter method imparts variable contrast in polymers in a multiphase system. Staining of polymers and polymer composite systems can be done either before or after thin sectioning. It is good to do staining prior to thin sectioning because staining usually hardens the polymer. This makes it easier to thin-section in an ultra-microtome. Small pieces of polymers are cut into small rectangular blocks (approximately 4×8 mm) and a small face (approximately 0.5×2 mm) is made and finished in an ultra-microtome. The block is then immersed in the stain for a given period of time at room temperature. After the staining period, the block is mounted back in an ultramicrotome chuck, very precisely aligned parallel to the knife, and thin sections are sliced. It is important that the first few thin sections are picked up for examination because the stain does not penetrate very deep. If the sections are to be stained after thin sectioning, then the initial step of staining as a block is skipped. The unstained block is thin sectioned. The thin sections are then stained by either floating the grid over a drop of stain, the section side in contact with the drop, or the sectioned material is kept in the vapors of the stain in an enclosed area for a fixed time.

12.5.3.1 Osmium Tetroxide (OsO₄)

An unsaturated polymer (by itself or as part of a multiphase system) is stained with osmium tetroxide (OsO_4). The stain reacts with the unsaturated portion of the polymer by opening the double bonds. This stain also fixes and hardens the polymer. The best example of osmium staining of a polymer is the butadiene rubber in ABS (acrylonitrile butadiene styrene) copolymer. Osmium stains the butadiene phase only. On TEM examination, the butadiene phase appears dark due to enhanced contrast (increased scattering imparted by osmium incorporation in the butadiene phase across the double bond). The osmium staining affects several polymers containing $-OH$, $-O-$, or $-NH_2$ groups also because most of these polymers contain some degree of unsaturation. The partial staining effect in these polymers imparts a light gray level when imaged in a TEM. The OsO_4 stain is generally used as a 2% aqueous solution. Also, it is used as a vapor stain by placing a crystal of OsO_4 in a closed dish next to the thin sections. This is due to the high vapor pressure of OsO_4.

12.5.3.2 Ruthenium Tetroxide (RuO₄)

Ruthenium tetroxide (RuO_4), like osmium tetroxide, is a very strong oxidizing agent and a good stain for the majority of polymeric materials. RuO_4 stains most polymers, including saturated polymers. There are only a few exceptions (e.g., PMMA, PVC, and PAN) that are not affected by RuO_4. The RuO_4 stain, like OsO_4, provides stability against high-energy electron beams. The stain is used in vapor form to either stain a block of the sample or thin sections. The staining time varies from a few minutes to 30 minutes or more, depending on the polymer and the application.

12.5.3.3 Other Staining and Contrast Enhancement Methods

There are several other staining methods used to stain polymers but those are used in very specific cases and thus are not elaborated upon here. Staining is a very important part of sample preparation of polymers for transmission electron microscopy. The staining by strong oxidizing agents such as OsO_4 and RuO_4 works for the majority of samples. There are certain other techniques such as plasma etching and chemical etching that are sometimes used for phase identification in multiphase systems.

12.6 Scanning Transmission Electron Microscopy

In scanning transmission electron microscopy (STEM), a fine electron probe scans thin electron transparent samples, the same as used for transmission electron microscopy. The electrons that emerge from the other side of the specimen are detected. The probe scans the specimen portion picked to be analyzed. The image is collected point by point and recorded on a film or digitally captured. The image is displayed on a cathode ray tube (CRT). The magnification is given by the ratio of the length of display line on the CRT to the length of the probe line on the specimen surface. There are no imaging lenses after the specimen. The resolution depends on the spot size of the probe. The feature is available (as an option) in the newer transmission electron microscopes, also called conventional transmission electron microscopes (CTEMs). In addition, there are dedicated STEM instruments with no TEM imaging capability. To obtain high resolution in a STEM instrument or in STEM mode in a CTEM, one needs to have a fine probe from a high-brightness source such as an FE source or, alternatively, an LaB_6 thermionic source. The size of the probe spot that determines resolution in a STEM cannot be made infinitely small. Thus, there is a need for a higher brightness source such as an FE that has 10^5 times more brightness and a probe size 10^{-4} times smaller than a conventional thermionic tungsten source. The enhancement in the resolving power of a STEM instrument is not only due to a very small high-brightness probe, but also to the fact that there are no image-forming lenses after the probe exits the specimen. Due to the absence of lenses after the specimen, there are no aberration effects of the imaging lenses that degrade the image resolution. Also, slightly thicker samples can be viewed in a STEM mode than in a CTEM mode.

Now we see what happens to the electrons that are transmitted when an electron probe scans a thin specimen in a STEM mode; refer to Figure 12.13:

1. A large number of incident electrons pass straight through the thin specimen without interacting with the sample.

2. Some incident electrons interact with the electron cloud around the atoms in the specimen. The incident electrons share their energies with the electrons of the specimen atoms and lose some energy. This energy loss changes the incident electrons from their original path. These are called inelastically scattered electrons.

3. Some incident electrons pass close to the nuclei of the specimen atoms and are repelled by the positively charged nucleus. Thus, they deviate through a much larger angle, sometimes through 180°. These are elastically scattered electrons and are called backscattered electrons.

If we collect each of the three kinds of electrons described above using three different detectors and display on three different CRTs, we obtain different information from each of these signals. The undeviated electrons, similar to the ones in a CTEM, give a Bright Field image. The second kind, called an inelastic signal, gives a reverse contrast called the *dark field image*. The third kind, the elastically scattered electrons, are also called backscattered electrons that have interacted with the nuclei. The higher Z elements will generate more of the elastic scatter; thus, this signal is an indication of the presence of higher Z elements. These three, simultaneously displayed images provide much more information than one can obtain from one CTEM image.

There are advantages and disadvantages to viewing polymeric materials by STEM. The three signals displayed simultaneously provide much more information than one CTEM image. Because there are no lenses after the high-energy electrons hit the specimen, there is no focus required with the change in magnification. Also, multiple signals can be combined digitally to obtain more information from the specimen. The beam hits only one part of the specimen at any one time. Thus, beam damage is confined to one area at a time. In CTEM, a much larger area is radiated at one time. One can image bright field, high magnification images directly. Also, one can image thicker specimens in a STEM mode. Dark field images from inelastically scattered electrons from noncrystalline polymeric materials have better resolution than in a CTEM. Because of the extremely small spot size in a STEM, it is possible to do micro-diffraction of any particular area of a specimen without causing radiation damage to a much larger portion of the sample.

If we compare a CTEM, a CTEM with STEM capability (CTEM/STEM), and a dedicated STEM, each instrument has its pros and cons. A CTEM is capable of providing much higher resolution than a CTEM with STEM capability. The dedicated STEM can provide higher resolution than a CTEM. However, the dedicated STEM instrument is more expensive due to the high brightness source that requires very high vacuum and very intricate electronics. One great advantage of a STEM, whether a CTEM/STEM or a dedicated STEM, is its ability to perform analytical electron microscopy (AEM) on a sample.

The STEM instrument has advantages over the CTEM for low Z materials such as polymers and biological samples. The ability to capture images obtained from separate signals, discussed earlier, and view slightly thicker sections, provides more information and flexibility.

12.7 Scanning Electron Microscopy

Scanning electron microscopy (SEM) is a technique used to study the surfaces of thick samples and bulk materials. The technique is similar to the light microscopy of the surfaces of thick materials. A high-energy electron beam is used that gives an advantage of much better spatial resolution due to the very short wavelength of electron radiation. As one can see from Figure 12.1, when a high-energy beam hits a solid sample, a variety of signals are released above the sample. In SEM, the signals such as backscattered electrons (BSEs), secondary electrons (SEs), characteristic x-rays, visible light, etc., can be harnessed with the use of suitable detectors, and a host of information about the material and its surface can be obtained.

The incoming high-energy primary electrons that strike the specimen release secondary electrons in the process. The SEs have relatively low energy (a few eVs to a few tens of eVs) and are released from the very top surface (≤100 nm) of the material. The SE signals used to form an image give topographical image of the surface. Other signals such as BSEs and x-rays are also collected to get additional information such as elemental information and the composition of the observed surface. A modern state-of-the-art SEM is capable of providing an image resolution of 1 nm, although resolutions of 5 to 10 nm can be obtained on a typical SEM. This resolution is much higher than that from a light microscope. Scanning electron microscopy has another great advantage; the image obtained by an SEM has a large depth of focus due to the short wavelength of the electron radiation used. A typical SEM view is shown in Figure 12.16.

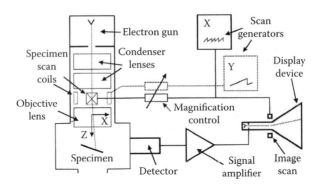

FIGURE 12.16
Schematic diagram of a scanning electron microscope. (*Source:* From *Physical Principals of Electron Microscopy*, Ray F. Egerton, Springer, 2005. With permission from the author and Springer Science and Business Media.)

A high-energy electron beam (typically 0.5 to 30 kV) is focused on to a fine spot by a combination of two condenser lenses. Then there is a final lens that demagnifies the spot into a probe 1 to 10 nm in diameter depending on the electron source (thermionic or FE) used. This final lens is called the objective lens (but the function is different than that of the objective lens in a TEM). The probe diameter determines the resolution of the SEM. This fine probe is scanned horizontally across the specimen in the x and y directions perpendicular to each other. The x scan moves across the specimen in a straight line, and the y direction moves incrementally down, thus covering the entire specimen area of interest. The x and y scans are in a synchronous way scanned on a CRT like a TV screen, and the SEM image is viewed on this screen. The ratio of the line scanned on the screen to the line scanned on the specimen gives the magnification. Because the image fills the entire TV screen, it is the scan on the image that is varied to change the magnification.

Electrons are negatively charged particles. Thus, when an electron probe hits a sample, the sample accumulates the negative charge if the material is not conductive. To avoid accumulation of negative charge on the surface of the sample, nonconductive samples are usually made conductive by depositing 10 to 20 nm of a conductive metal such as Au or Au/Pd alloy. Alternatively, one can use very low accelerating voltages (0.5 to 2.0kV), depending upon the sample material; this obviates the need for any conductive coating. In this scenario, the electron hitting the sample surface and those being ejected (SEs and BSEs, respectively) balance and thus there is no charge accumulation on the surface. This technique is now being widely used in viewing nonconducting materials such as polymers. The technique is called low-voltage scanning electron microscopy (LVSEM). The only requirement in LVSEM is the use of a high brightness source such as LaB_6 or an FE source. This technique has become very common in the past 10 years or so due to the availability of high brightness sources at lower cost.

12.8 Energy Dispersive X-Ray Analysis

When a high-energy electron beam strikes a specimen, x-ray photons are also emitted (see Figure 12.1 and Figure 12.17). Characteristic x-rays that are emitted have specific wavelengths and energies associated with them that are characteristic of the atoms in the

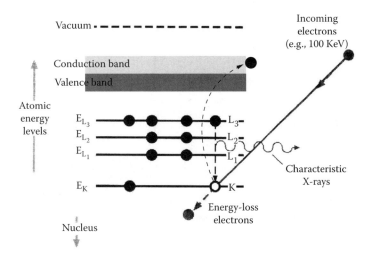

FIGURE 12.17
X-ray emission. When a high-energy electron strikes matter, an inner (K) shell electron is ejected from an atom. An electron from the L shell fills the vacancy in the K shell. In the process K x-ray emitted. The beam electron loses energy but the process continues on through the specimen. (*Source:* From *Transmission Electron Microscopy*, D.B. Williams and C.B. Carter, Springer, 1996. With permission from the authors and Springer Science and Business Media.)

specimen. When the x-ray wavelength is detected, the technique is called wavelength dispersive spectroscopy (WDS). When the x-ray energy is detected, the technique is called energy dispersive spectroscopy (EDS).

The EDS technique is more widely used on TEMs and SEMs because it is more compact and easy to use. The main advantage is the speed of data acquisition. Also, there is the fact that x-rays in a wide range can be detected and analyzed simultaneously. The complete EDS system is less expensive than the WDS system. Both techniques can do elemental analysis, beginning with element boron in the periodic table. A special thin window detector or a detector with no window is used for low Z elements from atomic number 4 to atomic number 10 (i.e., from B to Ne). EDS in most of the modern SEMs and STEMs is the more common and convenient way to detect x-rays from a specimen bombarded with high-energy electrons. The elemental information obtained by this technique helps in performing both qualitative and quantitative elemental analyses on a particular specimen, in combination with other techniques such as electron energy loss spectroscopy (EELS).

The x-ray micro-analyzer can be operated either in probe mode or in scan mode. In probe mode, a fine electron probe (spot) is focused on a specimen (such as a nanoparticle) and x-rays collected. In scan mode, a fine electron probe is scanned on a portion of a specimen and x-rays are collected from that area. In mapping mode, one can detect the presence or distribution of a particular element in a specimen. In this case, in scan mode, the detector is set to the x-ray spectral energy corresponding to one or more than one selected element and the probe is scanned in a raster across the desired area of the specimen. A map of the specimen area selected in light of the selected element or elements is generated on a display CRT and digitally captured. The areas rich in any particular element/elements are displayed bright in a dark background. Figure 12.18b shows an x-ray map of an area in Figure 12.18a. An electron micrograph of the entire area using the TEM mode or the STEM

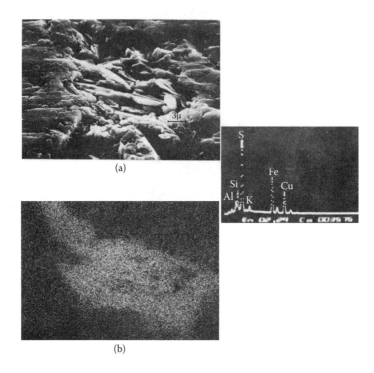

FIGURE 12.18
(b), K, Al, and Si map of an area in (a) showing area rich in these elements; the surrounding outer area is rich in Cu, Fe, and S. An x-ray spectrum is shown in the insert on the right.

mode is then taken. The two images are compared or superimposed to show the distribution of the selected elements in the specimen. Using the proper procedures, a quantitative elemental analysis of the sample can be performed. This feature provides a quantitative estimate of the concentration ratios of the elements present in a given area of the specimen. The quantitative elemental analysis procedure takes into account several factors, including the thickness of the specimen.

12.9   Electron Diffraction

Most inorganic materials such as metals, ceramics, oxides, etc. are made of regularly spaced arrangement of atoms in arrays or a lattice of rows and columns. These are called crystalline materials. A beam of electrons passing through such a material will be scattered in a preferential manner. An amorphous material has a random distribution of atoms; it does not scatter electrons in a preferential manner. If the illuminating electron beam is coherent, the scattered rays will interfere with one another. In a crystalline material this will result in a pattern of spots or rings of electrons. The intensity of the spots and rings and their pattern will depend on individual grains that scatter electrons to a different extent despite the specimen thickness and uniform composition. There is another factor that determines the intensity and the pattern; this is the orientation of the atomic lattices of rows and columns with respect to the incident beam. The pattern of spots and rings can be imaged and recorded by suitably energizing

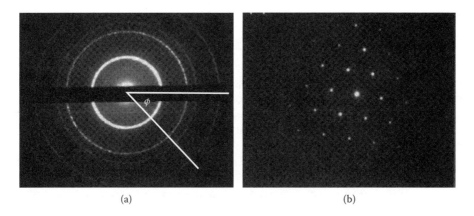

(a)   (b)

FIGURE 12.19
A typical Bi ring diffraction pattern from many crystallites (a), and a spot diffraction pattern from Bi crystallites (b). (*Source:* From *Physical Principals of Electron Microscopy*, Ray F. Egerton, Springer, 2005. With permission from the author and Springer Science and Business Media.)

various lenses in the microscope. The image is a diffraction pattern of the specimen area selected. One can pick a narrow selected area in the specimen using a suitable aperture to select and display the area of interest. This is called selected area diffraction (SAD). Crystalline materials give rise to either a ring pattern or a spot pattern. A ring pattern is indicative of a multi-crystalline nature of the specimen. That means various crystals are randomly oriented with respect to the beam (Figure 12.19a). A spot pattern is indicative of a single-crystal nature, which means that all the crystals in the viewing area of the specimen are equally oriented with respect to the beam (Figure 12.19b). Modern transmission electron microscopes have a built-in feature such that with the touch of a button, one can display a diffraction pattern for a given selected area in a specimen. The image is digitally captured and software is available to automatically analyze the diffraction pattern. One needs to plug in certain information, such as the accelerating voltage, camera length, etc. The resulting information gives insight into the crystalline structure of the specimen and its orientation. An electron diffraction pattern from an amorphous material gives rise to an extremely diffuse ring pattern. Thus, one can distinguish, using the diffraction mode, if the sample area is crystalline or amorphous.

12.10 Electron Energy Loss Spectroscopy (EELS)

When high-energy electrons traverse a thin specimen, forward scattering of the transmitted primary electrons takes place. The primary electrons lose energy primarily by the process of inelastic scattering. Forward scattering also takes place by the process of elastic scattering, where the primary electrons do not lose any energy or lose very little energy. The events resulting from inelastic or elastic scattering contain a host of information, some of which have already been discussed. We now discuss the primary electrons that have lost energy and that loss in energy is quantifiable. The process is called electron energy loss spectroscopy (EELS). In EELS, the spectrum obtained also contains information about atomic arrangement and chemical bonding. The technique complements EDS by providing better detection of low Z elements. Actually, one can get more information from EELS spectra than mere elemental information.

12.11 Analytical Electron Microscopy

The term "analytical electron microscopy" is used when both qualitative and quantitative information is obtained on a sample being analyzed. One can obtain such information simultaneously using a combination of imaging, spectroscopy, and diffraction techniques. A plethora of information from a sample can be obtained in a STEM-type instrument. This transforms a TEM/STEM instrument into an analytical electron microscope. This is very routinely being performed in research labs using a STEM instrument with a high brightness source such as an FE source, fitted with detectors to capture a SEM, TEM, STEM, EDS, electron diffraction, and EELS signals from the same area of the sample. By the combined use of these techniques, one can obtain not only the information about the internal and the external structure of the sample, but also the elements present and overall chemical composition, including chemical bonding information.

12.12 Applications

There has been a spectacular growth of materials using various polymers as the matrix material in combination with heterogeneous fillers. These fillers are usually in the shape of ductile fibers, such as carbon nanofibers, various kinds of clay, inorganic materials in the shape of nano-sized particles, etc. The heterogeneous combination of such fillers and a matrix material is usually called a polymer nanocomposite.

Figure 12.20 shows an example of two nanocomposites. In Figure 12.20a, there is a nanocomposite of 5% clay in low-density polyethylene (LDPE). The clay was observed to be well dispersed. Also, we could see good exfoliation of clay and evidence of good intercalation. The arrows indicate a few clay particles in the field of view. A lamellar structure of the semi-crystalline LDPE can be seen. The sample was cryo-sectioned at –120°C, stained with ruthenium tetroxide vapors, and viewed in JEOL 100CX TEM at 100-kV accelerating

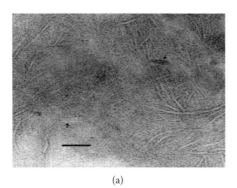
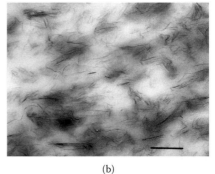

(a)   (b)

FIGURE 12.20
(a) This is a nanocomposite of 5% clay in low-density polyethylene (LDPE). The clay particles are shown with arrows. (The scale bar is 100 nm.) (b) A nanocomposite of 10% clay in polypropylene grafted maleic anhydride. The clay is well dispersed. The scale bar is 300 nm.

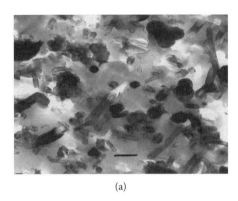
(a)

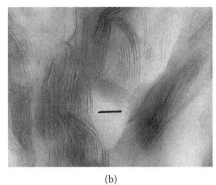
(b)

FIGURE 12.21
(a) A nanocomposite of 10% carbon fiber and 30% carbon nanofiber in 60% polycarbonate matrix. The sample is a strand from a capillary rheometer and sectioned 90° from the flow direction. (The scale bar is 200 nm.) (b) A sample of 2% clay in polyester matrix showing that the clay platelets are aligned parallel to each other. (Spacing between the clay platelets is between 15 and 30 nm.)

voltage. Figure 12.19b shows a nanocomposite of 10% clay in polypropylene grafted with maleic anhydride. The clay is well dispersed. The sample was prepared by cryo-sectioning at −40°C in a cryo-ultra-microtome. The thin sections were viewed in a JEOL 100CX TEM operated at 100-kV accelerating voltage.

Figure 12.21a shows a nanocomposite of 10% carbon fiber, 30% carbon nanofiber, and 60% polycarbonate matrix. The sample was an extruded strand from a capillary rheometer. Thin sections were prepared at room temperature. The carbon fiber and the carbon nanofibers are oriented in the flow direction. Thus, we see these fibers in cross-section. The sections were prepared from the face of the capillary perpendicular to the capillary flow. We can see that the carbon fibers and the carbon nanofibers are pretty good. Figure 12.21b shows clay platelets stacked parallel to each other. The spacing between the platelets is between 15 and 30 nm.

12.13 Conclusion

When preparing nanocomposite materials, scientists and researchers look for means to perform a structural characterization of the newly developed material. A detailed and accurate characterization of the material allows them to make correlations between the structure and its properties, what is called "structure-property relationship studies." The morphological and structural modifications at the nano or molecular level will relate to changes in the properties. The structure in a nanocomposite can easily be modified not only by the chemical nature of the additives and the matrix material, but also by the use of various processing conditions. To see the effect of these variables on the microstructure of nanocomposites, it is important to be able to visually see the microstructure and relate it to the amount and the nature of the additives. The microstructure of nanocomposites also affects the mechanical properties of the final product. Thus, the microstructure plays a very important role in the developmental process. For these reasons, electron microscopy has lately gained increasing importance. The use of electron microscopes and their development in the area of nanotechnology, including nanocomposites, has been pushed more by the needs of industry than by academic researchers.

Those using microscopy as a characterization tool are not always trained, dedicated microscopists. These are people who are specialists in fields such as materials science, chemical engineering, chemistry, physics, and others but their sole aim in using microscopy techniques is to obtain an in-depth knowledge in the research area they are pursuing by understanding the structure and morphology at the nano-level.

The newer, more modern microscopes are more automated, computer controlled, and more user-friendly. Despite all this, the skills of a trained, dedicated microscopist play a very significant role in the complete characterization of materials at a nano-level employing a host of electron microscopic and related techniques.

References

1. Egerton, R.F., *Physical Principals of Electron Microscopy*, Springer, 2005.
2. Goldstein, J.I. et al., *Scanning Electron Microscopy and X-Ray Microanalysis*, Plenum, 1992.
3. Meek, G.A., *Practical Electron Microscopy for Biologists*, John Wiley & Sons, 1976.
4. Murr, L.E., *Electron Optical Applications in Materials Science*, McGraw-Hill, 1970.
5. Sawyer, L.C. and Grubb, D.T., *Polymer Microscopy*, Chapman & Hall, 1987.
6. Williams, D.B. and Carter, C.B., *Transmission Electron Microscopy*, Springer, 1996.

13

Mechanical Properties of Clay-Containing Polymeric Nanocomposites

Leszek A. Utracki

CONTENTS

13.1 Introduction

Clay-containing polymeric nanocomposites (CPNCs) are used as either functional or structural materials. The former are employed in applications where specific functionality is
required; for example, electrical conductivity, electronic semiconductivity, magnetic properties, or biocompatibility. However, this chapter will deal exclusively with the structural,
large-volume CPNCs with a commodity or engineering polymer as the matrix, where the
mechanical behavior is the main concern (followed by gas–vapor barrier properties, flammability, etc.). In structural applications, the only nanosized particles of industrial interest
are layered, natural or synthetic 2:1-phyllosilicates (mainly montmorillonite [MMT], and
hectorite [HT]). It will be evident from the data in this chapter that for balanced optimal performance, a low clay concentration is required. For example, addition of 2-wt% of
organoclay to polyamide-6 (PA-6) increased the flexural modulus and strength by 26% and
tensile strength by 14%.[1] Similarly, dispersion of 2- and 5-wt% of vermiculite in polypropylene (PP) increased the tensile strength by 18 and 30%, the tensile modulus by 20 and
54%, and the storage modulus by 204 and 324%, respectively.[2]

In the absence of impurities and flaws, the mechanical performance of a material
depends on the magnitude of forces that bind atoms together. Unfortunately, in the real
world, flaws are always present, e.g., voids in crystalline homopolymers, for example, PP,
visible under acoustic microscope; surface imperfections in glassy materials (e.g., polystyrene [PS] or polymethylmethacrylate [PMMA]); or contaminating particles of dust, metal
chips from equipment wear, solid additives, and so forth. Evidently, incorporation of the
second, reinforcing phase significantly augments the possibilities of flaws—for example,
a common problem occurs involving the interphase and its dependence on processing
conditions. In CPNCs containing natural minerals, the common contaminants are grits
and amorphous silicates, such as quartz, silica, feldspar, gypsum, albite, anorthite, ortho-
clase, apatite, halite, calcite dolomite, sodium carbonate, siderite, biotite, muscovite, chlorite, stilbite, pyrite, kaolinite, hematite, and many others. The polymer-grade bentonites
are laboriously purified, but still they may contain < 5 wt% of nonsmectite impurities. The
impurities are particularly troublesome for low-permeability films. Because of poor adhesion of a polymeric matrix to grit particles, holes form at the interphase.[3]

This chapter is divided into three parts. The first provides the fundamental information
on the inherent properties of CPNCs that influence mechanical performance. The second
part summarizes the theoretical studies of the mechanical performance, and the third part
provides examples of the tensile and impact properties of CPNCs with selected polymers:
polyamides (PA), isotactic PP, and epoxy (EP).

13.2 Fundamentals

13.2.1 Thermodynamics

The mechanical performance of composites depends on the degree of dispersion of reinforcing solids and the quality of the interphasial interactions, and thus on thermodynamics. Fully exfoliated MMT with nonporous platelets has the specific surface area $A_{sp} = 750$
to 800 (theoretically 834) m²/g. A CPNC with nanosized clay platelets having > 40% atoms

on the surface should be treated as a macromolecular mixture of two species. The degree of clay dispersion (intercalation or exfoliation) depends on the system miscibility (equilibrium thermodynamics) and on the kinetics that control the distance from equilibrium conditions.

The self-consistent field (SCF) theory has been used to investigate the thermodynamic properties of four-component CPNCs, such as PP with preintercalated clay and a functional compatibilizer.[4] The Kim et al. simulation showed that intercalation and exfoliation is expected within limited ranges of independent variables. The presence of bare clay platelets, on the one hand, hinders the dispersion, but on the other, provides an opportunity for interactions with functional groups of a compatibilizer. The simulation identified the most influential factors that control free energy of the system.

Because of small thermal effects and slow diffusion rates, direct thermodynamic methods of measurement are practically impossible for CPNCs. However, good mean-field assessment of the system energetics might be derived from high-pressure dilatometry combined with the theoretically sound statistical thermodynamics of multicomponent systems, such as that derived by Simha and Somcynsky,[5] and over the years explored by Simha and his colleagues and collaborators.[6–11]

13.2.2 Molecular Adsorption

The high surface energy of crystalline solids causes adsorption and immobilization of molecules on the surface.[12–14] The thickness of the solidified layer on each side of the clay platelet ranges from 4 to 7 nm; thus, the solid content in a CPNC is that of (inorganic) clay multiplied by a factor of 9 to 13. Furthermore, the macromolecular mobility exponentially decreases with the reduction of distance from the clay surface; thickness of this gradient layer is about 100 to 120 nm. Evidently, these two layers of low mobility hamper clay dispersion. Judging by the reduction of polymer free volume, which results from clay addition, the adsorption is equivalent to reduction of the compounding temperature by approximately 50°C.[9] Compounding at higher temperatures would compensate for the immobilization problem, but poor thermal stability of the commonly used quaternary ammonium intercalants makes this approach impractical. The effects of the reduction of molecular mobility on mechanical properties of CPNCs as a result of the incorporation of clay has been discussed, for example, by Utracki and Simha[10] and Rao and Pochan.[15]

13.2.3 Degradation

The ammonium-ion-intercalated organoclays degrade at $T \geq 180°C$ by the Hofmann elimination[16] mechanism leaving the clay surface bare, which might cause reaggregation of clay platelets and loss of performance:

The reaction by-products are volatile amine and a vinyl group terminated on the longest paraffinic substituent. The vinyl terminal group might oxidize, forming peroxides capable of initiating free radical chain scission of the matrix polymer, which results in the reduction of mechanical performance. The thermal degradation of ammonium-intercalated organoclay is a painful reality for most CPNCs.

In addition to thermal degradation, thermomechanical degradation might cause chain scission and accelerate Hofmann elimination reactions, particularly when using a twin-screw extruder (TSE). The processes are significantly less severe in a single-screw extruder (SSE), but because SSE is a notoriously poor mixer, screw modification or add-on devices are required. Thus, for example, good performance was reported for CPNCs compounded in an SSE equipped with an extensional flow mixer (EFM).[17] The system provided good clay dispersion without the detrimental effects of thermomechanical degradation of the matrix polymer.

13.2.4 Maximum Packing Volume Fraction

By definition, the fully exfoliated clay platelets are randomly distributed in the matrix. If this is true, then their volumetric concentration should be less than the maximum volume fraction of their encompassed volumes, $\phi \leq \phi_m \approx 0.62$. For the monodispersed circular disks with diameter d and thickness t (the aspect ratio: $p = d/t$), the ratio of the encompassed to actual volume is: $V_{enc}/V = 2p/3$. For many commercial organoclays $p \approx 280$; thus, the maximum volume fraction for free rotation is $\phi_m \approx 0.0033$, which is equivalent to about 1-wt% of inorganic solid. In other words, exfoliation is possible only at low clay content; above approximately 1-wt%, the geometry progressively forces the platelets to form local stacks with gradually (with ϕ) decreasing interlayer spacing, d_{001}.

The above argument is valid for PNC (polymeric nanocomposites) with a molten or glassy matrix. When the matrix is semicrystalline (e.g., PP), clay is expelled from the crystalline lamellae into the mesocrystalline and amorphous domains during crystallization. Empirically, the following linear relation was found valid:[18]

$$d_{001} = d_{001}^0 - a_1 c_{clay} - a_2 X \tag{13.1}$$

The dependence held for PP-based CPNCs with $d_{001}^0 = 10.5 \pm 0.6$, $a_1 = 0.39 \pm 0.05$, $a_2 = 0.11 \pm 0.01$, and the correlation coefficient $r = 0.999$. Thus, infinitely diluted, amorphous PP-based CPNCs would be exfoliated ($d_{001} = 10.5$ nm), but as the clay concentration (c_{clay}) and the crystalline content (X) increase, the interlayer spacing decreases to the commonly measured values of $d_{001} \approx 3 - 3.5$ nm.

13.2.5 CPNCs with a Polymer Blend as the Matrix

It is a historical rule that systems are getting progressively more complicated. This has happened to polyvinyl chloride (PVC) formulations (with more than 30 components) and to polymer blends where mixtures of 6 polymers with at least 3 compatibilizers, fillers, and other additives are currently on the market.[19] CPNCs have started to evolve in that direction.

Thermodynamics provide the guiding rules for nano-reinforcing polymer blends. Consider the equilibrium free energy of mixing in a system composed of polymer-A, polymer-B, and a reinforcing solid, S. The free energy of mixing is given by

$$\Delta G_m = \Delta G_{AS} + \Delta G_{BS} - \Delta G_{AB} \tag{13.2}$$

The miscibility requires that $\Delta G_m < 0$. Starting with miscible blends where $\Delta G_{AB} < 0$, this condition is met when $\Delta G_{AS} + \Delta G_{BS} \ll 0$. However, starting with immiscible blends where $\Delta G_{AB} > 0$, the miscibility is easier to achieve, that is, $\Delta G_{AS} + \Delta G_{BS} \leq 0$.[20]

Thermodynamics also indicates the phase to which the nanofiller will migrate. In the case where $\Delta G_{AS} < \Delta G_{BS}$, the nanofiller will prefer polymer-A with which it forms more stable (lower free energy) associations. In reality, it appears that CPNC systems are more complicated as there is clay intercalant and most likely a compatibilizer or two. However, the above principles are readily extendable to more complex cases.

13.3 Theoretical Derivations

13.3.1 Micromechanical Models of the Tensile Modulus of CPNCs

The theories of mechanical behavior in a multiphase polymeric system have been discussed in several reviews.[21–28] One of the better-known expressions for the tensile modulus, E, of composites is that derived by Halpin and Tsai (H-T) for systems reinforced with continuous, infinitely long fibers:

$$E_r \equiv E_c/E_m = (1 + 2p\kappa\phi_f)/(1 - \kappa\phi_f)$$

$$\kappa = (E_R - 1)/(E_R + 2p); \quad E_R \equiv E_f/E_m \tag{13.3}$$

Here ϕ is the volume fraction, $\phi_m = 1 - \phi_f$; subscripts c, m, and f stand for composite, matrix, and filler, respectively; and p is the aspect ratio. The H-T model is simple and easy to use, as it contains two components and all the equation parameters are clearly specified. For small values of ϕ_f, Equation (13.3) can be expanded into Maclaurin's series and truncated at $0(\phi_f^2)$:

$$E_c/E_m = 1 + \kappa(2p + 1)\phi_f + 0(\phi_f^2) \cong 1 + \kappa^*\phi_f. \tag{13.4}$$

For the modulus in the stress and transverse direction, Equation (13.3) with the aspect ratio $p = \infty$ and $p = 0$ yields, respectively

$$\text{in stress:} \quad E_{c\parallel} = E_m\phi_m + E_f\phi_f$$

$$\text{transverse:} \quad 1/E_{c\perp} = \phi_m/E_m + \phi_f/E_f \tag{13.5}$$

In physics, the aspect ratio is usually defined in terms of the rotational spheroid axial ratio; that is, the principal axis of rotation divided by an orthogonal axis of rotation, such as for the prolate ellipsoids $p > 1$, and for oblate ellipsoids $p < 1$. Using this notation, expressions have been derived that predict particle behavior in the full range of the aspect ratio, that is, from fibers to flakes.[29] However, H-T Equation (13.3) does not belong in this category. The predicted dependence is plotted in Figure 13.1a for two values of $p = 250$ (fibers), and $p = 1/250$ (flakes). Evidently the dependencies are quite different, resembling the case of modulus in the stress and in the transverse directions (see Equation [13.5]). Thus,

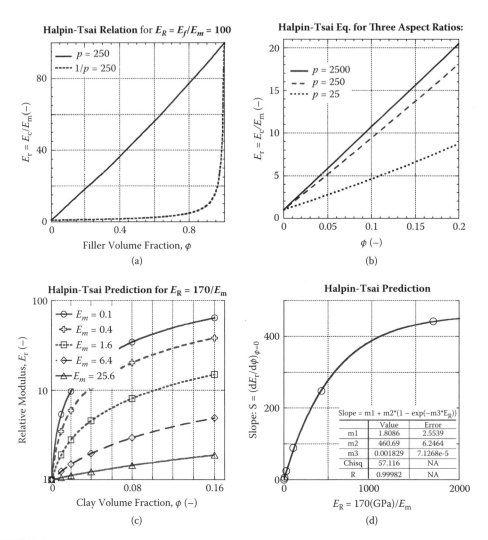

FIGURE 13.1
The relationships predicted by Halpin-Tsai Equation (13.3): (a) Relative tensile modulus for $E_R = E_f/E_m = 100$ and $p = 250$ or $1/250$; (b) relative tensile modulus for $E_R = E_f/E_m = 100$ and $p = 25$, 250, and 2500; (c) relative tensile modulus vs. filler content for five E_R; (d) the initial slope of the dependencies shown in C vs. E_R. See text.

empirically, H-T dependence reasonably predicts nanocomposite modulus reinforced with platelets. For this reason, the aspect ration in Equation (13.3) should be defined as $p = (fiber\ length)/diameter$ or $p = (platelet\ diameter)/thickness$.

The effect of p on E_r for $E_R = E_f/E_m = 100$ is displayed in Figure 13.1b. Equation (13.4) suggests that within the range of clay content of practical interest ($w \leq 5\text{-wt}\%$), the relative modulus should increase linearly with the clay concentration. Furthermore, the effect of the aspect ratio on stiffness is expected to be more important at low values, that is, it is more dramatic when p changes from 25 to 250 than when it increases from 250 to 2500 (for permeability effects the opposite would be true!). The dependencies in Figure 13.1c and Figure 13.1d illustrate a rarely noted H-T prediction, namely the effectiveness of filler addition on the modulus increase. In these two figures, the modulus of mica plate $E_f = 170$ GPa and matrix modulus $E_m = 0.1$ to 25.6 GPa were used; the former value represents an elastomeric matrix, whereas the latter modulus represents a rigid polymer. At constant clay loading of approximately 0.2 the relative modulus of an elastomer is about two orders of magnitude larger than that of a rigid polymer. In Figure 13.1d the initial slope

$$S = \lim_{\phi \to 0}(\partial E_r / \partial \phi)$$

is practically zero for $E_f \approx E_m$, and increases to 442 for $E_m = 0.1$ GPa.

Another micromechanical model, based on the mean-stress approach, is that from Mori and Tanaka.[30] It gives the best results for large aspect ratio fillers:

$$E_c / E_m = (1 - \phi_f \zeta)^{-1}$$

$$\zeta = [-2v_m A_3 + (1 - v_m)A_4 + (1 + v_m)AA_5]/(-2A) \tag{13.6}$$

where v_m is the matrix Poisson ratio, while A and A_i are the aspect ratio depended Eshelby coefficients. Equation (13.6) predicts slightly lower values of modulus than H-T Equation (13.3).

Brune and Bicerano[31] modified the H-T expression for predicting the modulus of intercalated CPNCs containing platelet stacks, each composed of N-clay layers with thickness t and the interlamellar gallery spacing, $s = d_{001} - t \cong d_{001} - 0.96$ nm, the variables p, ϕ, and E_r of Equation (13.3) were modified to read

$$p' = (p/N)/[1 + (1 - 1/N)(s/t)]$$

$$\phi' = \phi[1 + (1 - 1/N)(s/t)] \tag{13.7}$$

$$E'_r = \frac{E_r + (1 - 1/N)(s/t)}{1 + (1 - 1/N)(s/t)}$$

Because the dependence did not have the correct upper limit for the s/t ratio, the parameter N was replaced by

$$N' = N + (1 - N)(s/t)\frac{\phi}{1 - \phi} \tag{13.8}$$

Equation 13.7 and 13.8 predict a sharp reduction of modulus for the residual stacks. For example, stacks with $N = 2$ and $s/t = 2$ are predicted to reduce E_r by 47%. Thus, even a small reduction of exfoliation significantly lowers the modulus.

The authors also proposed a refinement to the predicted aspect ratio effects. The derived scaled modulus is

$$E_S \equiv \frac{E_c(p) - E_c(p = 1)}{E_c(p \to \infty) - E_c(p = 1)} = \frac{2(p - 1)}{(E_R - 1)(1 - \phi_f) + 1 + 2p} \; ; \; E_R \equiv \frac{E_f}{E_m} \tag{13.9}$$

For the clay content, $\phi_f < 5$ vol% the relation strongly depends on E_R and p, but it is insensitive to composition, and virtually traces Equation (13.3) dependence. Considering that the mica modulus $E_f \cong 170$ GPa (e.g., see Chen and Evans[32]) and the modulus of unoriented, unfilled engineering plastic ranges from 2 to 4 GPa, the clay platelets with $p \cong 200$ should increase the composite modulus to about 90% of the theoretically possible value for infinite platelet reinforcement.

Other modifications of the H-T equation were recently proposed by Yung et al.,[33] Chen and Evans,[34] and Utracki et al.[17] The latter authors considered the effect of the solidified polymer layer on clay platelets, which modifies the platelet modulus, aspect ratio, and the effective volume fraction. Assumption of the volumetric additivity yields the effective quantities:

$$E'_f = (E_f + \xi E_m)/(\xi + 1)$$

$$p' = p/(\xi + 1) \tag{13.10}$$

$$\phi'_f = \phi_f(\xi + 1)$$

where the prime (') indicates the modified parameters of Equation (13.3), and ξ is the thickness of the adsorbed polymer expressed in multiples of the clay platelet thickness, t. Algebraically, it has been shown that the effective relative modulus of exfoliated CPNCs is approximately equal to that calculated from H-T Equation (13.3), which assumes bare filler. Substituting Equation (13.10) into Equation (13.3) yields

$$\left. \begin{array}{l} E'_r = (1 + 2p'\phi'_f\kappa')/(1 - \phi'_f\kappa') \cong 1 + 2p\phi_f\kappa' \\[2mm] \kappa' = [E_R - 1]/[E_R + \xi + 2p] \\[2mm] E_f \gg E_m; \quad p \gg 1 \end{array} \right\} E'_r \cong E_r \tag{13.11}$$

Thus, the solidification of the polymer increases the volume of the reinforcing particles, but at the same time it reduces their aspect ratio and the ever-important filler modulus—these two tendencies compensate for each other. The agreement between the H-T prediction and experimental values of the relative modulus does not indicate absence of solidification. As will be discussed later in this chapter, the tensile strength is also sensitive to the solidified layer, and it directly provides information on its thickness.

Figure 13.1b indicates that within the low clay content, the relative modulus of CPNCs vs. concentration increases linearly. The simplest relation, based on Einstein's suspension model, is

$$E_r \equiv E_c/E_m = 1 + [\eta]\phi + O(\phi^2) \approx 1 + aw$$

$$a = [\eta]/(100\rho_f/\rho_m) \approx [\eta]/314 \tag{13.12}$$

$$\text{experimentally:} \quad [\eta] \approx 2.5 + 0.025(p^{1.47} - 1)$$

In Equation (13.12) the hydrodynamic volume of the reinforcing clay platelets, $[\eta]$, depends on the aspect ratio.[27] However, the aspect ratio also depends on the degree of dispersion—the larger the clay stacks, the smaller the effective aspect ratio. As shown in Figure 13.2, the

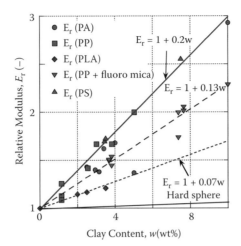

FIGURE 13.2
Relative tensile modulus vs. clay content for CPNCs of PA-6, PP, PLA, and PS with MMT, as well as for PP with flouromica (FM). See text.

lowest predicted value of the parameter $a = 0.008$ is for hard spheres with $p = 1$. For poorly exfoliated PP, PA-6 or polylactic acid (PLA), the values of E_r increase with the slope $a = 0.07$, giving $p = 93$. PP-based CPNCs with synthetic fluoromica followed Equation (13.12) with $a = 0.13$, equivalent to $p = 150$, i.e., close to $p = 132$ calculated from Zilg et al. data.[35] For the fully exfoliated, best-performing systems, Equation (13.12) was found to follow data with $a = 0.20$, equivalent to $p = 200$, i.e., close to the expected value for commercial organoclays with MMT. Thus, this simple dependence offers an insight into the nanofiller dispersion in CPNCs: Because the slope derived from the H-T equation is proportional to the a-parameter, the latter depends on $E_R = E_f/E_m$.

The linearity of the E_r vs. ϕ dependence of Equation (13.12) is based on the Einstein derivation for suspensions, often extended to the suspensions in the polymeric matrices.[6] The experimental relation between $[\eta]$ and p is valid at least up to $p = 300$. It is interesting to compare the relation between p and the initial slope in Equation (13.4) and Equation (13.12). The comparison is shown in Figure 13.3. It seems that the H-T relation predicts a small effect of aspect ratios for large, $p > 100$, values, expected for oriented, fiber-filled nanocomposites, for which Equation (13.3) was derived, but not for platelet-filled systems. On the other hand, the relation between $[\eta]$ and p is valid for infinitely diluted platelet suspensions, such as in exfoliated CPNCs, i.e., at $\phi \leq \phi_m \approx 0.93/p$. As the aspect ratio increases, the critical volume fraction of the onset of platelet overlap decreases, and the effective aspect ratio (of stacks that act as filler particles) is reduced.

It is important to recognize that the micromechanical relations presented so far are valid for idealized oriented CPNCs where the platelets are perfectly bonded to the matrix. In short, the considered systems are implicitly assumed to contain two bonded phases without interphase, thus without "adjustable" parameters.

The following derivation by Hui and Shia[36] still maintains the assumption of orientation, but incorporates the effect of interphase. For the tensile modulus of composites with oriented fibers or flakes (assuming that for both components Poisson's ratio $v_p = 0.5$, and

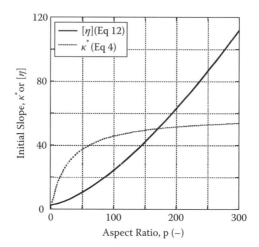

FIGURE 13.3
Initial slope of the relative modulus vs. volume fraction dependence as predicted by Equation (13.4) and Equation (13.12).

that there is perfect adhesion between the filler and matrix) the following expression was derived:

$$E_m/E_c = 1 - \frac{\phi}{4}\left[\frac{1}{\varsigma} + \frac{3}{\varsigma + \Lambda}\right]; \quad \text{where:}$$

$$\varsigma \equiv \phi + \frac{1}{E_r - 1} + 3(1 - \phi)\frac{(1 - g) - (gp^2/2)}{1 - p^2} \qquad (13.13)$$

$$\Lambda \equiv (1 - \phi)\frac{3(1 + 0.25p^2)g - 2}{1 - p^2}; \quad g \cong \pi/2p$$

Equation (13.13) predicts E_r values that are higher and more sensitive to p than those from the H-T theory. However, experimentation produced a discrepancy.[37] The authors postulated that the difference originated in imperfect bonding between the matrix and the filler, which affected the "effective" aspect ratio and clay volume fraction, both assumed to depend on the interfacial shear stress, σ_i. Fitting data to the derived set of equations led to low value of the effective friction coefficient (0.0932) within the interphase.

The three-phase approach to CPNCs was followed by Ji et al.[38] Here the filler particles (*f*) of any shape are randomly distributed in the matrix (*m*), with an interphase (*i*) that is particularly important for the nanoparticles. For large particles, the interphase contribution is negligible, and Takayanagi's relation is recovered. The three phases are connected to each other in series and in parallel. The realistic representation of a filled system is then reduced to three idealized regions, A, B, and C, connected in series. The general form of the derived relation is

$$\frac{1}{E_c} = \frac{1 - \beta}{E_m} + \frac{\beta - \varphi}{(1 - \alpha)E_m + \alpha E_{Bi}} + \frac{\varphi}{(1 - \alpha)E_m + (\alpha - \lambda)E_{Ci} + \lambda E_f} \qquad (13.14)$$

where α, β, φ, and λ are idealized dimensions of the interphase and filler regions determining the volume fraction of each component, that is, $V_f = \lambda\varphi$; $V_i = \alpha\beta - \lambda\varphi$; and $V_m = 1 - V_f - V_i$. The interphase modulus is defined for the model regions B (in series) and C (in parallel) as E_{Bi} and E_{Ci}, respectively. For CPNCs, these two contributions might be assigned to the interphases within and outside the clay stacks. The two types depend on the distance from the filler surface, linearly and exponentially, respectively:

$$\frac{1}{E_{Bi}} = \frac{\ln(E_{0i}/E_m)}{E_{0i} - E_m}; \quad E_{Ci} = [E_{0i} + E_m]/2 \quad \text{with:} \quad E_{0i}/E_m \equiv k \tag{13.15}$$

where E_{0i} is the interphase modulus on the filler surface at $\tau = 0$ and k is one of the model parameters. The platelet and the interphase shapes were assumed to be square, thus

$$\alpha = \beta = \sqrt{[2(\tau/t) + 1]V_f}; \quad \varphi = \lambda = \sqrt{V_f} \tag{13.16}$$

Substituting these relations into Equation (13.14) yields

$$\frac{1}{E_c} = \frac{1 - \sqrt{[2(\tau/t) + 1]V_f}}{E_m} + \frac{\sqrt{[2(\tau/t) + 1]V_f} - \sqrt{V_f}}{\{1 - \sqrt{[2(\tau/t) + 1]V_f}\}E_m + \sqrt{[2(\tau/t) + 1]V_f}\,(k - 1)E_m/\ln k}$$

$$+ \frac{\sqrt{V_f}}{\{1 - \sqrt{[2(\tau/t) + 1]V_f}\}E_m + \{\sqrt{[2(\tau/t) + 1]V_f} - \sqrt{V_f}\}(k + 1)E_m/2 + \sqrt{V_f}E_f}$$

$$\tag{13.17}$$

The interphase-characterizing parameters of Equation (13.17) are τ and k, related to its thickness and rigidity, respectively. Their values are determined by fitting the relation to experimental data. For PA-6/MMT the tensile data were fitted with $\tau = 7$ nm and $k = 12$. Since the thickness of the solidified organic layer on the clay surface ranges from 4 to 7 nm, the expected value would be $\tau = 9 - 15$ nm; thus, the fitted value of $\tau = 7$ nm is of the right order of magnitude, but small. By contrast, the experimental value of $k = E_{0i}/E_m = 12$ might be too high for solid CPNCs.

In short, by contrast with the simple two-phase models, the models that include the interphase offer more leeway to fit the data to equations, but with "adjustable" parameters that considerably differ from the expected values.

13.3.2 Prediction of Tensile Strength

Prediction of the tensile strength, σ, for a multiphase system is more difficult than for a modulus system. Strength involves transmission of stresses from one phase to another. Thus, for CPNCs, strength depends on concentration, degree of exfoliation, the type of bonding between clay platelets, and the matrix, as well as compatibilizers. The simplest case is for the two-component system with particles perfectly adhering to the matrix.

The tensile strength in the stress direction is expected to follow the volumetric rule of mixtures:[39]

$$\sigma_c = \sigma_m \phi_m + \sigma_f \phi_f$$

$$\sigma_r \equiv \sigma_c / \sigma_m = 1 + \phi_f (\sigma_R - 1) \tag{13.18}$$

$$\sigma_R \equiv \sigma_f / \sigma_m$$

In a recent publication,[17] the tensile strength of mica platelets, $\sigma_f = 240$ MPa, was assumed for MMT.[40] The matrix tensile strength, σ_m, was measured and the clay volume fractions in CPNCs were known; hence, $\sigma_r \equiv \sigma_f / \sigma_m$ could be predicted. Surprisingly, the value of the tensile strength calculated from Equation (13.18) was significantly smaller than measured, indicating the presence of a larger volume (by a factor of about 5 and 9 for the PP- and PA-6-based CPNCs, respectively) of reinforcing particles. However, these increases are reasonable considering that the thickness of the adsorbed polymer layer varies from about 4 to 7 nm.

It is noteworthy that Clegg and Collyer[41] found that Equation (13.18) overestimated the tensile strength of composites. To correct the disparity, they introduced the strength-reducing factor, F_R:

$$\sigma_c = \sigma_m \phi_m + F_R \sigma_f \phi_f ; \quad \text{where}$$

$$F_R = \left[1 - \frac{\tan h(u)}{u} \right] \Big/ 1 - \sec h(u) \tag{13.19}$$

$$u = p \sqrt{G_m \phi_f / [E_f (1 - \phi_f)]}$$

(G_m is the matrix shear modulus.) The expected value of the reducing factor is $0.666 \leq F_R \leq 0.900$.

Turcsáyi et al.[42] expressed the tensile strength by relations, which incorporate effects of the decreasing specimen load-bearing cross-section and interfacial interactions:

$$\sigma_r \equiv \sigma_c / \sigma_m = (1 - \phi_f)/(1 + 2.5\phi_f)\exp\{B\phi_f\}; \quad \text{where:}$$

$$B = (1 + A_{sp}\rho_f l)\ln(\sigma_i / \sigma_m) \tag{13.20}$$

In Equation (13.20) A_{sp} and ρ_f are the specific surface area and the density of reinforcing particles, while l and σ_i are the thickness and strength of the interphase. Thus, parameter B is a measure of the interactions between the matrix and the dispersed filler particles. For $\sigma_i < \sigma_m$, the parameter $B < 0$ and the relative tensile strength decreases with filler loading.

Examples of the relative tensile strength dependence on clay volume fraction are displayed in Figure 13.4.[43] As expected, the strength depends on the polymer–filler interactions, which in turn control the degree of clay dispersion. The X-ray defraction (XRD) data showed a high degree of C30B dispersion ($d_{001} \leq 10$ nm) and relatively poor dispersion of C10A. In spite of this, the relative strength for both CPNCs is the same up to approximately $\phi \leq 0.05$ (or about 10-wt%).

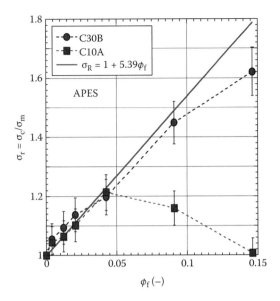

FIGURE 13.4
Relative tensile strength of biodegradable polyester of glycol + butanediol + succinic + adipic acids (APES; M_w = 60 Kg/mol) with Cloisite 30B (open circles) and 10A (filled squares). Data: Lee et al.[43] The straight line is the least squares fit to C30B data; ± 5% error bars are shown.

13.3.3 Creep

Creep is defined as the permanent deformation of material resulting from persistent application of stress below the elastic limit. The deformation depends on the material, load, temperature, pressure, and the loading duration. Creep might lead to unacceptable deformation and eventual failure.[44] There are several types of standard test methods for measuring the tensile, compressive, or flexural creep (e.g., see ASTM D-2990 or ISO 899). The schematic test curve for creep (at constant load) and its recovery after load removal before fracture is displayed in Figure 13.5. The numbers indicate regions of: 1 = instantaneous elastic deformation; 2 = primary (or transient) creep; 3 = secondary (or steady state) creep; 4 = tertiary creep (rapid deformation that might lead to necking and rupture); 5 = elastic recovery after unloading; and 6 = creep recovery.

The creep measurements of polymeric materials usually stop before the onset of strain hardening, which is relatively rare. Thus, the simplest analysis might be based on the empirical formula:[45]

$$\Delta\varepsilon = \varepsilon - \varepsilon_0 = B\sigma^m t^k; \quad \varepsilon_0 \cong E/\sigma \tag{13.21}$$

where strain $\varepsilon = \Delta L/L_0$ (L_0 is the initial length and ΔL is the specimen increase in length), ε_0 is the instantaneous elastic deformation, t is time, σ is the applied stress, and m and k are equation exponents. When the experiment is carried out under constant stress, the response of material to creep is characterized by the parameter k:

$$k = (\partial \ln \Delta\varepsilon / \partial \ln t)_{\sigma,T} = \dot\varepsilon(t/\Delta\varepsilon) \tag{13.22}$$

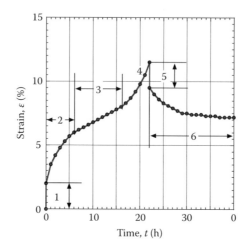

FIGURE 13.5
Strain vs. time with numbers indicating different regions: 1 to 4 represent creep under constant load, and 5 to 6 represent creep recovery after unloading. See text.

The values of k range from zero (purely elastic behavior) to one (purely viscous behavior); thus, k might be considered a relative measure of elastic and viscous contributions to creep. In the case in which the deformation is proportional to time under load, the parameter k is proportional to the linear strain rate, $\dot{\varepsilon}$.

Another method of creep data evaluation is by plotting a logarithm of the time to break, t_b, under constant load vs. temperature, T:

$$(\ln t_b)_\sigma = a_0 - a_1 T \tag{13.23}$$

In this dependence, a_i are the characteristic parameters of the equation, depending on the applied load and the material properties. In most cases the slope, a_1, is constant, and $a_0 = a_0(\sigma)$, which facilitates interpretation of the material behavior.

Since polymers are viscoelastic, the creep behavior is well described by the creep compliance:

$$J(t, \sigma(t), T) = \varepsilon(t, \sigma(t), T)/\sigma_0 \approx c\varepsilon(t, T) \tag{13.24}$$

where $c = \sigma(t)/\sigma_0$ is assumed to be constant. The advantage of this approach is the opportunity for generating master curves by employing the time–temperature or time–stress superposition principles.[46] The plot of $\log(J/J_0)$ vs. $\log(t)$ at constant T, σ or aging time, t_a, is frequently used to calculate the shift factor, a, the characteristic material response to the variable of interest: $v = v\{T, \sigma, t_a\}$. Its slope, $\mu = d\ln(a)/d\ln(v)$, is known as the *shift rate*, and its value also characterizes the material behavior.

The creep behavior of fiber reinforced polymer (FRP) composites has been reviewed by Scott et al.[47] The authors focused on the experimental techniques for (1) accelerated test methods, (2) linear and nonlinear viscoelastic theories for the modeling and prediction of the viscoelastic FRP response, as well as (3) the interrelations between creep and fatigue properties.

13.3.4 Fatigue Resistance of CPNCs

The tensile modulus and strength provide information on the material behavior at relatively low energy input under steady-state loading. Another type of mechanical test addresses the ultimate properties—how the material breaks. Three principal test methods used by fracture mechanics are impact tests, steady-state deformation (at stresses exceeding strength of the material), and fatigue.[48]

In the latter case, the fatigue crack grows when the stress intensity (cycling between σ_{min} and σ_{max}) is smaller than that needed for crack growth under steady-state loading. The cycling causes failure at stresses well below the tensile strength level discussed previously. Analysis of this behavior is done by determining the relationship between stress and the number of cycles to failure (S-N diagram), usually followed by studies of the crack-propagation mechanism (e.g., by fractography), and calculation of the fracture energy.

Fatigue crack growth will take place if (1) the stress amplitude, $\sigma_a = \sigma_{max} - \sigma_{min}$, exceeds a critical value; (2) the maximum stress intensity σ_{max} exceeds the critical value; and (3) there is insufficient energy dissipation during cycling. Above the threshold stress value, the crack growth rate, $\partial l / \partial N$, is a function of the stress amplitude, σ_a. To generate the S-N diagram, the specimen may be subjected to either constant stress amplitude or to the mean stress, respectively defined by Moet as[49]

$$\sigma_a = \sigma_{max} - \sigma_{min}; \quad \sigma_{mean} = (\sigma_{max} + \sigma_{min})/2 \tag{13.25}$$

Construction of the S-N diagram is time consuming, and in spite of its importance, is not frequently done.

The fatigue behavior of polymeric systems, with high energy dissipation and low thermal conductivity, is sensitive to test frequency, ν. Depending on the magnitude of ν, σ_a or the rate of energy input $(\partial E / \partial t) \propto \nu \times \sigma_a$ one of the following two mechanisms might take place:

1. At low rates of energy input, $(\partial E / \partial t) < (\partial E / \partial t)_{crit}$, the material is capable of dissipating the deformation energy; thus, the temperature increases only to a specific level, and the fracture proceeds by the crack initiation and propagation (FCP) mechanism.

2. At high rates of energy input, $(\partial E / \partial t) \geq (\partial E / \partial t)_{crit}$, there is an unbounded temperature increase and the specimen fails due to the thermal fatigue failure (TFF).

Crawford and Benham[50] proposed an empirical relation between the critical value of the stress amplitude, $\sigma_{a,crit}$, and the critical frequency, ν_{crit}:

$$(\partial E / \partial t) = (\partial E / \partial t)_{crit} \Rightarrow \sigma_{a,\,crit} = (c_1 - c_2 \log \nu_{crit})\sqrt{A/V} \tag{13.26}$$

where A and V are specimen surface area and volume, respectively, while c_i represents material parameters.

The FCP mechanism often starts with craze formation that, depending on the material and test conditions, leads to cracks or shear banding, which in turn proceeds to the crack propagation stage. The crack opens in either a single craze or a shear band. The initiation is best described in terms of the fracture mechanics stress at the crack tip, $\sigma_{tip}, \propto \sqrt{H/L}$, where H is the crack length and L is its radius of curvature. Growth of the crack length during the number of cycles in the fatigue test was found to be proportional to the ampli-tude of the stress concentration factors in the tensile mode, $\sigma K_I = \sigma_{I, max} - \sigma_{I, min}$:

$$\partial l / \partial N = c_o \Delta \sigma_I^m \qquad (13.27)$$

where c_o and m are characteristic constants of this "Paris law" equation. Continuum theo-ries that assume that the crack growth rate is proportional to the crack opening displace-ment predict the exponent value $m = 2$. Nguyen et al.[51] have developed a finite element model to simulate the fatigue behavior of a specimen under plane strain.

As reported in the cited monograph,[27] the effect of clay on fracture and fatigue behavior has not been widely studied. In CPNCs, the fracture proceeds mainly by the FCP mecha-nism, often with large volume expansion. The crack initiation is often caused by a min-eral contaminant, presence of clay aggregates, grid particles, or a nanoscale defect at the polymer–clay interface. As a consequence, in CPNCs the internal energy change per unit crack length, J_c, might be smaller than that in a neat polymeric matrix.[52]

13.3.5 Molecular Modeling of Mechanical Properties

During the last two decades, the understanding of solid–liquid interactions progressed rapidly due to the advancement of simulation techniques via the Monte Carlo (MC) or molecular dynamic (MD) methods, as well as the rapid progress of nuclear magnetic reso-nance (NMR) techniques.[14,53] Computations show that macromolecules are greatly affected by solid surface energy; their molecular arrangements, conformations, and dynamics are strongly perturbed with respect to the isotropic bulk. For example, macromolecular chain segments in layers near a solid surface are densely packed. The distorted Gaussian coils stretch for a distance of about 3/2 of the unperturbed radius of gyration, $\langle r_g^2 \rangle_0^{1/2}$.

MC simulations have been carried out on and off the lattice. On-lattice computations, with polymeric chains represented as random, self-avoiding walks, are more economic, but not as realistic as off-lattice methods.[54] MC and MD simulations have been used to study the static and dynamic properties of CPNCs. For example, mathematical modeling of organoclays and CPNCs has been used to describe the structure and energetics of organic molecules in the vicinity of 1:2 layered silicates.[55–57] The intercalant ion charge was found dislocated and spread over the stretched on-clay surface hydrocarbon molecules, which formed a low-mobility, immiscible, isolating layer between the clay and the matrix polymer.

Vacatello[58] carried out MC simulations for dense polymeric melt with solid, spherical nanoparticles. The model incorporated off-lattice approximation and conformational dis-tribution of the simulated chains, similar to that of a real polymer, with the Lennard-Jones 6-12 potential. The results showed that at the interface the polymer segments are densely packed in the form of ordered shells around the nanoparticles, analogous to the layer formation near the planar solid surface. The thickness of the shells was approximately 2 to 3 times the transverse diameter of the polyethylene iso-diametrical segments. The size of these was taken to be equal to about 3.5 $-CH_2-$ units, hence having the diameter of approximately 0.45 nm.[59] Some macromolecules visited several nanoparticle shells; thus, the nanoparticles behaved as physical crosslinks. Furthermore, the conformational distri-bution of the polymer was perturbed by the presence of these solid nanoparticles, e.g., the average dimension of chain segments was reduced.

Amorphous polymers with spherical nanoparticles were analyzed by equilibrium MD. The mechanical reinforcement was found to originate in the formation of a long-lived tran-sient polymer–particle network composed of macromolecular loops and bridges between the nanoparticles. It was found that the volume of the interphase should be significantly smaller than that of the bulk phase, and that the particle–polymer interactions must be strong. The reinforcement originates in the particles having the volume expanded by the adsorbed segments.[60]

These calculations successfully simulate the structure and dynamics of the polymer–solid surface system. The results, determined by topology and entropy, are applicable to diverse situations, including clay intercalation. The presence of preferential interactions between the polymer and a solid should not significantly change the computed structure, the order of interface shells, or the macromolecular conformation. Obviously, since the computations so far have been carried out for binary clay–polymer systems, the presence of a third component will affect the result. This would be expected for clays modified with quaternary onium ions having more than one type of substituent group. However, the overall image that emerges from these computations is expected to remain intact: a reduced segmental mobility near the high-energy clay surface, increasing toward the characteristic bulk mobility at a distance of approximately 100 nm.

Manias and Kuppa[61] provided a short topical review, focusing on the simulation of the structure and dynamics of PS macromolecules in slits. The MD results were compared to experimental results for CPNCs containing onium-modified fluorohectorite dispersed in a PS matrix; a reasonable agreement was found.

During the last few years the MC/MD computation was extended to simulation of the mechanical properties of nanocomposites. The adopted procedure is multiscale, starting with (1) atomic/molecular dynamics modeling of the nanoparticles and their interactions; (2) construction of a representative volume element (RVE); and (3) computation of the macroscopic behavior of the nanocomposite from the RVE, using either classical continuum expressions or a finite element method (FEM). Originally, Odegard et al. applied this approach to systems containing carbon nanotubes.[62] Later, the model was used for analyzing the mechanical properties of polyimide (PI) with dispersed silica particles, whose radius ranged from $r = 1$ to 1000 nm. For large particles ($r > 100$ nm) the computations reproduced the Mori-Tanaka predictions [i.e., Equation (13.5)], whereas for $r < 100$ nm, the model predicted increasing sensitivity of properties to the solid–matrix interactions.[63]

Sheng et al.[64] applied the multiscale modeling strategy to CPNCs: (1) at nanoscale, the interactions between the matrix and nanoparticles were computed; (2) at microscale, the clay particles were considered either exfoliated or intercalated, forming stacks; (3) at a length scale of millimeters, the structure was assumed to be of a matrix with high-aspect-ratio particles dispersed in it. The model allowed the matrix to be either amorphous or semi-crystalline. In the latter case the effects of polymer lamellae orientation and transcrystallization should be accounted for. Starting with the structural parameters (extracted from XRD and transmission electron microscopy [TEM]) and platelet modulus obtained from MD, a layer of matrix surrounding the solid particle was modeled. Next, the numerical or analytical models based on the "effective clay particle" were used to calculate the CPNC elastic modulus. The new model correctly predicted the elastic moduli for CPNCs with an MXD6 or PA-6 matrix. It is noteworthy that the model predicted only a moderate increase in the overall modulus for an exfoliated system over that of an intercalated system. The authors speculated that due to potential bending of exfoliated clay platelets, the full exfoliation might not lead to the highest modulus.

More recently, three-dimensional (3D) FEM was used to predict variation of the relative modulus with the orientation and volume fraction of clay platelets ($\phi_f \leq 0.1$).[65] The authors reported that the values predicted by the two-dimensional (2D) FEM computations were too low. The 3D FEM was found to be well approximated by the Mori-Tanaka

Equation (13.5). However, the modulus predicted by the later model was found to be too high for oriented CPNCs and too low for the random clay orientation.

Multiscale modeling was also applied to damage in ceramic and polymeric composites.[66] The fracture behavior in fully exfoliated CPNCs was recently modeled using MD.[67] The polymer was represented by a bead-spring linear chain interacting according to the Lennard-Jones 6-12 potential. The clay platelets were represented by thin squares with area $0.36 < r_g^2 >$. The density of clay platelets was assumed to be three times larger than that of polymeric segments. The system was subjected to tension in the z-direction. The simulation indicated that addition of clay platelets may improve the polymer fracture strength. The effect depends on the magnitude of polymer–clay interactions, the relaxation time of polymer chains, and the polymer glass transition temperature, T_g. For CPNCs with the matrix $T_g \leq$ RT (room temperature), addition of clay enhances the mechanical properties. However, when the matrix T_g is above RT (e.g., vitreous epoxy or polystyrene), addition of clay does not toughen the polymers. In this case, creation of a stress–relaxation mechanism might be necessary. This might be accomplished, for example, by addition of a suitable compatibilizer or a coupling agent, which will bond clay platelets to the matrix. Another successful method would involve incorporating an elastomeric toughener having an optimized drop size. However, these methods are expected to decrease the CPNC modulus.

The MC/MD findings might be confirmed by NMR. For example, the solid-state Magic Angle Spinning (MAS) NMR spectroscopy has been used to determine interactions between nuclei and the surroundings, and thus interpret the macroscopic behavior of PNC. The chemical shift, Cs, dipole–dipole interactions, and J-coupling are capable of elucidating the electronic structure surrounding the nuclei, internuclear distances, and orientation. More recently the 2D NMR spectroscopy has been developed for studying the geometry of the molecular dynamics and interactions.[68–70] The homo- and heteronuclear 2D correlated solid-state NMR provided invaluable information on the organic–inorganic complex structures.[71,72]

13.4 Experiments

13.4.1 Standard Testing of Mechanical Properties

Standard test methods are provided by the national standardization organizations, as well as international organizations such as the International Organization for Standardization (ISO). These methods are continuously upgraded by internal technical committees and the revisions are published every few years. The national and international methods may differ in detail, but the principles are the same, and thus they are considered equivalent. For example, the American Society for Testing and Materials (ASTM) standard D638 on "tensile properties of plastics" is equivalent to ISO 527. For testing CPNCs, the standard tests designed for the matrix polymer (thermoplastic or thermoset) should be used.

The principal test methods of mechanical properties for plastics are

1. Steady-state tensile, compressive, or flexural deformation to determine modulus, yield stress, stress at break, and their corresponding strains
2. Dynamic tests in shear (torsion) or elongation
3. Impact tests (Izod, Charpy, and their modifications)
4. Creep and fatigue tests

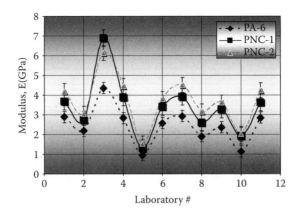

FIGURE 13.6
Tensile modulus as measured in 11 laboratories for three polymeric systems: PA-6, and its two CPNCs, PNC-1 and PNC-2.

There are many other properties assigned to "mechanical behavior," e.g., abrasion resistance, micro-indentation methods, environmental stress cracking, etc. However, they have rarely been used for characterizing the performance of CPNCs.

13.4.2 Reproducibility of Measurements

Recently, a round-robin test of three materials was carried out.[73] The pelletized commercial resins of homopolymer (PA-6) and its two CPNCs (PNC-1 and PNC-2) were dried, injection molded to ASTM specifications, vacuum packed, and distributed to 11 laboratories on three continents. The participants were asked to carry out tensile, flexural, and impact tests according to locally used specifications. An example of the results is presented in Figure 13.6.

For the three polymeric specimens (PA-6, PNC-1, and PNC-2) the reported (and confirmed) values of the tensile modulus are: E (GPa) ranged from 4.36 to 0.95, 6.89 to 1.31, and 6.13 to 1.53, respectively, i.e., with a spread factor of about 5. Even larger differences were reported for the elongation at break, ε_b (%), where the spread of values from one laboratory to the next ranged (e.g., for PA-6) from 21.5 to 494 %. The relative values, $E_r = E(PNC)/E(PA)$, showed a spread factor of 1.3, and a standard deviation of the average values within the required range of ±10%. This clearly indicates that the reason for the large scatter of absolute values was not the material, but rather the test procedure and/or data treatment.

Some mechanical parameters are prone to larger error than others, e.g., the elongation at break, tensile modulus, and impact strength notoriously show a large spread of values, while the tensile strength and the flexural data have better reproducibility. For example, the spread of the tensile strength and the flexural modulus values was by a factor of 1.4 and 1.5, respectively. The observation that the flexural modulus data are more reliable than the tensile strength data is not new—many commercial resin manufacturers do not list tensile modulus, but provide values for strength, flexural properties, etc.; see Table 13.1. It is noteworthy that for CPNCs, linearity has been observed between the tensile and flexural moduli and strengths.[74] Thus, while the mechanical test data from well-run laboratories are valid with about a ±10% limit, one has to be careful comparing values from different laboratories.

TABLE 13.1

Mechanical Properties of PA-6 and Its PNC

Property	ASTM	Units	Ube		Unitika	
			PA-6	PNC	PA-6	PNC
Tensile strength, σ	D-638	MPa	78	89	79	91
Tensile elongation at break, ε_b	D-638	%	100	75	100	4
Flexural strength, σ_f	D-790	MPa	108	136	106	155
Flexural modulus, F	D-790	GPa	2.80	3.52	2.84	4.41
Impact strength at RT, *NIRT*	D-256	J/m	64	49	48	44
HTD (18.56 kg/cm^2)	D-648	°C	75	140	70	172
HTD (4.6 kg/cm^2)	D-648	°C	180	197	175	193
H$_2$O permeability, P_{H_2O}	(JIS Z208)	g/m^2 24 h	203	106	—	—
Density, ρ	D-792	kg/m^3	1140	1150	1140	1150

Data from Ube Industries, Ltd., 2002, and Unitika Plastics, 2004.

TABLE 13.2

Mechanical Properties of PA-based CPNC

PA	Organoclay	Loading (wt%)	Properties	Ref.
-6	C15A[1]	0, 2, 4	E, σ, ε_b, tensile and flexural; *NIRT*	17
-6	MMT-ADA[2]	0, 2, 5	E, σ, ε_b; MD and TD and dynamic	75
-6	C15A, C30B[3]	0, 5	E, σ, ε_b; *NIRT*	77
-6	I.30TC[4]	0, 2, 5, 10	E, σ, ε_b	78
-6	Vermiculite	0, 2, 4, 6, 8	E, σ, ε_b, and dynamic	79
-6	MMT[5]	0, 1.7, 2.4, 2.7, 3.4, 3.8, 4.2, 6.5	E, σ, ε_b tensile and compression	76
-6	(RTP Co.)[6]	3	E, σ, ε_y v_s. strain rate; fatigue	80
-6	C30B	0, 5	E, σ, ε_y, ε_b, σ_{max} v_s. specimen thickness	81
-6	MMT; MMT-ODA	0, 2.8, 4.2, 6.7	tensile σ and flexural E, σ; *NIRT*	82
-6	MMT-ADA[1]	0, 2	E, σ, ε_y, ε_b, G', K_{IC}; process, aspect ratio	83
-6	I.30.TC and BT-ODA[7]	0, 5	E; effect of inorganic contaminants	84
-6	C30B[8]	0, 0.5, 1,5, 2.5, 5, 7	E, σ_b, ε_b, *toughness*	85
-6	(RTP Co.)[6]	0, 2.5, 5, 7.5	E, ε_y, ε_b	86
-6	MMT-3MDDA[9]	0, 4	E, σ, ε_b; *NIRT*	87
-6	C30B	0, 5	E, σ_b, ε_b	88
-6	MMT, C15A, C30B	0, 3, 5, 7	E, σ, ε_y, *impact force*	89
-6	C15A, C30B	0, 2.55, 4.11, 3.14, 5.07	E	90
-6	C15A	0, 2, 4	Tensile E, σ, ε_b and flexural F, σ_F; *NIRT*	74
-6 and -66	Somasif ME-100 and MEE[10]	0, 2.6, 5.5, 7.5, 10.3	E, ε_y, ε_b	91
-66	MMT-3MHDA	0, 1, 3, 5, 7, 10	E, σ, *NIRT*	92
-66	MMT-ODA or -AUDA	0, 2.1, 5.7	tensile E, σ, ε_b and flexural F, σ_F; *NIRT*	93
-66	MMT[11]	0, 2, 5, 10	E, ε_b	94
-66	I.34TCN[12]	0, 1, 2, 5	E from tensile and indentation	95
-66	Nanofil 919[13]	0, 1.6	E, σ, ε_y, creep	96
-66	C30B	0, 2, 4, 5	E, σ, ε_b	97
-12	I.34TCN[12]	0, 1, 2, 5	E, σ; *NIRT*, G'	98

Notes: Abbreviations are listed at the end of this chapter.
[1]C15A = MMT-2M2HTA; [2]CPNC produced by Ube; [3]C30B = MMT-MT2EtOH; [4]MMT-ODA; [5]MMT with 2M2ODA, M2EtOHODA, ADA, MHRODA, MHTODA; [6]Melt compounded product from RTP Co; [7]BT = non-purified bentonite with 63% MMT; [8]With 5-wt% of elastomers; [9]With 5, 10, 15 and 20-wt% of SEBS-*g*-MA elastomer; [10]Synthetic fluoromica with MCoco2EtOH; [11]MMT with 14% intercalant (*tert*-butyl *di*-methyl chlorosilane); [12]I.34TCN = MMT-MOD2EtOH; [13]Nanofil 919 = MMT-2MBODA, also TiO$_2$ particles with d = 21 and 300 nm diameters were used; [12]C30B was modified by addition of 0, 11.6 and 22.7-wt% of DGEBA epoxy (on the mass of organoclay).

13.4.3 Mechanical Properties of Polyamide-Based CPNCs

CPNCs based on PA-6 are the most common; they have been commercially available for at least ten years, and have been subjects of numerous technical publications. Traditionally, these CPNCs have been prepared by the reactive method, but at present they are produced also by melt compounding. Table 13.2 provides a summary of the recent reports on the mechanical properties of PA-based CPNCs; the data up to about 2003 can be found in the monograph.[27]

Over the years several factors affecting mechanical behavior have been investigated. For example, in Figure 13.7 effects of the relatively mild stretching of extruded sheet (by a 4:1 ratio) on the relative tensile modulus, E_r, tensile strength, σ_r, and elongation at break, ε_{br}, are displayed. The data demonstrate that orientation significantly alters the performance. All three functions are larger in the stress (machine direction, MD) than in the transverse direction (TD).[75] The tensile modulus of these commercial, reactively exfoliated CPNCs was similarly increased by stretching in TD and MD; σ_r was lower in TD than in MD, while $\varepsilon_{br} \leq 1.0$ was found for TD. These data suggest that the MMT platelets were predominantly oriented parallel to the sheet surface.

Figure 13.8 shows the relative compressive modulus: E_r vs. clay content for mildly oriented bars.[76] The properties were determined in the three orthogonal directions. As in tension, in compression the strongest enhancement is noted for MD. At low deformations, no significant difference between intercalated and exfoliated systems could be discerned.

13.4.3.1 Tensile Modulus and Strength

Mallick and Zhou[80] studied yield and fatigue of PA-6- and PP-based CPNCs. The measurements of the tensile modulus, yield stress, and strain were carried out at three temperatures and two strain rates. The results in reduced form are shown in Figure 13.9; the modulus and the yield stress decrease with T, whereas the yield strain increases. The effect

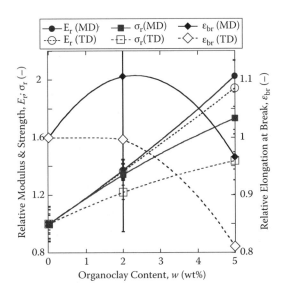

FIGURE 13.7
PA-6 with MMT-ADA; effect of sheet stretching by a 4:1 ratio on the relative tensile modulus, strength, and elongation at break as measured in the machine direction (MD) and the transverse direction (TD).[75]

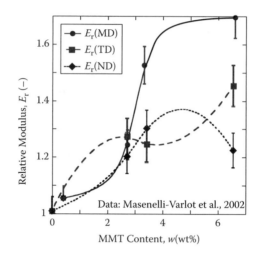

FIGURE 13.8

Relative compressive modulus vs. MMT content in MD, TD, and normal direction.[76] The reported errors are shown; the lines are only to guide the eye.

of the strain rate goes in the opposite direction—the strain rate is higher, the modulus and yield stress are higher, and the yield strain is lower. Similar dependencies were observed for the PP-based CPNCs.

Another potentially important variable for the tensile property determination is specimen thickness.[81] The reported relative tensile properties of injection molded PA-6 and its CPNCs containing 5-wt% of C30B (as determined following ISO standard 527-2) are displayed in Figure 13.10. The high clay loading accentuates the difference in behavior

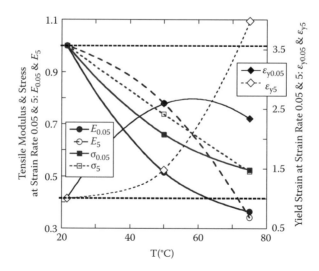

FIGURE 13.9

Tensile modulus, yield stress, and strain at the strain rates of 0.05 (solid symbol) and 5 (1/min) for PA-6 with 3-wt% clay.[80]

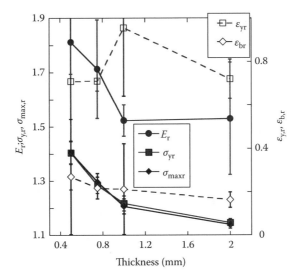

FIGURE 13.10
Effect of specimen thickness on the relative tensile properties of injection molded PA-6 with 5-wt% C30B.[81] The lines are only to guide the eye.

between PA-6 and its CPNCs. While the matrix polymer properties depend only slightly on thickness, those of the CPNCs are more sensitive. The most likely reason for this is the skin-core structure of the injection-molded specimens; in the skin, the clay platelets are oriented in the flow direction, which significantly improves the overall performance. The relative modulus, yield stress, and maximum stress (E_r, $\sigma_{y,r}$, $\sigma_{max,r}$, respectively) show a parallel decrease as the specimen thickness increases from 0.5 to 2 mm. Judging by these data the values are expected to decrease even further with an increase in thickness (n.b., ASTM D638 requires 3.26 ± 0.4-mm-thick injection-molded specimens). Considering the experimental error, the relative yield strain and strain at break ($\varepsilon_{y,r}$ and $\varepsilon_{b,r}$, respectively) are less affected by thickness.

Table 13.2 indicates that many of the CPNCs studied belong to PA with C30B-type organo-clay. These systems are known to exfoliate during melt compounding, and as such they constitute a new type of PA-based nanocomposites with properties that are slightly different than those prepared during polycondensation in the presence of clay.

The tensile modulus data from nine publications were converted to the relative modulus:

$$E_r = E_{CPNC}/E_{PA} \tag{13.28}$$

and are plotted in Figure 13.11 as a function of the (inorganic) clay content, w. There is a clear tendency for E_r to increase with w, but there is a bit of scatter at $w < 5$-wt%. The three lines in Figure 13.11 (from the top) are: $a = (E_r - 1)/w = 0.20$; $a = 0.14 \pm 0.02$; and $a = 0.11$. The highest dotted line corresponds to the best performance data.[27] Line 2 ($a = 0.2$) was calculated from Equation (13.28) using all values except these from Vlasveld et al.[91] (the error bar is indicated by the perpendicular double-headed arrow). Line 3 ($a = 0.11$) was calculated for data from Vlasveld et al.[91] It is noteworthy that in the cited monograph[27] the Somasif data followed Equation (13.12) with a similar slope, i.e., $a = 0.13$. Thus, in spite of

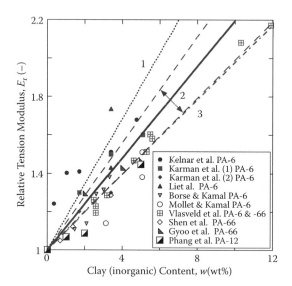

FIGURE 13.11
Relative tensile modulus of PA with C30B vs. the organic clay content. The legend indicates the sources of the data: References 85 to 90 are for PA-6 as the matrix; Reference 91 is for PA-6 and PA-66; References 95 and 97 are for PA-66; and Reference 98 is for PA-12. Lines 1 to 3 represent Equation (13.12) with the slope $a = 0.20$, $a = 0.14 \pm 0.00$, respectively.

the originally large aspect ratio (nominal $p \leq 6,000$),[53] the synthetic fluoromica does not increase the modulus as much as MMT with $p = 200 - 300$. The possible explanations are (1) high fragility of the synthetic fluoromica (FM) to attrition, which during compounding reduces the aspect ratio to approximaately $p \approx 140$;[35] (2) low modulus of the FM platelets, which on the one hand reduces the E_R, and on the other leads to folding and bending the platelets; and (3) the presence of nonplatelet-like contaminants, which reduce the effectiveness of FM as nanofiller.[53]

The reported relative modulus data by Kelnar et al.[85] are unexpectedly high at low clay loadings. The source of this enhancement is probably the method of specimen preparation for the tensile tests. In contrast with the standard test procedures, the authors used compression-molded 1-mm-thick specimens, which might have oriented the clay platelets. Statistically, PA-6 and PA-66 follow the same dependence, but the single set of data points for PA-12-based CPNCs[98] is low, which may indicate a lower degree of clay dispersion. As it can be seen in Figure 13.12, these two systems deviated from the common behavior of the tensile strength vs. clay content reported in three other publications.

It is noteworthy that the slope calculated from Equation (13.18) is $(\sigma_R - 1)/\phi_f = 0.0088$, whereas the experimental slope in Figure 13.12 (at the limit $w \to 0$) is 6.7 times larger. Thus, due to the solidification of the organic phase on the clay platelets, their volume was increased. This value is smaller than previously reported for the melt-compounded PA-6/C15A CPNC ($\phi_{effect}/\phi = 9.0$).[17] As there is no significant difference in the polymer molecular weights, the reduction of the solidified polymer might indicate a lower degree of clay dispersion. Since the surface force analyzer (SFA) experiments gave the solidified layer thickness of approximately 6 nm,[12,13] the current data suggest the presence of short stacks with about two platelets per stack.

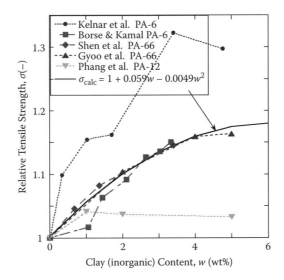

FIGURE 13.12

Relative tensile strength of PA with C30B vs. the organic clay content. The legend provides the sources of the data: References 85 and 89 are for PA-6 as the matrix; References 95 and 96 are for PA-66; and Reference 98 is for PA-12. The solid line $\sigma_{calc} = 1.002 + 0.059w - 0.0049w^2$ (r = 0.99) was calculated for the three systems.[89,95,97]

As evident from the summary in Table 13.2, most PA-based CPNCs are prepared with either the C15A or C30B type of organoclay. Dennis et al.[77] reported that while melt compounding of PA-6 with C30B resulted in easy exfoliation, compounding with C15A produced only intercalated compounds. Similar results were described by other authors compounding PA with the C30B-type and C15A-type organoclays.[82,99–107] It is enlightening to consider the mechanisms that lead to these different results. The structure of these organoclays and their properties are listed in Table 13.3.

Molecular modeling of PA/organoclay[56,57] indicated that the binding energy between PA-6 or PA-66 melt and bare MMT is the highest, linearly decreasing with the volume of intercalant. In short, neat MMT would readily disperse in PA if not for the minimal intergallery height of Na-MMT into which the PA chains must diffuse. When organoclay is dispersed in a PA matrix, the situation is reversed—the intergallery spacing is sufficient for the diffusion, but intercalant is immiscible with PA. Only a partial removal (e.g., by thermal degradation) offers a chance for the macromolecules to interact with the clay surface

TABLE 13.3

Comparison of C15A and C30B Properties

Property	C15A	C30B
Intercalant	2M2ODA	MT2EtOH
Interlayer spacing, d_{001} (nm), of organoclay	2.96	1.86
Intercalant content (CEC)	1.25	0.90
TGA wt loss (% at 250°C)	1.28	1.88
Thermal degradation during compounding (see text)	Less	More
Miscibility with PA-6	Very poor	Poor
Dispersion after compounding PA-6 with 2-wt% organoclay in a TSE	Intercalated, $d_{001} \approx 3.5$ (nm)	Exfoliated

and improve dispersion. The degradation and diffusion are rate-dependent, and when the degradation rate is higher than that of diffusion, clay reaggregation would take place. Thus, it is desirable that the diffusion rate is higher than that of degradation. Since C30B has a lower thermal stability than C15A[108–110] and smaller interlayer spacing, the diffusion rate of PA-macromolecules must be higher into the C30B intergallery space than into C15A. Now, the miscibility of PA with the two types of intercalant must be considered: 2M2ODA is immiscible with PA, whereas MT2EtOH has two polar hydroxyl groups that favorably interact with PA. As a result, CPNCs with C30B can be exfoliated readily, whereas those with C15A often remain intercalated.

Theoretically, there is a direct relation between the degree of clay dispersion and mechanical properties incorporated into most equations listed in Section 13.3.1 by means of the aspect ratio and in Equation (13.7) and Equation (13.8), directly via the number N of clay platelets in statistical stack. The latter relation predicts approximately 50% lower tensile modulus for CPNCs with short stacks ($N = 2$ to 3) in comparison to full exfoliation. Thus, one should expect that the performance of CPNCs prepared with C30B will be superior to those with C15A. However, there are two additional aspects: the semicrystallinity of PAs, and the asymptotic character of exfoliation. Crystallization of the matrix causes crowding of clay platelets within the noncrystalline regions, which reduces the degree of clay dispersion. The second aspect implies that, as the methods of CPNC preparation advance, a significant improvement in performance might be expected.

Figure 13.13 presents E_r vs. w for PA-6/C15A systems. The average value of the slope, $a = 0.128 \pm 0.009$, is comparable to that calculated for PA/C30B systems ($a = 0.14 \pm 0.2$). However, while in Figure 13.11 several data points are above the dotted line with $a = 0.2$, in Figure 13.13 only a couple of points approach it. The two circular points refer to compounding in an SSE with EFM;[17,74] the compounding resulted in full exfoliation for the

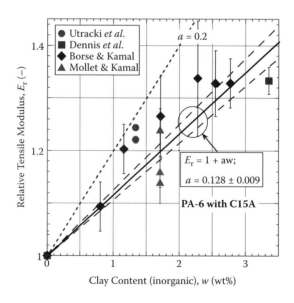

FIGURE 13.13
Relative tensile modulus of PA-6 with C15A vs. the inorganic clay content. The legend lists the source of data.[17,77,89,90] The solid black line with $a = 0.13 \pm 0.01$ was calculated for all data; the dotted line with $a = 0.2$ indicates the highest observed increase of stiffness.[27]

upper point. The square at $w = 3.35$-wt% represents the results from Dennis et al. [77] for the materials compounded either in an SSE or in diverse types of TSE; all values fall within the limit indicated by the error bar.[77] The three triangles at $w = 1.71$-wt% also differ only by the method of melt compounding in a TSE—the lowest point was obtained for the compounded material, the middle for compounded and compression-molded specimens, and the top one for material compounded using high melt intensity conditions and long residence time.[90] The sequence correlated with the surface density of particles—a measure of clay dispersion calculated from TEM micrographs. The diamonds were also obtained for a series of compositions compounded in a TSE at a high melt intensity and long residence time.[89] The data seem to indicate that for the PA-6/C15A system, there are two dependencies, the first for $w < 2.3$-wt%, and the second above this limit. Such a duality has not been observed for CPNCs with C30B, in which the modulus and stress tended to increase in the full range of clay content: $w \leq 10$-wt%. Thus, the compounding methods may significantly improve the CPNC performance, but the thermodynamic effects determine the achievable goal.

The relative tensile stress of PA-6 with C15A vs. the inorganic clay content dependence is displayed in Figure 13.14. The data are grouped around two straight lines with slopes, for the #1 broken line,[17,77] $a = 0.074 \pm 0.006$, while for the #2 solid line,[89] $a = 0.037 \pm 0.004$. Judging by these values, solidification increased the filler volume for the #1 and #2 data sets by a factor of 8.4 and 4.2, respectively. Thus, the average thickness of the adsorbed organic phase was about 4 and 2 nm, respectively, suggesting that the degree of exfoliation was about one-half for set #2 than that for set #1.

It has been shown that the tensile modulus of a PA-based CPNC linearly increases with tensile strength and vice versa.[17,74] Similarly, a correlation was observed between the tensile modulus and the notched Izod impact strength at room temperature (*NIRT*). However, in

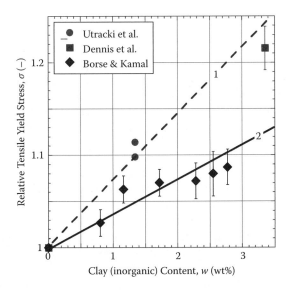

FIGURE 13.14
Relative tensile yield stress of PA-6 with C15A vs. the inorganic clay content. The legend indicates the source data.[17,77,89] The #1 broken line with $a = 0.074 \pm 0.006$ was calculated for data from refs[17,77] while the #2 solid green line with $a = 0.037 \pm 0.004$ data from.[89]

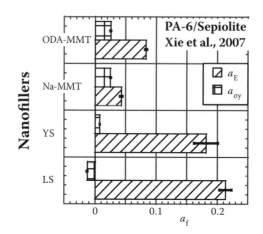

FIGURE 13.15

PA-6 with pre-intercalated MMT, Na-MMT and two sepiolites: YS and LS. The plotted quantities represent the effect of addition of 1-wt% of nanofiller on tensile modulus (E) and tensile strength at yield (σy). Data: Xie et al.[111]

the latter case the correlation existed only for the CPNC produced using the same method, i.e., the data points for materials prepared by melt compounding in a TSE clustered in one part of the graph, whereas those prepared either by reactive exfoliation or melt compounding in an SSE + EFM formed another group.[74] This difference could be related to the thermomechanical degradation of the matrix during energetic compounding in TSE.

Before closing this part it is important to recognize that good mechanical performance might also be obtained using other types of nanofillers. For example, recently Xie et al.[111] reported that incorporation of natural, nonintercalated sepiolites resulted in a significant improvement of the mechanical performance of PA-6.[111] The mineral is a hydrated magnesium silicate with the theoretical half unit-cell formula $Si_{12}O_{30}\text{-}Mg_8(OH, F)_4(OH_2)_4 x 8 H_2 O$. In spite of a similar composition and crystalline unit cell, the minerals have different morphologies: YS is needle-like, LS is platelet-like. As shown in Figure 13.15, PA-6 with either of the two sepiolites shows higher enhancement of properties than that with either Na-MMT or its preintercalated version. The enhancement parameter $a_f = [(f/f_o) - 1]/w$, where for modulus $f = E$, and for strength at yield $f = \sigma_y$; f_o represents the property of PA-6. However, sepiolite is less effective than MMT for increasing the heat distortion temperature.

13.4.3.2 Tensile Creep

Creep is one of the most important engineering properties, especially in applications where the product is under constant load (e.g., pressurized pipes). As described in Section 13.3.3, the rate of deformation under load is rapid after the initial loading and just prior to fracture. The phenomenon is related to the viscoelastic properties of the material, imposed load, temperature, pressure, and the time under load.

The effect of filler addition on creep depends on the system. For example, the creep reduction is significantly higher for flexible polymers than for rigid ones. The filler particle shape and concentration also affect creep. The best performance has been reported for fibrillar particles at concentration at which 3D interlocked structures were possible.[47] Thus, the creep deformation of semirigid liquid crystal polymer (LCP) with 40-wt% of $MgCO_3$, wollastonite, and glass fiber (GF) were reduced from 0.44% to 0.17, 0.14, and

TABLE 13.4

Tensile and Creep Properties of PA-66-based Nanocomposites

Composition	Tensile modulus, E (GPa)	Tensile strength, σ (MPa)	Creep ε (%) 200h (40 MPa load)	Creep rate 10^6 $\dot{\varepsilon}$ (1/h) (40 MPa load)	$\dot{\varepsilon}$ (1/h) Burgers model
PA-66	2.28 ± 0.04	75.8 ± 0.7	6.0	103	99.0
PA + 300 nm TiO$_2$	2.65 ± 0.32	72.5 ± 4.5	4.2	61.2	58.7
PA + Nanofil 919[1]	2.77 ± 0.16	73.0 ± 1.5	3.4	63.2	57.5
PA + 21 nm TiO$_2$ surface treated[2]	2.27 ± 0.09	75.6 ± 2.2	3.2	47.4	44.6
PA + 21 nm TiO$_2$	2.76 ± 0.11	71.7 ± 3.8	2.68	42.9	38.7

Note: At $T = 23°C$, $\sigma_0 = 40$ MPa.[96,115] [1] Nanofil 9[19] = MMT-[2]MBODA; [2] TiO$_2$ ($d = 21$ nm) treated with octyl-silane.

0.02%, respectively.[112] For nearly spherical particles of MgCO$_3$ ($d \approx 2.3$ μm), creep linearly decreased with loading: $\varepsilon = 0.445 - 0.00672w$ ($r^2 = 0.988$). Similarly, the room temperature creep of acrylic copolymer ($T_g \approx 50°C$) significantly decreases with 0 – 11-wt% TiO$_2$ loading.[113] Within this range of filler content, the creep constant, k, decreased from about 0.185 to 0.115. By contrast, incorporation of wood fibers into a vitreous PS matrix had a slight (if any) effect on creep.[114]

On the molecular scale, creep is related to the mobility of atoms or molecules forced to change their structure under stress. Since polymer molecular mobility depends on the free volume fraction, h, any factor that reduces h is expected to reduce creep. In CPNCs, reduction of h as a function of the degree of clay dispersion and its concentration has been well documented;[27] thus, it could be expected that in comparison to the matrix polymer, CPNCs would show reduced creep. Unfortunately, the creep resistance data are scarce.

Zhang et al. (2004) and Yang et al. [115] (2006) studied the influence of 1-vol% of different nanoparticles (see Table 13.4) on PA-66 creep.[96] The compositions were compounded in a TSE, and then injection molded into ASTM specimens for tensile and creep tests. As shown in Table 13.4, incorporation of nanoparticles reduced the creep strain and strain rate. Addition of Nanofil 919 (hardly appropriate organoclay for PA-66) resulted in the best tensile performance, but that of neat TiO$_2$ particles ($d = 21$ nm) was most effective for creep reduction.

The data in Table 13.4 illustrate good correlation among creep, creep rate, and tensile strength, but none with tensile modulus—the surface treated TiO$_2$ has a smaller modulus than that of the neat resin. Similarly, the Nanofil 919 engendered the highest modulus, but the creep value is higher than that for the compound with 21 nm of TiO$_2$ particles. It is noteworthy that the relative modulus of CPNCs with Nanofil 919 is 1.2, whereas ≥ 1.4 would be expected from the exfoliated system. Incomplete exfoliation might be responsible for slippage under load, and thus low tensile strength and a smaller-than-expected creep reduction related to it. The authors interpreted the extensive experimental data using the Burgers four-element constitutive model and the Findley empirical power-law model—both satisfactorily described the observed behavior; examples of $\dot{\varepsilon}$ calculation from Burgers model are listed in Table 13.4.

The physical aging and creep of PA-6-based CPNCs was studied by Vlasveld et al.[116] Several compositions were prepared using low- and high-molecular-weight polymers (coded PA-6L and PA-6H, respectively) compounded in a TSE with synthetic FM; Somasif ME100 and Somasif MEE (ME100 preintercalated with MCoco2EtOH). The ISO 527 standard specimens were injection molded, then aged at RT for 4 to 2450 hours. Creep tests were performed under constant load with 16 MPa, i.e., approximately 20% of the yield

TABLE 13.5

Creep Behavior of PA-6-based CPNCs at room temperature

Composition	J ($t_a = 4$ h; $t = 1000$s)	μ (-)
PA-6L ($t_a = 4$ h)	0.42	0.70
PA-6H ($t_a = 4$ h)	0.41	0.76
PA-6L + 2.7% Somasif MEE ($t_a = 4$ h)	0.26	0.82
PA-6L+ 6% Somasif MEE ($t_a = 4$ h)	0.23	0.82
PA-6L+ 11% Somasif MEE ($t_a = 6$ h)	0.18	0.83
PA-6L+ 10% Somasif ME-100 ($t_a = 4$ h)	0.38	0.84
Unitika M1030D ($t_a = 4$ h)	0.38	0.93

stress. For each composition a master curve was constructed by horizontally shifting the creep compliance data of specimens aged for different times, t_a. The shape of the CPNC curves was found to be similar to that of the matrix, PA-6L or PA-6H, but the magnitude of creep compliance was smaller. Two methods were used for assessing the effects of the nanocomposite: (1) a plot of compliance vs. time after 4 h of aging, and (2) calculating the shift rate parameter, $\mu = d \ln a_t / d \ln t_a$ (a_t is the shift factor and t_a is the aging time). The results are presented in Table 13.5. The molecular weight of PA-6 had little influence on the creep behavior at RT. Addition of Somasif MEE significantly reduced creep, but a relatively large amount of clay was required to reduce J by a factor of 2. For the long-term behavior, the shift rate parameter μ is more important and here the superiority of the commercial CPNC (Unitika 1030D) is clear. In the reactively prepared CPNC, the attrition of high-aspect-ratio clay platelets ($p \leq 6000$) is probably significantly smaller than that in a TSE compounding; thus, at relatively low clay loading its effect on creep is large.

13.4.3.3 Fatigue Resistance

The main difference in fatigue behavior between metals and polymers is the lack of sensitivity of metals to test frequency, while polymers with high energy dissipation and low thermal conductivity are sensitive to it. As discussed in Section 13.3.4, depending on the material characteristics (e.g., internal friction coefficient and thermal conductivity), frequency, v, and stress amplitude, σ_a, two mechanisms, FCP and TFF, have been identified. The boundary between these two is determined by the rate of energy input: $(\partial E / \partial t) \propto v \times \sigma_a$. Schematically, these two mechanisms are shown in Figure 13.16. At low frequency and/or low σ_a, the conventional FCP fracture mechanism is observed, but when the frequency and stress amplitude exceeded the critical values, the number of cycles to failure, N, dramatically decreased forming the TFF-branch characteristic for specific frequency.

The fatigue behavior of PA-6 and PA-6 with 2-wt% MMT-ADA (Ube 1015B and 1015C2, respectively) was examined by Bellemare et al.[117] The measurements were conducted at 5 Hz, $\sigma_{max}/\sigma_{mim} = $ const, with maximum stress $\sigma_{max} = 75$ or 57 MPa. As shown in Figure 13.17, the fatigue resistance of PA-6 and CPNC was found to be similar, the latter performing marginally better. In CPNCs the crack initiations occurred within the specimen, near inorganic particles, suggesting fracture initiation by micron-sized clay aggregates or contaminants containing variable amounts of Si, Mg, Ca, and Al. The mechanism of crack propagation in CPNCs had a higher tendency to follow the TFF mechanism than the matrix. After a similar number of cycles, $N = 10^4$ to 10^5, there was an onset of volume expansion in CPNCs, but volume reduction in PA-6. The critical maximum cyclic stress for PA-6 and CPNC was 60 and 78 MPa, respectively. The fractured surfaces showed an initial zone of crazing, followed by fatigue crack propagation, and then rapid fracture.

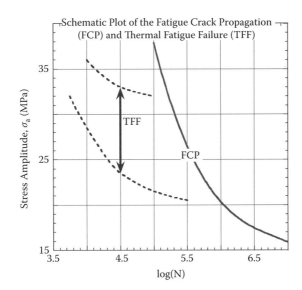

FIGURE 13.16
S-N plot for two mechanisms of polymer failure by FCP and by TTF. Adapted from Crawford and Benham.[50]

Mallick and Zhou[80] studied the fatigue behavior of PP- and PA-6-based nanocomposites. The materials were produced by RTP Co. by melt compounding with 5- and 3-wt% clay, respectively.[118] Injection-molded specimens were tested at 21.5°C in tension-tension mode at 0.5 and 1 Hz. Since the matrix specimens were not tested, the effect of organoclay cannot be assessed. The fatigue strength of the PA-6 CPNC was higher than that of

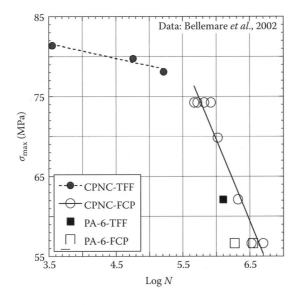

FIGURE 13.17
Maximum cyclic stress vs. the number of cycles to failure for PA-6 and its CNPC containing 2-wt% organoclay. Data from Bellemare et al.[117]

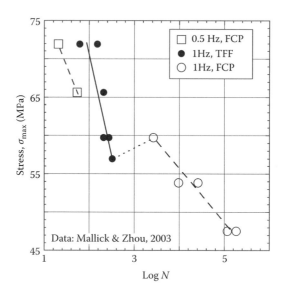

FIGURE 13.18
Maximum cyclic stress vs. the number of cycles to failure for PA-6-based CPNCs containing 3-wt% organoclay.
Data from Mallick and Zhou.[80]

the PP-based CPNC, i.e., 72 vs. 40 MPa. As shown in Figure 13.18, the PA-6 CPNC cycled at 1 Hz at high stress failed by the TFF mechanism and at low stress by the FCP mechanism. At the frequency of 0.5 Hz, FCP failure was reported at high stresses. As noted by Bellemare et al.,[117] Mallick and Zhou[80] also reported that fracture during fatigue tests was initiated at agglomerated nanoparticles.

13.4.3.4 *Impact Strength*

Several methods of impact stress determination have been proposed. The specimens might be notched or not, supported at both ends (Charpy-type), or attached by one (Izod-type), etc.

The Izod impact strength at room temperature (*NIRT*) is most commonly used. As described in ASTM D 256 (ISO 8256), the tests use "standardized" pendulum-type hammers, mounted in "standardized" machines, to break standard specimens with a milled notch that produces a stress concentration, increasing the probability of a brittle fracture. The results are reported in terms of energy absorbed per unit of specimen width (J/m) or optionally per unit of cross-sectional area (kJ/m²). The precision of the data within a laboratory ranges from 1 to 5%, and that among 25 laboratories from 5 to 41%.

Figure 13.19 shows that incorporation of organoclay into PA usually reduces impact strength. This is in accord with a common wisdom that rigidity is associated with brittleness. For example, the data from Wu et al.[82] are for compositions with PA-6, having lower or higher molecular weight (MW). As expected, the former has lower modulus and higher *NIRT*, while the properties of the latter are reversed. There are few exceptions to this rule, one being for PA-6 CPNC with 5-wt% of C15A.[77] In this case, the *NIRT* of PA-6 is cited as 37 while that for the CPNC ranges from 40 to 50, with a mean of 44 J/m. However, for this system there is also a negative correlation with tensile modulus. It seems that in this case, compounding in a TSE reduced MW, which in turn decreased the *E*–value and increased *NIRT*.

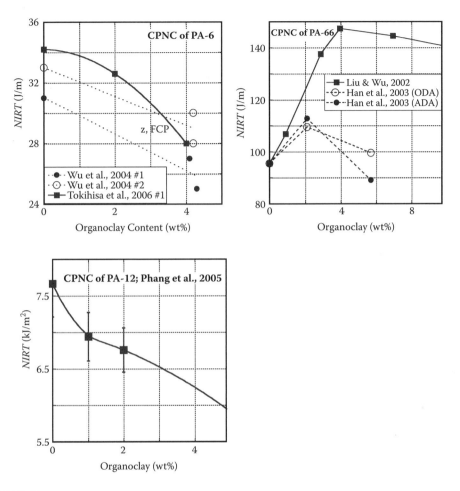

FIGURE 13.19
Concentration dependence of the notched Izod impact strength at room temperature (*NIRT*) for (clockwise) PA-6, PA-66 and PA-12-based CPNCs with different organoclays.

The PA-66-based CPNC systems studied by Liu and Wu[92] are more complex. The authors first prepared MMT-3MHDA, which prior to compounding with PA-66 was mixed in an internal mixer with 13-wt% of DGEBA epoxy. The improved impact strength came at a price of smaller-than-expected increase of modulus. In a sense, this treatment might be considered a form of Bucknall's toughening, by surrounding rigid particles with an elastic envelope.[119]

Another method of toughening relies on the formation of an elastomeric dispersion in the matrix, which modifies the system performance leaving the solid particles more or less unaffected. Here, the size of the dispersed drops is critical.[119] An example of this approach might be found in the work by Tjong and Bao,[87] which is illustrated in Figure 13.20. The data show that incorporation of 4-wt% of MMT-3MHDA improved the matrix modulus by 54% at a cost of reducing *NIRT* by 45%. Addition of maleated SEBS block copolymer to CPNCs affected both these properties to the extent that at about 10-wt% loading of SEBS-*g*-MA, those of the neat matrix are recovered—no gain and no loss, except the cost.

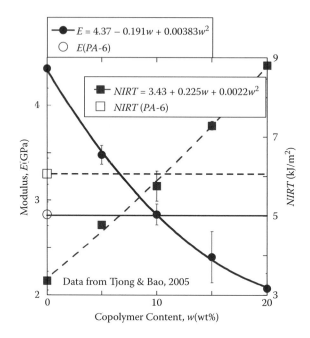

FIGURE 13.20

Tensile modulus and notched Izod impact strength at room temperature (*NIRT*) for PA-6 with 4-wt% organo-clay as functions of SEBS-g-MA content.[87] The horizontal lines indicate properties of neat PA-6.

Kelnar et al.[85] also tried to improve toughness by adding 5 or 10-wt% of elastomer to CPNCs of PA-6 with 5-wt% C30B. Table 13.6 summarizes their findings. It is worth noting that all the data fall into two groups; one highlighted, the other not. The highlighted group might be labeled "less reactive." Here the modulus and impact strength increase with the elastomer drop size. Within the reactive group the opposite is true—E_r increases with reduction of the drop size. Furthermore, σ_r increases with E_r twice as fast as in the

TABLE 13.6

Mechanical Properties of PA-6-based CPNCs Containing 5-wt% C30B and 5-wt% Elastomer

Elastomer		Relative CPNC properties		
Composition	Drop size (nm)	E_r	σ_r	$NIRT_r$
Neat PA-6	—	1	1	1
EPR	< 1000	1.59	1.13	2.18
u-SBR[1]	100–150	1.41	1.08	1.88
EPR-MA	60	1.38	1.15	2.73
SEBS-MA	70	1.35	1.11	2.48
PP-MA[2]	< 500	1.45	1.12	1.82
E-MA-GMA[3]	< 200	1.25	1.03	2.00
PE-GMA	< 100	1.28	1.03	1.52

Notes: Data from Kelnar et al.[85]. [1]Ultra-fine, vulcanized (non-reactive) styrene-butadiene rubber; [2]10-wt% PP-MA was used; . [3]E-MA-GMA is ethylene-methacrylate-glycidyl methacrylate copolymer.

less-reactive group. Elastomer presence rather than elastomer reactivity seems to be more important for $NIRT_r$.

The essential work of fracture (EWF) method has been used in studies of PA-66 toughness. The systems contained 1-vol% of spherical nanoparticles: 21 nm TiO_2, 13 nm SiO_2, and 13 nm Al_2O_3.[115] The crack initiation fracture toughness was investigated within the 23 to 120°C range. Double-edge-notched tension specimens with originally different ligament lengths were tested at a constant cross-head speed. The results were plotted as the specific crack initiation work of fracture, w_{ini}, vs. ligament length, l:

$$w_{ini} = w_{e,ini} + \beta w_{p,ini}\, l \qquad (13.29)$$

where $w_{e,ini}$ and $w_{p,in}$ are the essential and nonessential components, while β is the shape factor.

The authors focused their attention on the spherical nanoparticles as an earlier work on PP-based CPNCs suggested that nanoparticles with a high aspect ratio might have a negative effect on toughness due to stress concentrations at the sharp edges of the particle. By contrast, the spherical nanoparticles promote cavitations, thus improving toughness. However, at higher concentrations, the aggregates might be responsible for the crack initiation and propagation that reduce toughness. Plots of $w_{e,ini}$ and $w_{p,in}$ vs. T indicate that the nanoparticles enhanced the essential component of EWF, especially at $T > 80°C$, and caused reduction of the nonessential EWF part (in comparison to neat PA-66 behavior), i.e., above the β-transition. The best results were obtained for the system comprising silane-treated TiO_2, but it is not clear if the amelioration originated in better particle dispersion or the silane presence. Thus, according to EWF measurements, the presence of solid particles (at least within the concentration region of interest in CPNCs) reduces PA matrix toughness.

The material strength frequently is orders of magnitude lower than that determined for a freshly prepared, flawless specimen. It has been shown that the reason for such disparity is the presence of stress concentrators, i.e., surface scratches, void bubbles, and solid particles, especially in the form of aggregates. As mentioned earlier, the stress at the crack tip, σ_{tip}, depends on the crack length, H, crack tip radius, L, and an average stress in the material, σ:

$$(\sigma_{tip}/\sigma)^2 = 4H/L \qquad (13.30)$$

When the tip stress reaches the cohesive strength of the material, the crack starts to grow and σ_{tip} becomes proportional to the tensile modulus: $\sigma_{tip} \approx E/15$. ASTM D5045 (ISO 13586) describes the procedure for determining the characteristic parameter for stress concentration, the critical stress intensity factor under the plane-stress, Mode I conditions, K_{IC}, which characterizes the stress distribution at the crack tip—a measure of fracture toughness. Knowing the latter value, one may calculate the critical strain energy release rate:[120]

$$G_{IC} = \left(K_{IC}^2/E\right)\left(1 - v_P^2\right) \qquad (13.31)$$

where v_P is Poisson's ratio. Both K_{IC} and G_{IC} are characteristic material parameters, and since $K_{IC} \propto E$, they show similar dependencies.

The mechanical properties of the commercial PA-6 (PA1015B) and CPNC containing 2-wt% organoclay (PA1015BC2) from Ube were studied by Weon and Sue.[83] The specimens were extrusion-injection molded using a custom-designed equal channel angular extrusion (ECAE). The process reduced the platelet aspect ratio (from $p = 132 \pm 33$ to 78 ± 21)

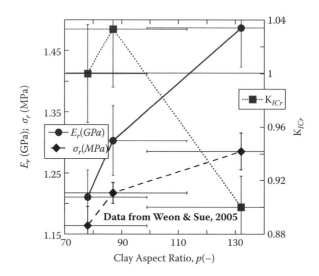

FIGURE 13.21

Tensile modulus and strength vs. the clay aspect ratio for Ube PA-6 and its CPNC with 2-wt% organoclay.[83] The dependence of K_{IC} is also shown. Experimental error bars are shown; the lines are to guide the eye.

and its orientation. The data are presented in Figure 13.21 as the tensile modulus, tensile strength, and K_{IC} vs. p. It is evident that modulus and strength increase with the aspect ratio, whereas K_{IC} tends to decrease. It is noteworthy that for the aspect ratios $p < 100$, the stress intensity factor of CPNCs is the same as that of the matrix polyamide: $K_{ICr} = 1$.

In summary, in PA-based CPNCs with intercalated or exfoliated clay particles, end-tethered or not, the fracture proceeds by the crazing-and-cracking mechanism, frequently with volume expansion (the shear banding fracture without volume expansion has not been observed in CPNCs). The crack initiation is usually a result of a mineral contaminant, a nanoscale defect at the polymer–clay interface, or the stress concentration on the tip of large aspect ratio clay particle or aggregate. According to Bureau et al.,[121] the internal energy change per unit crack length is lower in CPNCs than that in the neat PA-6, i.e., $K_{IC} =$ 4.9 ± 0.1 and 1.8 ± 0.2 for PA1015B and PA1015BC2, respectively. One part of the reason might be the clay-engendered change of PA-6 crystalline morphology (an increase of the γ-crystalline form); another reason might be a linear-elastic fracture in CPNCs, which does not undergo shear yielding (as PA-6 does).

13.4.4 Mechanical Properties of Polypropylene-Based CPNCs

By contrast with the previously discussed PA-based CPNCs, the ones with polyolefins (PO) are significantly more complex. Because of good clay–polyamide interactions,[56,57] exfoliation has been achieved during polymerization of ε-caprolactam in the presence of synthetic, non-intercalated clay.[122] In view of the nonpolar, hydrophobic character of PO, such a possibility does not exist for this matrix. Thus, the PO-based CPNC contains at least three components: the PO matrix, preintercalated clay, and at least one compatibilizer. The diversity of materials within each category, as well as the component concentrations and process variables, result in a nearly infinite combination of influences affecting the performance of CPNCs. These problems are discussed in Utracki's monograph[27] as well as in several reviews.[123,124]

13.4.4.1 Factors Affecting the Mechanical Performance of PP-Based CPNCs

In the case of PP-based CPNCs, crystallinity and diversity of crystalline morphologies augment the confusion. For example, there are conflicting reports on the influence of PP crystallinity on tensile modulus; in some publications a linear increase of modulus with the total crystallinity was reported (e.g., see Dell'Anno[125]), while in others the modulus was found to be independent of crystallinity.[74,126]

Initially, following the success with PA-based CPNCs, there were attempts to employ the reactive method for the preparation of PO-based CPNCs. For example, Heinemann et al.[127] polymerized olefins via metallocene catalysis in the presence of HT-2MBSA (interlayer spacing: d_{001} = 1.96 nm). After ethylene polymerization into high-density polyethylene (HDPE), the XRD peak shifted to d_{001} = 1.40 nm. Better dispersions were obtained for LLDPE-type ethylene-octene copolymers. As more octene was incorporated, the modulus decreased and the elongation at break increased. These nanocomposites performed better than those prepared by the melt compounding method. However, the postpolymerization melt processing might result in a reduction of the interlayer spacing and performance. Furthermore, the preintercalated clay was found to interfere with metallocene catalysts. As a result, the current tendency is to focus on the melt compounding method, assuring good thermodynamic interactions among all components of the system.

The influence of the methods of melt compounding on mechanical performance has frequently been studied. For example, Modesti et al.[128] investigated the influence of temperature and screw speed in TSE on the mechanical properties of PP with 6-wt% PP-MA (1% MAH), and 3.5 or 5-wt% MMT-2M2HTA. The best performance was obtained at a low temperature profile (T = 170-180°C) and high screw speed of 350 rpm. However, the effect of processing on performance was relatively small in comparison to that caused by the compatibilizer and organoclay content.

Several publications report that organoclay dispersion in a PP-matrix is determined within the melting zone. According to Lertwimolnun,[129] the interlayer spacing, d_{001}, after the melting zone, may remain constant or decrease (for high-shear screw configuration), but never increases. However, it should be stressed that d_{001} reflects the spacing in the organoclay stacks, not the global degree of clay dispersion, and in the absence of mechanical performance data the emphasis on the melting zone might be misleading. Wang et al.[130] stressed the role of shearing near the solidification temperature; the authors used a dynamic packing injection molding device, which closely resembles the old SCORIM system.[131] The process created a graded morphology, with large clay aggregates in the skin and good dispersion in the core.

The advantages of mixing in an extensional flow field have been recognized for decades and the EFM has been patented.[132] During the last few years the device attached to SSE has been successfully used for the manufacture of well-dispersed CPNCs based on PET or PA.[17,74] However, the results for the PP-based (or PS-based) CPNCs were not as spectacular. The data for PP-based CPNCs show that the degree of clay dispersion improves with the increase of the elongational stress, reaching a maximum at a convergent-divergent gap height of approximately 30 μm, where d_{001} reaches the highest value of 3.48 nm, and the number of particles in TEM micrographs doubles. However, these morphological refinements do not translate into significant improvements of the tensile modulus or strength. Only *NIRT* shows superiority of SSE + EFM compounding over that in TSE, which may be caused by in greater chain scission in the latter compounder.

Table 13.7 summarizes the recent efforts in optimization of the mechanical properties for the PP-based CPNCs. However, it should be noted that these systems are based on

TABLE 13.7

Mechanical Properties of PP-based CPNC

Organoclay	Compatibilizer	Concentrations (wt%)	Properties	Ref.
HT-, MMT-, mica-Q[1]	PP-MA (10%MAH) + DAAM[2]	3 org-clay; 3 DAAM,	σ, E, σ_F, NIRT, HDT	136
MMT-2M2ODA	PP-MA (1%MAH)	0–7 org-clay, 2 PP-MA	σ, NIRT	137
MMT-2M2ODA with or without fluoro-Si	Functionalized PP[3]	0–10 org-clay	E, σ, ε_b	138
MMT-ODA	PP-MA	0–50 org-clay rest PP-MA	E, σ, J_{IC}	139
ME-100-2M2EtOH	28% PP-MA	0–8.5 org-clay	F, NIRT	140
Bentonite-CPCl[4], MMT-2M2HTA, -2MHTL8, -2M2ODA	None	0, 1, 3, 5, 7, 10-vol%	E, σ.	141
MMT-2M2HTA	PP-MA (0.7%MAH) + etanol-steramice	5-wt% C15A; compatibilizer/clay = 1:1	F, σ_F, HDT	142
MMT-2M2HTA	PP-MA (0.7%MAH)	0–5 clay, 0–7 compatibilizer	E, σ, ε_b	143
MMT- and ME-100 with 2M2ODA	PP-MA (0.7%MAH)	7.5 PP-MA, 5 inorganic clay	E, σ, ε_b, ε_y, Charpy impact strength, HDT	144
ME-100 with 2M2HTA	8 PP-MA or PP-AA (0.5-8.2%MAH)	4.2 to 14.3% comp., 0.13–1.0 clay	E, σ, ε_b, NIRT	145
MMT, MMT-2M2HT, MMT-MT2EtOH	PP-MA (0.1 MAH), PP-AA (6% AA), PP-GMA	84 PP+ 12 compatibilizer + 4 clay	E, σ, ε_b	109
MMT-DA[5]	PP-MA/MMT-DA	75 PP + 25 PP-MA/MMT-DA	E, σ, ε_y	146
MMT-HA[6]	PP-MA (5 %MAH)	20 PP-MA, 0–2 clay	E, σ, ε_b	147
MMT-TAPTMS[7]	PP-MA, epoxy	9–27 PP-MA, 5–15 epoxy, 1–3 clay	E, σ.	148
MMT-ODA/APTES[8]	PP-MA(0.1% MAH)	PP-MA with 0, 2.5, 5, 7.5, 10% clay	Tensile and compressive stiffness	149
MMT-, HT-, synthetic HT- with ODA	PP-MA (0.6% MAH); PP-IT (0.7–1.8% IT[9])	3 compatibilizer, 1 clay	E, σ, ε_b	150
MMT-2M2HT	PP-MA	9 PP-MA, 5 clay	E, σ, ε_y	151
MMT-2M2HT	PP-MA (1% MAH)	4 and 8 PP-MA, 2 and 4 clay	E, σ, NIRT	18
MMT-2M2HT	PP-MA	0–3 PP-MA, 0–1 clay	E, σ, ε_b	152
MMT, ME-100, MMT-2M2HTA, MMT-2M2ODA	PEG, PE-PEG, PEG-PMMA, PEG-PDMS	0–5 MMT, 0–1 compatibilizer	E, σ, ε_b	153
MMT-2M2HT grafted with silane[10]	PP-MA (0.5%MAH)	PP:PP-MA = 4:1, 4 inorganic clay	E, σ	154
MMT-ODA	PP-MA (1%MAH), POE[11]	6 clay, 6 PP-MA, 0–20 POE	E, σ, F, σ_F, NIRT	155
MMT-2M2HTA,	PP-MA (3.8%MAH), or hydroxyamide[12]	5 PP-MA, 2 hydroxy-amide, 3 organoclay	E, σ. NIRT	156
MMT-2M2HT	PP-MA (1%MAH)	0–5 clay, PP-MA/clay = 3:1	E	157

Notes: Abbreviations are listed at the end of this chapter; [1]Q = quaternary ammonium ion; 3MHDA, 2MHDODA; [2]DAAM = di-acetone acryl amide; [3]PP-MS. PP-MA, PP-OH, PP-b-PMMA, [4]cetyl-pyridinum chloride, [5]DA = di-amido-dodecane; [6]α, ω-hydroxy amine; [7]TAPTMS = 3-amino propyl 3-methyl silane; [8]APTES = γ-aminopropyl 3-etoxy silane; [10]*i*-butyl 3-methoxy silane; [9]IT = itaconic acid; [11]POE = metallocene polyethylene-*co*-octene; [12]Paricin-220 or Paricin-285.

isotactic PP homopolymer with a glass transition temperature of $T_g \approx 0°C$. The presence of solids and high test frequency increases T_g.[133] As it has been noted in this chapter, the performance of nanocomposites with a glassy matrix (thermosets or thermoplastics) is notoriously difficult to improve. As Figure 13.1C and Figure 13.1D show, the improvement of mechanical properties by addition of organoclay is significantly greater when isotactic PP is replaced with a copolymer having a lower value of T_g.[18,134,135]

13.4.4.2 Tensile Modulus and Strength

Table 13.7 lists but a fraction of the publications on PP-based CPNCs. This partial summary of recent research provides evidence of the great diversity of systems that have been explored while attempting to reach the elusive goal of full exfoliation and enhancement of mechanical performance to the ultimate level. Different industries focus on different sets of performance characteristics for "ideal" materials. For example, in the automotive industry the balance between rigidity and toughness is critical—for flexible applications: $E = 1.5 \pm 0.15$ GPa *and* $NI(T = 0°C) = 600 \pm 100$ J/m, while for rigid applications: $E > 4$ GPa *and* $NIRT > 100$ J/m.

Oya et al.[136] examined the role of clay as well as a method of dispersion on mechanical properties of CPNCs containing 3-wt% of a nanofiller or talc. Synthetic FM, MMT, and MC were preintercalated with 2MHDODA. In method A the organoclay was dispersed in a toluene solution of di-acetone acrylamide (DAAM) with the free radical initiator *N,N'*-azo-bis(iso-butyro nitrile), AIBN. After polymerization of DAAM, a solution of PP-MA was added, the product was precipitated, washed, dried, and then melt compounded with PP. In method B, the CPNC was similarly prepared but without DAAM. In spite of a rather complicated procedure, only intercalation was obtained. The interlayer spacing, d_{001}, flex modulus, *F*, and *NIRT* are shown in Figure 13.22. The

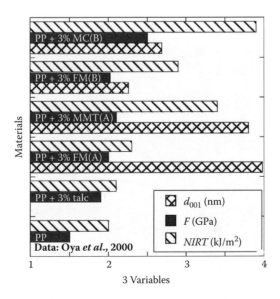

FIGURE 13.22
Interlayer spacing, flexural modulus, and *NIRT* for PP and PP with 3% talc or organoclay.[136]

best mechanical performance was obtained using MC (in spite of a small d_{001} value). Performance of CPNCs with MMT was not as good, and performance with talc the poorest. Polymerization of DAAM seems to crosslink the system, preventing enhanced expansion by PP-MA, and reducing performance. Clearly, d_{001} is not the only factor responsible for performance.

Manias et al.[138] prepared PP-based CPNCs with MMT-2M2ODA or MMT-ODA following two approaches: (1) incorporating these organoclays into functionalized PP or (2) modifying the organoclays and using these compounds with neat PP. The functionalized random copolymers (M_w = 200 kg/mol) contained approximately 0.5 wt% of *p*-methyl styrene (PP-MS), or PP-MS modified by attaching functional groups to it, i.e., –(CH$_2$)$_3$–OH (PP-OH) or –CH$_2$-maleic anhydride (PP-MA). Furthermore, two PP-*b*-PMMA block copolymers were employed. The organoclays were treated with semifluorinated silane. CPNCs were: (1) melt compounded in an extruder, (2) in an internal mixer, or (3) ultrasonicated in tri-chloro benzene. All three methods resulted in a similar morphology of CPNCs with intercalated tactoids and exfoliated platelets. Figure 13.23 illustrates the effect of the two preparation methods. Evidently, the use of functionalized PP resulted in more rigid materials, with the yield stress and the strain at break similar to that of neat PP. However, neither of these two approaches resulted in exfoliation; the interlayer spacing ranged from about 2.8 to 4.0 nm. Two years later, fully exfoliated CPNCs with –NH$_3^+$ end-terminated PP were described, but unfortunately neither Manias et al. nor any other study during the intervening years described mechanical performance.[158]

The commercial literature shows that organoclay manufacturers might incorporate up to 40% excess intercalant. Morgan and Harris[140] investigated the effect of purification

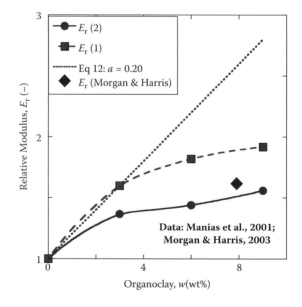

FIGURE 13.23

Relative modulus of PP and its CPNC prepared by mixing organoclay with functionalized PP (method #1) or by mixing neat PP with organoclay treated with semifluorinated silane.[138] The straight line represents Equation (13.12) with a = 0.2, valid for the best-performing CPNC.[27]

of an organoclay on mechanical performance. With increasing extraction time (using ethanol in a Soxhlet) the performance of CPNCs containing PP/PP-MA/organoclay = 66/26/8 changed, i.e., the modulus increased from 1.99 to 2.09 GPa (E_r is shown in Figure 13.23), and *NIRT* decreased from 42.7 to 37.4 J/m. Thus, as expected, extraction of the plasticizing excess intercalant makes the system more rigid (by 5%) and more fragile.

A similar problem was attacked from another angle—preintercalating clay with CPCl to 10, 30, 50, 70, and 100% of the cation exchange capacity (CEC).[141] As it could be expected from the published information (e.g., see Beal et al.[159]), the interlayer spacing increased with the amount of intercalant from d_{001} = 1.26 to 1.77 nm. The tensile modulus increased with clay loading—more for the nonintercalated than for the fully intercalated clay, i.e., E_r (10-vol%) = 1.45 and 1.31, respectively. The yield stress decreased with clay loading—again, less so for the nonintercalated than for the fully intercalated clay. The authors offered possible orientation effects as an explanation for this unexpected behavior.

Benetti et al.[142] studied the behavior of PP-based CPNCs with 5-wt% C15A and either PP-MA or ethanol-stearate. Two PP resins were used: low-viscosity injection molding grade and high-viscosity extrusion grade. It is interesting that larger improvement in properties was obtained for the former resin than for the latter:

- Low viscosity: $F_r \leq 1.50$, $\sigma_{Fr} \leq 1.13$, and $\Delta(HDT) = HDT_{CPNC} - HDT_{PP} \leq 40°C$
- High viscosity: $F_r \leq 1.40$, $\sigma_{Fr} \leq 1.35$, and $\Delta(HDT) = HDT_{CPNC} - HDT_{PP} \leq 30°C$

F_r is the relative flexural modulus, σ_{Fr} is the relative flexural yield stress, and HDT is the heat deflection temperature. For both resins, PP-MA better enhanced the modulus and HDT than did ethanol-stearate.

These results agree with E_r for CPNCs with different PP resins—average values of the parameter a in Equation (13.12) for the low- and high-viscosity resin were calculated as 0.33 and 0.11, respectively; thus, there was excellent enhancement for PP with MFR = 12, and poor for PP with MFR = 1.9 g/min.[18]

Figure 13.24 and Figure 13.25 display, respectively, the relative modulus and relative strength (at yield) for systems containing $\leq$ 50-wt% organoclay. The modulus enhancement is about one half of the expected best, i.e., a = 0.2. The tensile strength also increased in parallel to E_r.[139] The XRD data for the PP/PP-MA/MMT-ODA system indicated that clay dispersion decreased with concentration.[139] It is interesting that the relative strength of this system behaves in a manner similar to PP/PP-MA/MMT-2M2ODA.[137] However, better performance was reported for the system where organoclay preintercalated with α,ω-di-amino dodecane and $Ti(C_4H_8-OH)_4$, dispersed in PP/PP-MA.[146] The presence of di-ammonium might be responsible for bridging the gap between the two platelets and preventing full dispersion. The best modulus enhancement was observed at low clay loading (1.25-wt%), while the strength continued to improve up to 5-wt%.[146]

Table 13.8 lists values of the parameter a from Equation (13.12). Depending on the system and concentration, its values range from approximately 0.02 to 0.42. The highest values are systematically observed at low clay concentration in low MW PP under mild compounding conditions. By contrast, when dispersing organoclay in a high-MW viscous matrix, better performance is obtained for a higher clay concentration compounded at high stresses. These results support the proposed dual mechanism of clay dispersion—diffusion and peeling. Diffusion (D) might be sufficient in a matrix with relatively high chain mobility, but as MW increases, diffusion slows down with the inverse square of the

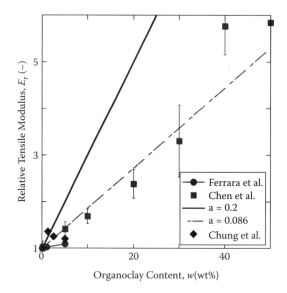

FIGURE 13.24
Relative modulus of PP and its CPNC as reported by Ferrara et al.,[143] Chen et al.,[139] and Chung et al.[146] The straight line represents Equation (13.12) with $a = 0.2$. The chain line is the fit to Chen et al. data ($a = 0.086$).

weight-averaged molecular weight: $D \propto 1/M_w^2$. Gianelli et al.[135] also reported better clay dispersion in a low-MW PP matrix.

In contrast to CPNC behavior with a PA matrix, in the ones based on PP, the modulus enhancement passes through a maximum at relatively low concentration, often at $w \leq 2$-wt% (e.g., see Figure 13.26). When several clays were examined, the synthetic

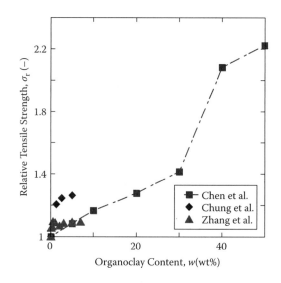

FIGURE 13.25
Relative tensile strength of PP and its CPNC as reported by Chen et al.,[139] Chung et al.,[140] and Zhang et al.[137]

TABLE 13.8

Relative Moduli of PP-based CPNCs

Organoclay	Compatibilizer	Composition (variable)	a [from Equation (13.12)]	Ref.
C15A	PP-1%MA	0, 2, 4 C15A in LCPP/HCPP (blends)	0.08–0.37	18
C20A	PP-0.5%MA	2% C20A + 4% PP-MA (compounding)	0.15–0.30 ± 0.01	126
NC(1)-NC(4)	PP-MA	ca. 5-wt% natural and synthetic (clays)	0.08–0.13	144
Somasif	PP-MA, PP-AA, PE-MA	3.5 organoclay + 2.1–4.3-wt% (compatibilizer)	≤ 0.09	145
MMT, C20A, C30B	PP-MA, PP-AA, PP-GMA	84 PP+ 12 compatibilizer + 4 clay (compatibilizer and clay)	MMT 0.02; C20A 0.09; C30B 0.04	160
MMT-HA[1]	PP-MA	0–2-wt% organoclay (intercalants)	0.19, 0.18	147
MMT-org	PP-MA + epoxy	1–3% clay + 9-27% PP-MA + 5–15% epoxy (composition)	≤ 0.14	148
MMT-ODA/ APTES	PP-MA(0.1% MAH)	PP-MA with 0, 2.5, 5, 7.5, 10% clay (global vs. local); see Figure 14.26.	a(global) ≈ 0.03, a(local) = 0.12-0.33	149
MMT-2M2HT	PP-MA	9 PP-MA, 5 clay (screw speed)	0.08–0.06 for 200–1000 rpm	151
MMT-2M2HT	PP-MA	0–3 PP-MA, 0–1 clay (clay and compatibilizer content for filaments)	For PP-MA = 1% a = 0.27-0.42	152
MMT, ME-100, C20A, N984[2]	PEG, PE-PEG, PEG-PMMA, PEG-PDMS	0–5 MMT, 0–1 compatibilizer (optimize compatibilizer)	a ≤ 0.18 at 5% MMT	153
MMT-2M2HT [3]	PP-MA (0.5%MAH)	PP:PP-MA = 4:1, 4 inorganic clay	a ≤ 0.07 at 4% clay	154
MMT-ODA	PP-MA, POE[11]	6 clay, 6 PP-MA, 0-20 POE	a ≤ 0.04 at 6% clay	155
C15A	PP-1%MA	0–5 clay, PP-MA:clay = 3:1	a ≈ 0.03 at ≤ 5% clay	157

Notes: Abbreviations are listed at the end of this chapter; [1]α,ω-hydroxy amines; [2]Nanofil 984 = MMT-2M2HTA; [3]silane grafted.

fluoromica (Somasif) gave the best result.[150] From between the organo-MMT, C20A (MMT-2M2HTA) engendered the best performance, e.g., the best performance at 4-wt% clay was reported for C20A, compatibilized with PP-0.1%MA (Polybond 3200).[160] The effects of processing and crystallinity were secondary, evident only at low clay contents. However, the orientation of clay platelets (e.g., in fiber spinning or biaxial stretching) had a significant effect on the modulus (see Galgali et al.[161])

13.4.4.3 Impact Properties

Figure 13.27 shows a plot of the tensile modulus vs. *NIRT*. The solid line represents the hyperbola $E = const./NIRT$. The data are for PP and its diverse CPNCs containing up to 4-wt% clay. It is noteworthy that the generally expected hyperbolic relation between

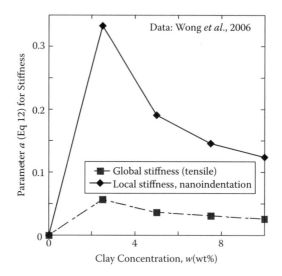

FIGURE 13.26
Parameter a of Equation (13.12), calculated for local and global stiffness in PP-based CPNCs with 0-10-wt% of MMT-ODA/APTES and PP-MA (0.1% MAH). See Wong et al.[149]

stiffness and brittleness is seldom observed within one set of data, where the influence of a specific variable (e.g., processing parameter, type of clay, type of intercalant) was studied. Often, the modulus and impact strength increase with the improved clay dispersion, i.e., with the reduced volume of clay aggregates.[18] Li et al.[126] studied the effect of processing for 2 wt% C20A; all the CPNC data had similar values of $NIRT \approx 30 \pm 1$ (J/m), but the moduli

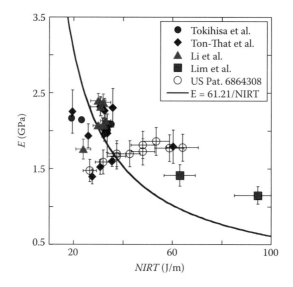

FIGURE 13.27
Tensile modulus vs. notched Izod impact strength at room temperature for PP-based CPNC. Data from Tokihisa et al.,[74] Ton-That et al.,[18] Li et al.,[126] Lim et al.,[155] and Rosenthal and Wolkowicz.[156]

concentrated around two values corresponding to a single-pass method and the master batch-method of compounding, i.e., $E = 2.1$ and 2.3 ± 0.1 GPa, respectively. The Rosenthal and Wolkowicz[156] data are for constant C20A loading of 3-wt% in CPNCs where the type of compatibilizer and its amount varies. Interestingly, the modulus is nearly constant ($E = 1.7 \pm 0.2$ GPa), while *NIRT* ranges from 27 to 64 (J/m). These results are encouraging, as they demonstrate that one might be able to develop CPNCs with a high modulus and good ductility by focusing on clay dispersion for the modulus and compatibilization for ductility.

13.4.5 Mechanical Properties of Epoxy-Based CPNCs

Both previously discussed CPNC systems, those with PA or PP as the matrix, represent thermoplastic nanocomposites with good and poor matrix–clay interactions, respectively. Both polymeric matrices are semicrystalline, with a higher crystallinity encountered in PP, which additionally made clay dispersion difficult. In this chapter it will be neither possible nor desirable to discuss nanocomposites with all possible diverse matrices. However, it would be appropriate to include a discussion of the amorphous matrix, either glassy (vitreous) or elastomeric. The epoxy-type matrices afford such systems; depending on the composition, either one of these two phases might be obtained.

As was the case for thermoplastic CPNCs, here the performance also depends on the composition and methods of preparation.[27] More recent publications suggest that high shear mechanical mixing produced better results than commonly used ultrasonics.[162] However, there was an optimum mixing speed above which the modulus decreased, probably due to clay platelet attrition.[163]

13.4.5.1 Tensile Properties

Wang et al.[164] reviewed preparation of amine-cured epoxy-based CPNCs in Tom Pinnavaia's laboratory. Curing diglycidyl ether of bisphenol-A (DGEBA) with a curing agent of increasing MW (e.g., polyether-di- or tri-amines, i.e., Jeffamine D-230, D-2000 or T-304, T-5000), either a glassy or elastomeric matrix could be obtained with the glass transition temperature varying from $T_g = 82$ to $-40°C$. The complex of preintercalated MMT-ODA or protonated-MMT (MMT-H$^+$) with a curing agent showed quite different interlayer spacing; for low-MW di- or tri-amines $d_{001} = 1.59$ to 1.88 nm, while for high-MW Jeffamines $d_{001} = 4.49$ to 9.78 nm. However, independent of the initial interlayer spacing, the cured epoxies with clay loading $\leq$ 15-wt% were fully exfoliated.[165] Examples of the tensile properties vs. clay content for the elastomeric epoxy CPNCs are displayed in Figure 13.28 and Figure 13.30, and those for the glassy epoxy in Figure 13.29 and Figure 13.31.

The dependence of E vs. clay content in Figure 13.28 shows that even at 15-wt% clay loading, the modulus continues to increase. It is noteworthy that protonated MMT-H$^+$ shows a higher modulus than the preintercalated MMT-ODA—the improvement is even more significant for tensile strength than for the modulus.[166,167]

In Figure 13.29 and Figure 13.31, data for CPNCs with a glassy epoxy matrix are presented from several sources.[168,169,172] It is noteworthy that in the glassy epoxy the modulus ranges from about 3 to 5 GPa, whereas that in the rubbery epoxy (see Figure 13.28 and Figure 13.30) from about 3 to 40 MPa. Thus, the glassy modulus is 2 to 3 orders of magnitude greater than the elastomeric one, and the effect of clay addition to glass is much smaller than that for elastomer.

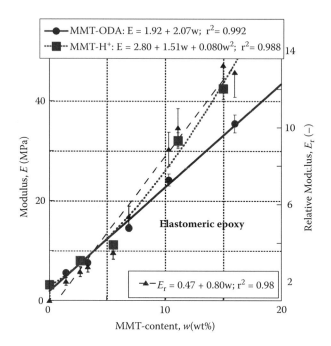

FIGURE 13.28
Tensile modulus vs. inorganic clay content for elastomeric epoxy-based CPNC. Data for epoxy/MMT-ODA from Lan and Pinnavaia.[166] Data for epoxy-MMT-H+ from Pinnavaia and Lan.[167]

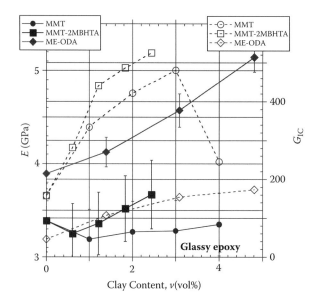

FIGURE 13.29
Tensile modulus and fracture energy vs. clay content for CPNCs in a glassy epoxy. Data for anhydride-cured epoxy with MMT from Zilg et al.[168] Data for ME-ODA from Kornmann et al.[169]

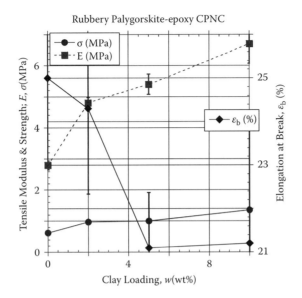

FIGURE 13.30
Tensile modulus, strength, and elongation at break vs. inorganic clay content for elastomeric epoxy-based palygorskite CPNCs. Data from Xue et al.[172]

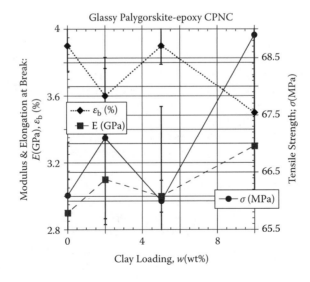

FIGURE 13.31
Tensile properties vs. inorganic clay content for glassy epoxy-based palygorskite CPNCs. Data from Xue et al.[172]

TABLE 13.9

Relative Tensile Strength, Modulus, and Elongation at Break Drop for Palygorskite Epoxy Compounds—Elastomeric and Glassy

Clay (wt%)	σ_r (rubber)	σ_r (glass)	E_r (rubber)	E_r (glass)	$\Delta\varepsilon_b$ (rubber)	$\Delta\varepsilon_b$ (glass)
0	1.00 (13%)	1.00 (32.8%)	1.00 (10%)	1.00 (14%)	0.00 (16%)	0.00 (8%)
2	1.56 (12%)	1.01 (3.5%)	1.71 (9%)	1.07 (16%)	0.0280 (16%)	0.077 (9%)
5	1.61 (14%)	0.998 (3.9%)	1.93 (11%)	1.03 (10%)	0.156 (14%)	0.00 (7%)
10	2.18 (13%)	1.04 (5.4%)	2.39 (14%)	1.14 (12%)	0.152 (21%)	0.103 (10%)

Note: The % error values are in parentheses. Data are from Xue et al.[172]

In Figure 13.29 the modulus with about 5-vol% (or 10-wt%) MMT-2MBHTA increased by $\leq 32\%$, i.e., the parameter of Equation (13.12) is $a = 0.032$.[168] A similar low value, $a \approx 0.029$, might be calculated from the modulus improvement of the high-functionality glassy epoxy with MMT-ODA.[170,171] By contrast, at similar clay loadings in the elastomeric epoxy (i.e., Figure 13.28), the parameter a ranges from 0.80 to 1.1. This enhancement is far better than that observed for glassy epoxy and any increase observed for the thermoplastic, semi-crystalline matrix nanocomposites: at 5-wt% clay loading, the thermoplastics' modulus increased by $\leq 100\%$, glassy epoxy by 15%, and elastomeric epoxy by 400 to 500%.

Additional comparative results of the mechanical CPNC behavior in elastomeric and a glassy epoxy matrix from Xue et al.[172] are shown in Figure 13.30 and Figure 13.31, respectively. It is noteworthy that the tensile strength in the elastomeric matrix increases with nanofiller loading from 0.6 to 1.35 MPa, while that in the glassy matrix is approximately two orders of magnitude larger, and is nearly invariant, ranging from $\sigma = 66$ to 68 (± 2) MPa. The strain at break shows the reverse tendency—while ε_b in the elastomeric matrix decreases with nanofiller loading from 25 to 21% that in the glassy epoxy is an order of magnitude smaller, i.e., $3.7 \pm 0.2\%$. It is also instructive to consider the relative values in Table 13.9. While in the elastomeric matrix the relative tensile strength and modulus increase with clay content, in the glassy system it remains about constant within the experimental error. Since silylation of palygorskite amounts to < 2-wt%, the parameter a in Equation (13.12) is about 0.13 for the elastomeric matrix, and 0.012 for the glassy one. In contrast to strain and modulus, the decrease of the elongation at break, $\Delta\varepsilon_b = [1-\varepsilon_b(\text{CPNC})/\varepsilon_b(\text{epoxy})]$, is about the same, i.e., decreasing by approximately 13% while the error is ± 10 to 21%.

Similarly, Messersmith and Giannelis[173] and Miyagawa et al.[174] report a significant difference in improved storage modulus below/above T_g by, respectively, 57/455% and 50/350%.[27] In these systems the clay concentrations and the test temperatures were different, but the enhancement in elastomeric material compared to that in the vitreous material was similar, i.e., by a factor of 7.9 and 7.0, respectively.

The tensile strength, σ, and dependence on clay content, w, in CPNCs with a glassy epoxy matrix is confusing. The experimental data scatter badly and a tendency to increase or decrease with clay loading has been reported. More consistent relationships were obtained when instead of clay content, the (proportional to it) modulus was used. Figure 13.32 provides an example of such plots with data from six recent publications.[175–181] In Figure 13.32 equations for the least-square straight lines are provided. The data from Fröhlich et al.[181] are divided into two groups: the pre-blends of organoclay with 67% of the three-block toughening copolymer (indicated as Fröhlich-2), and the rest (indicated as Fröhlich-1); good linear dependencies are observed for each set. The data from Ratna et al.[180] refer to the flexural modulus and strength. For the thermoplastic CPNCs it

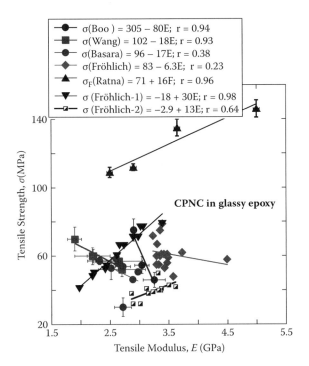

FIGURE 13.32
Tensile strength vs. modulus for CPNCs with a glassy epoxy as the matrix. The data from Ratna et al.[180] are for flexural modulus and strength. See text.

has been observed that while the tensile and flexural moduli are about the same, $E \approx F$, the flexural strength is significantly larger than the tensile, i.e., $\sigma_F \propto 1.73\sigma$.[74] If this empirical observation is also valid for epoxy-based CPNCs, then the data in Figure 13.32 fall within a small rectangle: $\sigma(MPa) \in [30, 85]$; $E(GPa) \in [2, 5]$.

A closer look at the Figure 13.32 data indicates that there is a certain regularity of behavior. The observed dependencies can be grouped in three categories:

1. where the strength increases with modulus,
2. where it decreases, and
3. where σ is nearly independent of E.

The three CPNCs that belong to Category 1 are compatibilized, and here the clay platelets are bonded to the matrix by an intercalant as an intermediary.[180,181]

The nanocomposites containing two α-zirconium phosphates, α-ZrP, with aspect ratio $p = 100$ and 1000 belong to Category 2.[175] The intercalant, 4-butyl ammonium hydroxide, might facilitate dispersion in water, but it does not react with the matrix. The other two systems in this category contain MMT preintercalated with quaternary ammonium ion without compatibilizer.[177,178] Again, the stress transfer from matrix to clay is severely hindered.

The "neutral" system of Category 3, studied by Fröhlich et al.,[179] comprises MMT, HT, fluoromica, and talc. The clays were preintercalated with phenolic imidazoline amines, which could chemically bond with epoxy. The modulus for all CPNCs ranges from 3.2 to

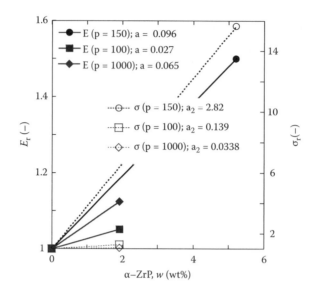

FIGURE 13.33
Relative tensile modulus and strength for PNC with α-ZrP. Data from Sue et al.[184] and Boo et al.[175]

3.6 GPa; hence within the expected experimental error of ± 10% (n.b., the higher modulus values in Figure 13.32 are for the epoxy/talc composites). The tensile strength shows a larger variation, from about 48 MPa for the system with 10-wt% of organo-MMT to 67 MPa for the one comprising 2.5-wt% organo-hectorite. The large quantity of intercalant apparently reduces the epoxy tensile strength (and T_g), but only slightly affects the low strain modulus.

The PNC of epoxy with α-ZrP requires a comment. This relatively recent synthetic layered mineral[182,183] can be prepared as large-aspect-ratio platelets, in situ preintercalated with ammonium ions.[184] Figure 13.33 illustrates early results as the relative tensile modulus and strength vs. the nanofiller content. The data from Sue et al.[184] are for α-ZrP with aspect ratio $p = 100$ to 200 (marked as 150) preintercalated with Jeffamine—a single composition was prepared. The modulus enhancement expressed as the Equation (13.12) gradient is respectable at $a = 0.096$. Even stronger enhancement was obtained for the relative tensile strength, $a_2 = 2.82$. The two other systems with $p = 100$ and 1000 were poorly intercalated intentionally; thus, there is hardly any enhancement for stress. Enhancement for the modulus was also weak: $a = 0.027$ for smaller platelets and $a = 0.065$ for the larger ones ($p = 1000$). As the micrographs showed, all three systems were fully exfoliated.

Boo et al.[176] prepared the epoxy/α-ZrP systems by diverse methods obtaining a full range of dispersion—from poor to exfoliation. At constant loading of 2-vol% of platelets with $p \approx 150$, the modulus increased from the neat value of 2.9 to 3.13 (poor dispersion), to 3.72 (moderate), and to 4.39 GPa (exfoliation). Thus, if enhancement of the modulus is desired, full exfoliation should be attempted. However, for K_{IC}, tensile strength, and elongation at break, fully exfoliated and moderately dispersed systems offered comparable performance.

Another type of synthetic layered nanofiller is the layered double hydroxide (LDH) nanofillers. These materials are relatively simple to prepare in preintercalated form.[53] While their use in PNC is at an early stage, the data are encouraging. However, LDH tends to lose adsorbed moisture at 50°C, with interlayer water loss at about 207°C.[185]

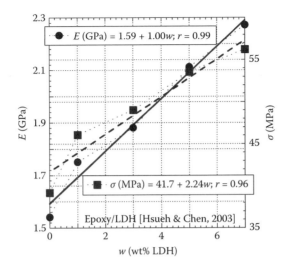

FIGURE 13.34
Tensile modulus and strength for epoxy/LDH PNC.[186]

The preparation and mechanical properties of the epoxy/LDH system were described by Hsueh and Chen.[186] The authors prepared LDH by the co-precipitation method. They incorporated up to 7-wt% of the product into DGEBA/Jeffamine D400 cured at 135°C. As Figure 13.34 shows, the addition resulted in large increases of both modulus and strength. Furthermore, the elongation at break (not shown) increased from 10% for neat epoxy to 15% at 3-wt% LDH, and then decreased to approximately 13% at the highest loading of 7-wt%.

As in the case of polymer blends, CPNCs are increasingly used as replacements for a homopolymer matrix in such multiphase polymeric systems as blends, composites, or foams.[27,187] A recent addition to a growing list of publications in this area comes from Norkhairunnisa et al.[188] The authors prepared (by hand) lay-up GF composites filled with amine-cured DGEBA epoxy-based CPNCs containing up to 5-wt% of Nanomer I.30E (MMT-ODA). The addition of organoclay increased the flexural modulus and strength by approximately 66 and 95%, respectively, at 3-wt% loading. It was reported that MMT-ODA was located at the epoxy–glass fiber interface, improving the GF-epoxy interactions.

13.4.5.2 Fracture Behavior

The epoxy-based CPNCs show improved stiffness, various changes of strength, and reduced elongation at break; in a glassy matrix, the higher the modulus, the lower the toughness. For example, incorporation of approximately 5-wt% organoclay caused reduction of the elongation at break by 50 to 75%.[179,175] It can be argued that because of the size of the clay platelets, the reduction of toughness is an inherent phenomenon in CPNCs; they cannot engender crack bridging and are too small to improve toughness via the matrix yield.[189] Studies of the fracture behavior have been conducted to identify mechanisms that might lead to improved toughness of glassy CPNCs. It has been observed that intercalated CPNCs show small increases in tensile modulus and strength, but the yielding behavior is modified by void formation within clay stacks or aggregates. The fracture energy of the intercalated CPNC may result in a 100% increase at 5-wt% clay loading.[190]

TABLE 13.10

Fracture Behavior of Epoxy CPNCs

Organoclay	Epoxy	Composition	Property	Ref.
I.30E = MMT-ODA	DGEBA, TGAP, TGDDM	0–10-wt% MMT-ODA; amine-cured	E, K_{IC}	170, 171
α-ZrP-TBA[1]	DGEBA amine-cured	0.7-vol% α-ZrP; p = 100 and 1000	E, σ, ε_b, K_{IC}	175
Na-MMT; M2HTA-MMT	DGEBA amine-cured	amine-cured + silane	K_{IC}, G_{IC}	177
Bentonite and ME-100[2] with imidazolineamine	DGEBA anhydride-cured	0–10-wt% organoclay	E, σ, ε_b, K_{IC}, G_{IC}	179
I.30E = MMT-ODA	DGEBA + HBE[3] amine-cured	0 % 15 HBE + 0 and 5 organoclay	F, σ_F, NIRT	180
ME-100 + M2EtOHDDA	DGEBA anhydride-cured	Toughener 0–13.5%; organoclay 0–7.5; 0–15% additives	T_g, E, σ, ε_b, K_{IC}	181
α-ZrP-monoamine[4]	DGEBA amine-cured	1.9-vol% (5.2-wt%) of α-ZrP-monoamine	E, σ, ε_b, K_{IC}	184
Na-MMT and I.28E = MMT-3MODA	DGEBA amine-cured	0–15-wt% organoclay	E, σ, K_{IC}, G_{IC}	190

Notes: [1]α-ZrP-TBA = tetra-butyl ammonium modified zirconium phosphate; [2]ME-100 = Somasif fluoro mica; [3]HBE = hyperbranched (11 branches) epoxy;. [4]monoamine = Jeffamine M715.

The critical stress intensity factor, K_{IC}, and the related [through Equation (13.32)] critical strain energy release rate, G_{IC}, have been determined for epoxy-based CPNCs. A partial summary of the published information is provided in Table 13.10, Figure 13.35, and Figure 13.36.

As shown in Figure 13.35, within the error of measurements, K_{IC} increases with clay loading (here expressed as mineral loading) parallel to the modulus. Such dependence has

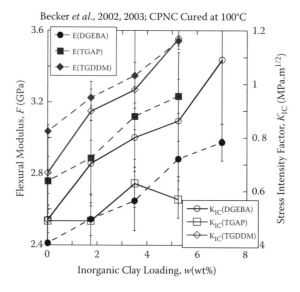

Becker *et al.*, 2002, 2003; CPNC Cured at 100°C

FIGURE 13.35

Flexural modulus and K_{IC} vs. inorganic clay content for three epoxies; see text.[170,171]

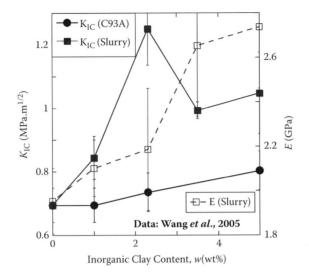

FIGURE 13.36
Tensile modulus and K_{IC} vs. inorganic clay content for DGEBA epoxy with C93A and Na-MMT treated with glycido-silane in a slurry process.[177]

been theoretically predicted. Of the three epoxies (cured at 100°C with *di*-ethyl-toluene *di*-amine, DETDA), the best performance was obtained for *tetra*-glycidyl-*di*-amino-*di*-phenyl methane (TGDDM), the worst for *tri*-glycidyl aminophenol (TGAP). In the latter system, K_{IC} did not show any enhancement. However, judging by the XRD data, both TGDDM and TGAP were intercalated ($d_{001} = 4.6$-5.0 nm) and had similar free volume (determined by the positron annihilation lifetime spectroscopy, PALS), whereas DGEBA, which shows intermediate behavior, was exfoliated and had the largest free volume. The moderate mechanical performance could be caused by the use of a bifunctional curing agent.[170,171]

As stated earlier in this chapter, K_{IC} and G_{IC} show similar dependencies on composition. According to Griffith, the energy release rate in brittle fracture is proportional to the surface energies of the created surfaces, $G_{IC} \propto 2\gamma$. Since the surface tension coefficient is expected to be constant, the increases of K_{IC} or G_{IC} must be related to the total surface area created during fracture; thus, in the studied epoxies, the higher the clay content, the higher K_{IC} or G_{IC} and the surface area. However, the increase of K_{IC} or G_{IC} is not universal for CPNCs, e.g., in system epoxy/α-ZrP, while modulus increases with nanofiller loading, K_{IC} decreases. It is noteworthy that the α-ZrP was well exfoliated, but in the absence of good interaction between epoxy and nanoparticles, the exfoliation was not sufficient for toughening.[175,184]

Wang et al.[177] prepared epoxy-based CPNCs using two procedures: (1) standard mixing epoxy with preintercalated M2HTA-MMT, and (2) mixing following a new slurry process. The latter involved pre-swelling Na-MMT in water, converting the aqueous suspension into an ethanol slurry, treating the swollen clay with glycido-silane, and then dispersing the treated clay in DGEBA and curing. The latter process yielded short stacks of clay platelets uniformly and randomly oriented within the matrix, which resulted in good performance. The clay-content dependence of the tensile modulus and K_{IC} is displayed in Figure 13.36. Extensive scanning electron microscopy (SEM) analysis showed that initiation and development of an extensive network of microcracks originated in short stacks dispersed in the epoxy matrix. A similar behavior (see Figure 13.29) was reported by Kornmann et al.[169]

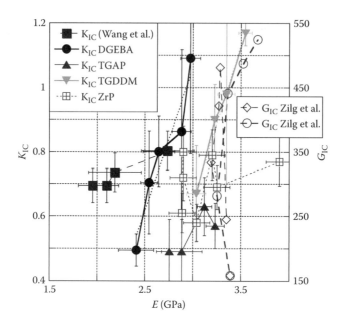

FIGURE 13.37

K_{IC} or G_{IC} vs. the modulus for an epoxy matrix with clay or α-ZrP as the nanofiller. The highest modulus was obtained for fully exfoliated α-ZrP having the aspect ratio $p = 100 - 200$.[184] The data from Wang et al.[177] are for intercalated CPNC with Na-MMT modified in the slurry process. The remaining data are for CPNC with three epoxies from Becker et al.[170,171]

These observations are also in agreement with the report by Fröhlich et al., [179] who noted that intercalated clay in anhydride-cured epoxy only slightly increased tensile modulus (i.e., by 3 to 10% at 7.5% inorganic clay loading), but significantly improved the toughness, increasing K_{IC} by up to 73%. When both stiffness and toughness are required, exfoliated clay and deformable dispersed microdomains of an elastomer might provide the desired performance, as suggested by the work by Ratna et al.[180]

Figure 13.37 demonstrates the relationships between K_{IC} or G_{IC} and the tensile modulus. The data are from the discussed polymeric nanocomposites, PNC, containing either clay or α-ZrP as the nanofiller.[170,171,177,184] It is noteworthy that the preliminary results for PNC with α-ZrP show that this nonoptimized system offers mechanical performance comparable to that obtained for clays. The dependencies seem to indicate an approximate linear increase of K_{IC} or G_{IC} with E, i.e., slope for DGEBA and TGDDM are 0.9 and 1.0, respectively. Evidently there are many departures from this theoretically expected correlation.

13.5 Summary and Generalization

This chapter examined the mechanical behavior of nano-reinforced polymers and attempted to trace the important influences. Three polymeric matrices were discussed: PA, PO, and EP thermosets. Natural or synthetic clays are the principal nanofillers, although zirconium phosphates (ZrP) and LDH were also discussed.

Because of the strong polarity of macromolecules, the PA-based CPNCs are the oldest and the easiest to produce. Exfoliation has been achieved for systems prepared by the reactive or melt-compounding method. Linear increase of the modulus with clay content up to approximately 10-wt% was obtained, but the elongation at break tended to peak at approximately 3-wt%. Depending on the type of system, the tensile strength was observed to increase along the second-order curve at least up to 6-wt%. Creep and fatigue tests have also been carried out; the internal energy change per unit crack length was found to be lower for CPNCs than for neat resins. The notched Izod impact strength at room temperature (*NIRT*) usually decreased with clay loading. Toughened PA-based CPNCs have been prepared either by modifying the organoclay or by dispersing an elastomer into the matrix. Evidently there is a trade-off—higher *NIRT* for reduced modulus.

While in PA-based CPNCs, as expected, the clay exfoliation and concentration affect the mechanical performance, the effect of the aspect ratio, p, remains uncertain. Most systems contain natural MMT with $p = 100$ to 300. By contrast, synthetic fluoromica is reported to have $p < 6000$. Theoretically, the modulus should increase with p nearly as fast as with concentration, but this has not been observed. The reason might be the fragility and/or low modulus of the synthetic platelets, which either bend or break during compounding; both were observed.

Preparation of the PP-based CPNCs is more complex and interpretation of the mechanical behavior is more difficult. Owing to immiscibility of PP with clays or organoclays, incorporation of at least one compatibilizer is necessary. Furthermore, the resin is highly crystalline and it concentrates clay platelets in noncrystalline domains, reducing the interlayer spacing. It is important to recognize that the relative modulus enhancement in PP-based CPNCs can be as high as that in PA-based systems, but while in the latter the increase is linear up to high clay content, in the former it is limited to organoclay loading of less than 3-wt%. Thus, the result for good enhancement of a PP-based CPNC modulus is low clay concentration, low polymer viscosity, optimized compatibilizer, and long residence time in a compounder. There is a general tendency of *NIRT* to increase with $1/E$. However, the data demonstrate that while the modulus depends on the clay content, *NIRT* is affected primarily by intercalants and compatibilizers. Thus, the desired rigid and tough CPNCs might be produced.

There were two reasons for discussing epoxy-based CPNCs: to examine the mechanical behavior improvement in amorphous nanocomposites, and to contrast this with the behavior of vitreous systems. Indeed, the enhancement observed for CPNCs with an elastomeric matrix was far better than the one seen in semicrystalline thermoplastic matrix nanocomposites, while the glassy epoxy behaved similarly to thermoplastic vitreous polymers such as PS. The highest observed values of CPNC relative modulus with 5-wt% clay (inorganic) are: $E_r = 1.15$, 2.0, and 5.5 for glassy, semicrystalline thermoplastic, and elastomeric matrices, respectively. In other words, at 5-wt% clay loading, the matrix modulus increased by up to 15, 100, and 450%, respectively. These differences of behavior are large and have many consequences.

To understand these increases of E one can ask two questions: (1) Is there a limiting value of E for polymeric materials, and (2) is the modulus and its enhancement related to the accessible matrix free volume?

Fundamentally, the ultimate modulus of a material is related to the interatomic bond strength and distance. Covalent bonds between C-C atoms are the strongest, leading to the highest modulus in diamonds.[191] Polymers have covalently bonded chains, but deformation affects not the backbone, but the intermacromolecular interactions. For van der Waals bonded systems, the molecular dynamics computations predict the modulus values $E_{max} \approx$ 4 – 5 GPa, comparable to those listed in this chapter. The E_{max} could be increased by replacing the van der Waals intermolecular bonds with stronger ones, i.e., hydrogen bonding, polar, or ionic. This requires chemical modification of the macromolecules, i.e., replacing weakly bonded polymers with strongly bonded ones, for example, PO by engineering or specialty polymer.[192,193]

There is a direct relation between the modulus and the compressibility parameter, κ:[194]

$$B = E/(3-6v_p) = 1/\kappa; \quad \kappa = -[\partial \ln V/\partial P]_T \quad\quad (13.32)$$

where B is the bulk modulus. The compressibility parameter depends on T and P through the hole fraction—a free volume quantity, $h = h(T, P)$. At ambient pressure, the Simha-Somcynsky (S-S) equation of state (eos) predicts that:[133,195]

$$\tilde{\kappa}_{P=0} = 0.202 + 1.80\tilde{h} + 6.42\tilde{h}^2 ; r^2 = 0.999 \quad\quad (13.33)$$

where the tilde indicates reduced variables.

Figure 13.38 displays the relation between the parameter a of Equation (13.12), and the effective hole fraction in the matrix, h_{eff}, defined as the free volume accessible to clay in the system. The latter parameter was computed from the matrix pressure-volume-temperature (PVT) data using the S-S eos.[133] Thus, for the molten or rubbery polymers it is numerically equal to the fraction computed from the eos: $h_{eff} = h$; for glasses h_{eff} is given by the nonfrozen fraction of holes; and for the semicrystalline polymers, h_{eff} corresponds to the portion associated with the noncrystalline phase.

The parameter a is a measure of the matrix modulus increase after incorporating 1-wt% of inorganic clay; selected values are the highest observed for the given type of CPNC. The

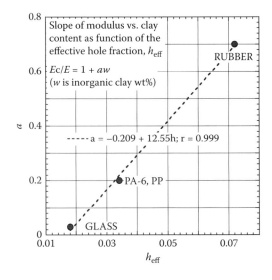

FIGURE 13.38
The correlation between the slope of the relative modulus vs. (inorganic) clay content a [see Equation (13.12)] and the effective hole fraction, h_{eff}, at $T = 300\ K$ and $P = 1$ bar.

graph suggests that the maximum increase in stiffness depends on the matrix; it is easy to increase E of elastomers, but difficult to do that for glassy polymers. As Figure 13.28 through Figure 13.31 show, E of elastomeric epoxy ranges from 2 to 43 MPa, whereas that of vitreous ones ranges from about 3 to 5 GPa, the latter values being close to the upper limit predicted for the van der Waals bonded systems.

One may ask whether it is possible to increase the elastic modulus of an elastomer all the way to the maximum value predicted for the van der Waals bonded system, i.e., if it would be possible to obtain $E_r \approx 2{,}000$. The Halpin-Tsai Equation (13.3) with the platelet modulus $E_f = 170$ GPa, and the aspect ratio of commercial MMT ($p = 295$)[17] predicts that the required volume fraction of a filler would need to be as high as $\phi_f = 0.777$; evidently, this value (required for increasing the matrix modulus by a factor of 2000) is unrealistic. However, if platelets with $p = 2500$ are used, the required clay loading decreases to $\phi_f = 0.302$.

In those calculations it was assumed that independent of the clay loading, the matrix remains elastomeric. However, since the glass transition temperature might increase with clay content, the $E_r \approx 2000$ might be obtained at significantly lower concentrations.

The correlation shown in Figure 13.38 illustrates the importance of molecular mobility on CPNC stiffness. Note that the considered modulus enhancements in these systems were the highest, corresponding to the maximum degree of clay dispersion, e.g., PA-6 and elastomer-based CPNCs were fully exfoliated. Thus, there is a coupled interaction between the molecular mobility and clay dispersions: mobility facilitates the dispersion, and the dispersed clay platelets influence the molecular mobility. The reduction is strong in the elastomeric system, moderate in the thermoplastic matrices, and negligible in the glassy ones. Adsorption and solidification of organic molecules on clay surfaces were found to depend on the clay content and matrix type. Incorporation of 2-wt% of organoclay in PS and in PA-6 resulted in the reduction of free volume by 4 and 15%, respectively.[10] Consequently, an even larger reduction must be expected for the elastomeric matrix. These observations might suggest that upon addition of clay, its influence on the matrix mobility is felt further from the clay surface in highly mobile systems (e.g., melts or elastomers) than in a low-mobility system (e.g., PS or vitreous epoxy).

The data presented in this chapter indicate that the principal mechanism for reinforcement by nanoparticles is solidification of the organic phase on crystalline nanofiller particles, which results in reduction of free volume, and increased stiffness and brittleness. Thus, as a result: (1) the highest degree of rigidity enhancement is obtained for full exfoliation, and (2) the efficiency of nanofiller particles is the highest for elastomeric systems and the poorest for vitreous ones. One may ask whether the use of micro- or even macrofillers with the latter matrix would be more appropriate. A recent review book clearly shows that this is not the case: On the graphs of impact energy vs. E_r, the data for microcomposites are systematically lower than those for PNC, and in addition, the weight factor makes such an alternative unattractive.[196]

Finally, one may wonder about the future research of PNC. There is a growing discontent with natural clays and their ammonium intercalants; they are immiscible with almost all polymers, are thermally unstable and toxic. This forces a search for alternatives—synthetic inorganic nanofillers, which on the one hand can be preintercalated during synthesis (e.g., LDH or ZrP), or on the other, may not require intercalation (e.g., sepiolite).[27,53] Before these materials will be adopted by the industry, much more must be known about their different aspects. One would like to know their mechanical and surface properties, i.e., strength, modulus, surface tension, charge density, etc. There are also a number of questions regarding the synthesis and optimization of the product and its cost of production. There is little doubt that interest in these materials is growing as is the demand for them.

13.6 Appendices

13.6.1 Abbreviations

AA	acrylic acid
ADA	ω-amino dodecyl (or lauric) acid, or 12-amino dodecyl acid
AEC	anion exchange capacity (meq/g)
AIBN	*N,N'*-azo-bis(iso-butyro nitrile)
ASTM	American Society for Testing and Materials
B	bulk modulus
BT	bentonite
C10A	Cloisite® 10A (MMT-2MBHTA)
C15A	Cloisite® 15A (MMT-2M2HTA)
C20A	Cloisite® 20A (MMT-2M2HTA)
C25A	Cloisite® 25A (MMT-2MHTL8)
C30B	Cloisite® 30B (MMT-MT2EtOH)
CB	carbon black
CEC	cation exchange capacity (meq/g)
CNT	carbon nanotubes
CPNC	clay-containing polymeric nanocomposite
Cs	NMR chemical shift
CTE	coefficient of thermal expansion
D	diffusion coefficient
DAAM	di-acetone acrylamide
DBS	dodecyl-benzene-sulfate
DDA	dodecyl ammonium chloride
DDS	4,4'-di-amino diphenyl sulfone
DETDA	di-ethyl-toluene di-amine
DGEBA	diglycidyl ether of bisphenol-A
DGEBE	diglycidyl ether of bisphenol-E
DTG	differential thermogravimetric analysis
ECAE	equal channel angular extrusion
EFM	extensional flow mixer
eos	equation of state
EP	epoxy
EWF	essential work of fracture
FCP	fatigue crack initiation and propagation mechanism
FEM	finite element method
FH	fluorohectorite
FM	fluoromica (e.g., Somasif ME-100)
FRP	fiber-reinforced polymer
FT-IR	Fourier transform infrared spectroscopy

GF	glass fibers
GMA	glycidyl methacrylate
HBE	hyperbranched (11 branches) epoxy
HDT	heat deflection temperature (*under load*)
HDTMOS	hexadecyl trimethoxy silane
H-T	Halpin and Tsai (model or equation)
HT	hectorite
I.30TC	Nanomer: MMT with ODA
I.34TCN	Nanomer: MMT with MOD2EtOH
ISO	International Organization for Standardization
IT	itaconic acid
LDH	layered double hydroxide
MAH	maleic anhydride
MAS	Magic Angle Spinning NMR spectroscopy
MB	master batch
MC	mica
MC	Monte Carlo computational method
MD	molecular dynamic computational method
MD	orientation in the stress (machine) direction
ME-100	Somasif ME-100 = synthetic fluoromica
MEE	Somasif ME-100 intercalated with MCoco2EtOH
MFR	melt flow rate
MMT	montmorillonite
MMT-H$^+$	protonated-MMT
MSDS	material safety data sheets
MW	molecular weight (unspecified average)
Na-FH	sodium-FH
Na-MMT	sodium-MMT
Nanofil 919	MMT with 2MBODA
Nanofil 984	MMT with 2M2HTA
ND	orientation normal to the stress (machine) direction
NI	notched Izod impact strength at other than room temperature
NIRT	notched Izod impact strength at room temperature
NMR	nuclear magnetic resonance
PA1015B	PA-6 from Ube (M_w = 21.7 kg/mol, density = 1.14 g/mL)
PA1015BC2	PA-6 based CPNC from Ube, containing approximately 2-wt% MMT-ADA
PALS	positron annihilation lifetime spectroscopy
PNC	polymeric nanocomposites
PTES	phenyl-tri-ethoxy silane
RT	room temperature
RVE	representative volume element
SCF	self-consistent field theory
SEM	scanning electron microscopy
S-N	diagram stress vs. number of cycles to failure in fatigue tests
SSE	single-screw extruder
TD	orientation transverse to the stress (machine) direction
TEM	transmission electron microscopy
TEOS	tetraethyl orthosilicate

TFF	thermal fatigue failure
TGA	thermogravimetric analysis
TGAP	tri-glycidyl aminophenol
TGDDM	tetra-glycidyl 4,4'-di-amino-diphenyl methane
TSE	twin screw extruder
VAMAS	Versailles Project on Advanced Materials and Standards
XRD	X-ray diffraction
α-ZrP (or ZrP)	alpha-zirconium phosphate; a synthetic layered mineral

13.6.2 Abbreviations for Intercalants

2M2DDA	di-methyl di-dodecyl ammonium chloride
2M2EtOH	di-methyl di ethoxy ammonium chloride
2M2HTA	di-methyl di-(hydrogenated tallow) ammonium chloride
2M2ODA	di methyl-di-octadecyl ammonium chloride
2M2TA	di-methyl di tallow ammonium chloride
2MBHTA	di methyl-benzyl-hexadecyl ammonium chloride
2MBODA	di methyl-benzyl-octadecyl ammonium chloride
2MBSA	di methyl-benzyl-stearyl ammonium chloride (2MBODA)
2MHDI	1,2-di-methyl-3-n-hexadecyl imidazolium bromide
2MHDODA	di methyl-hexadecyl-octadecyl ammonium chloride
2MHTL8	di-methyl hydrogenated tallow 2-ethylhexyl ammonium methyl sulfate
3BHDP	tributyl-hexadecyl phosphonium
3MDDA	tri-methyl dodecyl ammonium bromide
3MHDA	tri-methyl hexadecyl ammonium bromide
3MODA	tri-methyl octadecyl ammonium chloride
ADA	ω-amino-dodecanoic acid
APTES	γ-aminopropyl 3-etoxy silane
ATPS	amine-terminated PS
AUDA	11-aminoundecanoic acid
CPCl	cetyl-pyridinum chloride
DA	di-amido-dodecane
DDA	dodecyl ammonium chloride
HEODI	hydroxy-ethyl-di-hydro-imidazolinium chloride
M2EPG	methyl di-ethyl polypropylene glycol
M2EtOHODA	methyl *di*-ethoxy octadecyl ammonium chloride
MCoco2EG	methyl coco di polyethylene glycol
MCoco2EtOH	methyl coco *bis*-ethoxy ammonium chloride
MDD2EG	methyl dodecyl di polyethylene glycol
MHRODA	methyl hydrogenated rapeseed octadecyl ammonium chloride
MHT2EtOH	methyl hydrogenated tallow *bis*-ethoxy ammonium chloride
MHTODA	methyl hydrogenated tallow octadecyl ammonium chloride
MOD2EtOH	methyl octadecyl *bis*-ethoxy ammonium chloride
MT2EtOH	methyl tallow di-2-hydroxy ethyl ammonium chloride (Ethoquad T12)
ODA	octadecylamine
PEA	2-phenylethylamine
RDI	ricinyl di-hydro-imidazolinium chloride

TAPTMS	3-amino propyl 3-methyl silane
TBA	tetra-butyl ammonium chloride
TGDDM	tetra-glycidyl 4,4'-di-amino-diphenyl methane
W75	1-methyl-2-norstearyl-3-stearinoacid-amidoethyl-dihydro-imidazo-linium methyl-sulfate

13.6.3 Polymer Abbreviations

ABS	acrylonitrile-butadiene-styrene copolymer
E-MA-GMA	ethylene-methacrylate-glycidyl methacrylate copolymer
EP	epoxy
EPR	ethylene-propylene rubber
EPR-MA	maleated ethylene-propylene rubber
HDPE	high-density polyethylene (approximately 960 kg/m³)
HIPS	high-impact polystyrene
LDPE	low-density polyethylene (approximately 918 kg/m³)
LLDPE	linear low-density polyethylene
MXD6	semi-aromatic polyamide from *m*-xylylene *di*-amine and adipic acid
PA-12	polyamide-12
PA-6	poly-ε-caprolactam
PA-66	polyhexamethylene-adipamide
PAA	polyacrylic acid
PBT	polybutylene terephthalate
PC	bisphenol-A polycarbonate
PEG	polyethylene glycol
PE-GMA	polyethylene grafted with glycidyl methacrylate
PET	polyethylene terephthalate
PI	polyimide
PLA	polylactic acid
PMMA	polymethylmethacrylate
PO	polyolefin
POE	metallocene polyethylene-*co*-octene
PP	polypropylene
PP-*b*-PMMA	block copolymer of PP and PMMA
PP-MA	PP grafted with maleic anhydride (MAH)
PP-MS	PP copolymer with *p*-methyl-styrene
PP-OH	-OH modified PP copolymer with *p*-methyl-styrene
PS	polystyrene
PU	polyurethane
PVAl	polyvinyl alcohol
PVC	polyvinyl chloride
SBR	styrene-butadiene rubber
SEBS	styrene-ethylene/butylene-styrene tri-block polymer
SEBS-*g*-MA	SEBS grafted with maleic anhydride

13.6.4 Symbols

$\partial l / \partial N$	the crack growth rate in fatigue tests
a	parameter in Equation (13.12) and Equation (13.29)

a_i	equation parameters
A_{sp}	specific surface area (m^2/g)
c_{clay}, c_{cr}	clay concentration and its critical value
d	particle diameter
d_{001}, d_{001}^0	clay interlayer spacing and its value extrapolated to zero concentration (nm)
E, E_c	tensile (Young's) modulus, modulus of composite
$E_r = E_c/E_m$	relative tensile modulus
$E_R = E_f/E_m$	ratio of the filler-to-matrix tensile modulus
F	flexural modulus
F_R	filler strength reducing factor
G_{1c}	critical strain energy release rate (a measure of toughness)
G_m	matrix shear modulus
H	crack length
J	creep compliance
K_{1c}	critical stress intensity factor under plane-stress in Mode I
L	crack tip radius
L_a	clay platelet size (nm)
M_n, M_w	number- and weight-average molecular weight
N	number of clay platelets in a stack
N	number of cycles in fatigue test
$p = d/t$	aspect ratio (d is platelet diameter)
r	correlation coefficient
r_g	macromolecular radius of gyration
$\langle r_g^2 \rangle_o^{1/2}$	unperturbed radius of gyration
s	interlamellar gallery spacing, $s = d_{001} - platelet\ thickness$
t	clay platelet thickness (in MMT $t = 0.96$ nm)
T	temperature
t, t_b	time, time to break
T_c	crystallization temperature
T_g	glass transition temperature
T_m	melting point
$v = v\{T, \sigma, ta\}$	material variables
V, V_{enc}	volume and encompassed volume
w_{ini}	crack initiation work of fracture
$X(\%)$	crystallinity content
Y	Young's modulus
$\dot{\varepsilon}$	strain rate
$[\eta]$	intrinsic viscosity, a measure of hydrodynamic volume
$\varepsilon = \Delta L/L_0$	strain (L_0 is the initial length of a body and ΔL is its increase in length)
ΔG_m, ΔG_{AB}	free energy of mixing, and its value for the A-B system
ΔT_d	change of the degradation temperature due to clay presence
Δt_{max}	delay time to reach the peak temperature in flammability tests
α, β, φ, λ	idealized dimensions of the interphase and filler regions in Equation (13.14) through Equation (13.16)
ε_0	instantaneous elastic deformation
ε_b	elongation (or strain) at break

ϕ, ϕ_m	volume fraction and its maximum packing value
ϕ_f, ϕ_m	filler and matrix volume fraction
η_r	relative viscosity
κ	compressibility parameter
ν	frequency in fatigue tests
ν_P, ν_m	Poisson's ratio and its matrix value
σ	tensile strength
$\sigma_a = \sigma_{max} - \sigma_{min}$	stress amplitude in fatigue tests
σ_F	flexural strength
σ_f, σ_m, σ_c	filler, matrix, and composite tensile strength, respectively
σ_i	interfacial shear stress
ξ	thickness of adsorbed polymer in multiples of the clay platelet thickness, t

References

1. Ube Industries, Ltd., Nylon Resin Dept., *High performance nylon resin—technical data* (Molding applications), March 2000.
2. Tjong, S.C., Meng Y.Z., and Hay A.S., Novel preparation and properties of polypropylene-vermiculite Nanocomposites, *Chem. Mater.*, 14, 44–51, 2002.
3. Lan, T., Psihogios V., Bagrodia S., Germinario, L.T., and Gilmer J.W., Layered clay intercalates and exfoliates having low quartz content, US Pat. Appl., 0028870, 2002b.03.07, Appl. 2001.06.29, to AMCOL International Corp.
4. Kim K., Utracki L.A., and Kamal M.R., Numerical simulation of polymer nanocomposites using a self-consistent mean-field model *J. Chem. Phys.* 121, 10766–10777, 2004.
5. Simha, R. and Somcynsky, T., On the statistical thermodynamics of spherical and chain molecule fluids *Macromolecules* 2, 342–350, 1969; Somcynsky, T. and Simha, R., *J. Appl. Phys.*, 42, 4545–4548, 1971.
6. Simha, R., Jain, R.K., and Jain, S.C., Bulk modulus and thermal expansivity of melt polymer composites: statistical versus macro–mechanics, *Polym. Compos.*, 5, 3–10, 1984.
7. Papazoglou, E., Simha R., and Maurer, F.H.J., Thermal expansivity of particulate composites: interlayer versus molecular model, *Rheol. Acta.*, 28, 302–308, 1989.
8. Simha, R., Utracki, L.A., and Garcia-Rejon, A., Pressure-volume-temperature relations of a polycaprolactam and its nanocomposite, *Composite Interfaces*, 8, 345–353, 2001.
9. Utracki, L.A., Simha, R., and Garcia-Rejon, A., Pressure-volume-temperature relations in nanocomposites, *Macromolecules*, 36, 2114–2121, 2003.
10. Utracki, L.A. and Simha, R., Pressure-volume-temperature dependence of polypropylene/organoclay nanocomposites, *Macromolecules*, 37, 10123–10133, 2004.
11. Utracki, L.A. and Simha, R., Statistical thermodynamics predictions of the solubility parameter, *Polym. Int.*, 53, 279–286, 2004.
12. Israelachvili, J.N., *Intermolecular and surface forces with applications to colloidal and biological systems*, Academic Press, New York, 1985.
13. Luengo, G., Schmitt, F.-J., Hill, R., and Israelachvili, J.N., Thin film rheology and tribology of confined polymer melts: contrasts with bulk properties, *Macromolecules* 30, 2482–2494, 1997.
14. Hentschke, R., Molecular modeling of adsorption and ordering at solid interfaces, *Macromol. Theory Simul.* 6, 287–316, 1997.

15. Rao, Y.Q. and Pochan, J.M., Mechanics of polymer-clay nanocomposites, *Macromolecules*, 40, 290–296, 2007.
16. Hofmann, A.W., Beiträge zur Kenntnifs der flüchtigen organischen Basen; X. Uebergang der flüchtigen Basen in eine Reihe nichtflüchtiger Alkaloïde, *Liebigs Ann. Chem.* 78, 253–286, 1851.
17. Utracki, L.A., Sepehr, M., and Li, J. Melt compounding of polymeric nanocomposites, *Inter. Polym. Process.* 21, 1–14, 2006.
18. Ton-That, M.-T., Leelapornpisit, W., Utracki, L.A., Perrin-Sarazin, F., Denault, J., Cole, K.C., and Bureau, M.N., Effect of crystallization on intercalation of clay-polyolefin nanocomposites and their performance, *Polym. Eng. Sci.* 46, 1085–1093, 2006.
19. Utracki, L.A., Ed., *Polymer blends handbook*, Kluwer, Dordrecht, 2002.
20. Lipatov, Y.S., *Polymer reinforcement*, ChemTec Pub., Toronto, 1995.
21. Agarwal, B.D. and Broutman, L.J., *Analysis and performance of fiber composites*, John Wiley & Sons, New York, 1980.
22. Halpin, J.C., *Primer on composite materials analysis*, 2nd ed., Technomic Publ. Co., Lancaster, PA, 1992.
23. Hancock, M., *Particulate-filled polymer composites*, Longman, Harlow, U.K., 1995.
24. Hui, C.Y. and Shia, D., Simple formulae for the effective moduli of unidirectionally aligned composites, *Polym. Eng. Sci.*, 38, 774–82, 1998.
25. Tucker, C.L., III and Liang, E., Stiffness predictions for unidirectional short-fiber composites: review and evaluation, *Compos. Sci. Technol.*, 59, 655–671, 1999.
26. Colombini, D., *Apports de la modelisation mecanique a l'etude de matériaux polymeres multiphases*, PhD thesis INSA, Lyon, France, 1999.
27. Utracki, L.A., *Clay-containing polymeric nanocomposites*, RAPRA, Shawbury, U.K., 2004.
28. Young, R.J. and Eichhorn, S.J., Deformation mechanisms in polymer fibres and nanocomposites, *Polymer*, 48, 2–18, 2007.
29. Goldsmith, H.L. and Mason, S.G., in *Rheology theory and application*, Eirich, F.R., Ed., Academic Press, New York, 1967.
30. Mori, T. and Tanaka, K., Average stress in the matrix and average elastic energy of materials with misfitting inclusions, *Acta Metall. Mater.*, 21, 571–574, 1973.
31. Brune, D.A. and Bicerano, J., Micromechanics of nanocomposites, *Polymer*, 43, 369–387, 2002.
32. Chen, B. and Evans, J.R.G., Elastic moduli of clay platelets, *Scripta Mater.*, 54, 1581–1585, 2006.
33. Yung, K.C., Wang, J., and Yue, T.M., Modeling Young's modulus of polymer-layered silicate nanocomposites using a modified Halpin-Tsai micromechanical model, *J. Reinforced Plast. Compos.*, 25, 847–861, 2006.
34. Chen, B. and Evans, J.R.G., Nominal and effective volume fractions in polymer-clay nanocomposites, *Macromolecules*, 39, 1790–1796, 2006.
35. Zilg, C., Thomann, R., Mülhaupt, R., and Finter, J., Polyurethane nanocomposites containing laminated anisotropic nanoparticles derived from organophilic layered silicates, *Adv. Mat.*, 11, 49–52, 1999.
36. Hui, C.Y. and Shia, D., Simple formulae for the effective moduli of unidirectionally aligned composites, *Polym. Eng. Sci.*, 38, 774–782, 1998.
37. Shia, D., Hui, C.Y., Burnside, S.D., and Giannelis, E.P., An interface model for the prediction of Young's modulus of layered silicate elastomer nanocomposites, *Polym. Compos.*, 19, 608–617, 1998.
38. Ji, X.L., Jing, J.K., Jiang W., and Jiang, B.Z., Tensile modulus of polymer nanocomposites, *Polym. Eng. Sci.*, 42, 983–993, 2002.
39. McCrum, N.G., Buckley, C.P., and Bucknall, C.B., *Principles of polymer engineering*, Oxford Sci. Pub., Oxford, 1988.
40. Milewski, J.V. and Katz, H.S., Eds., *Handbook of reinforcements for plastics*, Van Nostrand Reinhold Co., New York, 1987.
41. Clegg, D.W. and Collyer, A.A., Eds., *Mechanical properties of reinforced thermoplastics*, Elsevier Appl. Sci. Pub., London, 1986.

42. Turcsáyi, B., Pukánszky, B., and Tüdös, F., Composition dependence of tensile yield stress in filled polymers, *J. Mater. Sci. Lett.*, 7, 160–162, 1988.

43. Lee, S.-R., Park, H.-M., Lim, H., Kang, T., Li, X., Cho, W.-J., and Ha, C.-S., Microstructure, tensile properties and biodegradability of aliphatic polyester/clay nanocomposites, *Polymer*, 43, 2495–2500, 2002.

44. Struik, L.C.E., *Physical aging in amorphous polymers and other materials*, Elsevier, Amsterdam, 1978.

45. Duncan, B.C. and Tomlins, P.E., Measurement of strain in bulk adhesive test pieces, *National Physical Laboratory Report* DMM(B) 398, 1994; Tomlins, P.E., Comparison of different functions for modeling the creep and physical ageing effects in plastics, *Polymer* 37, 3907–3913, 1996.

46. Jazouli S., Luo, W., Bremand, F,. and Vu-Khanh, T., Application of time–stress equivalence to nonlinear creep of polycarbonate, *Polym. Testing*, 24, 463–467, 2005.

47. Scott, D.W., Lai, J.S., and Zureick, A.-H., Creep behavior of fiber-reinforced polymeric composites: a review of the technical literature, *J. Reinforced Plast. Compos.*, 14, 588–617, 1995.

48. Van der Giessen, E. and Needleman, A., Micromechanics simulations of fracture, *Annul. Rev. Mater. Res.*, 32, 141–162, 2002.

49. Moet, A., Fatigue failure, in *Failure of plastics*, Brostow, W. and Corneliussen, R.D., Eds., Hanser Pub., Munich, 1986.

50. Crawford, R.J. and Benham, P.P., Some fatigue characteristics of thermoplastics, *Polymer*, 16, 908–914, 1975.

51. Nguyen, O., Repetto, A., Ortiz, M., and Radovitzky, R.A., A cohesive model of fatigue crack growth, *Int. J. Fract.* 110, 351–369, 2001.

52. Bureau, M.N., Denault, J., Cole, K.C., and Enright, G.D., The role of crystallinity and reinforcement in the mechanical behavior of polyamide-6/clay nanocomposites, *Polym. Eng. Sci.*, 42, 1897–1906, 2002.

53. Utracki, L.A., Sepehr, M., and Boccaleri, E., Synthetic, layered nano-particles for polymeric nanocomposites (PNCs), *Polym. Adv. Technol.*, 18, 1–37, 2007.

54. Binder, K., Ed., *Monte Carlo and molecular dynamics simulations in polymer science*, Oxford University Press, New York, 1995.

55. Hackett, E., Manias, E., and Giannelis, E.P., Molecular dynamics simulation of organically modified layered silicates, *J. Chem. Phys.*, 108, 7410–7415, 1998.

56. Tanaka, G. and Goettler, L.A., Predicting the binding energy for nylon 6,6/clay nanocomposites by molecular modeling, *Polymer*, 43, 541–553, 2002.

57. Brown, D., Mélé, P., Marceau, S., and Albérola, N.D., A molecular dynamics study of a model nanoparticle embedded in a polymer matrix, *Macromolecules*, 36, 1395–1406, 2003.

58. Vacatello, M., Monte Carlo simulations of polymer melts filled with solid nanoparticles, *Macromolecules*, 34, 1946–1952, 2001.

59. Flory, P.J., Yoon, D.Y., and Dill, K.A., The interphase in lamellar semicrystalline polymers, *Macromolecules*, 17, 862–868, 1984.

60. Sen, S., Thomin, J.D., Kumar, S.K., and Keblinski, P., Molecular underpinnings of the mechanical reinforcement in polymer nanocomposites, *Macromolecules*, 40, 4059–4067, 2007.

61. Manias, E., and Kuppa, V., Molecular simulations of ultra-confined polymers. Polystyrene intercalated in layered-silicates, in *Polymer nanocomposites*, Vaia, R. and Krishnamoorti, R., Eds., ACS Symp. Ser., Chapter 15, 2002.

62. Odegard, G.M., Gates, T.S., Nicholson, L.M., and Wise, K.E., Equivalent-continuum modeling with application to carbon nanotubes, *NASA/TM-2002-211454*; Equivalent continuum modeling of nano-structured materials, *Comp. Sci. Tech.*, 62, 1869–1880, 2002.

63. Odegard, G.M., Clancy, T.C., and Gates, T.S., Modeling of the mechanical properties of nanoparticle/polymer composites, *Polymer*, 46, 553–562, 2005.

64. Sheng, N., Boyce, M.C., Parks, D.M., Rutledge, G.C., Abes, J.I., and Cohen, R.E., Multiscale micromechanical modeling of polymer/clay nanocomposites and the effective clay particle, *Polymer*, 45, 487–506, 2004.

65. Hbaieb, K., Wang, Q.X., Chia, Y.H.I., and Cotterell, B., Modelling stiffness of polymer/clay nanocomposites, *Polymer*, 48, 901–909, 2007.
66. Talreja, R., Multi-scale modeling in damage mechanics of composite materials, *J. Mater. Sci.*, 41, 6800–6812, 2006.
67. Song, M. and Chen, L., Molecular dynamics simulation of the fracture in polymer-exfoliated layered silicate nanocomposites, *Macromol. Theory Simul.*, 15, 238–245, 2006.
68. Schmidt, C., Wefing, S., Blümich, B., and Spiess, H.W., Dynamics of molecular reorientations: direct determination of rotational angles from two-dimensional NMR of powders, *Chem. Phys. Lett.*, 130, 84–90, 1986.
69. Potrzebowski, M.J., What high-resolution solid-state NMR spectroscopy can offer to organic chemists, *Eur. J. Org. Chem.*, 1367–1376, 2003.
70. Usuki, A., Koiwai, A., Kojima, Y., Kawasumi, M., Okada, A., Kurauchi, Y.T., and Kamigaito, O., Interaction of nylon 6-clay surface and mechanical properties of nylon 6-clay hybrid., *J. Appl. Polym. Sci.*, 55, 119–123, 1995.
71. Sozzani, P., Bracco, S., Comotti, A., Mauri, M., Simonutti, R., and Valsesia, P., Nanoporosity of an organo-clay shown by hyperpolarized xenon and 2D NMR spectroscopy. *Chem. Commun.*, 1921–1923, 2006.
72. Comotti, A., Bracco, S., and Valsesia, P., Adsorption properties of nanoporous organoclay and NMR characterization, AIZ Workshop, Alessandria, Italy, September 1–2, 2006.
73. VAMAS, Technical Working Area on polymer nanocomposites meeting, NRCC/IMI, Boucherville, QC, Canada, June 11, 2007.
74. Tokihisa, M., Yakemoto, K., Sakai, T., Utracki, L.A., Sepehr, M., Li, J., and Simard, Y., Extensional flow mixer for polymer nanocomposites, *Polym. Eng. Sci.*, 46, 1040–1050, 2006.
75. Shelley, J.S., Mather, P.T., and DeVries, K.L., Reinforcement and environmental degradation of nylon-6/clay nanocomposites, *Polymer*, 42, 5849–5858, 2001.
76. Masenelli-Varlot, K., Reynaud, E., Vigier, G., and Varlet, J., Mechanical properties of clay-reinforced polyamide, *J. Polym Sci., Part B: Polym. Phys.*, 40, 272–283, 2002.
77. Dennis, H.R., Hunter, D.L., Chang, D., Kim, S., White, J.L., Cho, J.W., and Paul, D.R., Effect of melt processing conditions on the extent of exfoliation in organoclay-based nanocomposites, *Polymer*, 42, 9513–9522, 2001.
78. Mehrabzadeh, M. and Kamal, M.R., Polymer-clay nanocomposites based on blends of polyamide-6 and polyethylene, *Can. J. Chem. Eng.*, 80, 1083–1092, 2002.
79. Tjong, S.C., Meng, Y.Z., and Xu, Y., Novel preparation and properties of polypropylene-vermiculite nanocomposites, *J. Polym Sci., Part B: Polym. Phys.*, 40, 2860–2870, 2002.
80. Mallick, P.K. and Zhou, Y., Yield and fatigue behavior of polypropylene and polyamide-6 nano-composites, *J. Mater. Sci.*, 38, 3183–3190, 2003.
81. Uribe-Arocha, P., Mehler, C., Puskas, J.E., and Altstädt, V., Effect of sample thickness on the mechanical properties of injection-molded polyamide-6 and polyamide-6 clay nanocomposites, *Polymer*, 44, 2441–2446, 2003.
82. Wu, S., Jiang, D., Ouyang, X., Wu, F., and Jian, S., The structure and properties of PA6/MMT nanocomposites prepared by melt compounding, *Polym. Eng. Sci.*, 44, 2070–2074, 2004.
83. Weon, J.-I. and Sue, H.-J., Effects of clay orientation and aspect ratio on mechanical behavior of nylon-6 nanocomposite, *Polymer*, 46, 6325–6334, 2005.
84. García-López, D., Gobernado-Mitre, I., Fernández, J.F., Merino, J.C., and Pastor, J.M., Influence of clay modification process in PA6-layered silicate nanocomposite properties, *Polymer*, 46, 2758–2765, 2005.
85. Kelnar, I., Kotek, J., Kaprálková, L., and Munteanu, B.S., Polyamide nanocomposites with improved toughness, *J. Appl. Polym. Sci.*, 96, 288–293, 2005.
86. Karaman, V.M., Privalko, V.P., Privalko, E.G., Lehmann, B., and Friedrich, K., Structure/property relationships for polyamide-6/organoclay nanocomposites in the melt and in the solid state, *Macromol. Symp.*, 221, 85–94, 2005.

87. Tjong, S.C. and Bao, S.P., Impact fracture toughness of polyamide-6/montmorillonite nano-composites toughened with a maleated styrene/ethylene butylene/styrene elastomer, *J. Polym. Sci., Part B: Polym. Phys.*, 43, 585–595, 2005.
88. Li, L., Bellan, L.M., Craighead, H.G., and Frey, M.W., Formation and properties of nylon-6 and nylon-6/montmorillonite composite nanofibers, *Polymer*, 47, 6208–6217, 2006.
89. Borse, N.K. and Kamal, M.R., Melt processing effects on the structure and mechanical properties of PA-6/clay nanocomposites, *Polym. Eng. Sci.*, 46, 1094–1103, 2006.
90. Mollet, V.A. and Kamal, M.R., Quantitative characterization of exfoliation in polyamide-6/layered silicate nanocomposites, *J. Polym. Eng.*, 26, 757–782, 2006.
91. Vlasveld, D.P.N., Vaidya, S.G., Bersee, H.E.N., and Picken, S.J., Nanocomposite matrix for increased fibre composite strength, *Polymer*, 46, 3452–3461, 2005.
92. Liu, X. and Wu, Q., Polyamide 66/clay nanocomposites via melt intercalation, *Macromol. Mater. Eng.*, 287, 180–186, 2002.
93. Han, B., Ji, G., Wu, S., and Shen, J., Preparation and characterization of nylon 66/montmoril-lonite nanocomposites with co-treated montmorillonites, *Europ. Polym. J.*, 39, 1641–1646, 2003.
94. Yu, Z.-Z., Yan, C., Yang, M., and Mai, Y.-W., Mechanical and dynamic mechanical properties of nylon 66/montmorillonite nanocomposites fabricated by melt compounding, *Polym. Int.*, 53, 1093–1098, 2004.
95. Shen, L., Phang, I.Y., Chen, L., Liu, T., and Zeng, K., Nanoindentation and morphological studies on nylon 66 nanocomposites, *Polymer*, 45, 3341–3349; 8221–8229, 2004.
96. Yang, J.-L., Zhang, Z., Schlarb, A.K., and Friedrich, K., On the characterization of tensile creep resistance of polyamide 66 nanocomposites, *Polymer*, 47, 2791–2801; 6745–6758, 2006.
97. Gyoo, P.M., Venkataramani, S., and Kim, S.C., Morphology, thermal and mechanical properties of polyamide 66/clay nanocomposites with epoxy-modified organoclay, *J. Appl. Polym. Sci.*, 101, 1711–1722, 2006.
98. Phang, I.Y., Liu, T., Mohamed, A., Pramoda, K.P., Chen, L., Shen, L., Chow, S.Y., He, C., Lu, X., and Hu, X., Morphology, thermal and mechanical properties of nylon 12/organoclay nanocomposites prepared by melt compounding *Polym. Int.*, 54, 456–464, 2005.
99. Liu, L., Qi, Z., and Zhu, X., Studies on nylon 6/clay nanocomposites by melt-intercalation process, *J. Appl. Polym. Sci.*, 71, 1133–1138, 1999.
100. Shah, R.K. and Paul, D.R., Nylon 6 nanocomposites prepared by a melt mixing masterbatch process, *Polymer* 45, 2991–3000, 2004.
101. Varlot, K., Reynaud, E., Kloppfer, M.H., Vigier, G., and Varlet J., Clay-reinforced polyamide: preferential orientation of the montmorillonite sheets and the polyamide crystalline lamellae, *J. Polym. Sci., Part B: Polym. Phys.* 39, 1360–1370, 2001.
102. Yu, Z.-Z., Yang, M., Zhang, Q., Zhao, C., and Mai, Y.-W., Dispersion and distribution of organically modified montmorillonite in nylon-66 matrix, *J. Polym. Sci.: Part B: Polym. Phys.*, 41, 1234–1243, 2003.
103. Chavarria, F. and Paul, D.R., Comparison of nanocomposites based on nylon 6 and nylon 66, *Polymer*, 45, 8501–8515, 2004.
104. Davis, R.D., Bur, A.J., McBrearty, M., Lee, Y.-H., Gilman, J.W., and Start, P.R., Dielectric spectroscopy during extrusion processing of polymer nanocomposites: a high throughput processing/characterization method to measure layered silicate content and exfoliation, *Polymer*, 45, 6487–6493, 2004.
105. Nair, S.S. and Ramesh, C., Studies on the crystallization behavior of nylon-6 in the presence of layered silicates using variable temperature WAXS and FTIR, *Macromolecules* 38, 454–462, 2005.
106. Li, Y. and Shimizu, H., Effects of spacing between the exfoliated clay platelets on the crystallization behavior of polyamide-6 in polyamide-6/clay nanocomposites, *J. Polym. Sci. Part B: Polym. Phys.*, 44, 284–290, 2006.
107. Zhang, Y., Yang, J.H., Ellis, T.S., and Shi, J., Crystal structures and their effects on the properties of polyamide 12/clay and polyamide 6-polyamide 66/clay nanocomposites, *J. Appl. Polym. Sci.* 100, 4782–4794, 2006.

108. Lee, J.W., Lim, Y.T., and Park, O.Ok., Thermal characteristics of organoclay and their effects upon the formation of polypropylene/organoclay nanocomposites, *Polym. Bull.* 45, 191–198, 2000.

109. López-Quintanilla, M.L., Sánchez-Valdés, S., Ramos de Valle, L.F., and Medellin Rodriguez, F.J., Effect of some compatibilizing agents on clay dispersion of polypropylene-clay nanocomposites, *J. Appl. Polym. Sci.* 100, 4748–4756, 2006.

110. Sepehr, M., personal communication, 2006.

111. Xie, S., Zhang, S., Wang, F., Yang, M., Seguela, R., and Lefebvre, J.-M., Preparation, structure and thermomechanical properties of nylon-6 nanocomposites with lamella-type and fiber-type sepiolite, *Composites Sci. Technol.*, 67, 2334–2341, 2007.

112. Scaffaro, R., Pedretti, U., and La Mantia, F.P., Effects of filler type and mixing method on the physical properties of a reinforced semirigid liquid crystal polymer, *Eur. Polym. J.*, 32, 869–875, 1996.

113. Perrin, F.X., Nguyen, V., and Vernet, J.L., Mechanical properties of polyacrylic-titania hybrids— microhardness studies, *Polymer*, 43, 6159–6167, 2002.

114. Xu, B., Simonsen, J., and Rochefort, W.E., Creep resistance of wood-filled polystyrene/high-density polyethylene blends, *J. Appl. Polym. Sci.*, 79, 418–425, 2001.

115. Zhang, Z., Yang, J.-L., and Friedrich, K., Creep resistant polymeric nanocomposites, *Polymer*, 45, 3481–3485, 2004; Yang, J.-L., Zhang, Z., Schlarb, A.K., and Friedrich, K., On the characterization of tensile creep resistance of polyamide 66 nanocomposites, *Polymer*, 47, 2791–2801; 6745–6758, 2006.

116. Vlasveld, D.P.N., Bersee, H.E.N., and Picken, S.J., Nanocomposite matrix for increased fibre composite strength, *Polymer*, 46, 12539–12545, 2005.

117. Bellemare, S.C., Bureau, M., Dickson, J.I., and Denault, J., Fatigue-induced damage in PA-6/ clay nanocomposite, *SPE Techn. Pap.*, 48, 2235–2239, 2002.

118. Kharbas, H., Nelson, P., Yuan, M., Gong, S., Turng, L.-S., and Spindler, R., Effects of nano-fillers and process conditions on the microstructure and mechanical properties of microcellular injection molded polyamide nanocomposites, *Polym. Compos.*, 24, 655–671, 2003.

119. Bucknall, C.B., *Toughened plastics*, Applied Science Publishers, London, 1977.

120. Pascoe, K.J., General fracture mechanic, in *Failure of plastics*, Brostow, W. and Corneliussen, R.D., Eds., Hanser Pub., Munich, 1986.

121. Bureau, M.N., Denault, J., and Glowacz, F., Mechanical behavior and crack propagation in injection molded polyamide-6/clay nanocomposites, *SPE Techn. Pap.*, 47, 2125–2129, 2001.

122. Yasue, K., Tamura, T., Katahira, S., and Watanabe, M., Reinforced polyamide resin composition and process for producing the same, US Pat. 5,414,042, 09.05.1995, Appl. 28.12.1993, to Unitika Ltd.

123. Zheng, Q.H., Yu, A.B., Lu, G.Q., and Paul, D.R., clay-based polymer nanocomposites research and commercial development, *J. Nanosci. Nanotechnol.*, 5, 1574–1592, 2005.

124. Okada, A. and Usuki, A., Twenty years of polymer-clay nanocomposites, *Macromol. Mater. Eng.*, 291, 1449–1476, 2006.

125. Dell'Anno, G., *Development of a new class of hybrid reinforced thermoplastic composites based on nanoclays and woven glass fibres*, Tesi di laurea, Ingegneria Chimica, University of Pisa, 2004.

126. Li, J., Ton-That, M.-T., Leelapornpisit, W., and Utracki, L.A., Melt compounding polypropylene based nanocomposites containing Cloisite 20A, *Polym. Eng. Sci.*, 47, 1447–1458, 2006.

127. Heinemann, J., Reichert, P., Thomann, R., and Mülhaupt, R., Polyolefin nanocomposites formed by melt compounding and transition metal catalyzed homo- and copolymerization in the presence of layered silicates, *Macromol. Rapid Commun.*, 20, 423–430, 1999.

128. Modesti, M., Lorenzetti, A., Bon, D., and Besco, S., Effect of processing conditions on morphology and mechanical properties of compatibilized polypropylene nanocomposites, *Polymer*, 46, 10237–10245, 2005.

129. Lertwimolnun, M.W., *Realisation de nanocomposites polypropylene/argile par extrusion bivis*, Ph.D. thesis, Ecole des mines de Paris, Sophia-Antipolis, 2006.

130. Wang, K., Liang, S., Zhang, Q., Du, R., and Fu, Q., An observation of accelerated exfoliation in iPP/organoclay nanocomposite as induced by repeated shear during melt solidification, *J. Polym. Sci., Part B: Polym. Phys.*, 43, 2005–2012, 2005.

131. Allan, P.S. and Bevis, M.J., Process for molding directionally-orientatable material using shear force, US Pat., 4,925,161, 15.05.1990.

132. Nguyen, X.Q. and Utracki, L.A., US Patent 5 451 106 (1995); Utracki L.A., and Luciani, A., Canadian Patent Application 2 217 374 (1997); Utracki, L.A., Bourry, D., and Luciani, A., PCT Patent Application WO 99/16540 (1999).

133. Utracki, L.A., Pressure-volume-temperature of molten and glassy polymers, *J. Polym. Sci. Part B: Polym Phys.*, 45, 270–285, 2007.

134. Hong, C.H., Lee, Y.B., Bae, J.W., Jho, J.Y., Nam, B.U., and Hwang, T.W., Preparation and mechanical properties of polypropylene/clay nanocomposites for automotive parts application, *J. Appl. Polym. Sci.*, 98, 427–433, 2005.

135. Gianelli, W., Ferrara, G., Camino, G., Pellegatti, G., Rosenthal, J., and Trombini, R.C., Effect of matrix features on polypropylene layered silicate nanocomposites, *Polymer*, 46, 7037–7046, 2005.

136. Oya, A., Kurokawa, Y., and Yasuda, H., Factors controlling mechanical properties of clay mineral/polypropylene nanocomposites, *J. Mater. Sci.*, 35, 1045–1050, 2000.

137. Zhang, Q., Fu, Q., Jiang, L., and Lei, Y., Preparation and properties of polypropylene/montmorillonite layered nanocomposites, *Polym. Int.*, 49, 1561–1564, 2000.

138. Manias, E., Touny, A., Wu, L., Strawhecker, K., Lu, B., and Chung, T.C., Polypropylene/montmorillonite nanocomposites. Review of the synthetic routes and materials properties, *Chem. Mater.*, 13, 3516–3523, 2001.

139. Chen, L., Wong, S.-C., and Pisharath, S., Fracture properties of nanoclay-filled polypropylene, *J. Appl. Polym. Sci.*, 88, 3298–3305, 2003.

140. Morgan, A.B. and Harris, J.D., Effects of organoclay Soxhlet extraction on mechanical properties, flammability properties and organoclay dispersion of polypropylene nanocomposites, *Polymer*, 44, 2313–2320, 2003.

141. Pozsgay, A., Fráter, T., Százdi, L., Müller, P., Sajó, I., and Pukánszky, B., Gallery structure and exfoliation of organophilized montmorillonite: effect on composite properties, *Europ Polym. J.*, 40, 27–36, 2004.

142. Benetti, E.M., Causin, V., Marega, C., Marigo, A., Ferrara, G., Ferraro, A., Consalvi, M., and Fantinel, F., Morphological and structural characterization of polypropylene based nanocomposites, *Polymer*, 46 8275–8285, 2005.

143. Ferrara, G., Sartori, F., Costantini, E., and Di Pietro, F., WO 2006/131450, 2006.12.14, to Basell.

144. Leuteritz, A., Pospiech, D., Kretschmar, B., Willeke, M., Jehnichen, D., Jentzsch, U., Grundke, K., and Janke, A., Polypropylene-clay nanocomposites: comparison of different layered silicates, *Macromol. Symp.*, 221, 53–61, 2005.

145. Lew, C.Y., Murphy, W.R., McNally, G.M., Godinho, J.E., and McConnell, D.C., Compatibilisation protocol for the processing of polypropylene organoclay nanocomposites, *SPE Techn. Pap.* 1960–1964, 2005.

146. Chung, M.J., Jang, L.W., Shim, J.H., and Yoon, J.-S., Preparation and characterization of maleic anhydride-*g*-polypropylene/diamine-modified clay nanocomposites, *J. Appl. Polym. Sci.*, 95, 307–3111, 2005.

147. Jang, L.W., Kim, E.S., Kim, H.S., and Yoon, J.-S., Preparation and characterization of polypropylene/clay nanocomposites with polypropylene-graft-maleic anhydride, *J. Appl. Polym. Sci.*, 98, 1229–1234, 2005.

148. Chen, L., Wang, K., Toh, M.L,. and He, C., Polypropylene/clay nanocomposites prepared by reactive compounding with an epoxy based masterbatch, *Macromol. Mater. Eng.*, 290, 1029–1036, 2005.

149. Wong, S.-C., Lee, H., Qu, S., Mall, S., and Chen, L., A study of global vs. local properties for maleic anhydride modified polypropylene nanocomposites, *Polymer*, 47, 7477–7484, 2006.

150. Moncada, E., Quijada, R., Lieberwirth, I., Yazdani-Pedram, M., Use of PP grafted with itaconic acid as a new compatibilizer for PP/clay nanocomposites, *Macromol. Chem. Phys.*, 207, 1376–1386, 2006.

151. Peltola, P., Välipakka, E., Vuorinen, J., Syrjälä, S., and Hanhi, K., Effect of rotational speed of twin screw extruder on the microstructure and rheological and mechanical properties of nanoclay-reinforced polypropylene nanocomposites, *Polym. Eng. Sci.*, 46, 995–1000, 2006.

152. Joshi, M. and Viswanathan, V., High-performance filaments from compatibilized polypropylene/clay nanocomposites, *J. Appl. Polym. Sci.*, 102, 2164–2174, 2006.

153. Moad, G., Dean, K., Edmond, L., Kukaleva, N., Li, G., Mayadunnem R.T.A., Pfaendnerm, R., Schneider, A., Simon, G., and Wermter, H., Novel copolymers as dispersants/intercalants/exfoliants for polypropylene-clay nanocomposites, *Macromol. Mater. Eng.*, 291, 37–52, 2006; *Macromol. Symp.*, 233, 170–179, 2006.

154. Lee, J.W., Kim, M.H., Choi, W.M., and Park, O.O., Effects of organoclay modification on microstructure and properties of polypropylene–organoclay nanocomposites, *J. Appl. Polym. Sci.*, 99, 1752–1759, 2006.

155. Lim, J.W., Hassan, A., Rahmat, A.R., and Wahit, M.U., Rubber-toughened polypropylene nanocomposite: effect of polyethylene octene copolymer on mechanical properties and phase morphology, *J. Appl. Polym. Sci.*, 99, 3441–3450, 2006.

156. Rosenthal, J.S. and Wołkowicz, M.D., US Pat 6,864,308, 2005.03.08 to Basell.

157. Treece, M.A. and Oberhauser, J.P., Ubiquity of soft glassy dynamics in polypropylene/clay nanocomposites, *Polymer*, 48, 1083–1095, 2007.

158. Wang, Z.M., Nakajima, H., Manias, E., and Chung, T.C., Exfoliated PP/clay nanocomposites using ammonium-terminated PP as the organic modification for montmorillonite, *Macromolecules*, 36, 8919–8922, 2003.

159. Beall, G.W., Tsipursky, S., Sorokin, A., and Goldman, A., Intercalates and exfoliates formed with oligomers and polymers and composite materials containing same, US Pat., 5,760,121, 02.06.1998, to Amcol International Corp.

160. Lopez-Quintanilla, L., Sanchez Valdes, S., and Ramos de Valle, L.F., Effect of compatibilizing agents on clay dispersion of polypropylene-clay nanocomposites, *SPE Techn. Pap.*, 3314–3317, 2005.

161. Galgali, G., Agarwal, S., and Lele, A., Effect of clay orientation on the tensile modulus of polypropylene–nanoclay composites, *Polymer*, 45, 6059–6069, 2004.

162. Zunjarrao, S.C., Sriraman, R., and Singh, R.P., Effect of processing parameters and clay volume fraction on the mechanical properties of epoxy-clay nanocomposites, *J. Mater. Sci.*, 41, 2219–2228, 2006.

163. Oh, T.-K., Hassan, M., Beatty, C., and El-Shall, H., The effect of shear forces on the microstructure and mechanical properties of epoxy–clay nanocomposites, *J. Appl. Polym. Sci.*, 100, 3465–3473, 2006.

164. Wang, Z., Massam, J., and Pinnavaia, T.J., Epoxy-clay nanocomposites, in *Polymer-clay nanocomposites*, Pinnavaia, T.J. and Beall, G.W., Eds., J. Wiley and Sons, Ltd., New York, 2000.

165. Pinnavaia, T.J., and Lan, T., Flexible resin-clay composite, method of preparation and use, US Pat., 5,802,216, 01.09.1998, Appl. 07.07.1997, to Board of Trustees Michigan State University; Pinnavaia, T.J. and Lan, T., Hybrid nanocomposites comprising layered inorganic material and methods of preparation, US Pat., 5,853,886, 29.12.1998, Appl. 17.07.1996 to Claytec, Inc.

166. Lan, T. and Pinnavaia, T.J., Clay-reinforced epoxy nanocomposites, *Chem. Mater.*, 6, 2216–2219, 1994.

167. Pinnavaia, T.J. and Lan, T., Hybrid organic-inorganic nanocomposites and methods of preparation US Pat., 6,017,632, 25.01.2000, Appl. 08.20.1998; to Claytec, Inc.

168. Zilg, C., Mülhaupt, R., and Finter, J., Morphology and toughness/stiffness balance of nanocomposites based upon anhydride-cured epoxy resins and layered silicates, *Macromol. Chem. Phys.*, 200, 661–670, 1999.

169. Kornmann, X., Thomann, R., Mülhaupt, R., Finter, J., and Berglund, L.A., High performance epoxy-layered silicate nanocomposites, *Polym. Eng. Sci.*, 42, 1815–1826, 2002.

170. Becker, O., Varley, R., and Simon, G., Morphology, thermal relaxations and mechanical properties of layered silicate nanocomposites based upon high-functionality epoxy resins, *Polymer*, 43, 4365–4373, 2002.
171. Becker, O., Cheng, Y.-B., Varley, R., and Simon, G., Layered silicate nanocomposites based on various high-functionality epoxy resins: the influence of cure temperature on morphology, mechanical properties and free volume, *Macromolecules*, 36, 1616–1625, 2003.
172. Xue, S., Reinholdt, M., and Pinnavaia, T.J., Palygorskite as an epoxy polymer reinforcement agent, *Polymer*, 47, 3344–3350, 2006.
173. Messersmith, P.B. and Giannelis, E.P., Synthesis and characterization of layered silicate-epoxy nanocomposites, *Chem. Mater.*, 6, 1719–1725, 1994.
174. Miyagawa, H., Rich, M.I., and Drzal, T., Amine-cured epoxy/clay nanocomposites. I. Processing and chemical characterization, *J. Polym. Sci. Part B: Polym. Phys.*, 42, 4384–4390, 4391–4400, 2004.
175. Boo, W.-J., Sun, L., Warren, G.L., Moghbelli, E., Pham, H., Clearfield, A., and Sue, H.-J., Effect of nanoplatelet aspect ratio on mechanical properties of epoxy nanocomposites, *Polymer*, 48, 1075–1082, 2007.
176. Boo, W.-J., Sun, L., Liu, J., Moghbelli, E., Clearfield, A., Sue, H.-J., Pham, H., and Verghese, N., Effect of nanoplatelet dispersion on mechanical behavior of polymer nanocomposites, *J. Polym. Sci. Part B: Polym. Phys.*, 45, 1459–1469, 2007.
177. Wang, K., Chen, L., Wu, J., Toh, M.L., He, C., and Yee, A.F., Epoxy nanocomposites with highly exfoliated clay: mechanical properties and fracture mechanisms, *Macromolecules*, 38, 788–800, 2005.
178. Basara, C., Yilmazer, U., and Bayram, G., Synthesis and characterization of epoxy based nanocomposites, *J. Appl. Polym. Sci.*, 98, 1081–1086, 2005.
179. Fröhlich, J., Golombowski, D., Thomann, R., and Mülhaupt, R., Synthesis and characterization of anhydride-cured epoxy nanocomposites containing layered silicates modified with phenolic alkylimidazolineamide cations, *Macromol. Mater. Eng.*, 289, 13–19, 2004.
180. Ratna, D., Becker, O., Krishnamurthy, R., Simon, G.P., and Varley, R.J., Nanocomposites based on a combination of epoxy resin, hyperbranched epoxy and a layered silicate, *Polymer*, 44, 7449–7457, 2003.
181. Fröhlich, J., Thomann, R., and Mülhaupt, R., Synthesis and characterization of anhydride-cured epoxy nanocomposites containing layered silicates modified with phenolic alkylimidazolineamide cations, *Macromolecules*, 36, 7205–7211, 2003.
182. Clearfield, A., Ed., *Inorganic ion exchange materials*, CRC Press, Boca Raton, FL, 1982.
183. Alberti, G., Costantino, U., Allulli, S., and Tomassini, N., Crystalline $Zr(R-PO_3)_2$ and $Zr(R-OPO_3)_2$ compounds, *J. Inorg. Nucl. Chem.*, 40, 1113–1117, 1978.
184. Sue, H.-J., Gam, K.T., Bestaoui, N., Spurr, N., and Clearfield, A., Epoxy nanocomposites based on the synthetic α-zirconium phosphate layer structure, *Chem. Mater.*, 16, 242–249, 2004.
185. Camino, G., Maffezzoli, A., Braglia, M., De Lazzaro, M., and Zammarano, M., Effect of hydroxides and hydroxycarbonate structure on fire retardant effectiveness and mechanical properties in ethylene-vinyl acetate copolymer, *Polym. Degradation Stab.*, 74, 457–464, 2001.
186. Hsueh, J.-B. and Chen, C.-Y., Preparation and properties of LDH/polyimide nanocomposites, *Polymer*, 44, 5275–5283, 2003.
187. Becker, O. and Simon, G.P., Epoxy layered silicate nanocomposites, *Adv. Polym. Sci.*, 179, 29–82, 2005.
188. Norkhairunnisa, M., Azhar, A.B., and Shyang, C.W., Effects of organo-montmorillonite on the mechanical and morphological properties of epoxy/glass fiber composites, *Polym. Int.*, 56, 512–517, 2007.
189. Bucknall, C.B., Karpodinis, A., and Zhang, X.C., A model for particle cavitation in rubber-toughened plastics, *J. Mater. Sci.*, 29, 3377–3383, 1994.
190. Zerda, A.S. and Lesser, A.J., Intercalated clay nanocomposites: morphology, mechanics and fracture behavior, *J. Polym. Sci. Part B: Polym. Phys.*, 39, 1137–1146, 2001.
191. Van Krevelen, D.W. and Hoftyzer, P.J. *Properties of polymers—Their estimation and correlation with chemical structure*, Elsevier, Amsterdam, 1976.

192. Theodorou, D.N. and Suter, U.W., Atomistic modeling of mechanical properties of polymeric glasses, *Macromolecules*, 19, 139–154, 379–387, 1986.
193. Porter, D., *Group interaction modeling of polymer properties*, Marcel Dekker Inc., New York, 1996.
194. Hartmann, B., in *Handbook* of *Polymer Science and Technology*, Vol. 2, Cheremisinoff, N.P., Ed., Marcel Dekker, New York, 1989, pp. 101–125.
195. Utracki, L.A. and Simha, R., Analytical representation of solutions to lattice-hole theory, *Macromol. Chem. Phys., Molecul. Theory Simul.*, 10, 17–24, 2001.
196. Friedrich, K., Fakirov, S., and Zhang, Z., *Polymer composites; from nano- to macro-scale*, Springer, New York, 2005.

14

Mass Transport through Polymer Nanocomposites

Daniel De Kee and Kyle J. Frederic

CONTENTS

14.1 Introduction

Membranes are thin sheets of synthetic or natural material that can be made selectively permeable. During the past few decades, membranes have been used for performing increasingly complex material separations. The number of potential uses for membranes has risen due to improved manufacturing, which includes reducing the membrane thickness, producing more defect-free materials, and introducing nano-sized fillers to form nanocomposites. That is, nanocomposites are hybrid composite materials consisting of a polymer matrix with dispersed nanoparticles. Furthermore, nanocomposite membranes are now being "functionalized" by the type of nanoparticles chosen—their size, orientation, charges, responses to temperature, stress, etc.—all of which lead to impressive applications such as enhanced barrier properties for packaging, protective clothing, gas separations, and enhanced properties for controlled drug release, to name just a few. The particles are chosen with an understanding of the interactions between the membrane material and the components involved in the system. This chapter focuses on the barrier properties of polymer nanocomposites, on achieving enhanced properties, and on the theory used to model these systems.

The mass transport of small molecules through macromolecular materials is a major factor in improving the shelf life of products. For example, oxygen diffuses through tires, which causes the steel belt to rust, thereby reducing the tires' life. Another example includes bottles in which there is a need for selectively permeable materials. For example, as carbon dioxide diffuses out of soda bottles, the soda becomes flat. Also, as oxygen diffuses into beer bottles, it can react with the beer to make it stale. This process is so complex that no single model has been able to explain the observed data. Nevertheless, the entire process can be broken down into three steps. In the first step, the molecules absorb onto the surface of the polymeric material, then the molecules diffuse through the polymer, and finally the molecules desorb onto the downstream side of the material.

Fick's laws are commonly used to describe the mass transport of one component through another component. Fick's first law relates the flux to a concentration gradient, where the diffusion coefficient is the proportionality constant. Fick's second law relates the change in concentration with respect to time to the change in flux with respect to position. Mass transport processes that satisfy Fick's laws are referred to as Fickian diffusion or Case I diffusion and have been studied extensively.[1, 2] However, the transport of small solvent molecules through polymeric materials does not necessarily follow Fickian behavior. The exact reasons for this are still debated but it is widely accepted that polymer swelling plays a key role.

Overall, deviation from Fickian diffusion can arise from a number of factors that can increase the complexity of the diffusion process. These factors are discussed in the following section.

14.2   Factors Affecting Permeation

14.2.1 Temperature

Many groups have studied the effect of temperature on diffusion, and a strong dependence on the permeation is clear.[3–6] As temperature increases, the steady-state flux increases, while the breakthrough time decreases. Several parameters, including the diffusion coefficient, breakthrough time, and flux, obey Arrhenius-type relations as follows:

$$D = D_0 \exp\left(\frac{-E_D}{RT}\right)$$
(14.1)

$$t_b = t_0 \exp\left(\frac{-E_B}{RT}\right)$$
(14.2)

$$J = J_0 \exp\left(\frac{-E_J}{RT}\right) \qquad (14.3)$$

where D is the diffusion coefficient, t_b is the breakthrough time, and J is the flux, with their respective activation energies, E_D, E_B, and E_J.

Solubility is determined by many factors involving the chemical nature of the polymer and penetrant, all of which depend on temperature. Temperature can also be related to the solubility of the solvent in the polymer by the following Arrhenius-type relation:[7]

$$S = S_0 \exp\left(\frac{-H_S}{RT}\right) \qquad (14.4)$$

where S is the solubility and H_S is the heat of sorption. The extent of sorption is a result of the equilibrium between the chemical tendency of mixing and the elasticity of the polymer network, which tends to limit swelling. Therefore, depending on the chemical nature of the solvent and the polymer, the sorption process can be either endothermic or exothermic.

Permeability is determined by the solubility and the diffusion coefficient. The equation relating permeability to temperature is

$$P = D \times S = P_0 \exp\left(\frac{-E_P}{RT}\right) \qquad (14.5)$$

where P is the permeability, P_0 is constant for a given system, and E_P is the activation energy of permeation.

14.2.2 Permeant

The molecular structure of the permeant is an important factor in the mass transport process through any membrane. It is quite intuitive that the diffusivity will increase with decreasing molecular size. This can be explained by the free volume theory because more space is required for the diffusion of larger molecules.[8] The shape of the molecule displays a similar effect. For example, a linear molecule will have a higher diffusivity than a branched or cyclic molecule.[9] Guo and De Kee introduced an effective molecular volume to account for the molecular size and shape of permeants according to the following equation:[10]

$$d_{\text{eff}} = 0.6l_{\text{min}} + 0.3l_{\text{mid}} + 0.1l_{\text{max}} \qquad (14.6)$$

where d_{eff} is the diameter of an effective sphere, l_{min} is the minimum dimension, l_{max} is the maximum dimension, and l_{mid} is the third dimension. The dimensions of the molecule are in three orthogonal directions. The effective volume V_{eff} was calculated by using d_{eff} as the diameter of an effective sphere. This model proved that there is a strong dependence of solvent structure on permeation.

In addition, the interactions between the permeant and the polymer are significant in influencing the diffusion. For example, polarity is an important factor in permeation. Nonpolar fluids permeate faster through nonpolar membranes.

The permeation of mixtures is an important factor in many applications. Gas chromatography (GC) is a common technique employed to analyze the permeants.[11–13] When the components of the mixture have different permeabilities, the slower permeant's diffusion coefficient is affected by the concentration gradient of both permeants. An interesting case involves the permeation of a certain solvent through a polymer material in the presence of another solvent.[14] Solvent interaction was analyzed by Gagnard.[15] If the Flory-Huggins interaction coefficient of the solvents is positive (i.e., they do not mix well), the polymer matrix will experience a higher degree of swelling compared to contact with a pure solvent. This leads to a higher permeation or "positive synergy." The converse is also true; solvents that do mix well result in a lower permeation or "negative synergy."

14.2.3 Polymer Structure

As stated before, Fick's laws do not accurately describe mass transport through complex polymeric materials. The free volume theory comes into play again because the creation of space for the solvent molecules depends on the polymer chain's mobility as well as on the cohesive energy of the polymer. In addition, other factors such as crosslinking, additives such as plasticizers and fillers, and the degree of crystallinity will have an effect on the diffusion process.[16]

Even the simplest composite polymeric materials are frequently associated with complex structures.[17] Nanocomposite membranes have been shown to improve the mechanical, thermal, flame-resistant, as well as barrier properties.

Polymers may swell when in contact with certain solvents. The polymer network elongates as the solvent penetrates, and elastic restrictive forces develop to counteract the swelling process. Viscoelastic properties are associated with the polymer's response to changes in configuration. These responses/relaxation times distinguish the types of diffusion processes. Case I (Fickian) diffusion refers to the case where the characteristic diffusion time is substantially larger than the polymer relaxation time. Case II diffusion refers to the case where the diffusion time is much smaller than the polymer relaxation time. Case III (anomalous) diffusion refers to a situation where the two time scales are comparable.[18]

According to the free-volume theory, diffusion in polymers occurs as small molecules travel through available spaces between polymer chains.[19] In addition, the diffusion of the molecules is enhanced as their local concentration increases. This is called plasticization. A low-molecular-weight fraction is introduced that reduces the entanglement and bonding forces between the polymer molecules and increases their free volume and mobility.[20] As the amount of solvent increases to a critical value, a glassy polymer will change to a rubbery polymer and the diffusion coefficient will be larger than in the glassy region.

The presence of solvent can also lead to polymer crystallization.[21] Cornelis and Kander's study reported the results of diffusion experiments on poly(ether ether ketone) (PEEK) films with methylene chloride and tetrahydrofuran as solvents, which are known to plasticize and induce crystallization in PEEK.[21] Once the polymer crystallizes, the solvent that was previously absorbed is rejected, leading to high diffusion rates.

Many authors have reported on mass transport through polymers with complex internal structures.[22-25] The effect of fillers on the transport process through natural rubber was reported by Unnikrishnan and Thomas.[26] Essentially, permeation was found to decrease with smaller filler particles. As the particle size decreased, the overall mobility and flexibility of the polymer links also decreased. The presence of smaller particles reinforces the polymer matrix and does not allow for much rearrangement. The degree of crystallinity within the polymer also affects the permeation. There is a decrease in mass transport with increasing degree of crystallinity as the penetrating solvent molecules can only diffuse and absorb through the amorphous regions.

Drozdov et al.[27] performed permeation experiments of water molecules in vinyl ester/montmorillonite nanocomposite materials. The diffusion of water through a pure resin is Fickian; however, it becomes non-Fickian as the clay concentration is increased, which is attributed to the water being immobilized on the hydrophilic clay layers.

In contrast, Liu[28] performed permeation experiments with organic solvents through polymer/clay nanocomposite membranes. The diffusion of dichloromethane (DCM) through pure poly(dimethylsiloxane) (PDMS) is non-Fickian. It becomes more Fickian as the clay concentration increases, which is most likely due to reduced polymer chain mobility.

Another unexpected result was obtained by Merkel et al.[29] Normally, the addition of nanoparticles significantly reduces mass transport, but Merkel et al.'s results indicated an increase in the penetrant permeability through a high free-volume, glassy PMP-fumed silica system as the concentration of the nanoscale fumed silica increased. The increase in permeability is most likely a result of the increase in free volume in the PMP, induced by the fumed silica particles.

The effect of clays and compatibilizer on the dispersion, intercalation, and morphology has been studied for both micro- and nano-structures.[30] However, an appropriate mass transport model to better distinguish between polymers filled with nanoparticles versus polymers filled with microparticles is still outstanding. A significant difference should be observed in the mass transport through nanocomposites as compared to mass transport through microcomposites, depending on the size of the penetrating species. Species small enough to permeate through nano-sized channels could be selectively chosen to separate from larger species whose motion would be obstructed.

14.2.4 Mechanical Deformation

Dramatic structural changes may occur in polymeric materials due to external mechanical deformation. These structural changes affect the transport of molecules through the polymeric matrix. There are numerous reports on stress-enhanced transport, which include a decreased breakthrough time and an increase in the steady-state permeation rate, as is also associated with thermal effects.[31, 33] This makes sense intuitively because the polymer's thickness decreases as the externally applied mechanical stress increases. However, the structure of the polymeric material may change under mechanical deformation.

Drummond et al.[34] used atomic force microscopy (AFM) to study the structural changes due to external deformation. The initial extension caused the crystalline regions to orient along the axis of extension. Thus, for semicrystalline polymers such as polyethylene, the drawing process leads to a small increase in crystallinity. A change in crystallite size and shape can affect the tortuosity and consequently the diffusion. An accompanying change in chain mobility in the amorphous phase can also affect the diffusion process.[35] As fibers stretch, the molecules become more oriented and the system tends to "crystallize." The fibers become stronger, tougher, and more elastic than the nonoriented fibers. The orientation of the crystalline regions will also change during drawing. The orientation was viewed as decreasing the amount of excess free volume, bringing the polymer closer to equilibrium conditions.[36] Also, the orientation of nanoparticles plays a pivotal role in the properties of the system. If the nanoparticles are aligned perpendicular to the transport direction (with a large surface area exposed to the penetrant), the barrier properties will increase significantly. A controlled rate of transport can be achieved by controlling the nanoparticle orientation. Because stretching can help orient nanoparticles, it is crucial to understand how mass transport is affected by mechanical deformation.

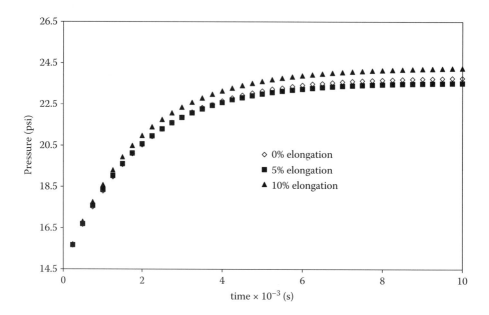

FIGURE 14.1

Pressure vs. time for oxygen through poly(dimethylsiloxane) (PDMS) at various degrees of mechanical deformation.

While several studies have shown the effects of external stress, few studies performed the permeation experiments while the membrane was actually under deformation.[37, 38] Hinestroza[39] developed an experimental technique to study the effect of mechanical deformation on the permeation through polymeric materials. This technique provided real-time monitoring of single-component as well as mixture transport. The setup consisted of an ASTM F739 permeation cell that was placed in the middle of a custom-built stretching device. Acetone and hexane were used as permeants through natural and butyl rubber membranes. The membranes were stretched uniaxially from 0 to 40% and biaxially from 5 × 5% to 40 × 40%. For all systems, a decrease in breakthrough time and an increase in steady-state flux were noted while measuring the diffusion through elongated membranes.

Liu and De Kee[40] also investigated the effect of mechanical deformation on the diffusion of dichloromethane through PDMS/clay nanocomposite membranes. While stretching the membrane decreases the thickness, it also packs the polymer chains, which could lead to an overall decrease in the diffusion coefficient. This study showed that at small deformations, the decrease in the free volume decreased the permeation, while at high deformation, a substantial decrease in thickness enhanced the permeation.

Current research in our (Frederic and De Kee[41]) laboratory involves performing gas permeation experiments while a membrane is actually stretched. The permeation of oxygen through pure PDMS and PDMS/clay nanocomposite membranes is being investigated. Figure 14.1 shows the effect of mechanical deformation on the permeation of oxygen through PDMS. At small deformations, the permeation actually decreases due to the packing of the polymer chains. However, the decrease in thickness again becomes the dominant factor and enhances the permeation. Figure 14.2 shows preliminary data on the effect

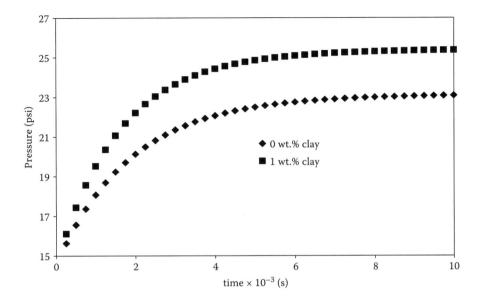

FIGURE 14.2
Pressure vs. time for oxygen through clay/ poly(dimethylsiloxane) (PDMS) nanocomposite membranes.

of the addition of nanoparticles into a polymer matrix. Uncharacteristically, a small addition in concentration of nanoparticles increased the permeation. This is most likely a result of nanoparticles being located between polymer chains and being oriented in such a way as to prevent chain packing. Further studies are underway to monitor trends as the nanoparticle concentration increases, as well as including the effect of mechanical deformation on a variety of viscoelastic systems.

Overall, mechanical deformation plays an important role in affecting the diffusion process, due to its effect on nanoparticle orientation. If the deformation is too pronounced, the membrane may lose its desired functionality.

14.3   Factors Enhancing Gas Barrier Properties

14.3.1 Exfoliation

To enhance gas barrier properties through polymer nanocomposites, the nanoparticles must present a large surface area in the direction of the permeation. The optimal case is to have large aspect ratio particles "blocking" permeant transport. In fact, a high length/ width is a key factor in maximizing a tortuous path. Therefore, the degree of dispersion is crucial and can be intercalated, exfoliated, or anywhere in between. The gas barrier properties can be maximized, if exfoliation is achieved. The concentration of nanoparticles is sufficiently small so that the effect on material clarity, color, weight, etc. is minimized. The clay particles must be exfoliated to obtain a nanocomposite material containing a dispersion of approximately 1 nm thick and 500 nm wide nanoclay particles.

The topology created by the dispersion/exfoliation of filler particles of substantial aspect ratio will affect the time scale of the transport process. The dimension of the diffusing molecules is another contributing factor. The gas barrier properties between composites and nanocomposites is significant.

14.3.2 Compatibility

Another important factor is the compatibility of the nanoparticles with the polymer itself. For example, in a polymer/clay nanocomposite, the clays are alumino-silicates and therefore must be functionalized to be compatible with the polymer. While this process can be quite complex, the simplest approach is to add compatibilizer or compatibilizer blends to the clay and polymer matrix.[42] The compatibilizer can be considered to "coat" the nanoclay and provide a suitable interaction with the polymer matrix.

14.3.3 Orientation

As mentioned previously, the nanoparticles must be oriented so that the flat surface of the nanoparticle is perpendicular to the direction of the permeation, in order to increase the surface area and maximize the barrier. This creates the tortuous path necessary to slow the transport. If the orientation can be controlled, the barrier properties can also be controlled. However, one of the main problems with this approach has been the reaggregation of the nanoparticles.

14.3.4 Reaggregation

Efforts must be focused on preventing aggregation of exfoliated particles during processing operations. If the particles aggregate, the flat surface is still perpendicular to the direction of the permeation but there is now less surface area to "block" the penetrant. The barrier properties are no longer optimal. Aggregation during extrusion, for example, could be due to the prevailing shearing forces and the self-diffusion of the clay particles. More work needs to be done in this area to limit or eliminate the reaggregation of the nanoparticles.[43]

14.4   Theoretical Modeling

Because it is prohibitively expensive and realistically impossible to test all combinations of permeants and polymer composites, the desirability of a predictive model becomes evident. Most of the theoretical studies used to model diffusion are based on Fick's laws. Further information on permeation through polymer nanocomposites can be found, for example, in Philip et al. and Nielsen.[44, 45] These provide popular theories that represent the data quite well. These laws state that the rate of transport through a unit area is proportional to the concentration gradient measured normal to the material. The actual driving force is a chemical potential gradient. The proportionality constant is the diffusion coefficient and is usually assumed to be constant. Another simplifying assumption is that equilibrium is immediately reached. However, most processes do not follow the simple, linear Fickian diffusion theory.

Long and Richman[46] and Richman and Long[47] proposed a two-stage diffusion process. The first stage involves instantaneous molecular rearrangements, while the second (slower stage) leads to the final equilibrium. Petropoulos and Roussis[48] considered an activity gradient as opposed to a concentration gradient as the driving force for diffusion.

Another popular assumption made to account for this apparent non-Fickian behavior was made by Crank and Park[49] and Crank and Park,[50] who considered a history-dependent diffusion coefficient that is a function of time as well as concentration.

All these theories partially help explain non-Fickian diffusion. However, their main limitations arise from the lack of consideration of interactions between the permeant and the polymer or the physical properties of the polymer.

Thomas and Windle[51, 52] stimulated research[53, 54] in the area by relating the diffusion process to the swelling rate of the polymer. As opposed to the simplified Fickian models mentioned previously, this model involves irreversible thermodynamics. The fluxes can be expressed in terms of a linear combination of chemical potential, temperature gradients, as well as stress distributions. This was a major advancement in the understanding of non-Fickian diffusion. This advanced model qualitatively predicts most characteristics of Case II diffusion.

Even more advanced models proposed that a convective flux associated with polymer stress exists. Usually, this was modeled by extending Fick's first law. Chan Man Fong[55] and Hinestroza[56] assumed that the chemical potential was a function of concentration and stress. While this model fits their experimental data reasonably well, it does not consider the interaction between polymer and solvent.

The situation becomes even more complex when dealing with polymer composites. Composition, miscibility, and phase morphology all affect the transport through polymer composite materials. The complex nature of the internal structure of a polymer composite material has not yet been fully explained.

More recently, a mesoscopic theory has been used to model non-Fickian diffusion behavior, which is the case for nanocomposite systems.[57, 58] This method uses the Hamiltonian mechanics formulation, including Poisson and dissipation brackets to derive the governing equations. Introducing dissipation brackets accounts for the irreversible processes. For the sake of simplicity, the following assumptions are made: (1) the system under consideration is composed of a solvent or simple fluid (small molecules) and a complex polymer fluid (large molecules); (2) the molecules under consideration are completely miscible, (3) isothermal conditions; (4) the density of the system is constant; (5) the system is in mechanical equilibrium; and (6) no bulk flow; that is, the mass flux is relatively small.

The procedure used to model diffusion through polymeric materials is as follows. First, the state variables to describe the overall system are chosen. In addition to the concentration of the penetrant and the flux as two state variables, the conformation tensor $\mathbf{m}$ has also been used to account for the structural changes of the polymer due to swelling or other external stress effects. The most fundamental and exact method uses a distribution function but the number of variables becomes too large to handle and several approximations must be made. The conformation tensor can be related to the distribution function ψ as:

$$m_{\alpha\beta}(t) = \langle \mathbf{RR} \rangle = \int \psi(\mathbf{R}, t) R_\alpha R_\beta d\mathbf{R} \qquad (14.7)$$

where $\mathbf{R}$ is a polymer segment end to end vector, and $\psi(\mathbf{R}, t)$ is a distribution function defined such that $\psi d\mathbf{R}$ is the number of segments per unit volume that have an end-to-end vector in the range $d\mathbf{R}$ about $\mathbf{R}$.

Second, the Poisson bracket and dissipation bracket formalism are applied so that the governing equations automatically satisfy the thermodynamic laws. Finally, the model is applied to a one-dimensional diffusion problem, considering a one-dimensional diffusion is valid when the thickness of the sample is much smaller than the other two dimensions. Although a one-dimensional model is used here, the tensorial formulations given in our earlier work[59, 60] allow for a higher-dimensional approach when required. That is to say, the one-dimensional model is a simplified version of the general case. Simplified models are associated with a smaller number of model parameters. Higher-dimensional models should be used in cases where the physics of the problem do not justify a one-dimensional approach or in case the simplified treatment does not produce quantitative agreement with the data. Clearly, if a one-dimensional model is satisfactory, a more general higher-dimensional version will work at least equally well, but this at the expense of the parameter economy. The governing equations for one-dimensional diffusion are

$$\rho\frac{\partial c}{\partial t} = -\frac{\partial J_s}{\partial x} \tag{14.8}$$

$$J_s = -\rho D\left(\frac{\partial c}{\partial x} + \Gamma\frac{\partial m_{11}}{\partial x}\right) \tag{14.9}$$

$$\frac{\partial m_{11}}{\partial t} = \frac{J_s}{\rho(1-c)}\frac{\partial m_{11}}{\partial x} - m_{11}\frac{\partial}{\partial x}\left(\frac{J_s}{\rho(1-c)}\right) - 2\lambda m_{11}\frac{\partial \varphi}{\partial m_{11}} \tag{14.10}$$

where J_s is mass flux of the solvent, m_{11} is the (1,1) component of the conformation tensor, λ is a phenomenological parameter, and φ is the Helmholtz free energy density. Equation 14.8 is Fick's second law. Equation 14.9 is an extension of Fick's first law. The second term on the right-hand side of Equation 14.9 contributes to changes in the flux due to structural changes of the polymer, which are described by Equation 14.10. Note that the solvent concentration in the membrane cannot be 1. That is, the factor $(1 - c)$ in the denominator cannot be zero. Γ relates the elastic and mixing parts of the free energy and is associated with the polymer's internal structural change in the diffusion process. For the case where the Hookean energy expression is used to derive the free energy, polymer swelling implies that Γ is negative and confirms that the elasticity of the polymer matrix decreases the mass flux.

The diffusion of small molecules through a composite membrane filled with impermeable objects such as clay particles is a practical problem. Cussler et al.[61] proposed equations for the flux through barrier membranes filled with flakes. Using Monte Carlo simulations of diffusion through impermeable flakes, they showed the tortuous paths produced by the flakes.[62] For transport through polymer nanocomposites with layered structured particles, Bharadwaj[63] proposed a tortuosity theory to describe the effect of these particles on the diffusion process.

A General Equation for the NonEquilibrium Reversible and Irreversible Coupling (GENERIC) formalism was applied by El Afif et al.,[64–66] who investigated the isothermal mass transport of a simple fluid through a blend of two immiscible Newtonian polymers. Two state variables were chosen to account for the size and shape of the polymer blend interface. These variables are an internal area density Q and an interfacial orientation density tensor $\mathbf{q}$. These two variables (Q and $\mathbf{q}$) were first introduced by Doi and Ohta[67] and have been used to study the rheological properties of immiscible blends. In addition, a nonlinear formulation that addressed the effects of the diffusion/interface coupling on the mass transport and on the morphology of the interface was derived. This new mass flux equation included a new term to account for the viscoelastic contribution of the interface, which leads to non-Fickian diffusion behavior.

Liu[68] and Liu and De Kee[69, 70] modeled mass transport through polymer-clay nanocomposites using an area tensor $\mathbf{A}$, first proposed by Wetzel and Tucker.[71] Liu[68] established this state variable to account for the internal, complex, interfacial structure of the nanocomposite. This area tensor can be related to the state variables, Q and $\mathbf{q}$, as follows:

$$Q = tr(\mathbf{A}) \tag{14.11}$$

$$\mathbf{q} = \mathbf{A} - \frac{1}{3} tr(\mathbf{A})\mathbf{I} \tag{14.12}$$

The structural state of the viscoelastic polymer cannot be ignored. To characterize polymer internal structural changes, the conformation tensor $\mathbf{m}$ was also included in Liu's model. Therefore, for the overall system that consists of one simple fluid (solvent) and one complex "fluid" (nanocomposite), the state variables chosen are the mass fraction, mass flux, conformation tensor, and area tensor. Applying the bracket/GENERIC formalism, the governing equations for one-dimensional diffusion were derived as follows:

$$\rho \frac{\partial c}{\partial t} = -\frac{\partial J_s}{\partial x} \tag{14.13}$$

$$J_s = -\rho D \left(\frac{\partial c}{\partial x} + E_{11} \frac{\partial m_{11}}{\partial x} + \Lambda_{\beta\gamma} \frac{\partial A_{\beta\gamma}}{\partial x} \right) \tag{14.14}$$

$$\frac{\partial m_{11}}{\partial t} = \frac{J_s}{\rho(1-c)} \frac{\partial m_{11}}{\partial x} - m_{11} \frac{\partial}{\partial x}\left(\frac{J_s}{\rho(1-c)} \right) - 2\lambda m_{11} \frac{\partial \varphi}{\partial m_{11}} \tag{14.15}$$

$$\frac{\partial A_{ij}}{\partial t} = \frac{J_s}{\rho(1-c)} \frac{\partial A_{ij}}{\partial x} + A_{1j}\partial_i \left(\frac{J_s}{\rho(1-c)} \right) + A_{1i}\partial_j \left(\frac{J_s}{\rho(1-c)} \right)$$

$$- \frac{1}{A_{kk}} A_{\alpha\beta} A_{ij}\partial_\beta \left(\frac{J_{s\alpha}}{\rho(1-c)} \right) - \lambda^A_{\alpha\beta ij} \frac{\partial \varphi}{\partial A_{\alpha\beta}} \tag{14.16}$$

where E_{11} is the (1,1) component of the tensor E that relates the elastic and mixing parts of the free energy, $\Lambda_{\beta\gamma}$ relates the interfacial interaction to the mixing part of the free energy, A_{ij} is the area tensor, $\lambda^A_{\alpha\beta ij}$ is a positive real-valued function of the state variables, and φ is the free energy density for the polymer-clay and solvent system. Note that Equation 14.13 is equivalent to Equation 14.8. Fick's second law has not changed. In addition, the introduction of the area tensor (A) to characterize the system has not affected the time evolution of the conformation tensor (m) (i.e., Equation 14.10 is equivalent to Equation 14.15). However, Equation 14.14 is a further extension of Fick's first law. The time evolution of the area tensor is given by Equation 14.16.

Nondimensionalizing the equations generates dimensionless parameters that characterize the physical properties of the polymer as well as the type of diffusion process. One dimensionless parameter relates the polymer elasticity and the mixing properties of the solvent and the polymer nanocomposite. Another relates the complex interface to the mixing properties. For swelling polymers, both the effect of the elasticity of the polymer matrix and the existence of complex interfaces will decrease the mass flux. The analysis also generates two dimensionless times: one dimensionless time involves the polymer relaxation time and the other relates to the interface relaxation time.

While complex, utilizing this approach reduces the number of arbitrary parameters and assumptions compared to traditional modeling techniques. In addition, the mesoscopic approach involves less phenomenological parameters than an analysis based on continuum mechanics. There is quantitative agreement between this model and experimental data.

This approach shows promise; but to fully predict the behavior of any nanocomposite material, the addition of characterizing tensors—to account, for example, for thermal and/or electrical effects—will be necessary. The necessary tensors will depend on the system. Therefore, by generalizing this method, there will be a model to predict a wide range of realistic situations.

Thus, following the approach of Liu[72] and Liu and De Kee,[73] the generalized governing equations for one-dimensional diffusion through nanocomposite membranes is as follows:

$$\rho \frac{\partial c}{\partial t} = -\frac{\partial J_s}{\partial x} \tag{14.17}$$

$$J_s = -\rho D \left(\frac{\partial c}{\partial x} + \sum_{i=1}^{N} X_{mn}^{(i)} \frac{\partial K_{mn}^{(i)}}{\partial x} \right) \tag{14.18}$$

where $K_{mn}^{(i)}$ is a generalized tensor to describe the system such as the conformation or area tensor and $X_{mn}^{(i)}$ is the companion coefficient. Again, Equation 14.17 depicts Fick's second law. Fick's first law is extended to include the required additional characterization (Equation 14.18). Unfortunately, the time evolution of each tensor needed cannot be generalized, but can only be derived once the tensors to be used are known. However, the approach is the same in each case.

The Fickian model should not be discarded. While it is less accurate for many cases, it should not be considered inadequate. A simple model may not predict transport perfectly, but may give sufficient results. That is to say, it may be beneficial to apply a simple model at first and then apply a more complex model if the results are truly misleading. Unfortunately, problems are different, and it is difficult to validate how complex a model should be used. For example, in neutron transport in nuclear reactor cores, while the Fickian model was known to be inadequate, it produced surprisingly accurate results. Also, Figure 14.5 shows that the Fickian prediction is significantly inaccurate. However, if all one is interested in is the steady-state permeation, then the Fickian model would not be misleading at all.

14.4.1 Limiting Cases

The preceding model can be simplified to cases where the polymer viscoelastic properties and/or the complex interface effect become less significant.

14.4.1.1 Model-c (*Fick's Laws*)

When the changes in the polymer's internal structure and the effect of complex interfaces are minimal, the only contribution to the mass flux is the mass concentration gradient.

$$\rho \frac{\partial c}{\partial t} = -\frac{\partial J_s}{\partial x} \tag{14.19}$$

$$J_s = -\rho D \left(\frac{\partial c}{\partial x} \right) \tag{14.20}$$

In Fick's law, it is assumed that the flux is a function of concentration only, which can describe some simple situations. However, many diffusion processes, such as those involving polymeric materials, do not always obey Fick's laws.

14.4.1.2 Model-c,m

When the complex interface is minimal or does not exist, all terms associated with area tensor **A** can be ignored. Now, the contribution to the mass flux is associated with the mass concentration gradient as well as the relaxation of the polymer internal structure.

$$J_s = -\rho D \left(\frac{\partial c}{\partial x} + \Gamma \frac{\partial m_{11}}{\partial x} \right) \tag{14.21}$$

$$\frac{\partial m_{11}}{\partial t} = \frac{J_s}{\rho(1-c)} \frac{\partial m_{11}}{\partial x} - m_{11} \frac{\partial}{\partial x} \left(\frac{J_s}{\rho(1-c)} \right) - 2\lambda m_{11} \frac{\partial \varphi}{\partial m_{11}} \tag{14.22}$$

Overall, there is quantitative agreement between this model and experimental data as shown in Figure 14.3.

14.4.1.3 Model-c,A

When the changes in the polymer's internal structure are not dominant, the terms associated with conformation tensor **m** can be ignored. Now, the contribution to the mass flux is associated with the mass concentration gradient as well as the complex interface effect.

$$J_s = -\rho D \left(\frac{\partial c}{\partial x} + \Lambda_{\beta\gamma} \frac{\partial A_{\beta\gamma}}{\partial x} \right) \tag{14.23}$$

$$\frac{\partial A_{ij}}{\partial t} = \frac{J_s}{\rho(1-c)} \frac{\partial A_{ij}}{\partial x} + A_{1j} \partial_i \left(\frac{J_s}{\rho(1-c)} \right) + A_{1i} \partial_j \left(\frac{J_s}{\rho(1-c)} \right)$$

$$- \frac{1}{A_{kk}} A_{\alpha\beta} A_{ij} \partial_\beta \left(\frac{J_{s\alpha}}{\rho(1-c)} \right) - \lambda^A_{\alpha\beta ij} \frac{\partial \varphi}{\partial A_{\alpha\beta}} \tag{14.24}$$

The last term in Equation 14.24 represents the effect of complex interfacial interaction due to the presence of nanoparticles. Overall, there is quantitative agreement between this model and experimental data as shown in Figure 14.4. However, the diffusion process is

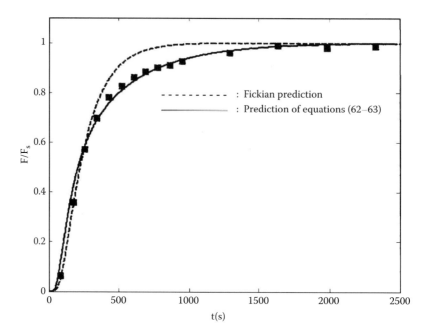

FIGURE 14.3
Normalized flux versus time for dichloromethane DCM through poly(dimethylsiloxane) PDMS.

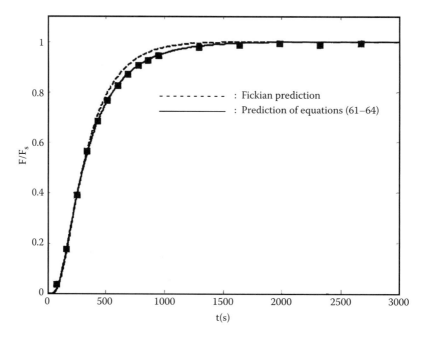

FIGURE 14.4
Normalized flux vs. time for dichloromethane (DCM) through poly(dimethylsiloxane) (PDMS) with 10 wt% clay.

nearly Fickian, with only small deviations near steady-state. This is likely due to the clay particles, which can limit the movement of the polymer chains.

Figure 14.3 and Figure 14.4 show results from Liu.[72] The liquid permeation measurements were obtained using a modified ASTM permeation cell (ASTM F-739-85). The major components of the apparatus are a modified permeation cell and a gas chromatograph that is connected to a personal computer for sample analysis.

A spherical 1-inch diameter, two-chamber modified ASTM F-739 glass cell was used to expose one side of the membrane to a challenging liquid. The challenge chamber has a volume of approximately 10 ml and is equipped with a stoppered nozzle that allows for liquid additions. The downstream (collection) chamber is equipped with inlet and outlet ports for nitrogen carrier gas flow. A Shimadzu GC-17A gas chromatograph (GC) with a flame ionization detector (FID) and a 30-m capillary column coated with Supelco Wax 10 is used to identify the concentration of the permeating species.

The system was secured using a Teflon ring and the membrane itself as a seal. The liquid solvent was introduced into the (challenging) upper chamber of the permeation cell. The start time for a permeation experiment is the time at which the solvent makes contact with the membrane. The concentration of the permeant was analyzed using the GC and saved on the computer.

14.5 Modeling of Diffusion through Polymer-Clay Nanocomposite Membranes

There are two levels of description of the area tensor based on the level of description of the volume and the bounded surface Ω. One level associated with discrete particles denotes an area tensor at the microscopic level. In the absence of particle (de)lamination, the trace of the area tensor is constant. The disadvantage at this level is that to fully describe the system, the number of variables becomes too large. At another level, one can consider the area tensor to be based on a finite volume. This volume is small enough to represent the interfacial interactions, but large enough to neglect microstructural details. This mesoscopic area tensor $\mathbf{A}$ is used here in subsequent modeling.

14.5.1 Poisson Bracket Description

Once the state variables are determined, the Poisson bracket formulation of Hamiltonian mechanics is applied to each component and to the overall system.[74, 75] First, an ideal system (S) (no energy dissipation) is considered.

The Poisson bracket description of state variable set S is

$$\{F,G\}^{(polymer)} = -\int_{\Omega} d^3x \left[F_\rho \frac{\partial}{\partial x_\alpha}\left(\rho G_{u_\alpha}\right) - G_\rho \frac{\partial}{\partial x_\alpha}\left(\rho F_{u_\alpha}\right) \right]$$

$$-\int_{\Omega} d^3x \left[F_{u_\beta} \frac{\partial}{\partial x_\alpha}\left(u_\beta G_{u_\alpha}\right) - G_{u_\beta} \frac{\partial}{\partial x_\alpha}\left(u_\beta F_{u_\alpha}\right) \right] \tag{14.25}$$

where F and G are arbitrary functionals. In addition,

$$\frac{dF}{dt} = \{F,H\} = \int_\Omega d^3x \left[\frac{\delta F}{\delta \rho} \frac{\partial \rho}{\partial t} + \frac{\delta F}{\delta u_\alpha} \frac{\partial u_\alpha}{\partial t} \right] \qquad (14.26)$$

and comparing terms yields the governing equations of the simple fluid for a reversible process:

$$\frac{\partial \rho_s}{\partial t} = -\partial_\alpha \left(u_{s_\alpha} \right) \qquad (14.27)$$

$$\frac{\partial u_{s_\alpha}}{\partial t} = -\partial_\beta \left(\frac{u_{s_\alpha} u_{s_\beta}}{\rho_s} \right) - \partial_\alpha p \qquad (14.28)$$

where the *Hamiltonian* $H = \int d^3r \left[\frac{u_\beta u_\beta}{2\rho} + h(\rho) \right]$ and the hydrodynamic pressure p is written as

$$p = -h + \rho_s \frac{\partial h}{\partial \rho_s} \qquad (14.29)$$

Note that the following notation will be used in all subsequent equations: $\partial_\alpha \equiv \partial / \partial x_\alpha$.

To obtain the Poisson bracket formulation for the nanocomposite system, we start from the f-level description.[76] By applying the relationship between f and $\mathbf{A}$, and following the chain rule, we obtain

$$\frac{\delta}{\delta f} = n_\alpha n_\beta \frac{\delta}{\delta A_{\alpha\beta}} \qquad (14.30)$$

where $\frac{\delta}{\delta f}$ is the Volterra functional derivative with respect to f. The Poisson bracket description for the interface contribution at the mesoscopic level is

$$\{F,G\}^{(\text{interface})} = \int d^3r A_{ij} \left[\partial_\alpha \left(\frac{\delta F}{\delta A_{ij}} \frac{\delta G}{\delta u_{N_\alpha}} \right) - \partial_\alpha \left(\frac{\delta G}{\delta A_{ij}} \frac{\delta F}{\delta u_{N_\alpha}} \right) \right]$$

$$- \int d^3r A_{\alpha j} \left[\frac{\delta F}{\delta A_{ij}} \partial_i \left(\frac{\delta G}{\delta u_{N_\alpha}} \right) - \frac{\delta G}{\delta A_{ij}} \partial_i \left(\frac{\delta F}{\delta u_{N_\alpha}} \right) \right]$$

$$- \int d^3r A_{\alpha i} \left[\frac{\delta F}{\delta A_{ij}} \partial_j \left(\frac{\delta G}{\delta u_{N_\alpha}} \right) - \frac{\delta G}{\delta A_{ij}} \partial_j \left(\frac{\delta F}{\delta u_{N_\alpha}} \right) \right]$$

$$- \int d^3r \frac{1}{A_{kk}} A_{\alpha\beta} A_{ij} \left[\frac{\delta F}{\delta A_{ij}} \partial_\beta \left(\frac{\delta G}{\delta u_{N_\alpha}} \right) - \frac{\delta G}{\delta A_{ij}} \partial_\beta \left(\frac{\delta F}{\delta u_{N_\alpha}} \right) \right] \qquad (14.31)$$

Note that in deriving this final equation, a fourth-order moment arises, and the quadratic closure approximation given in Doi and Ohta[77] has been used,

$$\int d^2 n n_\alpha n_\beta n_i n_j f(\mathbf{r}, \mathbf{n}, t) = \frac{1}{A_{kk}} A_{\alpha\beta} A_{ij} \tag{14.32}$$

The Poisson bracket description for a nanocomposite system can be expressed as

$$\{F, G\} = \{F, G\}^{(\text{polymer})} + \{F, G\}^{(\text{interface})} \tag{14.33}$$

where $\{F, G\}^{(\text{polymer})}$ is the Poisson bracket expression for the polymer matrix. Since

$$\frac{dF}{dt} = \{F, H\} = \int d^3 x \left(\frac{\delta F}{\delta \rho_N} \frac{\delta \rho_N}{\partial t} + \frac{\delta F}{\delta u_{N_\alpha}} \frac{\delta u_{N_\alpha}}{\partial t} + \frac{\delta F}{\delta m_{\alpha\beta}} \frac{\delta m_{\alpha\beta}}{\partial t} + \frac{\delta F}{\delta A_{ij}} \frac{\delta A_{ij}}{\partial t} \right) \tag{14.34}$$

it follows that

$$\frac{\partial \rho_N}{\partial t} = -\frac{\partial}{\partial x_\alpha} \left(u_{N_\alpha} \right) \tag{14.35}$$

$$\frac{\partial u_{N_\alpha}}{\partial t} = -\frac{\partial}{\partial x_\beta} \left(u_{N_\alpha} \frac{u_{N_\beta}}{\rho_N} \right) - \frac{\partial p_N}{\partial x_\alpha} - \frac{\partial \sigma_{N_{\alpha\beta}}}{\partial x_\beta} \tag{14.36}$$

$$\frac{\partial m_{\alpha\beta}}{\partial t} = -\frac{\partial}{\partial x_\gamma} \left(m_{\alpha\beta} \frac{u_{N_\gamma}}{\rho_N} \right) + m_{\gamma\beta} \frac{\partial}{\partial x_\gamma} \left(\frac{u_{N_\alpha}}{\rho_N} \right) + m_{\gamma\alpha} \frac{\partial}{\partial x_\gamma} \left(\frac{u_{N_\beta}}{\rho_N} \right) \tag{14.37}$$

$$\frac{\partial A_{ij}}{\partial t} = -\frac{u_{N_\alpha}}{\rho_N} \frac{\partial}{\partial x_\alpha} (A_{ij}) - A_{\alpha j} \frac{\partial}{\partial x_i} \left(\frac{u_{N_\alpha}}{\rho_N} \right) - A_{\alpha i} \frac{\partial}{\partial x_j} \left(\frac{u_{N_\alpha}}{\rho_N} \right) + \frac{1}{A_{kk}} A_{\alpha\beta} A_{ij} \frac{\partial}{\partial x_\beta} \left(\frac{u_{N_\alpha}}{\rho_N} \right) \tag{14.38}$$

The hydrodynamic pressure is

$$p = -h + \rho_N \frac{\delta h}{\delta \rho_N} + m_{\alpha\beta} \frac{\delta h}{\delta m_{\alpha\beta}} \tag{14.39}$$

and the extra stress tensor is

$$\sigma_{\alpha\beta} = -2 m_{\beta\gamma} \frac{\delta h}{\delta m_{\alpha\gamma}} + 2 A_{\alpha\gamma} \frac{\delta h}{\delta A_{\beta\gamma}} - \frac{1}{A_{kk}} A_{\alpha\beta} A_{ij} \frac{\delta h}{\delta A_{ij}} \tag{14.40}$$

where the Hamiltonian can now be expressed as

$$H = \int d^3 r \left[\left(\frac{u_{N_\alpha} u_{N_\alpha}}{2 \rho_N} \right) + h(\rho_N, \mathbf{m}, \mathbf{A}) \right] \tag{14.41}$$

Note that in Equation 14.39, the area tensor **A** does not appear explicitly. Equation 14.40 shows that the complex interface contributes to extra stress in the system.

Using relations between the state variables of the nanocomposite system (polymer and nanoparticles) and the overall system (polymer, nanoparticles, and solvent), the following governing equations for the overall incompressible system are obtained:

$$\rho \frac{\partial c}{\partial t} = -\frac{\partial J_{s_\alpha}}{\partial x_\alpha} \tag{14.42}$$

$$\frac{\partial J_{s_\alpha}}{\partial t} = \frac{\partial}{\partial x_\beta}\left(\frac{J_{s_\alpha}J_{s_\beta}}{\rho(1-c)}\right) - c\frac{\partial}{\partial x_\alpha}\left(\frac{\delta h}{\delta c}\right) \tag{14.43}$$

$$\frac{\partial m_{\alpha\beta}}{\partial t} = \frac{\partial}{\partial x_\gamma}\left(m_{\alpha\beta}\frac{J_{s_\gamma}}{\rho(1-c)}\right) - m_{\gamma\beta}\frac{\partial}{\partial x_\gamma}\left(\frac{J_{s_\alpha}}{\rho(1-c)}\right) - m_{\gamma\alpha}\frac{\partial}{\partial x_\gamma}\left(\frac{J_{s_\beta}}{\rho(1-c)}\right) \tag{14.44}$$

$$\frac{\partial A_{ij}}{\partial t} = \frac{J_{s_\alpha}}{\rho(1-c)}\frac{\partial}{\partial x_\alpha}(A_{ij}) + A_{\alpha j}\frac{\partial}{\partial x_i}\left(\frac{J_{s_\alpha}}{\rho(1-c)}\right)$$

$$+ A_{\alpha i}\frac{\partial}{\partial x_j}\left(\frac{J_{s_\alpha}}{\rho(1-c)}\right) - \frac{1}{A_{kk}}A_{\alpha\beta}A_{ij}\frac{\partial}{\partial x_\beta}\left(\frac{J_{s_\alpha}}{\rho(1-c)}\right) \tag{14.45}$$

We assume the system to be in mechanical equilibrium; that is,

$$\partial_\alpha p + \partial_\beta \sigma_{\alpha\beta} = 0 \tag{14.46}$$

where p is the (scalar) pressure,

$$p = p_s + p_p = -h + \rho_s\frac{\delta h}{\delta\rho_s} + \rho_p\frac{\delta h}{\delta\rho_p} + m_{\alpha\beta}\frac{\delta h}{\delta m_{\alpha\beta}} \tag{14.47}$$

The *Hamiltonian H* corresponding to h in Equation 14.47 is the one for the overall system. Note that the extra stress expression is the same as for the nanocomposite system. That is, the solvent introduces hydrodynamic pressure but does not change the extra stress.

Equations 14.42 through 14.45 are the governing equations for an ideal solvent-nanocomposite system. To describe nonequilibrium systems involving dissipation, the GENERIC formalism[78, 79] yields

$$\frac{\partial X}{\partial t} = L\frac{\delta\Phi}{\delta X} - \frac{\delta\Psi}{\delta(\delta\Phi/\delta X)} \tag{14.48}$$

where X is the set of independent state variables for the overall system, t denotes time, $\Phi = E - TS$ is the Helmholtz free energy, and E and S are the total energy and entropy, respectively. Ψ is the dissipative potential, and L is the Poisson operator. Under isothermal conditions, the specific GENERIC formulation is identical to the general bracket formulation:[80]

$$\frac{dF}{dt} = \{[F,\Phi]\} = \{F,\Phi\} - [F,\Phi] \tag{14.49}$$

where {[,]} is a generalized bracket. The first term (Poisson bracket) on the right-hand side of Equation 14.49 accounts for the reversible kinematics, and the second term (dissipative bracket) represents the irreversible kinematics.

Beris and Edwards[80] gave a general expression for the dissipative bracket. Here we assume the following expression for the dissipative bracket:

$$[F, \Phi] = \int d^3r \left[\left(\frac{\delta F}{\delta J_{s_\alpha}} \right) \rho \, c(1-c) \xi \left(\frac{\delta \Phi}{\delta J_{s_\alpha}} \right) \right] + \int d^3r \left[\left(\frac{\delta F}{\delta m_{\alpha\beta}} \right) 2\lambda \, m_{\alpha\gamma} \left(\frac{\delta \Phi}{\delta m_{\alpha\beta}} \right) \right]$$
$$+ \int d^3r \left[\left(\frac{\delta F}{\delta A_{\alpha\beta}} \right) \lambda^A_{\alpha\beta ij} \left(\frac{\delta \Phi}{\delta A_{\alpha\beta}} \right) \right]$$
(14.50)

where the third term on the right-hand side accounts for the complex interface of a nanocomposite. The last part accounts for the dissipation due to the complex interface. $\xi, \lambda, \lambda^A_{\alpha\beta ij}$ are positive, real-valued functions of the state variables. The corresponding dissipative potential due to the complex interface can be expressed as

$$\Psi^{interface} = \int d^3r \left[\left(\frac{\delta \Phi}{\delta A_{ij}} \right) \frac{\lambda^A_{\alpha\beta ij}}{2} \left(\frac{\delta \Phi}{\delta A_{\alpha\beta}} \right) \right]$$
(14.51)

Applying the dissipative bracket expression, we obtain the governing equations for the overall system as follows:

$$\rho \frac{\partial c}{\partial t} = -\frac{\partial J_{s_\alpha}}{\partial x_\alpha}$$
(14.52)

$$\frac{\partial J_{s_\alpha}}{\partial t} = \frac{\partial}{\partial x_\beta} \left(\frac{J_{s_\alpha} J_{s_\beta}}{\rho(1-c)} \right) - c \frac{\partial}{\partial x_\alpha} \left(\frac{\partial \varphi}{\partial c} \right) - \xi J_{s_\alpha}$$
(14.53)

$$\frac{\partial m_{\alpha\beta}}{\partial t} = \frac{\partial}{\partial x_\gamma} \left(m_{\alpha\beta} \frac{J_{s_\gamma}}{\rho(1-c)} \right) - m_{\gamma\beta} \frac{\partial}{\partial x_\gamma} \left(\frac{J_{s_\alpha}}{\rho(1-c)} \right)$$
$$- m_{\gamma\alpha} \frac{\partial}{\partial x_\gamma} \left(\frac{J_{s_\beta}}{\rho(1-c)} \right) - 2\lambda m_{\alpha\gamma} \frac{\partial \varphi}{\partial m_{\gamma\beta}}$$
(14.54)

$$\frac{\partial A_{ij}}{\partial t} = \frac{J_{s_\alpha}}{\rho(1-c)} \frac{\partial}{\partial x_\alpha} \left(A_{ij} \right) + A_{\alpha j} \frac{\partial}{\partial x_i} \left(\frac{J_{s_\alpha}}{\rho(1-c)} \right) + A_{\alpha i} \frac{\partial}{\partial x_j} \left(\frac{J_{s_\alpha}}{\rho(1-c)} \right)$$
$$- \frac{1}{A_{kk}} A_{\alpha\beta} A_{ij} \frac{\partial}{\partial x_\beta} \left(\frac{J_{s_\alpha}}{\rho(1-c)} \right) - \lambda^A_{\alpha\beta ij} \frac{\partial \varphi}{\partial A_{\alpha\beta}}$$
(14.55)

Here, the Helmholtz free energy Φ is given by

$$\Phi = \int d\mathbf{r} \left[\frac{\mathbf{J}_s^2}{2\rho c(1-c)} + \varphi(c, \mathbf{m}, \mathbf{A}) \right] \tag{14.56}$$

Note that

$$\partial_\alpha \left(\frac{\partial \varphi}{\partial c} \right) = \frac{\partial^2 \varphi}{\partial c^2} \partial_\alpha c + \frac{\partial^2 \varphi}{\partial c \partial m_{\gamma\beta}} \partial_\alpha m_{\gamma\beta} + \frac{\partial^2 \varphi}{\partial c \partial A_{\gamma\beta}} \partial_\alpha A_{\gamma\beta} \tag{14.57}$$

Equation 14.53 becomes

$$\frac{\partial J_{s_\alpha}}{\partial t} = \partial_\beta \left(J_{s_\alpha} \frac{J_{s_\beta}}{\rho(1-c)} \right) - \xi \left[J_{s_\alpha} + \rho D(\partial_\alpha c + E_{\beta\gamma} \partial_\alpha m_{\beta\gamma} + \Lambda_{\beta\gamma} \partial_\alpha A_{\beta\gamma}) \right] \tag{14.58}$$

The diffusion coefficient D is expressed as a function of solvent concentration c and the Helmholtz free energy density φ, which is related to the polymer structure via a conformation tensor $\mathbf{m}$ and to the interfacial characteristics via an area tensor $\mathbf{A}$. Tensor $\mathbf{E}$ relates the elastic and mixing parts of the free energy, which accounts for the effect of the polymer relaxation on the flux. Tensor Λ relates the interfacial interaction to the mixing part of the free energy.

By comparing the governing equations with the set of governing equations in El Afif et al.,[81, 82] the same equations for the conservations of mass (Equation 14.52) and momentum (Equation 14.53) are obtained. The difference is in the evolution equations for the complex interface because different state variables are chosen. This approach also allows for the recovery of Equation 14.7 in El Afif et al.,[81] via the trace of $\mathbf{A}$ in Equation 14.55. Note that the governing equations derived here are much simpler than those in El Afif et al.[81, 82]

For small values of the flux and for a situation where steady state is reached quickly, Equation 14.58 can be linearized as follows:

$$J_{s_\alpha} = -\rho D(\partial_\alpha c + E_{\beta\gamma} \partial_\alpha m_{\beta\gamma} + \Lambda_{\beta\gamma} \partial_\alpha A_{\beta\gamma}) \tag{14.59}$$

Again, this expression for the flux is an extension of Fick's first law. The last term represents the effect of interfacial interaction.

The governing equations that result for the area tensor will have many components that depend on the shape of the nanoparticles in the composite. Let us focus on a lamellar geometry. Note that Bharadwaj[83] used an order parameter (S) to reflect the effect of the orientation of nanoplatelets:

$$S = \frac{1}{2}(3\cos^2\theta - 1) \tag{14.60}$$

where θ represents the angle between the direction of the preferred orientation ($\mathbf{n}$) and the sheet normal unit vectors. S can range from 1 ($\theta = 0$), indicating that the nanoplatelets are oriented perpendicular to the diffusion direction, to $-1/2$ ($\theta = \frac{\pi}{2}$), indicating that the nanoplatelets are oriented along the diffusion direction. A value of 0 for S ($\cos^2\theta = 1/3$)

indicates random orientation of the nanosheets. It would be ideal to orient nanoparticles (clay particles, for example) in one of the two orthogonal directions compared to the direction of transport to produce a composite membrane that could be used as a barrier material.

In the case of one-dimensional diffusion, the nondimensional expressions for the governing equations are

$$\frac{\partial C^*}{\partial \theta^*} = -\frac{1}{\sqrt{m_{11}^*}} \frac{\partial J_s^*}{\partial X^*}$$
(14.61)

$$J_s^* = -\frac{1}{\sqrt{m_{11}^*}} \left(\frac{\partial C^*}{\partial X^*} + \Pi \frac{\partial m_{11}^*}{\partial X^*} + \Theta \frac{\partial A_{11}^*}{\partial X^*} \right)$$
(14.62)

$$\frac{\partial m_{11}^*}{\partial \theta^*} = \frac{c_{eq}}{(1 - C^* c_{eq})} \frac{J_s^*}{\sqrt{m_{11}^*}} \frac{\partial m_{11}^*}{\partial X^*} - \sqrt{m_{11}^*} \frac{\partial}{\partial X^*} \left(\frac{c_{eq}}{1 - C^* c_{eq}} J_s^* \right)$$

$$- \frac{1}{De_m} (1 - C^* c_{eq}) \left\{ \left[1 - c_{eq}(2 - c_{eq})C^* \right] m_{11}^* - 1 \right\}$$
(14.63)

$$\frac{\partial A_{11}^*}{\partial \theta^*} = -\frac{1}{\sqrt{m_{11}^*}} \frac{\partial}{\partial X^*} \left(\frac{c_{eq} J_s^*}{(1 - C^* c_{eq})} A_{11}^* \right) - \frac{1}{De_A} C^*$$
(14.64)

Figure 14.5 shows the modeling of anomalous water moisture uptake in a 2.5 wt% clay-vinyl ester nanocomposite (data from Drozdov et al.[84, 85]). The relative water uptake W is defined as the water mass gain (per unit mass of the sample) divided by the final moisture uptake. The sorption process is clearly non-Fickian.

Qualitatively, if the polymer swells, the volume of the system increases but the total surface area of the nanoparticles does not change in, say, a fully exfoliated system. In such a case, the area tensor components decrease. Thus, $\left(\Theta \frac{\partial A_{11}}{\partial X} \right)$ has a sign opposite to that of $\frac{\partial C}{\partial X}$. That is to say, the existence of the nanoparticles decreases the flux.

14.6 Conclusion

Many factors, including the temperature, nature of the diffusing molecules, physical properties and internal structure of the polymer, external mechanical deformation, thermal effects, etc., can contribute to mass transport, and thus make it quite complex. Adding nanoparticles to a polymer matrix can strengthen the resultant membrane and can significantly improve the barrier properties of the membrane. Controlling the exfoliation, compatibility, and orientation of nanoparticles will lead to improved barrier properties. It is essential to have a fundamental understanding of mass transport through membranes to produce polymer nanocomposites with improved properties.

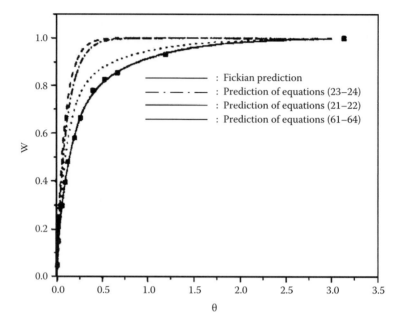

FIGURE 14.5

Normalized water mass uptake (W) vs. normalized time (θ) for a 2.5 wt% clay/vinyl ester nanocomposite.

This chapter described in detail the model derived by the Poisson bracket/GENERIC formalism. Applying a mesoscopic theory allows for an accurate characterization of the complex interface of a nanocomposite. Generalizing this approach generates a set of governing equations describing the time evolution of concentration, flux, and other variables that describe the system. In the limiting cases discussed previously, this model has been shown to quantitatively describe mass transport through nanocomposites quite well.

Investigations on mass transport will remain an important topic of scientific investigation due to the ever-increasing applications of polymer-nanocomposite systems.

While substantial progress has already been made toward understanding non-Fickian diffusion processes, further research is necessary to fully comprehend these complex processes.

Acknowledgment

Daniel De Kee gratefully acknowledges support via NASA grant NNC06AA02A.

References

1. Crank, J. and G.S. Park, *Diffusion in Polymers*. New York: Academic Press, 1968.
2. Crank, J., *The Mathematics of Diffusion*. Oxford: Oxford University Press, 1979.
3. Perkins, J.L. and M.J. You, *Am. Hyg. Assoc. J.*, 53, 77, 1992.

4. De Kee, D. et al., *J. Appl. Polym. Sci.*, 78, 1250, 2000.
5. Ghosh, P. and V.A. Juvekar, *J. Chem. Eng. Japan* 36, 711, 2003.
6. Bromwich, D. and J. Parikh, *J. Occup. Environ. Hyg.*, 3, 153, 2006.
7. Aminabhavi, T.M. and H.G. Naik, *J. Appl. Polym. Sci.*, 72,1291, 1999.
8. Crank, J. and G.S. Park, *Diffusion in Polymers*. New York: Academic Press, 1968.
9. Hayduk, W. and W.D. Buckley, *Chem. Eng. Sci.*, 27, 1997, 1972.
10. Guo, C.J. and D. De Kee, *Chem. Eng. Sci.*, 46, 2133, 1991.
11. Hinestroza, J., D. De Kee, and P.N. Pintauro, *Ind. Eng. Chem. Res.*, 40, 2183, 2001.
12. Hinestroza, J.P., Ph.D. dissertation, Tulane University, 2002.
13. Gagnard, C., Y. Germain, P. Keraudren, and B. Barriere,. *J. Appl. Polym. Sci.*, 90: 2727, 2003.
14. Hinestroza, J.P., Ph.D. dissertation, Tulane University, 2002.
15. Gagnard, C., Y. Germain, P. Keraudren, and B. Barriere,. *J. Appl. Polym. Sci.*, 90: 2727, 2003.
16. Crank, J. and G.S. Park, *Diffusion in Polymers*. New York: Academic Press, 1968.
17. Larson, R.G., *The Structure and Rheology of Complex Fluids*. New York: Oxford University Press, 1999.
18. Stastna, J. and D. De Kee, *Transport Properties in Polymers*. Lancaster, PA: Technomic Publishing, 1995.
19. Crank, J. and G.S. Park, *Diffusion in Polymers*. New York: Academic Press, 1968.
20. Deanin, R.D., *Polymer Structure, Properties and Applications*. Boston: Cahners Books, 1972.
21. Cornelis, H. and R.G. Kander, *Polymer*, 37, 5627, 1996.
22. Unnikrishnan, G. and S. Thomas, *J. Appl. Polym. Sci.*, 60, 963, 1996.
23. Drozdov, A.D., J.D. Christiansen, R.K. Gupta, and A.P. Shah, *J. Polym. Sci Part B: Polym. Phys.*, 41, 476, 2003.
24. Merkel, T.C. et al., *Chem. Mater.*, 15, 109, 2003.
25. Liu, R.Y.F., A. Hiltner, and E. Baer, *J. Polym. Sci., Polym. Phys.*, 42, 493, 2004.
26. Unnikrishnan, G. and S. Thomas, *J. Appl. Polym. Sci.*, 60, 963, 1996.
27. Drozdov, A.D., J.D. Christiansen, R.K. Gupta, and A.P. Shah, *J. Polym. Sci Part B: Polym. Phys.*, 41, 476, 2003.
28. Liu, R.Y.F., A. Hiltner, and E. Baer, *J. Polym. Sci., Polym. Phys.*, 42, 493, 2004.
29. Merkel, T.C. et al., *Chem. Mater.*, 15, 109, 2003.
30. Perrin-Sarazin, F., M.-T. Ton-That, M.N. Bureau, and J. *Polymer*, 46, 11624, 2005.
31. Li, Y., D. De Kee, C.F. Chan Man Fong, P. Pintauro, and A. Burczyk, *J. Appl. Polym. Sci.*, 74, 1584, 1999.
32. Hinestroza, J., D. De Kee, and P.N. Pintauro, *Ind. Eng. Chem. Res.* 40, 2183, 2001.
33. Hinestroza, J.P., Ph.D. dissertation, Tulane University (2002).
34. Drummond, K.M., R.A. Shanks, and F. Cser, *J. Appl. Polym. Sci.*, 83, 777, 2002.
35. Hardy, L., E. Espuche, G. Seytre, and I. Stevenson, *J. Appl. Polym. Sci.*, 89, 1849, 2003.
36. Liu, R.Y.F., A. Hiltner, and E. Baer, *J. Polym. Sci., Polym. Phys.*, 42, 493, 2004.
37. Hinestroza, J., D. De Kee, and P.N. Pintauro, *Ind. Eng. Chem. Res.*, 40, 2183, 2001.
38. Liu, R.Y.F., A. Hiltner, and E. Baer, *J. Polym. Sci., Polym. Phys.*, 42, 493, 2004.
39. Hinestroza, J., D. De Kee, and P.N. Pintauro, *Ind. Eng. Chem. Res.*, 40, 2183, 2001.
40. Liu, Q. and D. De Kee, *J. Non-Newtonian Fluid Mechanics*, 131, 32, 2005a.
41. Frederic, K. and D. De Kee, Gas permeation through nanocomposite materials, in preparation, 2008.
42. Zhong, Y., D. Janes, Y. Zheng, M. Hetzer, and D. De Kee, *Polym. Eng. Sci.*, 1101, 2007.
43. Dumont, M.-J., A. Reyna-Valencia, J.-P. Emond, and M. Bousmina, *J. Appl. Polym Sci.*, 103, 618, 2006.
44. Phillip, W.A., J. Razyev, M.A. Hillmyer, and E.L. Cussler, *J. Membr. Sci.*, 286, 144, 2006.
45. Nielsen, L.E., *J. Macromol. Sci., Part A*, 5, 929, 1967.
46. Long, F.A. and D. Richman, *J. Am. Chem. Soc.*, 82, 513, 1960.
47. Richman, D. and F.A. Long *J. Am. Chem. Soc.*, 82, 509, 1960.
48. Petropoulos, J.H. and P.P. Roussis, *J. Membr. Sci.*, 3, 343, 1978.
49. Crank, J. and G.S. Park, *Diffusion in Polymers*. New York: Academic Press, 1968.

50. Crank, J., *The Mathematics of Diffusion*. Oxford: Oxford University Press, 1979.
51. Thomas, N.L. and A.H. Windle, *Polymer*, 21, 613, 1980.
52. Thomas, N.L. and A.H. Windle, *Polymer*, 23, 529, 1982.
53. Cox, R.W. and D.S. Cohen, *J. Polym. Sci., Part B: Polym. Phys.*, 27, 589, 1989.
54. Durning, C.J., D.A. Edwards, and D.S. Cohen, *AIChE J.* , 42, 2025, 1996.
55. Chan Man Fong, C.F., C. Moresoli, S. Xiao, Y. Li, J. Bovenkamp, and D. De Kee, *J. Appl. Polym. Sci.*, 67, 1885, 1998.
56. Hinestroza, J. P., Ph.D. dissertation, Tulane University (2002).
57. El Afif, A. and M. Grmela, *J. Rheol.*, 46, 591, 2002.
58. Beris, A.N. and B.J. Edwards, *J. Rheol.*, 34, 55, 1990.
59. Liu, Q. and D. De Kee, *J. Non-Newtonian Fluid Mechanics*, 131, 32, 2005.
60. Liu, Q. and D. De Kee, *Rheol. Acta*, 44, 287, 2005.
61. Cussler, E.L., S.E. Hughes, W.J. Ward, and R. Aris, *J. Membr. Sci.*, 38, 161, 1988.
62. Falla, W.R., M. Mulski, and E.L. Cussler, *J. Membr. Sci.*, 119, 129, 1996.
63. Bharadwaj, R.K., *Macromolecules*, 34, 9189, 2001.
64. El Afif, A., R. Cortez, D.P. Gaver III, and D. De Kee, *Macromolecules* 36, 9216, 2003.
65. El Afif, A., D. De Kee, R. Cortez, and D.P. Gaver III, *J. Chem. Phys.*, 118, 10244, 2003.
66. El Afif, A., D. De Kee, R. Cortez, and D.P. Gaver III, *J. Chem. Phys.*, 118, 10227, 2003.
67. Doi, M. and T. Ohta, *Int. J. Chem. Phys.*, 95, 1242, 1991.
68. Liu, R.Y.F., A. Hiltner, and E. Baer, *J. Polym. Sci., Polym. Phys.*, 42, 493, 2004.
69. Liu, Q. and D. De Kee, *J. Non-Newtonian Fluid Mechanics*, 131, 32, 2005.
70. Liu, Q. and D. De Kee, *Rheol Acta*, 44, 287, 2005.
71. Wetzel, E.D. and C.L. Tucker III, *Int. J. Multiphase Flow*, 25, 35, 1999.
72. Liu, R.Y.F., A. Hiltner, and E. Baer, *J. Polym. Sci., Polym. Phys.*, 42, 493, 2004.
73. Liu, Q. and D. De Kee, *J. Non-Newtonian Fluid Mechanics*, 131, 32, 2005.
74. Beris, A.N. and B.J. Edwards, *Thermodynamics of Flowing System with Internal Microstructure*. New York: Oxford University Press, 1994.
75. El Afif, A., M. Grmela, and G. Lebon, *J. of Non-Newtonian Fluid Mechanics*, 86, 253, 1999.
76. El Afif, A., D. De Kee, R. Cortez, and D.P. Gaver III, *J. Chem. Phys.*, 118, 10227, 2003.
77. Doi, M. and T. Ohta, *Int. J. Chem. Phys.*, 95, 1242, 1991.
78. Grmela, M. and H.C. Ottinger, *Phys. Rev. E.*, 56, 6620, 1997.
79. Ottinger, H.C. and M. Grmela, *Phys. Rev. E.*, 56, 6633, 1997.
80. Beris, A.N. and B.J. Edwards, *Thermodynamics of Flowing System with Internal Microstructure*. New York: Oxford University Press, 1994.
81. El Afif, A., D. De Kee, R. Cortez, and D.P. Gaver III, *J. Chem. Phys.*, 118, 10244, 2003.
82. El Afif, A., D. De Kee, R. Cortez and D.P. Gaver III, *J. Chem. Phys.*, 118, 10227, 2003.
83. Bharadwaj, R.K., *Macromolecules* 34, 9189, 2001.
84. Liu, Q. and D. De Kee, *J. Non-Newtonian Fluid Mechanics*, 131, 32, 2005.
85, Drozdov, A.D., J.D. Christiansen, R.K. Gupta, and A.P. Shah, *J. Polym. Sci Part B: Polym. Phys.*, 41, 476, 2003.

15

Flammability Properties of Polymer Nanocomposites

Jin Zhu and Charles A. Wilkie

CONTENTS

15.1 Introduction

Polymer nanocomposites are one of the most important nanomaterials developed in the past decade. In nanocomposites, the dispersed particles in the polymer matrix have at least one dimension at the nanoscale (<100 nm).[1, 2] Nanocomposites can be classified into three categories according to the number of nanoscale dimensions of the nanofillers dispersed in polymer matrix: (1) lamellar, (2) nanofiber or nanotube, and (3) spherical polymer nanocomposites. The lamellar structural nanofillers include clays, graphite, and layered double hydroxides (LDHs). Carbon nanotubes are the best-known example of the two-dimensional systems but a variety of inorganic nanotubes also exist[3] and these do not appear to have been widely studied in combination with polymers. The best-known example of the three-dimensional materials is polyhedral oligomeric silsesquioxane (POSS).

In the presence of nanofillers in the polymer matrix, polymer nanocomposites exhibit enhanced properties compared to their microstructured counterparts. These properties include enhanced mechanical, barrier, and fire properties. Of these properties, fire retardancy has received special interest. Compared to conventional flame retardants, nanofillers can greatly improve fire retardancy at a small loading (typically less than 5 wt%). A recent book comprehensively reviews the topic of the flame retardancy of nanocomposites.[4]

The fire property of polymer nanocomposites has been studied for more than 10 years. Polymer nanocomposites were first found to exhibit fire retardancy by Giannelis[5] in 1996. Through cone calorimetry, Gilman[6,7] found that the peak heat release rate (PHRR) can be reduced by more than 50% in the presence of only 5 wt% nanoclay compared to the virgin polymers, while there is essentially no reduction for a microcomposite, also known as an immiscible system. It has been noted that the combination of nanocomposites with conventional flame retardants is necessary to achieve the flame retardancy requirement for industrial applications.[8]

This chapter focuses on the flame retardancy of polymer nanocomposites; the principal focus is on the clay-containing systems but some attention is also devoted to carbon nanotubes and to three-dimensional particles. This includes the mechanism and characterization of flame retardancy, fire properties of various polymer nanocomposites, and the flame retardancy of polymer nanocomposite together with conventional flame retardants.

15.2   Validation of Nanocomposite Formation

Nanocomposites are usually classified into three categories: (1) immiscible (also known as microcomposite), (2) intercalated, and (3) exfoliated (also known as delaminated).[1] In the immiscible system, the polymer does not penetrate between the clay layers and the clay acts as a traditional filler; there is no nanodispersion. In the other two cases, nanodispersion of the clay is achieved. Intercalated hybrids retain the registry of the clay layers but the distance between the layers increases. No registry is maintained in exfoliated nanocomposites and one clay platelet may have any orientation with respect to another; the separation between clay layers may be as large as 10 to 15 nm. A 100% exfoliated nanocomposite is difficult to achieve and probably has never been produced. Generally, a nanocomposite contains both intercalated and exfoliated morphologies and it is not unusual for there also to be some immiscible component.

The inorganic clay is not used directly to make the polymer-clay nanocomposites because this is a hydrophilic material that will not show good compatibility with the organic polymer. Rather, one must first ion-exchange the clay to insert an organophilic ion, usually an ammonium ion, into the gallery space so that it will be compatible with the polymer. There have been a few studies on the solubility parameter and its effect on miscibility of the polymer and the clay.[9,10] The commercially available, organically modified clays contain either one or two long carbon chains, along with either alkyl groups or hydroxyethyl groups. The presence of the hydroxyethyl group imparts some polarity to the surfactant and this makes it more attractive for somewhat polar polymers.

There are a number of techniques that have been used to evaluate nanocomposite formation but the most common is the combination of x-ray diffraction (XRD) and transmission electron microscopy (TEM). XRD provides the d-spacing, if the d-spacing has increased and maintains its intensity; this is a good indicator that nanodispersion has been achieved. In some instances, the peak seems to vanish upon nanocomposite formation and some investigators want to take this as an indication that the clay layers are now widely dispersed and that very good nanodispersion has been achieved. It is also possible that the clay is simply disordered and that there is no nanodispersion. For this reason, one should never rely on only XRD measurements to evaluate the morphology of a nanocomposite.[11] The usual complement to XRD is TEM, which will provide an image of the clay in the polymer. Usually, two images are provided, one at somewhat low magnification, which enables one to see how well dispersed the clay layers are, and one at higher magnification, which enables one to see the individual layers and evaluate their orientation. Because TEM examines only a very small fraction of the material and one then extrapolates to say that this is representative of the rest of the material, the use of this technique is fraught with problems. To truly understand the morphology of a material, one must do a statistical sampling of the material and likely use image analysis also to be sure of the conclusions. The majority of systems will consist of at least two, and sometimes all three, of the possible morphologies.

A bulk measurement of morphology is greatly needed to provide good information on these systems. Techniques that have been used include infrared spectroscopy,[12] nuclear magnetic resonance spectroscopy,[13] atomic force microscopy,[14] and rheological measurements.[15–18]

It is widely felt that exfoliated nanocomposites exhibit the best performance. While this may be true for barrier and mechanical properties, it is not true for flame retardancy applications. In the majority of investigations, no difference was observed between an intercalated and an exfoliated morphology. There is certainly a drive to obtain exfoliation but this is not necessarily important for fire applications.

15.3 Mechanism of Flame Retardancy of Polymer Nanocomposites

Several mechanisms have been proposed to explain the flame retardancy of polymer nanocomposites. The first reported mechanism, proposed by Gilman et al.,[19] was that the accumulation of clay at the surface of the degrading material functions as a barrier to both mass transport of degradation products and thermal transfer of energy from the heat source to the polymer that underlies the clay. Char accumulation during the burning process can be observed in Figures 15.1[20] and 15.2.[21] Char is not formed upon burning of pure polyamide 6, but a significant amount of char was formed for polyamide 6 with 2 or 5 wt% clay. If the char is stiff and continuous, it will be more effective in enhancing flame retardancy.[22]

The migration of clay onto surface was also confirmed by XRD and attenuated total reflectance Fourier transform infrared (ATR-FTIR) measurements on the isothermally heated samples.[23] There are a few possible mechanisms for the accumulation of char at the surface.[20] One mechanism is due to the force of the volatile products produced in the bulk of the material. The numerous rising bubbles during combustion push the nanoparticles outward from the burning area. Another mechanism is due to recession of polymer resin from the surface during pyrolysis, leaving the de-wetted nanoparticles behind. The decomposition of the organic modifier in the clay will lead to aggregation of the clay. Another suggestion is that the surface free energy of the clay is such that the clay rises to the surface at relatively low temperatures, even as low as 200 to 250°C.[23]

It should also be pointed out that it has recently been suggested that it may not even be necessary for the clay to rise to the surface to function as a barrier. The thermal degradation products of a nanocomposite are different from those of the virgin polymer, in most cases. This has been explained by noting that the clay retains the degrading radicals for enough time to permit radical recombination reactions to occur. This means that polymers will be reformed, although different in small ways from the starting polymer, and the time of burning will be spread out. Thus, the clay platelet may function as a barrier wherever it is in the material.[24–29]

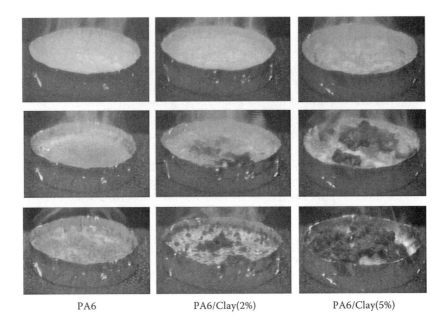

PA6   PA6/Clay(2%)   PA6/Clay(5%)

FIGURE 15.1
Char formation of nylon 6 clay nanocomposites during combustion. (*Source:* Reprinted from T. Kashiwagi, R.H. Harris, Jr., X. Zhang, R.M. Briber, B.H. Cipriano, S.R. Raghavan, W.H. Awad, and J.R. Shields, *Polymer*, 44, 881, 2004. Copyright 2004. With permission from Elsevier.)

An additional mechanism for enhanced flame retardancy is paramagnetic radical trapping by iron or other paramagnetic species that are present in the clay.[30] The barrier mechanism can be effective at high amounts of clay; but at very low amounts, there is not enough material to form a good barrier. Yet, even at 0.1% clay in the polymer, there is good evidence for flame retardancy.[31] The iron is present due to the isomorphous substitutions that occur in the clay. It appears that radical trapping is only important at quite low levels of clay and that as the amount of clay increases, the barrier mechanism is dominant.

FIGURE 15.2
Char formation build-up on the surface during burning of polymer nanocomposites. (*Source:* Reproduced with permission from Süd Chemie from a presentation given by P. Andersen at SPI.)

15.4   Characterization of Fire Retardancy of Polymer Nanocomposites

There are several methods to evaluate the fire properties of polymeric materials, depending on the type of information desired. There are three indirect methods: (1) cone calorimetry (ASTM E 1354), (2) radiative gasification,[32] and (3) limiting oxygen index (LOI) (ASTM D2863, ISO 4589). The UL-94 (ISO 9772 and 9773, ASTM D635) test is a direct burning method that is used for commercial products to evaluate and qualify a material. Both cone calorimetry and radiative gasification provide information on the heat release rate while the oxygen index measures the ease of extinction and the UL protocol evaluates the ease of ignitability. One must be very careful not to compare data obtained from one test with data from a different test protocol, because each measures a different aspect of fire retardancy. It should not be expected that if a material performs well by one evaluative technique that it will also show good performance by another technique. It is well known that almost all nanocomposites show good reductions in the peak heat release rate when evaluated by cone calorimetry but will completely burn in the UL-94 protocol, retaining their shape, and will not show improvement in the oxygen index.

Cone calorimetry is a quantitative laboratory method to evaluate fire performance of polymeric materials; a schematic diagram of the instrument is shown in Figure 15.3.[33, 34]

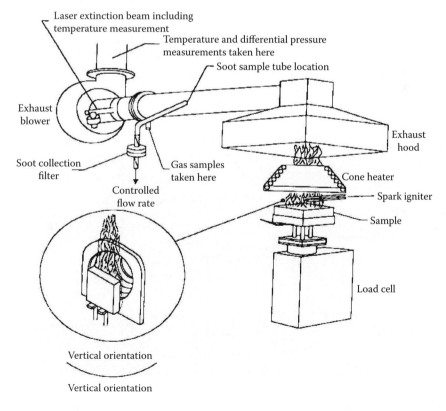

FIGURE 15.3
The cone calorimeter. (*Source:* From http://doctorfire.com/cone.html.)

The fire performance parameters obtained include the time to ignition, the entire heat release rate as a function of time and especially its peak value and the heat of combustion, the rate of mass loss, the smoke that is produced, and the production of CO and CO_2. The test is on the small scale, normally 20 to 50 g of sample is used. The entire heat release rate (HRR) curve is obtained but the main focus is usually on the peak heat release rate (PHRR). The test is very effective to evaluate flame retardancy but there is no necessary correlation between the results of this test and other evaluations of fire performance. Cone calorimetry is usually considered the method of choice for the bench-scale evaluation of flame retardancy.

Radiative gasification is a variation on the cone calorimeter in which the heat source is applied in a nitrogen atmosphere; the radiative gasification device is schematically shown in Figure 15.4.[35] Because the pyrolysis is conducted in nitrogen, flaming does not occur and thus smoke is eliminated, which enables the presence of a video camera to view the pyrolysis process. This technique, which is not widely used, enables the viewing of the degrading polymer so one can see and identify the processes that are occurring.

The limiting oxygen index (LOI) is measured by controlling the concentration of oxygen in a synthetic nitrogen-oxygen atmosphere during the burning of a specimen of polymeric material. The minimum oxygen concentration required to sustain burning of a sample is recorded. Generally speaking, if the LOI is near 20 (close to the concentration of oxygen in air), then it will burn easily. The higher the value can be raised, the more likely it is that some level of flame retardancy has been obtained. There is no magic value at which one can say that flame retardancy has been achieved. In some cases, an LOI of 30 may be acceptable and in other cases a much higher value may be required.

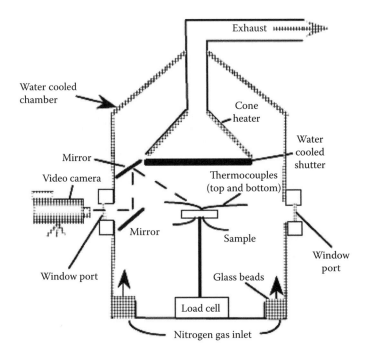

FIGURE 15.4
Schematic of gasification device. (*Source:* Gilman, J.W. et al. NIST, *Flammability of Polymer Clay Nanocomposites Consortium: Final Report,* November, 2003, http://www.nist.gov/pcnc/nano_consort_report_final_24.pdf.)

TABLE 15.1

UL-94 Flammability Ratings Summary

UL-94 ratings	Description of Ratings
5VA Surface Burn	Burning stops within 60 seconds after five applications of 5 seconds each of a flame (larger than that used in Vertical Burn testing) to a test bar. **Test specimens may not have a burn-through (no hole).** **This is the highest (most flame retardant) UL-94 rating.**
5VB Surface Burn	Burning stops within 60 seconds after five applications of five seconds each of a flame (larger than that used in Vertical Burn testing) to a test bar. **Test specimens may have a burn-through (a hole).**
V-0 Vertical Burn	Burning stops within 10 seconds after two applications of 10 seconds each of a flame to a test bar. **NO flaming drips are allowed.**
V-1 Vertical Burn	Burning stops within 60 seconds after two applications of 10 seconds each of a flame to a test bar. **NO flaming drips are allowed.**
V-2 Vertical Burn	Burning stops within 60 seconds after two applications of 10 seconds each of a flame to a test bar. **Flaming drips are allowed.**
H-B Horizontal Burn	Slow horizontal burning on a 3-mm thick specimen with a burning rate of less than 3″/min or stops burning before the 5″ mark. H-B rated materials are considered "self-extinguishing." **This is the lowest (least flame retardant) UL-94 rating.**

Source: From Fire Testing Technology Ltd., http://www.fire-testing.com/html/instruments/ul94ad.htm; see Reference 36.

The UL-94 test is a widely accepted, small-scale method test in industry for flammability of plastic materials;[36] this method is usually used to qualify materials. There are three different burning methods: (1) surface, (2) vertical and horizontal burning, and (3) seven UL-94 tests (94HB, 94V, 94VTM, 94-5V, 94HBF, 94HF, and Radiant Panel).[36] The test results are ranked according to burning methods and phenomena. Only polymeric materials that achieve a certain level of flame retardant effect receive a rating; those that cannot achieve a rating are designated as "not classified" (nc). The UL flammability ratings are summarized in Table 15.1.

15.5 Flame Retardancy of Polymer Nanocomposites

Although polymer nanocomposites show improved flame retardancy, different kinds of nanofillers have different effects on improvement in flame retardancy. The fire properties of polymer nanocomposites are related to the types of nanofillers, types of polymeric resins, and the nanostructures obtained. In the following, we review the flame retardancy of polymer nanocomposites with various nanofillers.

15.5.1 Flammability of Polymer Clay Nanocomposites

Clay is either an aluminosilicate or a silicate with a layered structure. The clays contain an octahedral sheet sandwiched between two tetrahedral sheets. Montmorillonite (MMT)

$[M_x(Al_{4-x}Mg_x)Si_8O_{20}(OH)_4]$ is the most commonly used clay because it is ubiquitous in nature and can be obtained in high purity at low cost[1] and it also exhibits very rich intercalation chemistry so that it can be easily organically modified. In the following discussion, the clay used to form the polymer nanocomposites is montmorillonite, unless otherwise noted. More information can be found in the recent book on polymer-clay nanocomposites.[37] A review on the range of synthetic clays has recently appeared.[38]

The flame retardancy of polymer clay nanocomposites has received significant attention because flame retardancy was one of the first desirable properties noted. Nearly all polymers exhibit improved flame retardancy, as evaluated by cone calorimetry, upon the incorporation of clay. The following summarizes the research progress on clay nanocomposites based on different polymers.

15.5.1.1   Flammability of Styrenic Nanocomposites

The flammability of polystyrene-clay nanocomposites has been thoroughly studied. It was reported that the peak heat release rate was reduced by more than 40% upon incorporation of as little as 0.1% mass fraction compared with virgin polystyrene.[31] The onset temperature of the degradation (TGA) also increased by about 40°C. The fire properties of polystyrene-clay nanocomposites prepared by bulk polymerization using both ammonium- and phosphonium-modified clays have been investigated.[39] The fire properties of these polymer nanocomposites were little affected by different nanostructures, although an exfoliated nanocomposite showed the highest thermal stability. Figure 15.5 shows the heat release rate of polystyrene nanocomposites with different nanostructures. The intercalated nanocomposite (PS-OH16) exhibited a little lower PHRR than exfoliated (PS-VB16) or mixtures of intercalated and exfoliated (PS-P16). The peak heat release rate was reduced by 27 to 58%, depending on the amount of clay that was present, and the mass loss rates were also

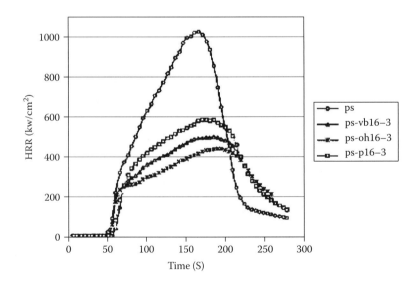

FIGURE 15.5
Heat release rate of polystyrene nanocomposites with different nanostructures. Note: PS-VB16-3 is an exfoliated nanocomposite; PS-OH16-3 is an intercalated nanocomposite; PS-P16 is a mixture of both exfoliated and intercalated nanocomposite. (*Source:* Reprinted with permission from *Chem. Mater.*, 13, 3774–3780 (2001). Copyright 2001, American Chemical Society.).

significantly reduced in the presence of the clay. Silicon-methoxide-modified clays and their polystyrene nanocomposites have also been studied.[40] Although the silicon-methoxide offers the possibility of reaction between the methoxide and a clay hydroxyl group to link together the cation and the clay, the nanocomposites did not show any difference in flammability compared to other examples of polystyrene nanocomposites.

In addition to modification with ammonium or phosphonium cations, other methods can be used to modify the clay so that it can be incorporated into a polymer. The flame retardancy of polystyrene nanocomposites using crown-ether-modified clays did not show the same improvement as seen when ammonium-cation-exchanged clay nanocomposites were used. The PHRR of nanocomposites was reduced by 25% to 30% compared with the normal 50% to 60% reduction in PHRR observed for other polystyrene nanocomposites. This suggested that there may also be some immiscible component present.

A new class of organically modified clays, oligomerically modified clays, was pioneered by Su[42–44] using oligomeric cations of styrene, methacrylate, and butadiene. The polystyrene nanocomposites based on this clay exhibited an exfoliated nanostructure upon melt-blending in a Brabender mixer. The flammability was quite similar to those of small-cation-modified clay nanocomposites. A tropylium-substituted clay was prepared by Zhang,[45] and its polystyrene nanocomposites exhibited a poorer reduction in the PHRR than that seen for the typical ammonium-substituted clays. Only very imperfectly nanodispersed clay was obtained in this case and that is most likely the reason for the poorer fire performance.

The flammability of polystyrene nanocomposites with different types of layered silicates was studied by Gilman.[46] This study indicated that the type of layered silicate, nanodispersion, and processing conditions had an influence on the fire properties. There were different effects on flame retardancy between montmorillonite and fluorohectorite. Fluorohectorite had no effect on PHRR while montmorillonite showed a 60% reduction in PHRR, compared to virgin polystyrene. The flammability of polystyrene nanocomposites was also affected by the clay loading and the polymer melt viscosity, according to the study of Morgan.[47, 48] While the viscosity of the polystyrene nanocomposite played a role in lowering the PHRR, the clay loading had a larger effect. The clay was also able to catalyze carbonaceous char formation and reinforce the char, which is believed responsible for the improved flame retardancy of nanocomposites.

High-impact polystyrene (HIPS) and acrylonitrile-butadiene-styrene terpolymer (ABS) nanocomposites have been studied. The HIPS nanocomposites based on the oligomeric-styrene-modified clay[42] exhibited about a 40% reduction in PHRR, which is comparable to that of other polystyrene nanocomposites. The reductions for the ABS nanocomposites were about 25%; the reductions for ABS is lower because the stability of some of the radicals produced upon degradation is lower, due to the presence of the nitrile, so the polymer degrades more rapidly.[28] Very similar reductions in the peak heat release rate have also been observed for nanocomposites in which graphite is the nanodimensional material (this is discussed in a subsequent section).

15.5.1.2 Flammability of Olefinic Clay Nanocomposites

The flame retardancy of polyolefins was improved by the incorporation of clay. Studies[49] on poly(propylene-*graft*-maleic anhydride)-layered silicate nanocomposites using montmorillonite or fluorohectorite indicated that a clay-reinforced carbonaceous char formed during combustion of the nanocomposites. This is particularly significant for polymers that produce little or no char when burned alone; reductions in the heat release rates of up to 70% to 80% were achieved in this system. Wang[66] studied the flammability of polypropylene nanocomposites using polypropylene-*graft*-maleic anhydride (PP-g-MA) as a compatibilizer. The typical values of PHRR reduction range from 11% to 34%, which are lower than what is seen when PP-g-MA is used directly to make the nanocomposites and the presence of a substantial immiscible portion in this system was found and explains the observation. The use of the oligomeric styrene-modified clay[42] enables the preparation of polypropylene nanocomposites in a Brabender mixer without the need for maleic anhydride; the resulting material was a mixed immiscible-intercalated-exfoliated nanocomposite and the reduction in PHRR was 35%.

The flammability of polyethylene clay nanocomposites was studied using melt-blending in a Brabender mixer,[50] either in the presence or absence of maleic anhydride. It was found that the reduction in PHRR was 30% to 40%. The reduction in PHRR and in the mass loss rate was about the same whether maleic anhydride was present or absent. Similar reductions in PHRR were observed when the oligomeric styrene-containing clay was used.[42] Similar flammability results were found in polyethylene-clay nanocomposites prepared by the melt intercalation technique directly from sodium montmorillonite in the presence of a reactive compatibilizer, hexadecyltrimethylammonium bromide. The nanocomposite showed a 32% reduction in PHRR,[51] while a microcomposite showed the same combustion behavior as pure PE.

The fire retardancy of poly(methyl methacrylate) (PMMA) clay nanocomposites had been studied by Zhu et al.[52] The reduction in PHRR was in the range of 20% to 28% for PMMA nanocomposites, less effective than in the polystyrene system, and no difference was observed between intercalated and exfoliated systems. PMMA nanocomposites showed a slightly longer time to ignition compared with neat PMMA;[53] for most polymer systems, the time to ignition of nanocomposites are shorter than for the virgin polymer. Typical heat release rate curves for a mixed intercalated-exfoliated poly(methyl methacrylate) nanocomposite at two different clay amounts are shown in Figure 15.6.[54]

The flame retardancy of polyvinyl chloride (PVC) clay nanocomposites was studied by Wang.[55–57] The nanocomposites were prepared by melt-blending with organically modified clay or sodium clay with or without plasticizers. The plasticized PVC exhibits lower

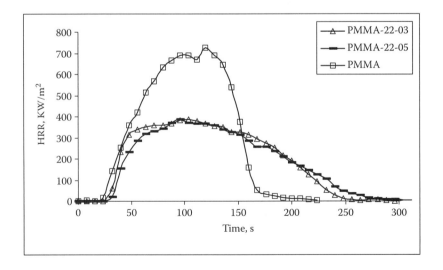

FIGURE 15.6
Heat release rate curves for PMMA and its nanocomposites with 3% and 5% clay. (*Source:* Reprinted from *Thermochim. Acta*, 435, 202–208, 2005, X. Zheng, D.D. Jiang, and C.A. Wilkie. Copyright 2005. With permission from Elsevier.)

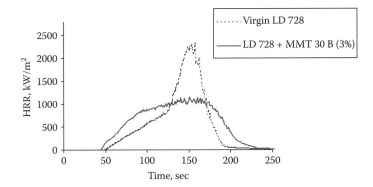

FIGURE 15.7
Heat release rate for EVA and its nanocomposite.

fire retardancy than the polymer because the plasticizer adds fuel, but PVC is difficult to process without the addition of plasticizers. Addition of clay can reduce the amount of plasticizers and further improve the flame retardancy of PVC. The concern with PVC is not the reduction in PHRR as PVC already shows a small enough heat release rate; rather, the problem with PVC is smoke generation and its reduction. The addition of clay did not increase the smoke and may in some cases cause some reduction.

The combustion behavior of an EVA/fluorohectorite nanocomposite has been studied by Zanetti.[32, 58] The nanocomposite showed acceleration of EVA deacetylation and delayed volatilization of the resulting polyene; the overall heat release rate was much lower than in the case of EVA. Accumulation of the silicate on the surface of the burning specimen may create a protective barrier to heat and mass transfer that is, however, much more effective for the nanocomposite than for the immiscible composite. A typical heat release rate plot for an EVA nanocomposite is shown in Figure 15.7. Nondripping behavior in vertical combustion was found only in the case of the nanocomposite, which reduces the hazard of fire spread to surrounding flammable materials. Studies of clay dispersion on flammability of the EVA nanocomposites were investigated by Morgan[59] and it was found that both intercalated and exfoliated nanocomposites had a similar effect on flammability. Particle sizes of pristine montmorillonite also had an effect on the flammability of EVA nanocomposites.[60] The PHRR of the nanocomposite loaded with the smaller size particles was lower than that for the larger particle size.

15.5.1.3 Flammability of Other Polymer Clay Nanocomposites

The polyamide-6-clay nanocomposite was the first discovered polymer nanocomposite. The flammability of polyamide 6 nanocomposites has been studied by Gilman et al,[61, 62] and the cone calorimeter was used as the principal tool to evaluate the exfoliated polyamide-6-clay nanocomposites. The PHRR is reduced by 63% in a polyamide-6-clay nanocomposite containing a 5% mass fraction of clay. The nanocomposites not only showed improved flame retardancy, but the mechanical properties also improved (compared with virgin polyamide).

The fire properties of vinyl ester nanocomposites have been studied. The PHRR of vinyl ester nanocomposites showed a reduction on the order of 25% to 40% compared with virgin resin.[63–65] In addition to the change in PHRR, there was a change in the mass loss rate while the heat of combustion, soot, and carbon monoxide yields remained unchanged.

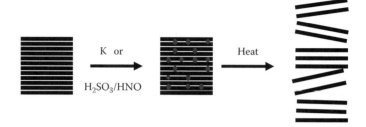

FIGURE 15.8
Formation of expanded graphite.

The thermal stability and fire properties were improved in other polymer systems such as poly(o-methoxyanilin),[66] poly(-caprolactone),[67–69] Poly(ether imide),[70] cyanate ester resins,[71] polycarbonate,[72] unsaturated polyester, and phenolic resins.[73] The flammability of polyurethane-clay nanocomposites was studied by Berta et al.[74] The presence of the clay brought about a reduced heat release rate and prevented dripping during combustion, but clay alone could not achieve nonburning behavior; to accomplish this, a conventional flame retardant must be added. Nanocomposite technology has been used to improve fire retardancy of fabrics by Bourbigot.[75] Both nanocomposite yarns and coatings were used for making fire-retarded textiles using nanocomposite technology.

15.5.2 Flammability of Polymer Graphite Nanocomposites

Because char formation in the process of combustion for polymer clay nanocomposites plays a vital role, the rigidity of the char is very important. The stronger the char layers that form, the more effective the role that the char plays in barrier formation. Studies have indicated that the clay formed a graphite-type char during the combustion of nanocomposites,[76] which leads to the possibility of developing polymer-graphite nanocomposites. If the graphite can be nanodispersed in the polymer matrix, the char formed during the combustion should be stronger and more effective. Two kinds of graphites have been exploited to form polymer-graphite nanocomposites: (1) expanded graphite and (2) graphite oxide. The expanded graphite can be synthesized by intercalation of either potassium or sulfuric acid in the graphite. When the expandable graphite (graphite-sulfuric acid) is heated, it will rapidly expand up to 2000 times to form nanolayers as depicted in Figure 15.8.

Polystyrene-graphite nanocomposites have been prepared by *in situ* polymerization of styrene in the presence of potassium graphite.[77] The d-spacing increased from 5.54Å in KC_8 to 15.1Å in the nanocomposite. The best reduction in the peak heat release rate (i.e., 43%) suggests that good nanodispersion has been obtained and the TEM images also confirm that good nanodispersion was obtained.

Uhl[78] has examined the fire properties of polystyrene nanocomposites with sulfuric acid intercalated graphite. The reduction in the PHRR was 50%; this value is in the same range as seen for montmorillonite systems and indicates that nanocomposite formation has occurred and that graphite may be used in place of clays for fire retardancy. The time to ignition for graphite-containing nanocomposites is routinely lower than that for clay-based systems, which is a large disadvantage for these systems. A typical heat release rate plot for a polystyrene-graphite nanocomposite is shown in Figure 15.9.

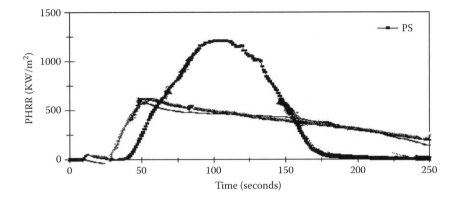

FIGURE 15.9
Heat release rate curves for virgin polystyrene and its graphite nanocomposite. The upper curve is virgin polystyrene while the lower one (showing a decreased heat release rate) is the nanocomposite. (*Source:* Reprinted with permission of John Wiley & Sons, Inc. from *Polymers for Advanced Technologies*, 16, 533, 2005, F.M. Uhl, Q. Yao, and C.A. Wilkie.)

The flame retardancy of polyamide-graphite nanocomposites was studied by Uhl.[79] The results showed that typical reductions in PHRR range from 50% to 65% were achieved, very similar to that seen for clay systems. The mass loss rate was also reduced and there was only a small increase in smoke production.

Graphite oxide can also be used to make polymer nanocomposites;[80] graphite oxide is obtained by the oxidation of graphite. Oxidation causes the formation of functional groups on the graphite, with the formation of a negative charge on the graphite layers, which therefore requires the presence of some counter ion; graphite oxide has an intercalation chemistry similar to that of clay. The ion exchange method can be used to surface-modify the graphite oxide. Using the emulsion polymerization technique, both polystyrene nanocomposites[81] and styrene-butyl acrylate nanocomposites[82] have been prepared in the presence of graphite oxide, and a mixed intercalated/delaminated material was obtained. A well-dispersed styrene nanocomposite has been achieved by Uhl,[80] who showed that the nanocomposite could only be obtained by *in situ* polymerization, as the graphite oxide thermally decomposes at the temperature at which melt-blending occurs. The heat release rate curve for a polystyrene-graphite oxide formulation is shown in Figure 15.10.

15.5.3 Flammability of Polymer-Double Layered Hydroxide Nanocomposites

Layered double hydroxides (LDHs) are synthetic minerals with positively charged brucite-type layers of metal hydroxides.[83] An LDH contains exchangeable anions located in the interlayer spaces that compensate for the positive charge on the surface, just the reverse of the cations in the gallery space for montmorillonite. Compared with the cation exchange capacity in montmorillonite, the anion exchange reaction in LDHs is more difficult due to their high selectivity for carbonate anion (CO_2 must be carefully and completely excluded during the preparation) and the large anion exchange capacity. Organic sulfates and sulfonates and carboxylates are widely used to modify LDHs before polymer intercalation.

The flame retardancy of epoxy-LDH nanocomposites was found to be much better than conventional clay-based nanocomposites or ATH-based traditional flame retardant microcomposites. LDH-filled epoxy nanocomposites showed self-extinguishing behavior in the

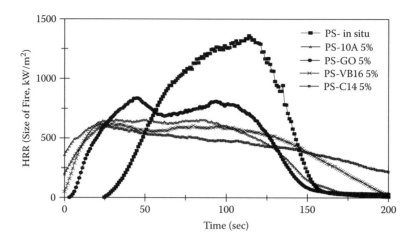

FIGURE 15.10
Heat release rate curves for polystyrene and polystyrene—graphite oxide nanocomposites. (*Source:* Reprinted from *Polymer Degradation and Stability*, 84, 215–226, 2004, F.M. Uhl and C.A. Wilkie. Copyright 2004. With permission from Elsevier.)

horizontal UL-94 HB test;[84] this is the first report of a self-extinguishing nanocomposite containing only nano-additives. Table 15.2 provides a comparison of the linear burning rates in a UL-94 HB test for the LDH-based epoxy nanocomposites and conventional nanocomposites or microcomposites. Compared to neat epoxy resin, the epoxy composite with montmorillonite (nanocomposite) showed a bit lower linear burning rate, but it did burn. The epoxy composite with magnesium-aluminum LDH (immiscible composite) also showed behavior similar to neat epoxy resin with a slightly lower burning rate. However, epoxy with LDH (nanocomposite), modified either with 3-aminobenzenesulfonic acid or 4-toluenesulfonic acid, demonstrated a self-extinguishing behavior. The cone calorimeter analyses confirmed the better performance of LDH-based nanocomposites. The LDH-based epoxy nanocomposites showed a decreased PHRR, up to 51% compared with neat epoxy, while montmorillonite based nanocomposites showed only a 27% reduction. The excellent flame retardancy of LDH-based nanocomposites is probably due to the nano-intumescent behavior of organically modified LDH. The char formed after combustion in LDH nanocomposites was intumescent and more compact, while the char in the montmorillonite nanocomposite showed a fragmented structure, which would be less efficient to protect the resin underneath from heat or flame.

TABLE 15.2

Linear Burning Rates of Epoxy Samples Obtained from UL-94HB test

Epoxy Samples	Linear Burning Rates (mm/min) and Notes
Neat epoxy resin	22.1 and burned completely
Epoxy resin with Cloisite 30B (modified montmorillonite)	20.0 and burned completely
Epoxy resin with aluminum trihydroxide (ATH)	19.1
Epoxy resin with magnesium-aluminum LDH	18.6 and burned completely
Epoxy with LDH modified by 3-aminobenzenesulfonic acid	Self-extinguishing
Epoxy with LDH modified by 4-toluenesulfonic acid	Self-extinguishing

Source: Adapted from Zamarano, M. et al., *Polymer*, 46, 9314, 2005.

In more recent work, nanocomposites of an LDH with polyethylene, polystyrene, and EVA have been examined.[85] The LDH is not as well dispersed in the polymers but the reduction in the PHRR is much larger than what would be expected for a microcomposite. These systems require further investigation to better understand how they resemble, and how they differ from, the montmorillonite nanocomposites.

15.5.4 Flammability of Polymer Nanotube or Nanofiber Nanocomposites

The two-dimensional nanoscale fillers include carbon nanotubes (single-walled and multi-walled), carbon nanofibers, and other inorganic nanofibers or nanotubes. Carbon-nanotube-based polymer nanocomposites have been intensively studied by Kashiwagi and others. The flammability of nanocomposites was measured with a cone calorimeter in air and a gasification device in a nitrogen atmosphere. Kashiwagi[86] reported that poly(methyl methacrylate) (PMMA) nanocomposites with 0.5 wt% single-walled carbon nanotubes (SWNTs) showed more than 50% reduction in the PHRR compared with neat PMMA. The flame retardancy of the nanocomposites was related to the concentration and dispersion of the SWNT; the flame retardancy of the nanocomposites was poor with less than 0.2 wt% SWNT. The nanocomposites with poorly dispersed SWNTs showed much higher PHRR values than the nanocomposites with well-dispersed SWNTs. It was proposed[86] that the SWNTs formed a continuous network that acts as a heat shield to slow the thermal degradation of PMMA. The heat release rate curves for PMMA and its nanocomposites with poorly dispersed and well-dispersed SWNT are shown in Figure 15.11.

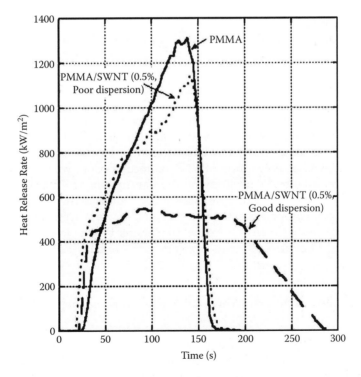

FIGURE 15.11
Heat release rate curves for PMMA and a nanocomposite in which SWNT is poorly dispersed and one in which it is well dispersed. (*Source:* Reprinted from *Polymer*, 46, 471–481, 2005, T. Kashiwagi, F. Du, K.I. Winey, K.M. Groth, J.R. Shields, S.P. Bellayer, H. Kim, and J.F. Douglas., Copyright 2005. With permission from Elsevier.)

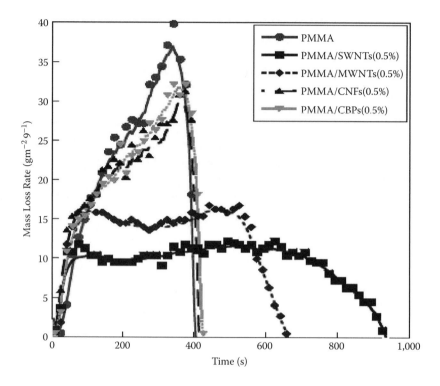

FIGURE 15.12

Mass loss rate of PMMA nanocomposites with various carbon nanofillers. (*Source:* Reprinted with permission from *Nature Materials*, 4, 928–933, 2005, T. Kashiwagi, F. Du, J.F. Douglas, K.I. Winey, R.H. Harris, Jr., and J.R. Shields. Copyright 2005.)

The flammability of polymer nanocomposites based on multiwalled carbon nanotubes (MWNTs) was studied in polypropylene[87, 88] and poly(ethylene-co-vinyl acetate).[89] The thermal stability and flame retardancy were greatly improved compared with neat polymers. The fire behavior of polyamide-MWNT nanocomposites was also shown to exhibit improved flame retardancy.[90] The study on the flammability of PMMA nanocomposites with various carbon nanofillers indicated that SWNTs showed the biggest improvement in flame retardancy and carbon black showed the smallest improvement (see Figure 15.12).[22] From the char residue structure of the cone calorimeter test, it was found that the SWNTs and MWCTs formed better continuous network protective layers than carbon nanofibers and carbon black.

15.5.5 Flammability of Polymer Three-Dimensional Nanoparticle Nanocomposites

Here, the nanoparticles are considered as those in which all three dimensions are at the nanoscale (<100 nm). The flammability of these types of polymer nanocomposites has not been extensively studied, possibly because some of these particles do not form a strong residue and POSS (polyhedral oligomeric silsesquioxane) is expensive. Yang[91] reported that the flammability of polycarbonate nanocomposites based on nano-silica and nano-alumina was improved compared with virgin polycarbonate, but polycarbonate/silica showed less change in oxygen indices and less impact in peak heat release rate reduction when compared to polycarbonate/alumina nanocomposites. This suggested that the performance of nanocomposites depends greatly on the interfacial interaction between nanoparticles and polymer matrix. Morgan[92] reported the thermal and flammability properties of a silica-PMMA nanocomposite. Silica nanoparticles were used as a fire retardant additive for PMMA; silica gel was more effective than nano-silica in flame retardancy enhancement, probably because silica gel *in situ* formed a silica network to cover the entire sample surface.[93] Qu et al.[94] reported that the flame retardancy of phosphorus-containing poly(ethylene terephthalate) (PET) was greatly improved by the incorporation of $BaSO_4$ nanoparticles, improving anti-dripping behavior and reducing the heat release rate. Phosphorus-containing epoxy-based hybrid nanocomposites were obtained from bis(3-glycidyloxy) phenylphosphine oxide, diaminodiphenylmethane, and tetraethoxysilane via an *in situ* sol-gel process.[95] The nanoscale silica showed a limiting oxygen index (LOI) of 44.5.

POSS reinforced epoxy vinyl ester resins have been investigated.[96] The POSS nanocomposites exhibit better fire retardancy than the pure resin, giving reduced smoke and heat release rate and an increased time to ignition. A patent reports a large reduction in the peak heat release rate for several materials that contain POSS.[97] POSS has also been examined in textiles and the time to ignition increases, but there is no reduction in the PHRR.[98] There have recently appeared several studies on the thermal stability of POSS-containing polymers, which may lead to their investigation in flame retardant applications.[99–101] The mechanism of POSS-reinforced composites exhibiting better fire retardancy was possibly due to two effects:[96] (1) reduced volatilization of fuel and (2) the formation of oxidatively stable, nonpermeable surface chars.

15.6 Flammability of Polymer Nanocomposites with Conventional Flame Retardants

Although polymer nanocomposites exhibit improved flame retardancy, they will not achieve a rating in the UL 94 V protocol, which is required for industrial applications. To achieve an acceptable UL-94 V rating, the nanocomposites must be combined with conventional flame retardants. Studies on the combinations of nanocomposite formation with conventional flame retardants have received significant attention recently, and this topic has been recently reviewed.[102] Synergy between conventional halogen-based fire retardants and clays has been found.[103] The combination of bromine-based flame retardants and nanocomposite formation can lead to a UL-94 V-0 rating at an amount of decabromodiphenyl oxide (DB) that is lower than the amount required when the flame retardant is used alone.

An intercalated nanocomposite is formed when a nanoclay is melt-blended with polypropylene-*graft*-maleic anhydride, decabromdiphenyl oxide, and antimony trioxide (AO). The nanocomposite shows a lower PHRR than does the virgin polymer; the PHRR is reduced further when antimony oxide or decabromodiphenyl oxide is added. When both additives are present, synergy is observed and the reduction in PHRR is much larger than expected from the independent components. Synergy between nanocomposites and halogenated flame retardants was also found in polystyrene.[104]

Self-extinguishing PMMA-clay nanocomposites with DB and AO had been prepared by melt-blending using a Brabender mixer.[105] It was found that the polymers with only the flame retardants or the clay could not achieve the UL-94 V-0 rating, but all polymers with both components were self-extinguishing. The addition of clay can effectively avoid dripping during the burning

TABLE 15.3

UL-94 V0 Test and LOI Value of PMMA and PMMA/DB/AO/Clay Composites

Materials	UL-94 V0 Test	LOI (%)
PMMA	Dripping and burning (failed)	16.9
PMMA/DB/AO (75/20/5)	Dripping and burning (t_1 = 16 s failed)	23.3
PMMA/DB/AO/Cloisite 20A (70/20/5/5)	Passed (t_1 = 1S, t_2 = 1S)	25.6

Source: From Si, M. et al., *Polym. Degrad. Stab.*, 92, 86, 2007.

test (see Table 15.3 and Figure 15.13). The synergy was due to char formation promoted by the clay in the condensed phase and free radical capture in the vapor phase by the DB and AO.

Synergy between phosphorus flame retardants and nanocomposites has been studied by combining polystyrene nanocomposites with phosphorous flame retardants, such as resorcinol diphosphate (RDP). The combination of the clay and the phosphate exhibited higher flame retardancy than polystyrene with only RDP or clay.[8] The combination of vinyl esters and either RDP or tricresyl phosphate (TCP) with clay or polyhedral oligomeric silsesquioxanes (POSS) also showed synergistic effects.[65] In another styrenic system,[106] the clay alone gives a 56% reduction in PHRR while the phosphate alone gives a reduction of 20% to 40%. The synergistic combination gives a 92% reduction in PHRR.

An alternative method to this synergy has been to incorporate the phosphorus onto the ammonium cation that is used to modify natural clay. Because the phosphate is attached to the clay, it will be uniformly distributed throughout the polymer once the clay forms

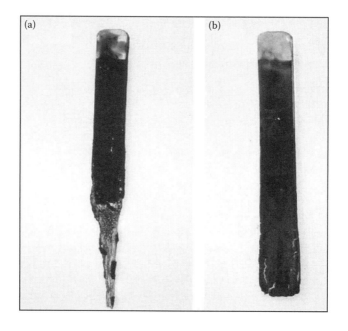

FIGURE 15.13
Optical images of specimens after UL-94 V0 test: (a) PMMA/DB/AO (75/20/5), and (b) PMMA/DB/AO/Cloisite 20A (70/20/5/5). (*Source:* Reprinted from *Polymer Degradation and Stability*, 92, M. Si, V. Zaitsev, M. Goldman, A. Frenkel, D.G. Peiffer, E. Weil, J.C. Sokolov, and M.H. Rafailovich, Self-extinguishing polymer/organoclay nanocomposites, pp. 86–93, Copyright 2007. With permission form Elsevier.)

the nanostructure, and it was hoped that this would enhance the fire retardancy.[107] The maximum reduction in PHRR achieved by this approach was 81%.

An improvement in fire retardancy was achieved by preparing polymer nanocomposites using intercalation of poly(vinyl alcohol) or polyamide 6 with 5% kaolinite.[108] This enabled the establishment of a concept to combine the intercalation of kaolinite with phosphorous fire retardants: trimethylphosphate, triphenylphosphate, or triphenylphosphine in acrylo-nitrile-butadiene-styrene copolymer. A synergistic effect of using the fire retardant nano-composites and addition of epoxy resin and silane coupling agent as co-fire retardants was also confirmed for the acrylonitrile-butadiene-styrene (ABS) copolymer compounds based on 2,6-dimethylphenol-resorcinol bis-(diphenylphosphate) (DMP-RDP).[109] LOI values as high as 44.8 were achieved for certain formulations.

Two different nanofillers used together can synergistically improve the thermal and flame retardant properties of EVA due to the strengthening effect on char formation by multi-walled or single-walled carbon nanotubes.[110, 111] EVA nanocomposites combining alumina trihydrate (ATH) with organoclay exhibited higher fire performance than either of them alone as a flame retardant,[112–114] where the clay can strengthen the char formation. Synergy was also found in polyethylene combining ATH and nanoclay.[115] Beyer et al.[116, 117] combined a nanofiller with alumina trihydrate to improve the fire retardancy properties of EVA nano-composites. The PHRR is 200 kW/m² when 65% ATH and 35% EVA are combined. When 5% of the ATH is replaced by an organoclay, the PHRR falls to 100 kW/m². If it is sufficient to maintain the PHRR at 200 kW/m², one can use 50% EVA, 45% ATH, and 5% organoclay and the mechanical properties are improved. Fire retardant nanocomposites were synthesized by melt-blending EVA copolymers with nanoclays.[118] Synergistic effects were also found with the combination of the TiO_2 nanoparticle and montmorillonite in PMMA.[119] The syn-ergistic effects lead to an improvement in PMMA thermal stability together with higher ignition times, reduced heat released, and a significant increase in the total burning time.

EVA combining an intumescent flame retardant (ammonium polyphosphate, APP) with polyamide 6 nanocomposites showed improved flame retardancy compared to the case where some other material was used as the carbonific.[120] Intumescent formulations, using either polyamide 6 or a polyamide 6-clay nanocomposite as the carbonization agent, were studied in an EVA-based intumescent formulation.[121] Using the polyamide 6-clay nano-composite instead of pure polyamide 6 has been shown to improve the fire properties of the intumescent blend and the mechanical properties are also improved. The role played by the clay in the improvement of the fire retardant (FR) performance was studied using FT-IR and solid-state NMR.[122] The clay allowed thermal stabilization of a phosphorocarbo-naceous structure in the intumescent char, which can act as a protective barrier.

Epoxy-LDH nanocomposites also exhibit synergistic effects with the intumescent addi-tive APP.[84] For example, to achieve the UL-94 V-0 rating for 3-mm-thick samples, 30 wt% APP was necessary in epoxy resin, whereas only 16 to 20 wt% of APP was required when 4% organically modified LDH was added.

15.7 Concluding Remarks

Nanotechnology has a splendid future as one of most promising methods to improve the flame retardancy of polymeric materials. Although the nanofiller alone cannot achieve the fire properties required in industrial applications, the presence of the nanofiller will enhance other properties, such as mechanical and barrier properties, and, in combination with other materials (flame retardants), can lead to flame retarded solutions.

Due to the thermal instability of organoclays, it is most likely that advances in the near term will occur with those polymers that can be processed below about 200°C. As new surfactants are developed and become commercialized, flame-retardant nanocomposite compositions that can be used for polymers that must be processed at higher temperatures may be developed. At the present time, the expense of carbon nanotubes and POSS makes it unlikely that these will be commercialized except in niche markets where the nanodimensional material offers additional advantages beyond enhanced flame retardancy. The nanocomposite does offer many advantages—enhanced barrier, flame, and mechanical properties—so these will be used in some instances to achieve another advantage and the enhanced flame retardancy will come along as a "bonus."

References

1. Alexandre, M. and Dubois, P. Polymer-layered silicate nanocomposites: preparation, properties, and uses of new class of materials, *Mater. Sci. Eng.*, R28, 1, 2000.
2. Sinha Ray, S. and Okamoto, M. Polymer/layered silicate nanocomposites: a review from preparation to processing, *Prog. Polym. Sci.*, 28, 1539, 2003.
3. Halford, B. Inorganic menagerie, *Chem. Eng. News*, 30, 2005.
4. Morgan, A.B. and Wilkie, C.A., Eds. *Flame Retardant Polymer Nanocomposites*, Wiley, 2007.
5. Giannelis, E.P. Polymer layered nanocomposite. *Adv. Mater.*, 8, 29, 1996.
6. Gilman, J.W. et al. In *Chemistry and Technology of Polymer Additives*. Al-Malaika, S., Golovoy, A., and Wilkie, C.A. Eds.; Blackwell Scientific, 1999, p. 249.
7. Gilman, J.W. et al. In *Fire Retardancy: The Use of Intumescence*, Le Bras, M., Camino, G., Bourbigot, S., and Delobel, R., Eds. Royal Society of Chemistry, Cambridge, 1998, **p.** 203.
8. Zhu, J., Uhl, F., and Wilkie, C.A. Recent studies on thermal stability and flame retardancy of polystyrene-nanocomposites, *Fire & Polymer III, ACS Books*, Oxford University Press, G. Nelson and C.A. Wilkie, Eds. 2001, p. 24–36.
9. Ishida, H., Campbell, S., and Blackwell, J. General approach to nanocomposite preparation, *Chem. Mater.*, 12, 1260, 2000.
10. Jang, B.N., Wang, D., and Wilkie, C.A. The relationship between the solubility parameter of polymers and the clay-dispersion in polymer/clay nanocomposites and the role of the surfactant, *Macromolecules*, 38, 6533, 2005.
11. Morgan, A.B. and Gilman, J.W. Characterization of polymer-layered silicate clay nanocomposites by transmission electron microscopy and x-ray diffraction: a comparative study, *J. App. Polym. Sci.*, 87, 1329, 2003.
12. Ijdo, W.L., Kemnetz, S., and Benderly, D. An infrared method to assess organoclay delamination and orientation in organoclay polymer nanocompsoites, *Polym. Eng. Sci.*, 46, 1031, 2006.
13. Bourbigot, S. et al. Investigation of nanodispersion in polystyrene-montmorillonite nanocomposites by solid state NMR, *J. Polym Sci Part B: Polym Phys.*, 41, 3188, 2003.
14. Zhu, J. et al. Silicon-methoxide-modified clays and their polystyrene nanocomposites, *J. Polym. Sci. Part A: Polym. Chem.*, 40, 1498, 2002.
15. Wagener, R. and Reisinger, T.J.G. A rheological method to compare the degree of exfoliation of nanocomposites, *Polymer*, 44, 7513, 2003.
16. Zhao, J., Morgan, A.B., and Harris, J.D. Rheological characterization of polystyrene-clay nanocomposites to compare the degree of exfoliation and dispersion, *Polymer*, 46, 8641, 2005.
17. Treece, M.A. and Oberhauser, J.P. Soft glassy dynamics in polypropylene-clay nanocomposites, *Macromolecules*, 40, 571, 2007.

18. Treece, M.A. and Oberhauser, J.P. Ubiquity of soft glassy dynamics in polypropylene-clay nanocomposites, *Polymer*, 48, 1083, 2007.
19. Gilman, J.W. Flammability and thermal stability studies of polymer layered-silicate clay nanocomposites, *Appl. Clay Sci.*, 15, 31, 1999.
20. Kashiwagi, T. et al. Flame retardant mechanism of polyamide 6-clay nanocomposites, *Polymer*, 45, 881, 2004.
21. Anderson, P. *NPE Educational Conference*, Nanotechnology applications for polymers, 2006.
22. Kashiwagi, T. et al. Nanoparticle networks reduce the flammability of polymer nanocomposites, *Nature Mater.*, 4, 928, 2005.
23. Lewin, M. et al. Nanocomposites at elevated temperatures: migration and structural changes, *Polym. Adv. Tech.*, 17, 226, 2006.
24. Jang, B.N. and Wilkie, C.A. The thermal degradation of polystyrene nanocomposites, *Polymer*, 46, 2933, 2005.
25. Jang, B.N. and Wilkie, C.A. The effect of clay on the thermal degradation of Polyamide 6 in polyamide 6/clay nanocomposites, *Polymer*, 46, 3264, 2005.
26. Costache, M.C., Jiang, D.D., and Wilkie, C.A. Thermal degradation of ethlene-vinyl acetate copolymer nanocomposites, *Polymer*, 46, 6947, 2005.
27. Jang, B.N., Jiang, D.D., and Wilkie, C.A. The effects of clay on the thermal degradation behavior of polystyrene-co-acrylonitrile, *Polymer*, 46, 9702, 2005.
28. Jang, B.N., Costache, M., and Wilkie, C.A. The relationship between thermal degradation behavior of polymer and the fire retardancy of polymer/clay nanocomposites, *Polymer*, 46, 10678, 2005.
29. Costache, M.C. et al. The thermal degradation of polymethyl methacrylate nanocomposites with montmorillonite, layered double hydroxides and carbon nanotubes, *Polymers Adv. Tech.*, 17, 272, 2006.
30. Zhu, J. et al. Studies on the mechanism by which the formation of nanocomposites enhances thermal stability, *Chem. Mater.*, 13, 4649, 2001.
31. Zhu, J. and Wilkie, C.A. Thermal and fire studies on polystyrene-clay nanocomposites, *Polym. Int.*, 49, 1185, 2000.
32. Zanetti, M. et al. Cone calorimeter combustion and gasification studies of polymer layered silicate nanocomposites, *Chem. Mater.*, 14, 881, 2002.
33. Babrauskas, V. Fire test methods for evaluation of fire-retardant efficacy in polymeric materials. In *Fire Retardancy of Polymeric Materials*. Grand, A.F. and Wilkie, C.A., Eds. Marcel Dekker, New York, 2000, p. 81.
34. Barrauskas, V. Release rate apparatus based on oxygen consumption, *Fire and Materials*, 8, 81, 1984.
35. Gilman, J.W. et al. NIST, Flammability of polymer clay nanocomposites consortrium: final report, November, 2003, http://www.nist.gov/pcnc/nano_consort_report_final_24.pdf.
36. Fire Testing Technology Ltd., http://www.fire-testing.com/html/instruments/ul94ad.htm
37. Utracki, L.A. *Clay-Containing Polymeric Nanocomposites*, Rapra Technology Limited, Shropshire, U.K., 2004. 2 volumes, 786 pages.
38. Utracki, L.A., Sephr, M., and Boccaleri, E. Synthetic, layered nanoparticles for polymeric nanocomposites PNCs, *Polym. Adv. Tech.*, 18, 1, 2007.
39. Zhu, J. et al. Fire properties of polystyrene-clay nanocomposites, *Chem. Mater.*, 13, 3774, 2001.
40. Zhu, J. et al. Silicon-methoxide-modified clays and their polystyrene nanocomposites, *J. Polym. Sci., Part A: Polym. Chem.*, 40, 1498, 2002.
41. Yao, H. et al. Crown ether-modified clays and their polystyrene nanocomposites, *Polym. Engr. Sci.*, 42, 1808 2002.
42. Su, S., Jiang, D.D., and Wilkie, C.A. A novel polymerically-modified clays permit the preparation of intercalated and exfoliated nanocomposites of styrene and its copolymers by melt blending, *Polym. Degrad. Stab.*, 83, 333, 2004.
43. Su, S., Jiang, D.D., and Wilkie, C.A. Methacrylate modified clays and their polystyrene and polymethyl methacrylate nanocomposites, *Polym Adv. Tech.*, 15, 225, 2004.

44. Su, S., Jiang, D.D., and Wilkie, C.A. Polybutadiene modified clay and its polystyrene nanocomposites, *J. Vinyl Add. Tech.*, 10, 44, 2004.

45. Zhang, J. and Wilkie, C.A. A carbocation substituted clay and its styrene nanocomposite, *Polym. Degrad. Stab.*, 83, 301, 2004.

46. Gilman, J.W. et al. Flammability properties of polymer-layered-silicate nanocomposites. polypropylene and polystyrene nanocomposites, *Chem. Mater.*, 12, 1866, 2000.

47. Morgan, A.B. et al. Flammability of polystyrene layered silicate clay nanocomposites: carbonaceous char formation, *Fire and Materials*, 26 247, 2002.

48. Morgan, A.B. et al. *Abstracts of Papers—American Chemical Society*, 2000, 220th PMSE-064.

49. Morgan, A.B. et al. Flammability properties of polymer-clay nanocomposites: polyamide-6 and polypropylene clay nanocomposites, *ACS Symp. Series*, 797 Fire and Polymers, 2001, p. 9.

50. Zhang, J. and Wilkie, C.A. Preparation and flammability of polyethylene-clay nanocomposites, *Polym. Degrad. Stab.*, 80, 163, 2003.

51. Wang, S. et al. Preparation and flammability properties of polyethylene/clay nanocomposites by melt intercalation method from Na⁺ montmorillonite, *Mater. Lett.*, 57, 2675, 2003.

52. Zhu, J. et al. Thermal stability and flame retardancy of polymethyl methacrylate-clay nanocomposites. *Polym. Degrad. Stab.*, 77, 253, 2002.

53. Jash, P. and Wilkie, C. A. Effects of surfactants on the thermal and fire properties of polymethyl methacrylate/clay nanocomposites, *Polym. Degrad. Stab.*, 88, 401, 2005.

54. Zheng, X., Jiang, D.D., and Wilkie, C.A. Methyl methacrylate oligomerically-modified clay and its poly methyl methacrylate nanocomposites, *Thermochim. Acta*, 435, 202, 2005.

55. Wang, D. and Wilkie, C.A. PVC/clay nanocomposites: preparation, thermal and mechanical properties, *J. Vinyl Add. Tech.*, 7, 203, 2001.

56. Wang, D. et al. Melt blending preparation of PVC-sodium clay nanocomposites, *J. Vinyl Add. Tech.*, 8, 139, 2002.

57. Wang, D. and Wilkie, C.A. Preparation of PVC-clay nanocomposites by solution blending, *J. Vinyl Add. Tech.*, 8, 238, 2002.

58. Zanetti, M., Camino, G., and Mulhaupt, R. Combustion behavior of EVA/fluorohectorite nanocomposites. *Polym. Degrad. Stab.*, 74, 413, 2001.

59. Morgan, A. B. et al. *Fire Safety Developments: Emerging Needs, Product Developments, Non-Halogen FR's, Standards and Regulations, Papers, [Conference]*, Washington, D.C., Publisher: Fire Retardant Chemicals Association, Lancaster, PA. March 2000, p. 12–15, 25–41.

60. Tang, Y. et al. Preparation and flammability of ethylene-vinyl acetate copolymer/montmorillonite nanocomposites. *Polym. Degrad. Stab.*, 78, 555, 2002.

61. Gilman, J.W., Kashiwagi, T., and Lichtenhan, J.D. Nanocomposites: a revolutionary new flame retardant approach, *SAMPE J.*, 33, 40, 1997.

62. Gilman, J.W., Kashiwagi, T., and Lichtenhan, J.D. Nanocomposites: a revolutionary new flame retardant approach, *Int. SAMPE Symp. Exhib.*, 42 Evolving Technologies for the Competitive Edge, Book 2, 1078, 1997.

63. Gilman, J.W. et al. Flammability studies of polymer layered silicate nanocomposites: polyolefin, epoxy, and vinyl ester resins, in *Chemistry and Technology of Polymer Additives*, S. Al-Malaika, A. Golovoy, and C.A. Wilkie, Eds. Blackwell Scientific, Oxford, 1999, p. 249.

64. Shah, A. P. et al. Flammability and other characteristics of vinyl ester/clay, vinyl ester/nomex/clay, and vinyl ester/glass fiber/clay nanocomposites, *Proc. Int. Conf. on Fire Safety*, 33, 18 2001.

65. Chigwada, G. et al. Fire retardancy of vinyl ester nanocomposites: synergy with phosphorus-based fire retardants, *Polym. Degrad. Stab.*, 89, 85, 2005.

66. Yeh, J.-M. and Chin, C.-P. Structure and properties of poly-*o*-methoxyaniline-clay nanocomposite materials, *J. Appl. Polym. Sci.*, 88, 1072, 2003.

67. Lepoittevin, B. et al. Poly-ε-caprolactone/clay nanocomposites prepared by melt intercalation: mechanical, thermal and rheological properties, *Polymer*, 43, 4017, 2002.

68. Pantoustier, N. et al. Poly-ε-caprolactone layered silicate nanocomposites: effect of clay surface modifiers on the melt intercalation process, *e-Polymers [online computer file]*, Paper No. 9, 2001.

69. Pantoustier, N. et al. Biodegradable polyester layered silicate nanocomposites based on poly-ε-caprolactone, *Polym. Engr. Sci.*, 42, 1928, 2002.

70. Lee, J., Takekoshi, T., and Giannelis, E.P. Fire retardant polyetherimide nanocomposites, *Mater. Res. Soc. Symp. Proc., 457 Nanophase and Nanocomposite Materials II*, 513, 1997.

71. Gilman, J.W., Harris, R., Jr., and Hunter, D. Cyanate ester clay nanocomposites: synthesis and flammability studies, *Int. SAMPE Symp. Exhib., 44 Evolving and Revolutionary Technologies for the New Millennium*, Book 2, 1408, 1999.

72. Stretz, H.A. et al. Flame retardant properties of polycarbonate/montmorillonite clay nanocomposite blends. *Polym. Prep. Am. Chem. Soc., Div.Polym. Chem.*, 422, 50, 2001.

73. Lee, J. and Giannelis, E.P. Synthesis and characterization of unsaturated polyester and phenolic resin nanocomposites, *Polym. Prep. Am. Chem. Soc., Div. Polym. Chem.*, 382, 688, 1997.

74. Berta, M. et al. Effect of chemical structure on combustion and thermal behaviour of polyurethane elastomer layered silicate nanocomposites, *Polym. Degrad. Stab.*, 91, 1179, 2006.

75. Bourbigot, S. et al. Nanocomposite textiles: new route for flame retardancy, *Int. SAMPE Symp. Exhib.*, 47, 1108, 2002.

76. Gilman, J.W. and Kashiwagi, T. Nanocomposites: a revolutionary new flame retardant approach, *SAMPE J.*, 33, 40, 1997.

77. Uhl, F.M. and Wilkie, C.A. Polystyrene/graphite nanocomposites: effect on thermal stability, *Polym. Degrad. Stab.*, 76, 111, 2002.

78. Uhl, F.M., Yao, Q., and Wilkie, C.A. Formation of nanocomposites of styrene and its copolymers using graphite as the nanomaterial, *Polym. Adv. Tech.*, 16, 533, 2005.

79. Uhl, F.M. et al., Expanded graphite/polyamide-6 nanocomposites, *Polym. Degrad. Stab.*, 89, 70 2005.

80. Uhl, F.M. and Wilkie, C.A. Preparation of nanocomposites from styrene and modified graphite oxides, *Polym. Degrad. Stab.*, 84, 215 2004.

81. Ding, R. et al. Preparation and characterization of polystyrene/graphite oxide nanoocomposite by emulsion polymerization, *Polym. Degrad. Stab.*, 81, 473, 2003.

82. Zhang, R. et al. Flammability and thermal stability studies of styrene-butyl acrylate copolymer/graphite oxide nanocomposite, *Polym. Degrad. Stab.*, 85, 583, 2004.

83. Ruiz-Hitzky, E. Organic–inorganic materials: from intercalation to devices, *Mol. Cryst. Liq. Cryst.*, 161, 433, 1988.

84. Zamarano, M. et al. Preparation and flame resistance properties of revolutionary self-extinguishing epoxy based on layered double hydroxides, *Polymer*, 46, 9314, 2005.

85. Costache, M.C. et al. The influence of carbon nanotubes, organically modified montmorillonites and layered double hydroxides on the thermal degradation of polyethylene, ethylene-vinyl acetate copolymer and polystyrene, manuscript in preparation.

86. Kashiwagi, T. et al. Flammability properties of polymer nanocomposites with single-walled carbon nanotubes: effects of nanotube dispersion and concentration, *Polymer*, 46, 471, 2005.

87. Kashiwagi, T. et al. Thermal degradation and flammability properties of polypropylene/carbon nanotube composites, *Macromol. Rapid Commun.*, 23, 761, 2002.

88. Kashiwagi, T. et al. Thermal and flammability properties of polypropylene/carbon nanotube nanocomposites, *Polymer*, 45, 4227, 2004.

89. Beyer, G. Carbon nanotubes as flame retardants for polymers, *Fire Mater.*, 26, 29,1 2002.

90. Schartel, B. et al. Fire behaviour of polyamide 6/multiwall carbon nanotube nanocomposites, *Eur. Polym. J.*, 41, 1061, 2005.

91. Yang, F., Yngard, R., and Nelson, G.L. Thermal stability and flammability of polymer/silica nanocomposites prepared via extrustion, *Polym. Mater. Sci. Eng.*, 91, 86, 2004.

92. Morgan, A.B. et al. Thermal and flammability properties of a silica-PMMA nanocomposite, *Polym. Mater. Sci. Engr.*, 83, 57, 2000.

93. Kashiwagi, T. et al. Thermal and flammability properties of a silica-polymethyl methacrylate nanocomposite, *J. Appl. Polym. Sci.*, 89, 2072, 2003.

94. Qu, M.H. et al. Flammability and thermal degradation behaviors of phosphorus-containing copolyester/$BaSO_4$ nanocomposites, *J. Appl. Polym. Sci.*, 102, 564, 2006.

95. Hsiue, G.H., Liu, Y.L., and Liao, H.H. Flame-retardant epoxy resins: an approach from organic-inorganic hybrid nanocomposites, *J. Polym. Sci., Part A: Polym. Chem.*, 39, 986, 2001.

96. Gupta, S.K. et al. Polyhedral oligomeric silsesquioxane reinforced fire retarding epoxy vinyl ester resins, *Int. SAMPE Symp. Exhib.*, 47, 1517, 2002.

97. Lichtenham, J.D. and Gilman, J.W. Preceramic additives as fire retardants for plastics U.S. Patent No. 6,362,279, March 26, 2002.

98. Bourbigopt, A. et al. Polyhedral oligomeric silsesquixones: application to flame retardant textiles, in *Fire Retardancy of Polymers New Applications of Mineral Fillers*, Eds., Le Bras, M., Wilkie, C.A., Bourbigot, S., Duquesne, S., and Jama, C. Royal Society of Chemistry, Cambridge, 2005, p. 189.

99. Fina, A. et al. Octaisobutyl POSS thermal degradation, in *Fire Retardancy of Polymers New Applications of Mineral Fillers*, Eds., Le Bras, M., Wilkie, C.A., Bourbigot, S., Duquesne, S., and Jama, C. Royal Society of Chemistry, Cambridge, 2005, p. 202.

100. Fina, A. et al. Polyhedral oligomeric silsesquioxanes POSS thermal degradation, *Thermochim. Acta*, 440, 36, 2006.

101. Fina, A. et al. Polypropylene metal functionalized POSS nanocomposites: a study by thermogravimetric analysis, *Polym. Degrad. Stab.*, 91, 1064, 2006.

102. Morgan, A.B. Flame retarded polymer layered silicate nanocomposites: a review of commercial and open literature systems, *Polym. Adv. Technol.*, 17, 206, 2006.

103. Zanetti, M. et al. Fire retardant halogen-antimony-clay synergism in polypropylene layered silicate nanocomposites, *Chem. Mater.*, 14, 189, 2002.

104. Chigwada, G. et al. Synergy between nanocomposite formation and low levels of bromine on fire retardancy in polystyrene, *Polym. Degrad. Stab.*, 88, 382, 2005.

105. Si, M. et al. Self-extinguishing polymer/organoclay nanocomposites, *Polym. Degrad. Stab.*, 92, 86, 2007.

106. Chigwada, G. and Wilkie, C.A. Synergy between conventional phosphorus fire retardants and organically-modified clay can lead to fire retardancy of styrenics, *Polym. Degrad. Stab.*, 81, 551, 2003.

107. Zheng, X. and Wilkie, C.A., Flame retardancy of polystyrene nanocomposites based on an oligomeric organically-modified clay containing phosphorus, *Polym. Degrad. Stab.*, 81, 539, 2003.

108. Zaikov, G.E. and Lomakin, S.M. New polymer flame retardant compositions for ABS and nylon 6. Part 1, *Oxidation Commun.*, 22, 556 1999.

109. Kim, J. et al. Studies on the thermal stabilization enhancement of ABS; synergistic effect of triphenyl phosphate nanocomposite, epoxy resin, and silane coupling agent mixtures, *Polym. Degrad. Stab.*, 79, 201, 2002.

110. Beyer, G. Filler blend of carbon nanotubes and organoclays with improved char as a new flame retardant system for polymers and cable applications, *Fire Mater.*, 29, 61 2005.

111. Peeterbroeck, S. et al. Polymer-layered silicate-carbon nanotube nanocomposites. Unique nanofiller synergistic effect, *Composites Sci. Technol.*, 64, 2317, 2004.

112. Zhang, X.G. et al. The investigation of interfacial modification for flame retardant ethylene vinyl acetate copolymer/alumina trihydrate nanocomposites, *Polym. Degrad. Stab.*, 87, 411, 2005.

113. Chuang, T.H. et al. Thermal properties and flammability of ethylene-vinyl acetate copolymer/montmorillonite/polyethylene nanocomposites with flame retardants, *J. Polym. Res.*, 11, 169 2004.

114. Beyer, G. Flame retardant properties of EVA-nanocomposites and improvements by combination of nanofillers with aluminium trihydrate, *Fire Mater.*, 25, 193, 2001.

115. Zhang, J. and Wilkie, C.A. Fire retardancy of polyethylene-alumina trihydrate containing clay as a synergist, *Polym. Adv. Technol.*, 16, 549, 2005.

116. Beyer, G. Flame retardant properties of EVA-nanocomposites and improvements by combination of nano-fillers with aluminum trihydrate, *Fire Mater.*, 25, 193, 2002.

117. Beyer, G. Flame retardancy of nanocomposites based on organoclays and carbon nanotubes with aluminium trihydrate, *Polym. Adv. Technol.*, 17, 218, 2006.

118. Alexandre, M. et al. Preparation and properties of layered silicate nanocomposites based on ethylene-vinyl acetate copolymers, *Macromol. Rapid Commun.*, 22, 643, 2001.
119. Laachachi, A. et al. Thermal degradation and flammability of polymethyl methacrylate containing TiO_2 nanoparticles and modified montmorillonite, in *Fire and Polymers IV: Materials and Concepts for Hazard Prevention*, Eds. C.A. Wilkie and G.L. Nelson, American Chemical Society Symposium Series # 922, Oxford University Press, Oxford, 2006, p. 36.
120. Bourbigot, S. et al. Recent advances for intumescent polymers, *Macromol. Mater. Eng.*, 289, 499, 2004.
121. Dabrowski, F. et al. The use of clay in an EVA-based intumescent formulation. Comparison with the intumescent formulation using polyamide-6 clay nanocomposite as carbonisation agent, *J. Fire Sci.*, 19, 219, 2001.
122. Bourbigot, S. et al. T. PA-6 clay nanocomposite hybrid as char forming agent in intumescent formulations, *Fire Mater.*, 24, 201, 2000.

16

Electrical Properties of Nanoparticle-Filled Polymers

Elliot B. Kennel

CONTENTS

16.1 Introduction

By the late 1980s, multiwalled carbon nanotubes (referred to as "fibrils™" by Hyperion Catalysis) and carbon nanofibers (Applied Sciences Inc.) were being manufactured in several gram size quantities, which was sufficient for testing composite samples. Although the spectacular estimates of enhanced strength and modulus from single-walled nanotubes (SWNTs) probably had not been appreciated at that time, it was supposed by many researchers that enhanced mechanical properties might be attainable by adding carbon nanomaterials of different types to polymer composites.

At the same time, it was recognized that the high cost of many types of nanomaterials would present a barrier to commercialization, unless truly exceptional mechanical property enhancements could be achieved.

On the other hand, it was soon realized that electrical properties could also be enhanced by the addition of carbon nanomaterials, and in some cases very significant modifications in electrical properties could be attained with even a few percent addition by weight. Moreover, the small characteristic dimensions of nanoreinforcements suggested that unique electromagnetic behavior might result.

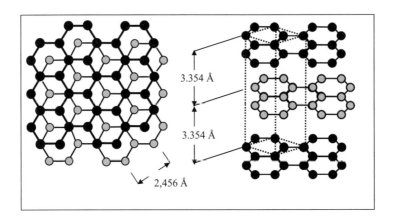

FIGURE 16.1
Idealized graphite structure.

16.2 Intrinsic Electrical Resistivity of Carbon Nanofilaments

The bulk electrical resistivity of planar graphite is highly isotropic, owing to its structure. Electrical charge transfer along the graphene planes occurs with low scattering losses, whereas charge transfer perpendicular to the graphene planes necessarily involves considerable scattering losses, resulting in low electron mobility.

Although a perfect single crystal of graphite would be very difficult to manufacture, the ideal graphite lattice can be approximated by materials such as Highly Ordered Pyrolytic Graphite (HOPG), in which the carbon atoms are arranged in offset layers of hexagons, as shown in Figure 16.1.

As shown in Table 16.1, the in-plane electrical resistivity of HOPG approaches metallic levels, about 4.8×10^{-7} Ω-m. By comparison, the electrical resistivity of titanium is about 4.0×10^{-7} Ω-m and that of copper is about 1.72×10^{-8} Ω-m. Yet in the perpendicular direction, the electrical resistivity of HOPG is only 0.002 Ω-m. The ratio of in-plane to perpendicular electrical resistivity is thus 1:4200.

TABLE 16.1

Approximate Properties of Highly Oriented Pyrolytic Graphite

Property	Value
Density	2.25 g/cm³
Thermal Conductivity (in-plane)	1700–2000 W/(mK)
Thermal Conductivity (perpendicular)	8 W/(mK)
Electrical Conductivity (in-plane)	2.1×10^{6} S/m
Electrical Resistivity (in-plane)	4.8×10^{-7} (Ω-m)
Electrical Conductivity (perpendicular)	5×10^{2} S/m
Electrical Resistivity (perpendicular)	0.002 (Ω-m)

Source: From Matsubara, K., Sugihara, K., and Tsuzuku, T., *Phys. Rev. B*, 41, 969, 1990.

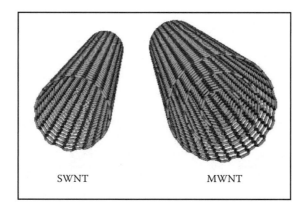

FIGURE 16.2
Idealized single-walled and multi-walled nanotubes. (From Immunology et Chime Therapeutiques, CNRS-UPR 9021.)

16.2.1 Electrical Resistivity of Carbon Nanotubes

The multiwalled nanotube (MWNT; Figure 16.2) might be imagined as planar graphene sheets wrapped into cylindrical shells, at least from the standpoint of electrical properties. Hence, an electrical resistivity similar to that of bulk pure graphite might be expected. However, owing to the extremely small size of an individual nanotube, the precise value of bulk electrical resistivity of a single MWNT is a matter of some controversy. Nevertheless, intrepid researchers have been able to isolate individual MWNTs by various means and apply a four-point probe to make such measurements.

For at least some MWNTs, electrons can be ballistically conducted between the concentric tubes.[1, 2] Ballistic conduction refers to a state, similar to superconductivity, in which electrons can be transmitted a substantial distance without collisional energy losses.

The single-walled nanotube (SWNT) bears special mention (Figure 16.2). The normal solid-state derivation of properties such as electrical resistivity and band gap is based on the assumption of a crystalline lattice. Because the SWNT contains only a single layer of carbon atoms, these basic assumptions are suspect, opening up an entirely new field of scientific inquiry that, needless to say, goes beyond the needs of polymer nanocomposites.

Nevertheless, a few key points can be highlighted. First, in a molecular conductor such as a carbon nanotube, the possibility of ballistic conduction exists. Ballistic conduction refers to an electron transport mode by which electrons can be transported, probably in the hollow core region of the nanotube, in a trajectory that avoids scattering.[3, 4] Inside the SWNT, extremely high currents are achieved with very little if any voltage losses. These properties are likely to result in the manufacture of ingenious electronic devices on a molecular scale, and thus will represent an important research and development opportunity for many years to come.[5] However, from the standpoint of polymer nanocomposites, it is not obvious how high electron currents would be transferred from one nanotube to the next; that is, electron transport would presumably be limited by the properties of the dielectric matrix material rather than the properties of the nanoscale additives.

In addition, the electron-conducting character of SWNTs can have fundamentally different electron conduction modes, depending on their chirality. Chirality refers to the way in which the graphene cylinder is twisted. Some examples are shown in Figure 16.3.

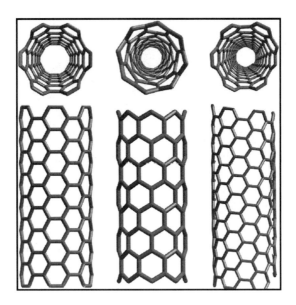

FIGURE 16.3
Single-walled nanotubes. From left, armchair (n,0), zigzag (n,n), and chiral (n,m). (Illustration courtesy T.S. Fisher, Purdue University.)

The values of n and m determine the chirality, or "twist" of the nanotube, as illustrated in Figure 16.4. SWNTs exhibit metallic conductivity when the value $n - m$ is divisible by three. Otherwise, the nanotube is semiconducting. In addition, the chiral vector (n,m) can also be used to estimate the diameter of the corresponding carbon nanotube using the following relationship:

$$d = (n^2 + m^2 + nm)^{1/2} \, 0.0783 \text{ nm} \tag{16.1}$$

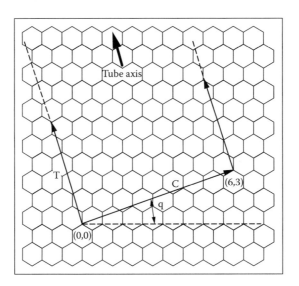

FIGURE 16.4
Chiral vector diagram for a (6,3) nanotube.

16.2.3 Electrical Resistivity of Carbon Nanofibers

Describing the electrical resistivity of carbon nanofibers based on the anisotropic resistivity of planar graphite is complicated because the graphene planes are wrapped into truncated conic sections, as shown in Figure 16.5. This model was proposed by Endo of Shinshu University.[6] In the Endo model, the conic sections create an angle of about 20° with respect to the fiber axis, which represents the direction of electron transport. For the moment, the possibility of ballistic conduction is neglected, and consideration is given to the electrical conduction only through the bulk of the nanofiber.

In planar graphite, the electrical conduction path can be at an angle θ with respect to the orientation of the graphene planes, as depicted in Figure 16.6. The resistance would be given by

$$R = \rho\beta \tag{16.2}$$

where R has the units of resistance per unit area normal to the conduction direction, ρ is the resistivity in units of Ω-m, and L is the length of the straight-line conduction (either give units of all parameters or none) path. However, the conduction path can also be constructed in terms of the in-plane and perpendicular directions:

$$R = \rho_\alpha L \cos\theta + \rho_\kappa L \sin\theta \tag{16.3}$$

Hence,

$$\rho(\theta) = \rho_\alpha \cos\theta + \rho_\kappa \sin\theta \tag{16.4}$$

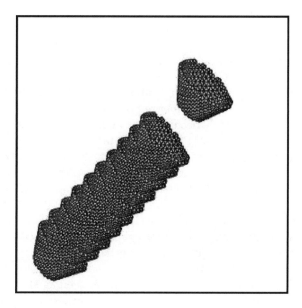

FIGURE 16.5
Carbon nanofiber: Endo stacked conic section model. (*Source:* Yanagisawa, T. and Higaki, S. [Nagareyama, JP], U.S. Patent No. 7014829, Carbon fiber, and filter and absorbent using the same, Filing Date: 03/18/2002.)

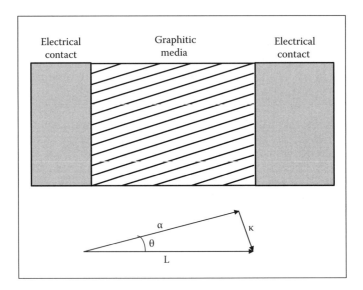

FIGURE 16.6
Idealized geometry for electrical conduction at an angle θ with respect to the orientation of graphene planes.

For $\rho_\alpha = 4.8 \times 10^{-7}$ Ω-m, $\rho_\kappa = 0.002$ Ω-m and θ of 20°,

$$\rho(\theta = 20^\circ) \approx 6.8 \times 10^{-4} \quad \text{Ω-m} \tag{16.5}$$

A similar result for bulk resistivity can be obtained in cylindrical coordinates. Thus, somewhat surprisingly, it appears that the electrical resistivity of an ensemble of stacked conic sections (i.e., a carbon nanofiber) appears to be about three orders of magnitude higher than a multiwalled carbon nanotube.

It would seem, then, that carbon nanofibers might be very poor electrical additives for polymer composites. However, in many cases, the production process for carbon nanofibers results in a layer of pyrolitically deposited carbon on the outer layers of the stacked-cup geometry, as shown in the transmission electron micrograph of Figure 16.7 below. Two distinct regions can be observed. The dark bands in the inner region are actually artifacts caused by electron diffraction effects, and thus are evidence of a high degree of crystallinity. The lighter, outer region is the pyrolytic layer, which is in the form of turbostratic carbon. Turbostratic carbon consists of layers of carbon atoms, but with little relationship between the layers. Hence, the true graphitic crystal structure is not present. Turbostratic carbon can be annealed ("graphitized") to reduce defect density and become more like true graphite. Graphitization is a commercial process that generally takes place at a temperature of 2500 to 3100°C in an electrically heated Acheson-type furnace. Thus, the outer section of the nanofiber could behave in a manner similar to planar graphite and might exhibit graphitic electrical conductivity.

Other means of enhancing the electrical properties of carbon nanofibers can also be considered. Intercalation compounds are formed by adding dopant atoms to the graphite lattice.[7] Commonly used intercalants are bromine as well as alkali metals and certain acids. Intercalated graphite can achieve electrical conductivity comparable to that of the best metals. The drawbacks of this procedure, however, are that the process is not easily controlled, generally involves chemically aggressive materials, and has only limited stability in the presence of air or moisture.

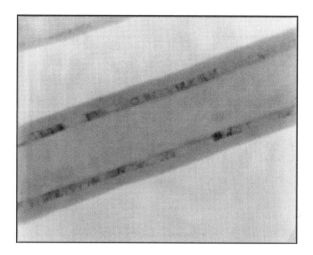

FIGURE 16.7
Transmission electron micrograph of a carbon nanofiber, showing an inner layer of highly ordered carbon, surrounded by a chemical vapor deposited layer of turbostratic carbon. (*Source:* Courtesy of Applied Sciences Inc.)

Boronation is similar to intercalation. However, in this case, boron is substitutionally bonded to the graphene lattice, and thus is quite stable at temperatures of interest for polymer fabrication. Some enhancement of electrical resistivity is achievable through this process.[8] The boronation process is self-limiting; however, because in addition to adding charge carriers, the presence of boron atoms in the graphene lattice results in the formation of additional scattering centers, thus reducing the electron mobility in the lattice.

Metallization is the most straightforward means by which electrical resistivity can be enhanced. Nickel is a commonly used coating and is applied via electroless plating. Nickel-coated carbon nanofibers can be prepared by electroless deposition.[9] This can be accomplished by first treating carbon nanofibers with polyacrylic acid to assist in dispersion prior to the nickel electroless deposition. The carbon nanofibers were coated homogeneously with nickel by an electroless deposition process using an electroplating bath containing sodium hypophosphite as a reducing agent, and resulted in a nickel-coated carbon nanofiber material. Similar processes can be used with copper, silver, and other metals.[10, 11]

Yet another emerging possibility for nanocomposite fabrication might be to use pure metal nanowires.[12] Such nanowires have been produced in small quantities from various techniques, but at the time of this writing were not available on a commercial scale or at commercially relevant prices.

16.3 Relationship between Thermal and Electrical Conductivity

In metals, thermal conductivity and electrical resistivity are related by the well-known Wiedemann-Franz law:

$$k\rho = LT \tag{16.6}$$

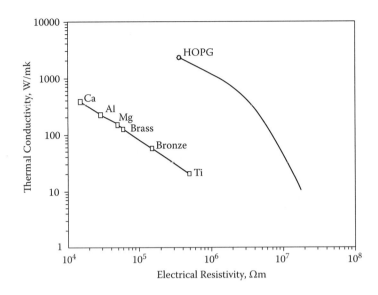

FIGURE 16.8
Relationship between thermal conductivity and electrical resistivity for different grades of carbon/graphite, with the Wiedemann-Franz correlation for metals included for comparison. (*Source:* Kowalski, I.M., *32nd Int. SAMPE Symposium and Exhibition,* 1987, p. 953–963.)

where k is thermal conductivity in W/mK, ρ is the electrical resistivity as before, T is the absolute temperature, and L is the Lorenz number equal to 2.44×10^{-8} WΩ/K^2. However, the Wiedemann-Franz law is derived for materials electron conduction and thus does not apply directly to lattice thermal conductors such as graphite.

The character of graphite is such that the thermal and electrical conductivity both improve dramatically, depending on the degree of order in the lattice—that is, the extent to which the matrix is thermally annealed. This is shown in Figure 16.8, which is based on pitch-derived graphite fibers but which is likely representative of bulk graphites derived from a variety of processes.

Thus, although pure annealed graphite is comparable to a poorly performing metal from the standpoint of electrical resistivity, high-grade graphite surpasses metals in thermal conductivity performance. As shown in Table 16.1, the thermal conductivity of HOPG (highly ordered pyrolytic graphite) can be as high as 2000 W/mK in the planar direction. By comparison, the thermal conductivity of copper is only 401 W/mK. On the other hand, the thermal conductivity through the planes is a mere 8 W/mK.

In the case of MSNTs or SWNTs, in which the direction of thermal conduction is usually along the axis of the tube, it can be anticipated that HOPG-like conduction modes might be present. In actuality, the situation is not so simple. Because of the unique molecular structure of SWNT, as semiconducting SWNTs can also exist, depending on the chirality of the particular nanotube. However, for the purpose of affecting the bulk electrical properties of polymers, the semi-metal case is the more relevant one.

Once again, however, the case of a stacked-cup geometry is different, owing to the fact that thermal transport must occur between layers. Thus,

$$\frac{1}{k(\theta)} = \frac{1}{k_\alpha} \cos\theta + \frac{1}{k_\kappa} \sin\theta \tag{16.7}$$

For $\kappa_\alpha = 2000$ W/mK, $\kappa_\kappa = 8$ W/mK, and θ of $20°$,

$$k(\theta = 20°) \approx 23 \quad \frac{W}{mK} \tag{16.8}$$

In analogy to the electrical conductivity application, a coating of pyrolytic carbon followed by thermal annealing would be advantageous for enhanced thermal conductivity.

Intercalation compounds and boronated material do not create enhanced thermal conductivity in carbon/graphite materials because thermal conduction is primarily by lattice conduction (phonon transport) rather than electron energy.

16.4 Percolation Theory

The previous section discussed the intrinsic bulk electrical resistivity of the different types of nanomaterials. However, one of the unique attributes of nanomaterials is that they often behave very differently than bulk materials, owing to effects that are size dependent. The electrical conductivity of mixtures of nano-additives and insulating polymers is an example of such nonlinear behavior. The discussion of why this is true will begin with the Rule of Mixtures (ROM) and its limitations, and how it specifically breaks down in the case of nano-additives.

The Rule of Mixtures assumption is often used to estimate the properties of macro-composites. Rule of Mixtures analysis assumes that any material property of a mixture or composite will often be a linear average of the properties of the components, weighted according to the mass percentage; that is, for some property ζ_n of a mixture with n separate components,

$$\zeta_n = \sum_{i=1}^{i} a_i \zeta_i \tag{16.9}$$

where the a_i represent suitable weighting factors, such as mass fraction, and the ζ_i are appropriate to the ith constituent.

Needless to say, it might be mentioned from the outset that ROM is far from correct for many well-known situations in materials science. For example, the entire science of metallurgical alloying is based on the premise that synergistic effects can result when different metals are combined.

Electrical resistivity is another property that departs strongly from the ROM approach, even at the macroscale. Empirically it is found that a small amount of additive does not affect the resistivity very much.

Percolation theory is based on the notion that additives in a dielectric matrix have a random chance of linking together to create regions that are more conductive than average. At low volumetric additions, the particles may not link up, and the resistivity is extremely high.

This can be visualized by assuming that a composite is created from some dispersed, highly conductive filaments in a dielectric matrix. The filaments are randomly oriented and have some probability of being in electrical contact or not, as shown in Figure 16.9.

When additional reinforcing filaments are added, eventually by chance the filaments contact each other and form a conductive path as illustrated in Figure 16.10. Thus, in the idealized situation, there is a critical concentration of fibers at which the composite can suddenly become conductive. This is known as the *percolation threshold*.

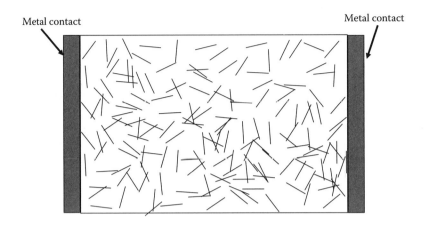

FIGURE 16.9
A composite sample with conductive filaments, no percolation.

Based on this notion, the percolation threshold depends mainly on the length (L) of the filament and the concentration of the filaments, but not the diameter (D). This holds important implications for nanocomposites, because the L/D ratio of nanofilaments can be extremely high. For example, a nanofilament with a diameter of 100 nm and a length of 100 μ would represent an L/D ratio of 1000. Hence, it might be supposed that the percolation threshold for nanocomposites could occur at a low concentration of reinforcements, with an abrupt transition between nonconducting and Rule-of-Mixtures conducting states.[13]

Additives with smaller dimensions are typically effective at low concentrations, whereas additives with larger diameters (e.g., metallic fillers) can achieve low resistivity for high material loadings.[14]

The resistivity as a function of additive concentration can often approach the ROM value when the concentration reaches two or three times the percolation point.

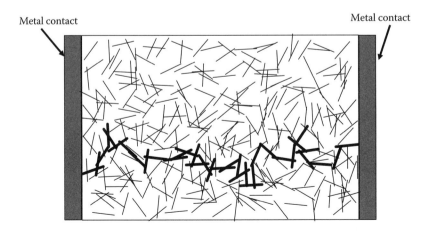

FIGURE 16.10
A conductive path in a composite with filamentary additives.

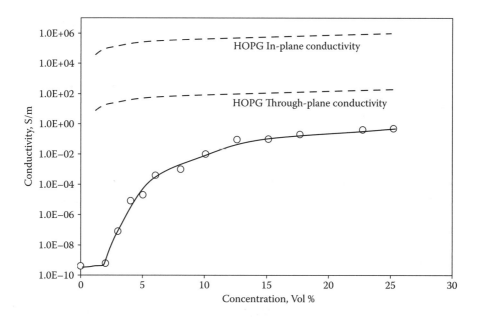

FIGURE 16.11

Percolation plot of SWNT-poly-ethyl-methacrylate composite conductivity, based on data of Grimes et al. (*Source:* Ounaies, Z., Park, C., Wise, K.E., Siochi, E.J., and Harrison, J.S., *Compos. Sci. Technol.*, 63, 1637, 2003.)

A percolation plot is shown in Figure 16.11, based on the results of Grimes et al.[15] for an SWNT composite in poly(ethyl-methacrylate). Also shown on the same graph are two curves corresponding to ROM behavior. The upper curve (corresponding to low resistance) assumes that the electrical conductivity is equal to the value for in-plane graphitic material. The middle curve is an ROM approximation for a material with the electrical conductivity value corresponding to the perpendicular direction. The results show percolation behavior with an apparent mass loading at the percolation point between 5 and 10 mass percent.

The bulk resistivity, far from approaching the ROM curve for semi-metallic conductivity, is even higher than expected for graphite in its poorest-conducting orientation. That is, this is similar to the type of result that one might anticipate for a micro-additive having electrical resistivity some three orders of magnitude higher than graphite.

The nanoscale is thus sort of a two-edged sword, as the potential for attractive properties resulting from nanoscale reinforcements is countered by the difficulties in dispersing nanomaterials uniformly. In addition, electron transport from one nanotube to the next is problematic. In terms of percolation theory, one of the inherent assumptions is that random contact between conducting filaments produces an effective conducting interface. Yet, on the atomic level it is not obvious how electrons would transition from a conducting state in one nanotube, travel a short distance (possibly through a polymer interface), and contact another conducting nanotube. Indeed, the study of quantum mechanical tunneling and electron hopping effects promises to be a fertile field of academic study in upcoming decades. But practical objections to using SWNT additives as a means to enhance dc electrical conductivity appear quite formidable.

These issues are made clear by the sequence of Figures 16.11 through 16.15. Curves corresponding to data points have been added to guide the reader's eye. The dashed curves correspond to ROM predictions based on electrical conductivities of Highly Oriented Pyrolytic Graphite (HOPG), and are added as a reference comparison. The upper curve corresponds to the in-plane-conductivity of HOPG (2×10^6 S/m), and the lower curve corresponds to the through-the-plane electrical conductivity of HOPG (500 S/m).

Figure 16.11 shows the percolation effect in SWNT in a poly(ethyl-methacrylate) composite, based on measurements by Grimes et al.[15] Weight percent was converted to volume percent assuming an SWNT effective density of 1.11 g/cm³, after Ericson et al.,[16] and matrix density of 1.12, after Kamiya et al.[17] The SWNT nanocomposites exhibit orders-of-magnitude lower electrical conductivity than one would expect for HOPG in the through-the-plane direction. Yet percolation effects are evidenced at very low concentrations. For example, Barrau et al.[18] report percolation threshold of a mere 0.1%. This suggests that indeed the percolation effect is strongly enhanced by nanoscale filaments. Given that the percolation effect is clearly present, this suggests that the poor performance is likely due to poor conductivity in the interface between neighboring conducting filaments.

At some point in the future, a conductive highly adherent polymer might succeed in overcoming the interface problem, which could permit SWNT nanocomposites to achieve a several order-of-magnitude enhancement in electrical conductivity as a function of nanotube volume fraction. However, for the near future, SWNT nanocomposites seem to be plagued by both high cost and low performance, at least for this application.

Figure 16.12 shows an analogous percolation effect in MWNTs from Hyperion Catalysis Inc., according to data obtained from Pötschke et al.[19] Composites produced from these commercial fibrils have electrical conductivity some two orders of magnitude higher than their SWNT counterparts. To be sure, the matrix material was considerably different, but on the other hand there is no empirical evidence that some other matrix material would reverse the conclusion that MWNTs produce higher electrical conductivity than SWNT additives when the concentration is at least several percent by unit volume.

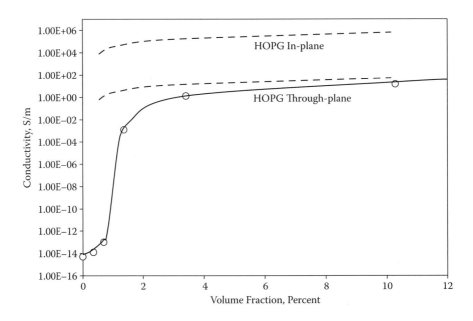

FIGURE 16.12
Percolation plot of Hyperion Catalysis Inc. MWNT in a polycarbonate matrix.

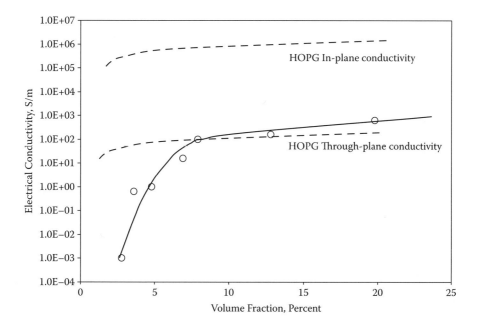

FIGURE 16.13
Percolation plot for heat-treated Applied Sciences Inc Pyrograf™-III nanofibers in a polypropylene matrix, showing percolation density at about 3%. The curve is added to guide the reader's eye.

Data from Burton et al.[20] on carbon nanofibers in polypropylene are presented in Figure 16.13. The magnitude of electrical conductivity as a function of volume fraction appears commensurate with the MWNT data, and in fact may be somewhat higher.

At volume loadings of some 30%, percolation can occur with graphitic carbon black, and the conductivity advantage is likely lost.

In all cases, it is possible to achieve lower electrical conductivity if the nanofibers are poorly dispersed. This begs the question of whether the conductivity might be enhanced if dispersion methods are significantly improved.

It also might be mentioned that the cited references are among the very best achievements in the nanocomposite field, and it should not be inferred that electrical conductivity enhancement is always uniform. Poor mixing technique, in particular, can result in failing by orders of magnitude to achieve the reported values. This issue is discussed further in Section 16.5.

Both MWNTs and CNFs (carbon nanofibers) are potential candidates for metallic overcoating to enhance the electrical conductivity, as well as to alter the interface characteristics. Electroless processes have succeeded in producing nickel coatings on MWNTs as well as CNF.[21] Processes such as electron beam evaporation have been used successfully to coat nanomaterials with metals such as gold and silver, although obviously this would be considered rather exotic for most polymer filler applications.[22] Similarly, particles of metals such as platinum have been attached to carbon nanomaterials with the possible application as catalysts.[23] Such exotic materials and processes may have excellent commercial applications as catalysts, sensors, or electronic devices, but would be less likely to have applicability as nanofillers, even at a fraction of a percent due to their high costs.

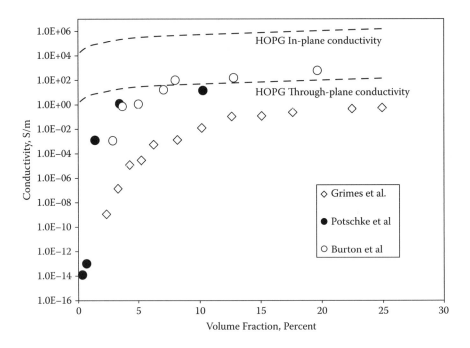

FIGURE 16.14
Composite of data on SWNT (Grimes et al.), MWNT (Potschke et al.), and CNF (Burton et al.).

In summary, percolation theory seems qualitatively able to describe conductivity enhancement in polymers due to nanoscale additives. However, the main fault is that the magnitude of enhancement is grossly overpredicted, likely due to the failure to adequately account for effects in the polymer-carbon interface that inhibit electron transport.

The interfacial resistance problem is the dominant effect on nanocomposite electrical performance, and lowers the electrical conductivity by several orders of magnitude. Consequently, SWNT nanocomposites have not achieved the same levels of conductivity as their cousins, the MWNT nanocomposites and CNF nanocomposites (Figure 16.14).

An interesting area of research in the future would be to combine nanoreinforcements with an adherent polymer matrix with intrinsic conductivity in the range of approximately 10^{-3} S/m or higher. If the interface problems can be overcome by developing an adherent polymer, it might be possible to enhance the electron transport to the point where semimetallic conductivity might be attainable in a polymer system.

In addition to the currently unavoidable difficulty of the interfacial resistance problem, there are other processing conditions that can affect the performance of nanocarbon composites. In particular, reinforcement material can agglomerate in a polymer melt, and thus not achieve uniform dispersion. This would likely have the effect of causing percolation effects to appear at higher additive concentrations than expected. The problem of agglomeration, although shared with nanoclay-reinforced nanocomposites, has a different character in the case of carbon nanomaterials. In the case of nanoclays, it is desired—and is often difficult—to achieve a high degree of exfoliation and dispersion in order to attain optimal properties. In the case of carbon nanomaterials, however, high shear is thought to be destructive to the nanomaterials. What is required, then, is an advanced capability to disperse nanomaterials with low shear and thus low damage to the nanoreinforcement materials.

16.4.1 Multicomponent Reinforcements

From the standpoint of improving electrical conductivity, it may well be cost-effective to consider standard fillers such as metal particles or carbon black, which might then be augmented by the inclusion of carbon nanofilaments. For the sake of discussion, these nanofilaments could be carbon nanofibers having an average diameter of some 100 nm. One might imagine that if a metallic macrofiller is used (i.e., one with lower bulk electrical resistivity compared to nanofibers, and a characteristic dimension in the range of microns), just below the percolation threshold, the inclusion of nanofibers might help close the open circuit between adjacent macroparticles. One might further imagine that even smaller size domain material (e.g., MWNTs having a characteristic diameter of some 10 nm) employed at an even smaller concentration could augment the ability of the first to succeed.

This can be expressed mathematically, similar to a fractal system. As an example, consider the family of mixtures with a concentration C_{total} of conductive additives. The conductive additives can, in turn, comprise several different additives with different size domains; that is,

$$C_{tot} = C_1 + C_2 + C_3 + \cdots + C_n \qquad (16.10)$$

where n is the total number of additives. The index $i = 1$ refers to the additive with the largest size domain, $i = 2$ refers to the additive with the second-largest size domain, etc.

The basic strategy is to select additive concentrations for $i = 2, 3, \ldots n$ that are close to the percolation thresholds $\rho_{2p}, \rho_{3p} \ldots$ for each additive. This may result in enhanced performance compared to the case in which a single additive with a single size domain is used. The intended effect is to reduce percolation effects in the largest scale additive, so that the ROM approximation is more nearly obeyed.

In addition, it can be observed that in general the smaller dimension materials are the more expensive ones. Hence, it is probably advantageous to minimize their use.

The net effect is shown in Figure 16.15. When additives 2 and 3 are included, the percolation effects are largely removed, and the modified resistivity of the composite as a function of the largest size domain $\rho'(C_1)$ is thus lowered for concentrations $C_1 < C_{1p}$, as signified by the thick dashed line in the figure.

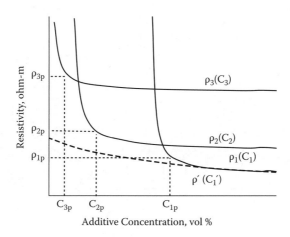

FIGURE 16.15
Resistivity as a function of additive concentration for additives with different size domains.

As a practical matter, meaningful reductions in resistivity can be accomplished via synergistic combinations of reinforcements with different size domains—for example, with the judicious addition of a fraction of a percent nanofiller(s), combined with a few percent of metallic microscale fillers.

16.5 Dispersion of Nanoscale Additives

The workhorse of the polymer compounding industry is the twin-screw extruder (Figures 16.16 and 16.17). Yet experience shows that shear forces created in such units is sufficient to cause significant breakage in carbon nanofilaments (Figure 16.16). In addition, it may be suspected that forces that are sufficient to cause breakage in some nanofilaments may also result in creating lattice damage at other sites. The excellent properties associated with nanofilaments are generally associated with maintaining a high degree of long-range order in the molecular lattice. Defects in the lattice would likely represent charge scattering centers and thus would contribute to degrading electrical properties.

Likely, these nanoscale damage mechanisms have prevented widespread success with the use of nanofilaments for electrical resistivity reduction. Conversely, the use of lower shear mixing methods in the laboratory has resulted in impressive results that have not yet been successfully transitioned to commercial production scale.

The answer to this dilemma likely lies in the development of new compounding techniques with an advanced generation of twin-screw extruders and other compounding machines.[23] In newer machines, advances in screw design have been accompanied by the ability to profile both temperature and shear in the barrel of the extruder itself, and to introduce additives (e.g., nanomaterials) at different points in the barrel to tailor the shear conditions (Figure 16.17). The key for nanocomposite compounding, then, will translate into maximizing the uniformity of dispersion while minimizing damage to the nano-additives.

Yet another enhancement is possible if there is an opportunity to align the nanotubes. For example, by extruding nanomaterials in a polymer melt, the result can be an alignment in the direction of extrusion.[24] The aligned materials then can offer a one-dimensional enhanced conductivity.

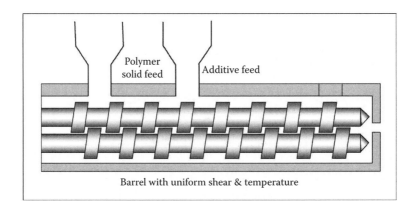

FIGURE 16.16
Older-generation twin-screw extruders generally damage carbon nanofibers during compounding.

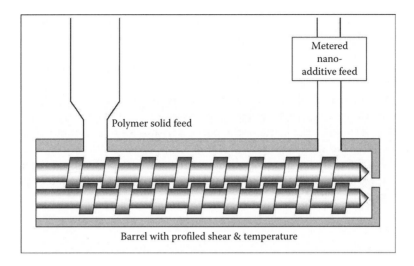

FIGURE 16.17
Modern twin-screw machine with features relevant for effective nanofiber dispersion.

16.6 Electrostatic Spray Painting

One of the first applications of carbon nano-additives for the purpose of enhancing electrical conductivity in a polymer nanocomposite was electrostatic spray painting. In this case, the automotive industry imposed a second requirement for an optically smooth surface so that the painted object would have the reflective, glossy appearance that the automobile-buying public has come to expect.

Electrostatic spray painting is a commercial process used to achieve uniform paint coatings with minimal overspray through the action of electrostatic attraction between the target and charged paint. In addition to the obvious economy of minimizing waste, electrostatic paint spray is particularly attractive because it minimizes environmental emissions of volatile organic compounds (VOCs).

Typically, paint particles are negatively charged by an ionizing electrode and impinge upon a workpiece that is electrically grounded. This results in the creation of an electrostatic field that draws the paint droplets to the workpiece, minimizing overspray and even drawing droplets to the rear of the workpiece, a phenomenon referred to as "wrap."

The transfer efficiency is the ratio of sprayed paint applied to the workpiece divided by the total amount consumed in the process. The typical transfer efficiency for an electrostatic paint spray system is 75%. Hence, for many situations, electrostatic paint spraying is highly economical and carries a minimal environmental impact.

A drawback of the process is that it requires an electrically conductive target. Thus, components fabricated using nonconductive polymers might be unsuitable for the electrostatic spray painting process.

Hence, it follows that the creation of conductive polymer nanocomposites can be attractive because it could permit, for example, injection-molded parts to be painted via the electrostatic spray painting process. Yet another requirement in many cases is the ability to produce a smooth finish, which might militate against the use of electrically conductive macro-additives such as chopped carbon fiber.

These factors favor the use of carbon nanomaterials to produce polymers with acceptable electrical conductivity with low mass addition, while also permitting the formation of a smooth surface.

GE Plastics with the assistance of Hyperion Catalysis Inc. produced a multiwalled nanotube-based NORYL GTX 990EP conductive resin, which was used in an electrostatic spray process by United Technologies Automotive.[25] Starting in October 1997, electrostatic discharge spray-painted nanocomposite components were used to produce Ford Taurus and Mercury Sable mirror housings, one of the earliest instances of polymer carbon nanocomposites to have achieved true commercial status. Compared to the case of mirror housings that ordinarily utilized hand-spray application, electrostatic paint spraying increased the application rate, from approximately 80 mirror housings coated per gallon of paint to about 290 housings coated per gallon.

16.7 Electrostatic Discharge Mitigation

Electrostatic discharge (ESD) is a rapid transfer of electrostatic charge between two objects, usually resulting when two objects at different potentials come near contact with each other. Materials such as polymers can acquire large static potentials if they are highly dielectric.

ESDs are serious problems in certain situations. For example, dielectric materials can accumulate static charge, meaning that care must be taken in using such materials near vapors of gasoline or other potentially explosive vapors or gases. Hence, polymer materials that are less susceptible to electrostatic discharge could find widespread use for handling combustible liquids, gases, and vapors.

Electronic equipment including computer components is likewise susceptible to upset or damage in the presence of a static shock. Likewise, materials that can avoid static discharge would be in demand for use in electronic devices.

The rate of charge dissipation is controlled by the time constant

$$\frac{dq}{dt} \sim \frac{1}{\tau} \sim \frac{\sigma}{\varepsilon} \qquad (16.11)$$

where τ is the time constant of charge dissipation (inversely proportional to the rate of dissipation), σ is the conductivity of the material (in S/m), and ε is the permittivity of the (units for all or none) material. The permittivity of a perfect metal is infinite, so the addition of conductive additives both increases the numerator and the denominator of the equation. However, in the case of nanomaterials, it is possible to increase the conductivity by orders of magnitude, while the permittivity increases more slowly. Thus, nano-additives can be especially attractive for electrostatic discharge applications compared to conventional conductivity enhancement.

16.8 Alternating Current Effects

Although many of the applications of nanocomposites are based on the steady-state direct current behavior of nanocomposites, frequency-dependent effects may also be important for certain applications. These applications can include radar cross-section modification,

TABLE 16.2

Materials as Described by Che et al.

Sample	Description
A.	Mainly MWNT
B.	Mainly Fe contained in carbon nanocages
C.	Mainly Fe nanowires contained in MWNT
D.	Same as B, but with α-Fe
E.	Same as C, but with α-Fe
F.	Pure iron sheet

Source: Che, R.C., Peng, L.M., Duan, X.F., Chen, Q., and Liang, X.L., *Adv. Mater.*, 6, 401, 2004.

lightning strike mitigation, electromagnetic pulse (EMP) mitigation, electromagnetic interference (EMI) mitigation, antennas, waveguides, etc.[26]

That nanomaterials should have unique impedance characteristics is ensured by their small dimensions and close-coupling. This is confirmed by experimental studies, including those of Che et al.[27] Che et al. carried out microwave reflection studies at 2 to 18 GHz, and found a strong dependence on the type of materials used as additives.

The materials were prepared according to the descriptions in Table 16.2. The researchers apparently have the capability to pyrolytically implant iron in the cores of the nanotubes. The structure formed thereby is regarded as a nanocomposite in its own right.

In Figure 16.18, it can be seen that the ability to attenuate microwaves is strongly enhanced by the presence of iron in the nanomaterials, and, in particular, by α-Fe. Through this

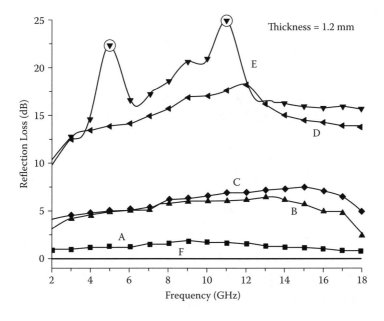

FIGURE 16.18
Che et al. data for microwave reflection loss from 2 to 18 GHz. (*Source:* Che, R.C., Peng, L.M., Duan, X.F., Chen, Q., and Liang, X.L., *Adv. Mater.*, 6, 401, 2004.)

simple example, it is certainly plausible that microwave attenuation might be even stronger for other metallic nano-additives as they are developed.

Thus, while the applicability of nanocomposites to the complex frequency domain has been well established, it is a much more difficult task to derive generalized performance goals for the polymer industry to respond to. Requirements for complex permittivity vary on a case-by-case basis, and there is no single specification or even representative goal that can be used at present to guide the launch of new master batches. Hence, the development of appropriate composites is likely to be accomplished on a product-specific basis.

16.9 Summary

The electrical conductivity of carbon nanotubes and other nanostructures is not easily quantified. It is argued that true multiwalled nanotubes may offer electrical characteristics similar to planar graphite. In the case of single-walled nanotubes, enhanced electrical conduction modes (e.g., ballistic conduction, with near zero losses) probably exist. Other single-walled nanotubes are semiconducting. On the other hand, carbon nanofibers (believed to resemble "stacked cups" with a characteristic conic angle of 20°) would offer poor conductivity unless the nanofibers are enhanced with a coating of either tubular pyrolytic carbon or some other conductive material. The ability to coat nanoreinforcements with conductive metal layers may result in significant improvement in electrical performance of future carbon nanocomposites.

Many of these effects offer exciting possibilities for electronic devices. (However, in the case of polymer compounding, for the most part nanomaterials are simply fillers. This author does not think anybody would use expensive nanomaterials as fillers.) In this case, the ability to add electrical conductivity through nanofillers can be important for specialized applications.

In any case, the composite electrical conductivity performance achieved in real nanocomposites has generally been much lower than would be expected based on rule-of-mixture analysis. Problems such as poor dispersion and adherence can likely be overcome by improved mixing techniques. On the other hand, the evidence is strong that the interface between nanomaterial and polymer matrix is highly resistive for many combinations of interest. Future research is aimed at the ability to produce adherent polymers with enhanced electrical caducity, which would result in greatly enhanced electrical performance for nanocomposites.

References

1. Poncharal, P., Berger, C., Yi, Y., Wang, Z.A., and de Here, W., Room temperature ballistic conduction in carbon nanotubes, *J. Phys. Chem. B*, 106, 12104, 2002.
2. Frank, S., Poncharal, P., Wang, Z.L., and de Heer, W.A. Carbon nanotube quantum resistors, *Science*, 280, 1744, 1998.
3. Dresselhaus, M.S., Dresselhaus, G., and Eklund, P.C. *Science of Fullerenes and Carbon Nanotubes,* London: Academic Press, 1996.
4. An, K.H. et al., Electrochemical properties of high power supercapacitors using single walled carbon nanotube electrodes, *Adv. Func. Mater.*, 11, 387, 2001.

5. Natori, K., Kimura, Y., and Shimizu, T., Characteristics of a carbon nanotube field-effect transistor analyzed as a ballistic nanowire field-effect transistor, *J. Appl. Phys.*, 97, 034306, 2005.
6. Kim,Y.A. , Hayashi, T., Naokawa, S., Yanagisawa, T., and Endo, M., Comparative study of herringbone and stacked-cup carbon nanofibers, *Carbon*, 43, 14, 3005, 2005.
7. Gaier, J.R. and Berkbile, S., Electrical characterization of pristine and intercalated graphite fiber composites, Extended Abstracts: *24th Biennial Conference on Carbon* sponsored by the American Carbon Society, 1999.
8. Lake, M.L., Fabrication and Characterization of Boronated Carbon Fibers and Electrical Conductos, Argonne National Laboratory Contract 7202401, 1987.
9. Arai, S., Endo, M., Hashizume, S., and Shimojima, Y., Nickel-coated carbon nanofibers prepared by electroless deposition, *Electrochem. Commun.*, 6, 1029, 2004.
10. Silvain, J.F., Richard, P., Douin, J., Lahaye M., and Heintz, J.M., Electroless coating process of carbon nano fibers by copper metal, *Mater. Sci. Forum*, 534-536, 1445, 2007.
11. Rawal, S.P., Kustas, F.M., and Rice, B., Evaluation of carbon nanofiber-based coatings and adhesives, *38th ITC*, Dallas, Texas, November 6–9, 2006.
12. Graff, A., Wagner, D., Ditlbacher, H., and Kreibig, U., Silver nanowires, *The European Physical Journal D – Atomic, Molecular, Optical and Plasma Physics*, 34, 263, 2005.
13. Sandler, J.K.W., Kirk, J.E., Kinloch, I. A., Shaffer, M.S.P., and Windle, A.H., Ultra-low electrical percolation threshold in carbon-nanotube-epoxy composites, *Polymer*, 44, 5893, 2003.
14. Last, B.J. and Thouless, D.J., Percolation theory and electrical conductivity, *Phys. Rev. Lett.*, 27, 1721, 1971.
15. Grimes, C.A., Mungle, C., Kouzoudis, D., Fang, S., and Eklund, P.C., The 500 MHz to 5.50 GHz complex permittivity spectra of single-wall carbon nanotube-loaded polymer composites, *Chem. Phys. Lett.*, 319, 460, 2000.
16. Ericson, L.M. et al., Macroscopic, neat, single-walled carbon nanotube fibers, *Science*, 305, 1447, 2004.
17. Kamiya, Y., Mizoguchi, K., Hirose, T., and Naito, Y., Sorption and dilation in poly(ethyl methacrylate)-carbon dioxide system, *J. Polym. Sci. Pt. B: Polym. Phys.*, 27, 4, 879.
18. Barrau, S., Demont, P., Peigney, A., Laurent, C., and Lacabanne, C., DC and AC conductivity of carbon nanotubes-polyepoxy composites, *Macromolecules*, 36, 5187, 2003.
19. Pötschke, P., Fornes, T.D., and Paul, D.R., Rheological behavior of multiwalled carbon nanotube/polycarbonate composites, *Polymer*, 43, 3247, 2002.
20. Burton, D.J., Glasgow, D.G., Kwag, C., Lake, M.L., Carbon nanofiber treatment effects on composite properties, *Materials Research Society Proceedings*, 702, 185–192, 2002.
21. Kong, F.Z. et al., Continuous Ni-layer on multiwall carbon nanotubes by an electroless plating method, *Surf. Coatings Technol.*, 155, 33, 2002.
22. Chin, K.C., Gohel, A., Chen, W.Z., Elim, H.I., Ji, W., Chong, G.L., Sow, C.H., and Wee, A.T.S., *Chem. Phys. Lett.*, 409, 85, 2005.
23. Guo, D.-J. and Li, H.-L., Electrochemical synthesis of Pd nanoparticles on functional MWNT surfaces, *Electrochem. Commun.*, 6, 999, 2004.
24. Wildi, R.H. and Maier, C., *Understanding Compounding*, Hanser Gardner Publications, July 1998.
25. Du, F., Fischer, J.E., and Winey, K.I., Effect of nanotube alignment on percolation conductivity in carbon nanotube/polymer composites, *Phys. Rev. B*, 72, 121404(R), 2005.
26. Purdue University Clean Manufacturing Technology and Safe Materials Institute Case Study, United Technologies Automotive, April 1998.
27. Grimes, C.A., Mungle, C., Kouzoudis, D., Fang, S., and Eklund, P.C., The 500 MHz to 5.50 GHz complex permittivity spectra of single-wall carbon nanotube-loaded polymer composites, *Chem. Phys. Lett.*, 319, 460, 2000.
28. Che, R.C., Peng, L.M., Duan, X.F., Chen, Q., and Liang, X.L., Microwave absorption enhancement and complex permittivity and permeability of Fe encapsulated within carbon nanotubes, *Adv. Mater.*, 6, 401, 2004.

17

Thermal Conductivity of Polymer Nanocomposites

Sushant Agarwal and Rakesh K. Gupta

CONTENTS

17.1 Introduction

Since their invention more than two decades ago, polymer nanocomposites have become a new and important class of materials that not only offer superior performance as compared to conventional composite materials, but also provide multifunctional properties that extend their scope of application to new areas.[1] For example, the addition of organically modified nanoclays to polymers results in improved mechanical properties such as stiffness, but the incorporation of the nanofiller also provides a better mass transfer barrier and improved flame retardant properties.[1] In another instance, although most thermoplastics and thermosets are inherently thermally and electrically insulating materials, if carbon nanotubes or nanofibers are used as fillers, one can obtain a material that not only has superior mechanical properties, but that also has high electrical and thermal conductivities.[2]

The utility of such versatile materials becomes readily apparent if one considers that one of the most successful usages of thermoplastics and their composites has been as a replacement for metals and ceramics in automotive and aerospace applications. Thermoplastics are preferred over metals and ceramics because of their low density, which results in very high stiffness-to-weight and strength-to-weight ratios. Another advantage of plastics over metals is the ease of fabrication of complex parts that can be manufactured on a large scale using methods such as injection molding. However, still wider application of thermoplastic materials as a replacement for metals and ceramics has been constrained by the fact that polymeric materials have very low electrical and thermal conductivities; in fact, plastics are widely used as insulating materials. Typically, the thermal conductivity of thermoplastics is approximately 0.2 W/mK, whereas the thermal conductivity of most metals is more than 100 times this value. Therefore, in recent years there has been considerable interest in improving the thermal properties of polymers for a variety of applications.[3] Most of these applications are in the area of heat removal and heat management. In some areas they can replace metals and ceramics, while in other cases they can replace nonconductive plastics.[4] The major commercial markets for thermally conductive plastics are commercial and industrial equipment; microelectronic applications, such as heat sinks in electronic devices; tubing for heat exchangers; enclosures for electrical appliances; casings for small motors; and heat exchangers used in corrosive environments. Other applications include automotive light reflectors, laser-diode encapsulation, fluorescent ballasts, heat sensors, and switches.[3–5]

Just as the stiffness of a generally pliant thermoplastic can be improved by adding stiff fillers such as glass fibers, the thermal conductivity of polymers can be enhanced by incorporating thermally conductive materials such as metals, carbon-based materials, and ceramics.[3] The most commonly used metal fillers are copper, steel, silver, gold, and aluminum, and these are generally used in the form of powders or flakes. Traditional carbon-based fillers are graphitic carbon fibers and carbon powders. Ceramic fillers include silicon carbide, boron nitride, aluminum nitride, and alumina. The thermal conductivity of graphitic carbon fibers is about 600 W/mK, for copper 400 W/mK, for boron nitride 275 W/mK, for aluminum nitride 200 W/mK, and for alumina about 30 to 40 W/mK. It has been suggested that for a polymer composite to be used in thermal management applications where the main barrier to heat transfer is the convective heat resistance, the composite should have a minimum thermal conductivity in the range of 1 to 10 W/mK.[3] Note that some filled commercial grades having 100-W/mK thermal conductivity are also available.[6] For this extent of thermal conductivity enhancement, however, very high loadings, of up to 50 vol%, must be used, and this results in increased density of the composites, loss in mechanical properties, processing difficulties, and increased cost. However, it is expected that by using nanofillers, such as high thermal conductivity carbon nanotubes, a substantial improvement in the thermal conductivity can be obtained at very low loading levels. The effectiveness of using carbon nanotubes is illustrated in Figure 17.1, which shows the thermal conductivity enhancement for a silicone rubber matrix containing silicon carbide and aluminum nitride on the one hand and multiwalled carbon nanotubes on the other.[7] It is seen that the same amount of thermal conductivity enhancement can be obtained using only 2 vol% carbon nanotubes as by using 30 vol% silicon carbide or aluminum nitride. This happens partly because carbon nanotubes have a higher thermal conductivity as compared to the ceramic fillers and partly because the nanotubes can

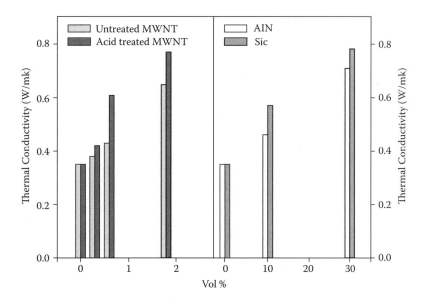

FIGURE 17.1
Comparison of thermal conductivity enhancement of silicone resin using silicon carbide (SiC) and aluminum nitride fillers vs. nanofillers such as multiwalled carbon nanotubes (MWCNTs). (*Source:* Reprinted with permission from Lee, G.-W. et al., *J. Mater. Sci.*, 40, 1259, 2005.)

form a conducting network within the polymer. Figure 17.1 also shows that acid treatment of carbon nanotubes helps to further increase thermal conductivity, possibly by promoting adhesion between the filler and the matrix and by promoting contacts among the nanotubes.

Carbon nanomaterials such as nanotubes or nanofibers are considered the most suitable additives for thermal conductivity enhancement because theoretically they can have thermal conductivity values as high as 6000 W/mK.[8] It is very difficult to verify these numbers, however, because of experimental difficulties in measuring the thermal conductivity of individual nanotubes. However some attempts have been made in this regard but the reported values vary quite widely. Thus, while Kim et al.[9] reported a value of 3000 W/mK for individual multiwalled carbon nanotubes (MWCNTs), Shioya et al.[10] reported 950 W/mK while some others[11–13] have reported values as low as 200 W/mK. In the case of individual single-walled carbon nanotubes (SWCNTs), Fujii et al.[14] have reported 2000 W/mK as the thermal conductivity. Some other nanomaterials that can be used as additives are expanded nanographite (EG) platelets and nanodiamonds, which can have thermal conductivities of nearly 2000 W/mK.[5] Table 17.1 summarizes data on the thermal conductivity of polymer nanocomposites with a variety of thermoplastic, thermoset, and elastomeric matrices. The nanofillers employed have been mostly carbon-based materials. The sections that follow present methods of making thermally conductive polymer nanocomposites and then describe the different techniques used to make thermal conductivity measurements. Also discussed are the theoretical equations that have been developed to predict the thermal conductivity of nanocomposites, and, finally, the effects of various parameters on thermal conductivity are examined.

TABLE 17.1

Polymer Nanocomposites for Thermal Conductivity Enhancement

Nanofiller	Matrix	Concentration Range	Effect on Thermal Conductivity	Maximum Thermal Conductivity Observed (W/mK)	Reference
SWCNT	Epoxy	0–2 wt%	~120% increase	0.44	Biercuk et al.[15]
SWCNT	Epoxy	0–1 vol%	~70% increase	0.36	Bryning et al.[16]
SWCNT	Epoxy	0–1.5 wt%	Almost doubles	~0.26	Song and Youn[17]
SWCNT	PMMA epoxy	0–7 wt%	>100% increase	0.61	Du et al.[18]
SWCNT buckypaper	Epoxy	25–30, 50 vol%	~20 times increase, directional effect	42	Gonnet et al.[19]
SWCNT	PMMA	0–9 vol%	~250% increase	~0.5	Guthy et al.[20]
SWCNT	PE	0–20 vol%	~7-fold increase	~4.5	Haggenmueller et al.[21]
SWCNT buckypaper	PEEK	——	~30-fold increase	7.6	Song et al.[22]
MWCNT	PP	0–20 wt%	~100% increase	~0.55	Kashiwagi et al.[23]
MWCNT	Silicone	0–3.8 wt%	65% enhancement	1.8	Liu et al.[24]
MWCNT	Silicone rubber	0–3 wt%	More than 3 times increase	~0.8	Lee et al.[7]
MWCNT	PP	0–2 wt%	~2-fold increase	0.55	Kim et al.[25]
MWCNT	CF-Epoxy, hybrid	0–10 wt%	~57% increase	397	Kim et al.[26]
CNT	PDMS rubber	2 wt%	~70% increase	0.191	Liu and Fan[27]
CNT	EPDM rubber	5 wt%	~3-fold increase, directional effect	0.7	Kim et al.[29]
CNF	PP	0–25 vol%	~30-fold increase, directional effect	~5.5	Kuriger and Alam[29]
CNF	Nylon 11	0–8 wt%	Almost 3 times increase, directional effect	~0.6	Cummings et al.[30]
CNF	PP	0–40 vol%	~17-fold increase, directional effect	3.5	Enomoto et al.[31]
EG	Nylon 66, HDPE	15, 20 wt%	5 to 10 times increase	~4.1	Fukushima et al.[32]
EG	Epoxy	1 vol%	<6% improvement	0.22	Hung et al.[33]
EG	Silicone rubber	0–30 wt%	Almost 3 times increase	~0.35	Mu and Feng[34]
MWCNT, CNF, EG	Polyetherimide	0–50 wt%	Up to 20 times increase, directional effect	6.8	Ghose et al.[35]
CB, SWCNT, DWCNT, MWCNT	Epoxy	0–0.6 vol%	Marginal increase	0.25	Gojny et al.[36]
Nanodiamond	PMMA	0–10 wt%	Marginal increase	~0.35	Tyler et al.[37]
Nanodiamond, onion-like nanocarbon	PDMS	0–2 wt%	~15% increase	——	Shenderova et al.[38]
Nano- and micro-alumina powder	Epoxy	50 wt%	Increases on increasing the micro-content	~0.65	Fan et al.[39]
Nano CaCO₃	HIPS	0–2.5 wt%	Up to 4 times increase	——	Mishra and Mukherji[40]
Nanoclay	Polyurethane foam	0–5 wt%	Thermal conductivity decreases	——	Seo et al.[41]

Key: SWCNT—single-walled carbon nanotube, MWCNT—multiwalled carbon nanotube, DWCNT—double-walled carbon nanotube, CNT—carbon nanotube, CNF—carbon nanofiber, EG—expanded or exfoliated nano-graphite, CB—carbon black, EPDM—ethylene propylene diene rubber, HDPE—high-density polyethylene, HIPS—high impact polystyrene, PDMS—polydimethyl siloxane, PE—polyethylene, PEEK—polyetherether ketone, PMMA—poly(methyl methacrylate), PP—polypropylene.

17.2 Methods of Making Polymer Nanocomposites for Thermal Conductivity Improvement

Various methods are used to formulate polymer nanocomposites with enhanced thermal conductivities, depending on whether the polymer matrix is thermoplastic or thermoset. Often, the physical shape of the nanofillers—whether long tubes, platelets, or particulates— also influences the choice of processing method. However, the objective in each case is to prepare nanocomposites to achieve the best enhancement in desired properties. For thermal conductivity enhancement, especially with the use of carbon nanotubes or nano-fibers, a number of characteristics in the processing method are desirable. The processing method should be able to de-aggregate and disperse the nanotubes uniformly throughout the matrix without causing fracture, which reduces their aspect ratio. The mixing should result in good interfacial bonding between the matrix and the nanofiller. In addition, if required, the processing method should also promote alignment of the nanomaterial in the desired direction to tailor the alignment-dependent properties of nanocomposites. While a general discussion on preparing polymer nanocomposites can found in references, the following discussion is restricted to methods used for making thermally conductive nano-composites.[2, 42, 43]

For thermoplastics, melt-mixing remains the most popular and practical way of incorporating nanofillers in the polymer matrix. Both extruders and batch mixers are used for this purpose. For example, Kuriger and Alam[29] used a twin-screw extruder to disperse vapor-grown carbon nanofibers in a polypropylene matrix. Furthermore, they used a converging die at the exit of the extruder to align the fibers in the flow direction. However, care must be exercised to ensure that good dispersion is achieved without causing breakage of the nanofibers. Kim et al.[28] used a Banbury batch mixer to compound carbon nanofibers in a rubber matrix. Solvent processing can also be employed for making thermoplastic nanocomposites, especially if small samples are desired or if any surface coating on the nanofiller is degraded at melt processing temperatures. For example, Guthy et al.[20] first dispersed SWCNTs in dimethylformamide using sonication and then poly(methyl methacrylate) (PMMA) was also dissolved. Finally, water was added to the mixture, which caused the precipitation of the PMMA trapping the SWCNTs within the polymer. In all cases, nanocomposite blends are either compression molded or injection molded to obtain samples for thermal conductivity applications. Injection molding is, however, preferred because it gives more reproducible results. For thermosets, the nanofillers are first mixed with the uncured resin using a high shear mixer or ultra-sonic mixing to disperse the nanofillers and then appropriate cross-linking agents are added.[33, 26] The mixture is then poured into molds and allowed to cure to obtain the final product.

17.3 Measurement of Thermal Conductivity

As an increasing number of thermally conductive plastic composites are available or are under development for ever-demanding requirements and new applications, the accurate and repeatable measurement of thermal conductivity is essential in order to compare different materials and in designing various applications. The thermal conductivity of a material in one-dimensional conductive heat transfer is given by the well-known Fourier's law of heat conduction:

$$k = -\frac{q/A}{\Delta T/L} \tag{17.1}$$

where q is the rate of heat transfer through a cross-section of area A, and ΔT is the temperature difference that exists over thickness L of the material.

Many methods exist to measure the thermal conductivity of polymer composites, some of which have been adopted as ASTM standards. These methods can be broadly classified into two categories: (1) steady-state methods and (2) transient methods. In what follows, both categories of methods are briefly described and their respective merits and limitations discussed. More detailed discussion can be found in References 44 through 46.

17.3.1   Steady-State Methods

17.3.1.1   ASTM D5470

ASTM D5470 and its variations are the most popular methods of measuring thermal conductivity of thin conductive plastics or composites, though this technique was originally meant for electrical insulation materials.[47] In this method, a thin and flat test specimen is exposed to a known amount of uniaxial steady-state heat flux by sandwiching the specimen between two metal bars. A schematic diagram of the setup is shown in Figure 17.2a. Heat flux is provided by a resistive heating element. At equilibrium or steady state, heat flux is measured either by measuring the temperature gradient along the metal bars or by

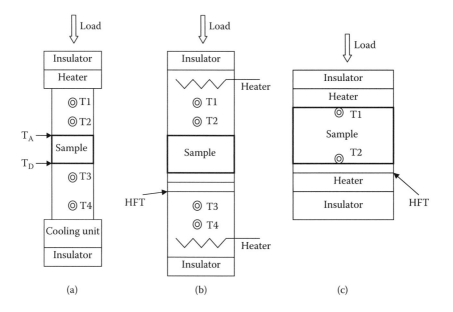

FIGURE 17.2
Schematic diagrams of steady state test methods for thermal conductivity measurement: (a) ASTM D5470, (b) ASTM E1530, and (c) ASTM F433.

measuring the electric power provided to the heating element. The entire unit is properly insulated to minimize heat loss and to ensure one-dimensional heat flow. Furthermore, the specimen is compressed between metal bars at a stress of at least 3 MPa to minimize contact resistance between the metal surface and the specimen. Still, there is resistance to heat flow due to boundary contact resistance. Here, this contact resistance can be separated from the thermal resistance of the sample by making measurements on samples of various thicknesses. At steady state, the thermal impedance, θ, defined as the ratio of the temperature difference across the assembly to the heat flux through the assembly, is given by

$$\theta = (T_A - T_D) \times \frac{A}{q} \qquad (17.2)$$

where A is the cross-sectional area, q is the rate of heat flow, T_A is the temperature in the upper rod in contact with the sample, and T_D is the corresponding temperature in the lower rod in contact with the sample. The thermal impedance is then related to the thermal conductivity of the sample by

$$\theta = \frac{L}{k} + R_i \qquad (17.3)$$

where L is the thickness of the sample, k is the thermal conductivity, and R_i is the interfacial resistance between the sample and the metal surfaces, which is assumed to be constant. θ is plotted versus L, and then a straight line is fitted to the data. The slope of the best-fit line is the reciprocal of the thermal conductivity of the sample. The intercept on the ordinate is the boundary or interfacial resistance.

Although this method remains a convenient and popular way of measuring thermal conductivity, the reproducibility and reliability of the data obtained from this method remain a matter of concern. Wide variations in reproducibility and repeatability have been reported using this method.[44] These errors arise due to the deviations in the apparatus and methodology from some of the underlying idealized assumptions. If the system is not properly insulated, there may be two-dimensional heat transfer resulting in not all the energy passing through the sample and thus leading to error in the calculation of the heat flux. Another major source of error is the assumption of a linear relationship between the thermal impedance and the sample thickness, an assumption that may not hold if the sample is not properly compressed between the metal bars due to misalignment or uneven surfaces. In such a case, the intercept does not necessarily represent the true interfacial resistance. It is also found that the boundary layer resistance decreases rapidly with the load of compression and then reaches a limit.[45] Care must also be exercised in determining the thickness of the sample as it may change under the compressive load, especially in the case of soft elastomeric composites.

17.3.1.2 ASTM E1530

ASTM E1530 is also a steady-state method of measuring the thermal conductivity of solid materials; however, it is a comparative method in the sense that it determines the thermal conductivity of the test sample by comparing the heat flow of calibration samples of known thermal conductivities with the unknown test sample.[48] Another major difference is the use of a heat flux transducer (HFT), which produces a voltage signal as a function of heat flux, to measure the heat flux, q, through the sample. Here again, samples are sandwiched between metal bars that are embedded with thermocouples to measure the temperature gradients. Heaters are installed to control the heat flow and a heat flux transducer is installed to measure the heat flux. A schematic of the setup is shown in Figure 17.2b. Sometimes, the entire setup is surrounded with a heater to minimize heat loss, hence the term "guarded heater method." The following equations are used to set up the procedure:

$$\theta_s = \frac{N(T_A - T_D)}{q/A} - R_i \tag{17.4}$$

$$\theta_s = \frac{L}{k} \tag{17.5}$$

Here, N is HFT calibration constant. In this procedure, calibration samples of various thicknesses and known thermal conductivities are used to obtain a plot of θ_s versus $(T_A - T_D)/(q/A)$ in such a way that the range covers the expected value of the unknown sample. The slope of this plot is N and the intercept is R_i. The same procedure is repeated for the test sample. Assuming that N and R_i are constant, θ_s is read from the curve and Equation 17.5 is used to calculate the thermal conductivity of the test sample. Major sources of error in this method are those caused by the assumption that interfacial resistance R_i remains constant for the calibration samples and the test sample, because in reality interfacial resistance is a strong function of surface properties and hardness of the samples. If calibration materials are hard substances such as metals or glass and the test sample is a soft elastomer, the error in estimating the contact resistance could be significant.[46] Thus, it is advisable that the calibration samples be similar to the test samples for good accuracy.

17.3.1.3   ASTM F433-02

This standard is also a steady-state method; it uses a comparative technique to measure the thermal conductivity.[49] As a matter of fact, it is used more for comparing the thermal conductance of comparable materials under controlled conditions. Figure 17.2c shows a schematic of the setup for this test method. Here, the test sample is compressed between heater plates and a heat flux transducer. The thermocouples are placed on the upper and lower surfaces of the sample to measure the surface temperatures directly. The device is first calibrated with a standard sample, for which the thermal conductivity is given by

$$k_s = N\phi_s \frac{L_s}{\Delta T_s} \tag{17.6}$$

Here N is the calibration constant for the HFT and ϕ_s is the heat flux reading from HFT, so that $N\phi_s$ is the heat flux through the sample. L_s is the sample thickness and ΔT_s is the temperature difference across the sample. A similar equation can be written for the test sample measurements:

$$k = N\phi \frac{L}{\Delta T} \tag{17.7}$$

Assuming N is constant, the thermal conductivity of the test sample can be calculated from

$$k = k_s \frac{\phi}{\phi_s} \frac{L}{L_s} \frac{\Delta T_s}{\Delta T} \qquad (17.8)$$

Because in this method the temperature difference across the sample is measured directly by placing the thermocouples on the sample surfaces, it is more susceptible to errors caused by surface contact resistance. Here, the contact resistance occurs not only between the thermocouple and sample surface, but also between the heaters and the sample surface. It is difficult to maintain a uniform contact between samples and the thermocouples, and sensitivity to the size and location of the sensors may cause variations in interfacial resistance from sample to sample. In this method also, it is advisable to use a material similar to the test sample as the calibration material.

17.3.2   Transient Methods

17.3.2.1   Transient Hot-Wire Method

The transient hot-wire technique is an absolute method of determining the thermal conductivity of a material without the need for any calibration standards. In this method, a thin metal wire is embedded in the material whose thermal conductivity is to be determined, and electric power is applied. The wire works both as a heat source and as a temperature sensor. When electric power is applied to the wire, some of the energy dissipates through the surrounding material by conduction, while the remainder goes to raise the temperature of the wire. As the wire temperature increases, however, the resistance of the wire increases. A measurement of the electrical resistance then gives the instantaneous temperature of the wire. This changing temperature can be related to the thermal conductivity of the surrounding material by an energy balance where it is assumed that the wire is an infinitely thin and long heat source embedded in an infinite medium.

To describe the method briefly, a constant current or voltage is applied at time $t = 0$ to a very thin and long wire made of a metal such as platinum that has been embedded in the sample medium. Then the change in resistance of the wire is monitored with respect to time using a digital data collection device. The change in the resistance of the wire is converted to the temperature change via the known resistance-temperature relationship. The thermal conductivity of the surrounding medium is then given by

$$k = \frac{Q}{4\pi l (dT / d \ln t)} \qquad (17.9)$$

where Q is the electric power supplied to the wire, l is the length of the wire, T is the temperature of the wire, and t is the time. This relationship holds only under certain conditions and is applicable only during a very brief period of heating time, which may last only a couple of seconds. Before using this method, it is advised to follow a more extensive discussion of the theory and practice of this method, which is available in References 50 through 54. One of the major concerns is that a special modification must be made if the test sample is electrically conductive. For example, a nanocomposite made from carbon nanofibers will be both electrically and thermally conductive. In that case, the wire must be coated with a thin layer of an insulating material to prevent loss of electrical current.[53]

In addition to the above concern, a major disadvantage of this method is that special sample preparation steps must be taken before the measurements can be made. The wire must be properly embedded in the sample material, and this may require compression-molding a thin long wire within the polymer matrix. The wire is usually very fragile as it is only few micrometers in diameter and thus very difficult to handle. Second, the matrix should have good thermal contact with the wire surface; any detachment of the matrix from the wire will result in erroneous results. However, this method is well suited for making measurements on molten polymers.

17.3.2.2 Transient Hot-Disk Method

This method is similar to the hot-wire method except that a thin flat round disk is used as a heat source, and a temperature sensor is used in place of a wire.[55] In other, slightly different versions, the technique is also known as the transient hot-strip (THS) or transient plane-source (TPS) method.[56] This method is more versatile, in that it also allows the measurement of specific heat and thermal diffusivity of the samples. A very thin metal foil such as a 10-μm thick spiral-pattern nickel disk is embedded between very thin layers of an insulating material such as Kapton™. This heat sensor is tightly sandwiched between two layers of the test sample. A constant current is applied to the nickel disk and the resistance-temperature change of the disk is monitored as a function of time. The following equation is plotted:

$$\Delta T = \frac{P_0}{\pi^{3/2} r k} D(\tau) \tag{17.10}$$

Here, ΔT is the temperature rise, P_0 is the electric heat generated, r is the radius of the disk, and k is the thermal conductivity that is being measured. τ and $D(\tau)$ are functions that are defined in References 55 through 57. The slope of Equation 17.10 can be used to obtain the thermal conductivity. In this method, one must be careful to ensure good thermal contact between the sample and the heating element to reduce errors.

17.3.2.3 Laser Flash Method

This method is also a transient method and is used to measure the thermal diffusivity (α) of the material from which the thermal conductivity can be easily obtained by the following equation:

$$\alpha = \frac{k}{\rho c_p} \tag{17.11}$$

where ρ is the density and c_p is the specific heat of the sample. This method was developed by Parker et al.,[58] who used a high-energy optical source as a heating source. However, the technique was later modified to allow the use of a laser beam as the heat source. Now this method has been adopted as ASTM E1461 to measure the thermal diffusivity of both solid and liquid materials, including composites.[59]

To briefly describe this method, one surface of a thin flat sample is exposed to a high-intensity energy pulse for a short period of time using a radiant energy source such as a laser beam or halogen lamp. The energy of the beam is absorbed by the sample, which results in a corresponding temperature rise on the rear surface of the sample. This temperature rise is monitored as a function of time using either a thermocouple, infrared detector, or optical pyrometer until the temperature reaches a maximum value.[59] From that, the half-time $t_{1/2}$ (i.e., the time to reach half of the maximum temperature) is calculated. For a sample of thickness l, the thermal diffusivity is given by[59]

$$\alpha = \frac{0.1388l^2}{t_{1/2}} \tag{17.12}$$

from which the thermal conductivity can be easily calculated.

There have been some studies to compare results obtained from the various methods of thermal conductivity measurement. Fukushima et al.[32] measured the thermal conductivity of exfoliated graphite nanocomposites of nylon and HDPE (high-density polyethylene) using hot-flash and hot-wire techniques, and they found that the measured values were in good agreement with each other. Miller et al.[60] used guarded heat flow and transient plane source methods to measure the thermal conductivity of nylon/carbon-fiber composites and found good agreement between the two techniques. In the commercial and research areas, some of these methods have been modified to suit particular needs. Especially in the case of testing thermal interface materials (TIMs) intended for heat management applications, it is felt that reliable data on thermal conductivity are difficult to obtain because the different test methods and test conditions are not representative of the intended applications.[44] In thermal management systems, the boundary layer resistance of the TIM, heat sink, and heat source play a crucial role in determining the performance of the TIM. Therefore, for thermal management systems, the entire heat dissipation package is sometimes tested for thermal performance rather than just the isolated thermal interface material. In summary, one must take into account the end usage of the conductive composite when selecting a method for thermal conductivity measurement.

17.4 Models to Calculate the Thermal Conductivity of Polymer Nanocomposites

Many theoretical and semi-theoretical models are available to represent the effective thermal conductivity of conventional polymer composites in which large-size fillers have been dispersed in a polymer matrix. Pragelhof et al.[61] provide a comprehensive listing and a review of such models to predict the thermal conductivity of conventional composites. Although some of these models can be applied to predict the thermal conductivity of polymer nanocomposites, each model must be carefully analyzed to ascertain its applicability as some models are only applicable in the dilute concentration range and some are meant just for spherical inclusions. Simple models such as the Rule of Mixtures (series model) give the upper bound, whereas the parallel model gives the lower bound of the thermal conductivity of a nanocomposite. As might be expected, experimental observations suggest that real values for nanocomposites fall somewhere in between these two limits. In the following section, some of the methods for predicting the thermal conductivity that are relevant to polymer nanocomposites are described.

Nielsen's model is often used to predict the thermal conductivity of nanocomposites.[62] Whereas in simple models, such as mixture rules, only the concentration of the two components and their respective thermal conductivities are considered, Nielsen's model also takes into account the shape, packing, and alignment of the filler material with respect to the direction of heat flow to predict the effective thermal conductivity of the composite. According to Nielsen's model, the thermal conductivity of the composite is given by[62]

$$k_c = k_m \left[\frac{1 + AB\phi}{1 - B\psi\phi} \right] \qquad (17.13)$$

where

$$B = \frac{k_f/k_m - 1}{k_f/k_m + A} \qquad (17.14)$$

and

$$\psi = 1 + \left(\frac{1 - \phi_m}{\phi_m^2} \right)\phi \qquad (17.15)$$

where A is a factor akin to Einstein's coefficient but also depends on the alignment of fibers with respect to the heat flow direction and the aspect ratio. For example, for isotropically oriented long fibers ($l/d > 15$), A is 8.38. Similarly, for heat flow parallel to the fibers, A is $2l/d$, while for heat flow perpendicular to fibers, it is 0.5. ϕ_m is the maximum packing fraction for the particles in the matrix. For three-dimensional random packing of fibers, this quantity is 0.52, whereas for uniaxial random packing it is 0.82. Values of A and ϕ_m can be found in Reference 62 for other packing situations. From the above equations it is clear that the effective thermal conductivity of the nanocomposite depends on the fiber aspect ratio and thermal conductivity of the nanofiller. The fiber aspect ratio of isolated nanotubes or nanofibers is very large (>1000); however, in reality, they are highly aggregated and not well-dispersed in the polymer matrix. Second, they undergo severe attrition during mixing operations, which further reduces their aspect ratio. Thus, the fiber aspect ratio must be determined experimentally. This model also suggests that the thermal conductivity of the nanocomposite becomes insensitive to the thermal conductivity of the nanofiller if $k_f/k_m \gg 1$ for isotropic distribution of the fibers. However, it has a very strong influence if fibers are aligned in the direction of heat flow. Figure 17.3 shows experimental thermal conductivity values of SWCNT-PMMA nanocomposites as a function of filler loading and its comparison with predictions from Nielsen's model.[20] The aspect ratio of the nanotubes was determined experimentally using an atomic force microscope, and it was found to be approximately 26, much less than that expected for an individual SWCNT. Using the theoretical thermal conductivity of SWCNTs of 6000 W/mK and the measured aspect ratio of about 26, Nielsen's model is found to overpredict the thermal conductivity. Reducing the theoretical aspect ratio to 20 gives a better fit to the data. However, an equally good fit is obtained if one uses 10 W/mK as the thermal conductivity of the carbon nanotubes. This points to the effect of thermal interfacial resistance between the fibers and the polymer matrix on the thermal conductivity of the nanocomposite. The thermal interfacial resistance arises due to imperfect bonding between the nanoparticle and the polymer matrix or due to phonon acoustic mismatch between the nanoparticles and the polymer.[63] This is also known as Kapitza resistance, R_K, and it is defined as the temperature drop across the interface for a given heat flow over the distance known as the Kapitza radius, ak.[64]

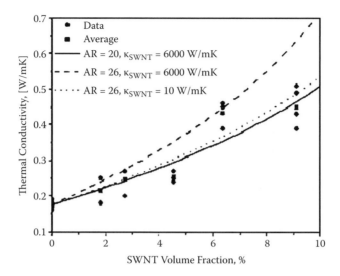

FIGURE 17.3
Thermal conductivity as a function of SWCNT volume fraction in a PMMA matrix. Lines represent predictions using Nielsen's model. The figure shows how theoretical predictions depend on the aspect ratio (AR) and thermal conductivity of the nanotubes. (*Source:* Reprinted with permission from Guthy, C. et al., *J. Heat Transfer*, 129, 1096, 2007.)

The Kapitza resistance can be very high in value ($\sim 10^{-7}$ to 10^{-9} m^{2}K/W), and thus it has a very significant impact on the effective thermal conductivity of a nanocomposite.[65, 66] It is due to the effect of the interfacial resistance that a polymer nanocomposite shows a much lower thermal conductivity than what one would expect from Nielsen's model using the correct aspect ratio and the correct (high) thermal conductivity of the nanofibers. Nan et al.[64, 66] have developed a model using an effective medium approach that takes into account the interfacial resistance between the nanofillers and the polymer matrix. While the derivations for the model predictions can be found in Reference 64, here we merely give the expressions for spherical, ellipsoidal, and cylindrical inclusions.

For spherical nanoparticles, the thermal conductivity of the nanocomposite is given by

$$\frac{k_c}{k_m} = \frac{k_f(1 + 4a_k/d) + 2k_m + 2\phi[k_f(1 - 4a_k/d) - k_m}{k_f(1 + 4a_k/d) + 2k_m - \phi[k_f(1 - 4a_k/d) - k_m]} \tag{17.16}$$

where *d* is the particle diameter and ak is the Kapitza radius defined by

$$a_k = R_k km \tag{17.17}$$

For the case of isotropically oriented ellipsoidal particles, which includes platelets, the nanocomposite thermal conductivity can be expressed as

$$\frac{k_c}{k_m} = \frac{3 + \phi[2\beta_{11}(1 - L_{11}) + \beta_{33}(1 - L_{33})}{3 - \phi[2\beta_{11}L_{11} + \beta_{33}L_{33}]} \tag{17.18}$$

Here, for $i = 1, 2,$ or $3,$

$$\beta_{ii} = \frac{k_{ii} - k_m}{k_m + L_{ii}(k_{ii} - k_m)} \tag{17.19}$$

$$L_{11} = L_{22} = \left\{ \frac{p^2}{2(p^2 - 1)} - \frac{p}{2(p^2 - 1)^{3/2}} \cosh^{-1} p \text{ for } p > 1 \right. \tag{17.20}$$

$$L_{11} = L_{22} = \left\{ \frac{p^2}{2(p^2 - 1)} - \frac{p}{2(1 - p^2)^{3/2}} \cosh^{-1} p \text{ for } p < 1 \right. \tag{17.21}$$

$$L_{33} = 1 - 2L_{11} \tag{17.22}$$

$$k_{ii} = \frac{k_f}{1 + \frac{\gamma L_{ii} k_f}{k_m}} \tag{17.23}$$

and

$$\gamma = (2 + 1/p) 2a_k / l \text{ for p} > 1 \tag{17.24}$$

and

$$\gamma = (1 + 2p) 2a_k / d \text{ for p} < 1 \tag{17.25}$$

Using this method, Hung et al.[33] were able to fit the thermal conductivities of nanocomposites of exfoliated graphite nanoplatelets and epoxy by adjusting various parameters because the thermal conductivity of nanoplatelets is difficult to obtain.

For nanocomposites containing carbon nanotubes, the corresponding results are given by[66]

$$\frac{k_c}{k_m} = \frac{3 + \phi(\beta_x + \beta_z)}{3 - \phi\beta_x} \tag{17.26}$$

where

$$\beta_x = \frac{2(k_{11} - k_m)}{(k_{11} + k_m)} \tag{17.27}$$

and

$$\beta_z = \frac{k_{33}}{k_m} - 1 \tag{17.28}$$

Here, ϕ is the volume fraction of the nanotubes, and k_{11} and k_{33} are equivalent thermal conductivities of a nanotube in the transverse and longitudinal directions in the presence of a

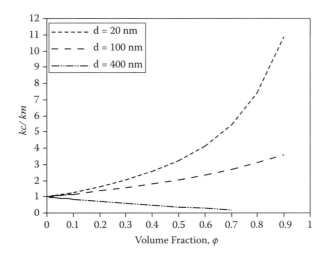

FIGURE 17.4
Predicted thermal conductivity values using the model of Nan et al. for nanocomposites made from spherical nanoparticles having a thermal conductivity of 3000 W/mK in a matrix of thermal conductivity equal to 0.4 W/mK. The Kapitza resistance has been assumed to be $8 - 10^{-8}$ m²K/W to compare with Figure 17.5.

layer of interfacial resistance barrier. The k values are given by

$$k_{11} = \frac{k_f}{1 + \frac{2a_k}{d}\frac{k_f}{k_m}} \quad \text{and} \tag{17.29}$$

$$k_{33} = \frac{k_f}{1 + \frac{2a_k}{l}\frac{k_f}{k_m}} \tag{17.30}$$

where d and l are diameter and length of the nanotubes, respectively.

The model of Nan et al. can be used to show the effect of nanoparticle shape on the thermal conductivity of nanocomposites.[64, 66] Figure 17.4 shows predicted values of relative thermal conductivity for nanocomposites containing spherical particles of thermal conductivity of 3000 W/mK in a polymer matrix of 0.4 W/mK thermal conductivity. The Kapitza resistance was assumed to be 8×10^{-8} m²K/W and constant for all three particle sizes of 20, 100, and 400 nm. Figure 17.5 shows predictions of this model for thermal conductivity of a nanocomposite containing carbon nanotubes of thermal conductivity of 3000 W/mK in a matrix of 0.4 W/mK thermal conductivity.[66] In comparing Figure 17.4 with Figures 17.5a and 17.5b, it can be seen that long nanofibers provide a much higher thermal conductivity enhancement than spherical nanoparticles for the same loading level. In fact, due to the very high Kapitza resistance, there is even a decline in the thermal conductivity with increasing volume fraction for the smallest spherical particles. Furthermore, Figure 17.5a shows the effect of nanotube diameter on the relative thermal conductivity of the nanocomposite. It is seen that as the diameter of the nanotube decreases, this leads to lower enhancement in the thermal conductivity. Figure 17.5b shows that better thermal conductivity enhancement is observed for larger aspect ratio nanotubes. Figure 17.5c further shows the dramatic effect

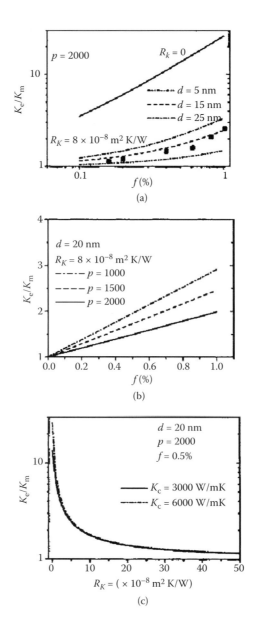

FIGURE 17.5
Theoretical predictions for nanocomposite thermal conductivity using the model of Nan et al. for nanofibers with an aspect ratio, p, of 2000 showing effects of interfacial thermal resistance (R_K) and the nanofiber diameter: (a) for constant R_K, a smaller particle size shows less enhancement, (b) a high aspect ratio results in better enhancement, and (c) if R_K is large, thermal conductivity enhancement is insensitive to the intrinsic thermal conductivity (K_C) of the nanofiller. (*Source:* Adopted from Nan, C.-W. et al. *Appl. Phys. Lett.*, 85, 3549, 2004. With permission.)

of interfacial resistance on the thermal conductivity enhancement. This figure shows that for large interfacial resistance, the thermal conductivity is insensitive to the thermal conductivity of the nanofiller. Separately, one notes that, due to increasing values of surface area per unit volume with decreasing size, it becomes increasingly difficult to incorporate large amounts of ultrafine particles into a polymer matrix.

Every et al.[67] also developed a model for particle-loaded composites, taking into account interfacial resistance. For $k_f/k_m \gg 1$, the model results given by[63]

$$k_c = \frac{k_m}{(1-\phi)^{\frac{3(1-\alpha)}{(1+2\alpha)}}} \quad (17.31)$$

in which α is the Biot number defined as

$$\alpha = \frac{R_b k_m}{d} \quad (17.32)$$

This model also shows that increasing the interfacial resistance or decreasing the particle size of the nanofiller has an adverse impact on the thermal conductivity enhancement of the nanocomposites.

The models described above assume that the nanofiller particles are completely surrounded with the polymer matrix and that they do not touch each other. Foygel et al.[68] considered the fact that, at very high aspect ratios, nanotubes start to form percolating clusters even at very low concentrations. Using Monte Carlo simulations and taking into account contact resistance between the particles, they arrived at the following relationship for the electrical or thermal conductivity:

$$k_c = k_0 |\phi - \phi_c|^t \quad (17.33)$$

where

$$k_0 \cong (R_0 L \phi_c^t)^{-1} \quad (17.34)$$

and k_0 is the preexponential factor that depends on the length of the nanotube, L, and the resistance of the nanotube or the contact resistance between the tubes, R_0, and percolation limit ϕ_c. Usually ϕ_c is estimated and then k_0 and t are obtained by data fitting. A knowledge of these quantities then yields the contact resistance. Conversely, if the contact resistance is known, k_0 can be independently estimated. The model of Nan et al., which is based on an effective medium approach, and Foygel's percolation model can be used in conjunction with each other to model the thermal conductivity over a wide range of nanofiller concentrations. Haggenmueller et al.[21] used the model of Nan et al. to fit thermal conductivity data at low filler concentrations in order to obtain the interfacial resistance and then used the percolation model of Foygel et al. to fit data at higher concentrations.

From the models described above, it can be concluded that to obtain the best enhancement in the thermal conductivity (1) there must be good thermal contact between the nanofillers and the matrix to reduce the interfacial boundary resistance, (2) particles should be large, and (3) particles should maintain their high aspect ratio. Some of these observations are also borne out by the experimental results discussed in the following sections.

17.5 Effect of Nanofiller Concentration

Increasing the nanofiller loading should result in an increase in the thermal conductivity of the nanocomposite. Indeed, this is exactly what is observed experimentally for very low levels of nanofiller loading (less than ~1 vol%) where more than a 100% increase in

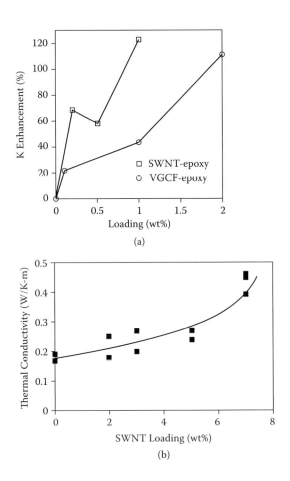

FIGURE 17.6
Effect of nanofiller concentration on thermal conductivity: (a) epoxy nanocomposite with SWCNT and CNF, and (b) PMMA nanocomposite with SWCNT. (*Source:* From Biercuk, M.J. et al., *Appl. Phys. Lett.*, 80, 2767, 2002; Du, F., *J. Polym. Sci., Part B: Polym. Phys.*, 44, 1513, 2006. With permission.)

the thermal conductivity, relative to the matrix, has been observed.[15, 18, 24, 36, 38] Figure 17.6 shows a very significant thermal conductivity enhancement for epoxy nanocomposites as a function of filler loading for nanofillers such as SWCNTs and carbon nanofibers, and PMMA nanocomposites with SWNTs.[15, 18] However, it has also been observed that, beyond a certain limit, upon increasing the fiber loading further, the enhancement in thermal conductivity tends to plateau or even decrease with increasing concentration.[20, 26, 38, 40, 69] This happens because, with an increase in the loading level, the interfacial thermal resistance increases, which, as previously discussed, has a dramatic negative effect on the thermal conductivity. Second, at higher loading levels, fibers usually undergo severe breakage during processing due to high stresses encountered in mixing operations. This results in a much reduced aspect ratio that further reduces the nanocomposite thermal conductivity. Concentration effects have been summarized in Table 17.1, which reviews observations made with various polymer nanocomposites prepared for thermal conductivity enhancement, and it also includes the maximum thermal conductivity value obtained for each of these systems.

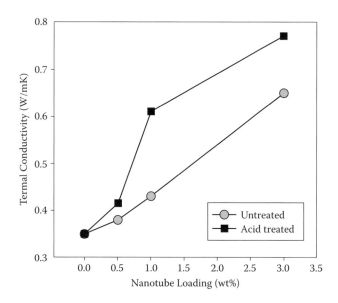

FIGURE 17.7
Effect of surface modification of carbon nanotubes by acid treatment on the thermal conductivity of nanocomposites of silicone resin. (*Source:* From Lee, G.-W. et al., *J. Mater. Sci.*, 40, 1259, 2005. With permission.)

17.6   Effect of Surface Treatment of Nanofillers

It is clear from the previous discussion that to realize the full potential of nanofillers as thermally conducting additives, the interfacial boundary resistance between the polymer matrix and the nanofiller must be minimized. Recent theoretical studies have shown that the introduction of covalent bonding between the carbon nanotubes (CNTs) and the polymer matrix can result in reduced interfacial resistance.[70, 71] The difficulty, however, is that CNTs have very low chemical reactivity and are largely chemically inert. However, CNT surfaces can be functionalized by chemical surface treatment to facilitate covalent bond formation between the polymer matrix and the surface of nanotubes. A review of CNT chemistry relating to surface fictionalization can be found in Reference 72.

Lee et al.[7] treated MWCNTs with nitric acid to functionalize the surface before mixing them in silicone resin. Acid treatment can introduce hydroxyl, carboxylic, or carbonyl groups on the CNT surface, which helps in bonding with the polymer matrix.[27, 72] Thermal conductivity measurements showed that the acid-treated nanotubes resulted in better enhancement of the thermal conductivity as shown in Figure 17.7. Hung et al.[33] showed that the thermal conductivity of nanocomposites made with acid-treated graphite nanoplatelets was higher than that of the untreated nanographite nanocomposites. Kim et al.[25] treated MWCNTs with nitric acid and potassium hydroxide solutions and found an almost doubling of the thermal conductivity compared to untreated nanotube nanocomposites. Gojny et al.[36] used amino-functionalized double-walled and multiwalled carbon nanotubes for epoxy nanocomposites; however, no improvement in the thermal conductivity was observed. Note that surface treatment helps in better dispersion of the fibers, which also results in better thermal conductivity enhancement as shown in Figure 17.8 for an MWCNT and epoxy nanocomposite.[17]

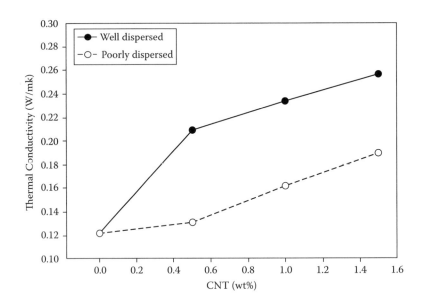

FIGURE 17.8
Effect of carbon nanotube dispersion on the thermal conductivity of epoxy nanocomposites. (*Source:* Adopted from Song, Y.S. and Youn, J.R., *Carbon*, 43, 1378, 2005. With permission.)

17.7   Effect of Fiber Orientation

Carbon nanotubes and nanofibers are the most popular nanofillers used to improve thermal conductivity of nanocomposites. Generally, fibers are distributed isotropically in the polymer matrix; however, by appropriate techniques, they can be aligned in any desired direction to tailor the alignment-dependent properties of the nanocomposites such as conductivity or mechanical properties. There are many ways of aligning nanofibers in a polymer matrix, and different orientations result, depending on the processing techniques employed. These techniques include extrusion, injection molding, fiber-spinning, electro-spinning, filtration, and magnetic force field alignment. Fibers can be aligned parallel or perpendicular to the direction of heat flow. A nice review of dispersing and aligning carbon nanotubes in polymer matrices is given in Reference 73.

To achieve the best results, nanotubes should be well-dispersed and aligned in the direction of heat flow, as suggested by theoretical models. If fibers are aligned along the heat flow direction, then energy travels along the fiber axis over a longer distance. This results in less resistance to heat flow, which leads to a higher thermal conductivity. Indeed, the longer the fiber (or higher the aspect ratio), the better are the results obtained.

Kuriger and Alam[29] aligned vapor grown carbon nanofibers in a polypropylene matrix by extruding the melt blend through a narrow die. Then the aligned fibers were compression-molded in the longitudinal or transverse direction with respect to the heat flow. Figure 17.9 shows thermal diffusivity values of the longitudinal and transverse samples. It can be seen that the longitudinal samples have a much higher thermal diffusivity than the transverse samples, and this property increases very rapidly with nanofiller loading in the longitudinally aligned samples. Enomoto et al.[31] aligned carbon nanofibers in a polypropylene matrix using an injection-molding technique. They compared the thermal

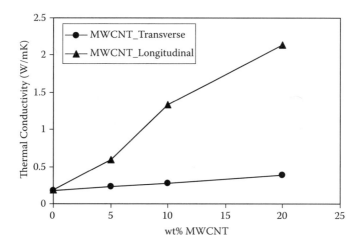

FIGURE 17.9

Effect of fiber alignment with respect to the direction of heat flow on the thermal conductivity of the nanocomposites of Ultem™ 1000 polyimide with MWCNT. (*Source:* Plot was prepared with the data from Ghose, S., et al., *High Perform. Polym.*, 18, 961, 2006.)

conductivities of highly aligned samples to low orientation samples and also to samples containing carbon black where there was hardly any alignment. It was found that the highly oriented samples gave the highest thermal conductivity. At 50 wt% loading, a dramatic, 17-times increase in the thermal conductivity was observed. Ghose et al.[35] prepared nanocomposites of Ultem™ brand polyetherimide with vapor-grown carbon nanofibers, MWCNT, and expanded graphite layers. Ribbons of the nanocomposites were extruded to align the nanofillers in longitudinal and transverse directions, while an internal mixer was used to obtain isotropic samples. Thermal conductivity measurements showed the following trend with respect to fiber orientation: transverse < isotropic < longitudinal. At a 40 wt% level, for the longitudinal case, an almost 40-fold increase in the thermal conductivity was observed for expanded graphite compared to only a twofold increase for transverse and 12-fold increase for the isotropic distribution. Gonnet et al.[19] used magnetically aligned SWNT buckypapers as the nanofillers in an epoxy resin. They also found that aligning the fibers in the direction of heat flow resulted in the best thermal conductivity enhancement.

17.8 Use of a Hybrid Mixture

As observed previously, nanofillers have very high interfacial thermal boundary resistance due to very high specific surface area that results in less than expected thermal conductivity enhancement in the nanocomposites. It has been shown that for a given filler, if the size of the particles decreases, then thermal conductivity also decreases. Devpura et al.[74] showed that the thermal conductivity of an alumina-polypropylene composite decreased when the alumina particle size was reduced from 10 to 0.2 μm. Zhou et al.[75] and Fan et al.[39] also observed a similar effect when, in hybrid mixtures of micro- and nano-alumina

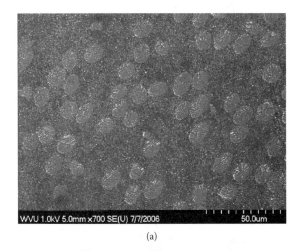

(a)

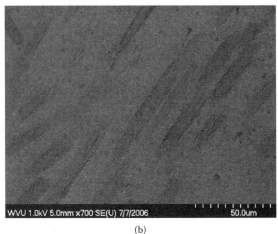

(b)

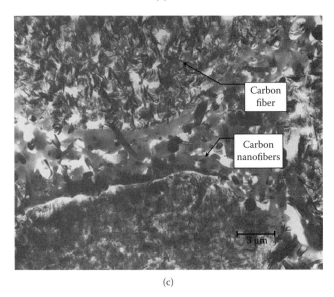

(c)

FIGURE 17.10
SEM of (a) vertically aligned and (b) horizontally aligned carbon fiber and nanofiber hybrid mixture in a polycarbonate matrix, and (c) TEM of a vertically aligned sample showing that the interstitial space between carbon fibers is filled with carbon nanofibers.

particles, the nanoparticle content was increased. This resulted in reduced thermal conductivity of the composites. However, the use of a hybrid mixture of microfillers, such as large aspect ratio graphitic carbon fibers, with low aspect ratio fillers such as SiC or carbon black, resulted in improved thermal conductivity relative to single-component composites.[76, 77] Kim et al.[26] used MWCNTs as an additive to conventional carbon fiber phenolic resin composites and found that a 57% increase in thermal conductivity resulted by adding just 7 wt% of nanotubes. It is suspected that carbon nanotubes create a conductive path between the micro-carbon fibers, thus resulting in increased thermal conductivity.

Agarwal et al.[69] used a hybrid mixture of carbon nanofibers and thermograph carbon microfibers to increase the thermal conductivity of a polycarbonate resin. Hybrid mixtures of carbon nanofibers and carbon fibers were melt-compounded with polycarbonate in an internal mixer. Then extrusion through a narrow capillary was used to obtain strands in which fibers were aligned along the axis. These strands were then carefully compression-molded to obtain disks in such a way that fibers were aligned either parallel to or perpendicular to the direction of heat flow. Figures 17.10a and 17.10b show scanning electron micrographs of the hybrid nanocomposites in which fibers have been aligned in these two directions. Figure 17.10c shows a transmission electron micrograph of the parallel-aligned fibers. These figures show that excellent alignment of the microfibers was achieved; however, the nanofibers are not aligned properly. This can be expected as nanofibers are aggregated and irregular in shape as opposed to microfibers, which are mostly straight cylinders. Figure 17.10c also shows that the nanofibers are dispersed in the interstitial region between the microfibers and seem to form connecting links between them. Thermal conductivity measurements were made on these samples using a method based on the steady-state ASTM D5470 method, and these results are shown in Figure 17.11. In the hybrid mixture, while the total filler content was kept at 40 vol%, the relative amount of nano- and microfibers was varied. It can be seen that an almost 50-fold increase in the thermal conductivity took place for the aligned hybrid mixture nanocomposites in which fibers are aligned in the direction of heat flow. Thermal conductivity enhancement for isotropic and perpendicularly aligned fibers is much less. Thus, it can be concluded that a hybrid mixture with fiber alignment in the direction of heat flow gives the best thermal conductivity enhancement. In addition, good contact between the fibers also results in better thermal conductivity.

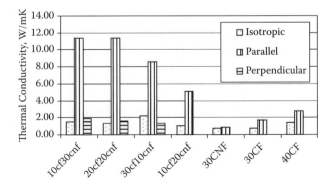

FIGURE 17.11
Effect of fiber alignment and concentration (vol%) on the thermal conductivity of polycarbonate nanocomposites containing a hybrid mixture of carbon fibers (CFs) and carbon nanofibers (CNFs).

17.9 Concluding Remarks

The thermal conductivity of polymers can be enhanced by orders of magnitude using high thermal conductivity nanofillers, especially in conjunction with larger-sized fillers. A wide variety of polymer materials such as thermoplastics, thermosets, and elastomers can be used as the matrix material. Carbon-based nanofillers such as nanotubes and nanofibers are particularly effective in this regard. Compared to conventional fillers, the use of even a very low level of nanofiller loading can result in substantial improvement in thermal conductivity. However, the full potential of these nanofillers in increasing the thermal conductivity has not been realized because of the phenomenon of interfacial thermal resistance, which imposes a severe barrier to the heat flow between the matrix and the nanofiller. This, to a great extent, negates the effect of the inherently high thermal conductivities of the nanofillers. Future work must be directed toward finding new ways to reduce the thermal boundary resistance. Another challenge is to ensure good dispersion of the nanofibers in the polymer matrix without causing the breakage of the fibers as high aspect ratio fibers are more effective in enhancing the thermal conductivity. In summary, it can be concluded that thermally conductive polymer nanocomposites are emerging as a new class of materials and their applications will only continue to increase in the future.

References

1. Okada, A. and Usuki, A., Twenty years of polymer-clay nanocomposites, *Macromol. Mater. Eng.*, 291, 1449, 2006.
2. Breuer, O. and Sundararaj, U., Big returns from small fibers: A review of polymer/carbon nanotubes composites, *Polym. Compos.*, 25, 630 (2004).
3. Finan, J. M., Thermally conductive thermoplastics, *Plast. Eng.*, 56, 51, 2000.
4. Sherman, L.M., Plastics that conduct heat, *Plast. Technol.*, 47, 52, 2001.
5. Zweben, C., Emerging high-volume applications for advanced thermally conductive materials, *SAMPE 2004, Materials and Processing Technology, 60 Years of SAMPE Progress*, May 16–20, Long Beach, CA, 49, 4046.
6. Miller, J., Thermally conductive thermoplastics, *Adv. Mater. Processes*, 61, 34, 2003.
7. Lee, G.–W., Lee, J.T., Lee, S.–S., Park, M., and Kim, J., Comparisons of thermal properties between inorganic filler and acid-treated multiwall nanotube/polymer composites, *J. Mater. Sci.*, 40, 1259, 2005.
8. Berber, S., Kwon, Y.K., and Tomanek, D., Unusually high thermal conductivity of carbon nanotubes, *Phys. Rev. Lett.*, 84, 4613, 2000.
9. Kim, P., Shi, L., Majumdar, A., and McEuen, P.L., Thermal transport measurements of individual multiwalled nanotubes, *Phys. Rev. Lett.*, 87, 215502-1, 2001.
10. Shioya, H., Iwai, T., Kondo, D., Nihei, M., and Awano, Y., Evaluation of thermal conductivity of a multiwalled carbon nanotube using the ΔV_{gs} method, *Jpn. J. Appl. Phys.*, 46, 3139, 2007.
11. Choi, T. –Y., Poulikakos, D., Tharian, J., and Sennhauser, U., Measurement of the thermal conductivity of individual carbon nanotubes by the four point three-ω method, *Nano Lett.*, 6, 1589, 2006.
12. Yang, D.J., Zang. Q., Chen, G., Yoon, S.F., Ahn, J., Wang, S.G., Zhou, Q., Wang, Q., and Li, J.Q., Thermal conductivity of multiwalled carbon nanotubes, *Phys. Rev. B*, 66, 165440, 2002.
13. Hone, J., Llaguno, M.C., Biercuk, M.J., Johnson, A.T., Batlogg, B., Benes, Z., and Fischer, J.E., Thermal properties of carbon nanotubes and nanotube-based materials, *Appl. Phys. A*, 74, 339, 2002.

14. Fujii, M., Zhang, X., Xe, H., Ago, H., Takahashi, K., and Ikuta, T., Measuring the thermal conductivity of a single carbon nanotube, *Phys. Rev. Lett.*, 95, 065502, 2005.
15. Biercuk, M.J., Llaguno, M.C., Radosavljevic, M., Hyun, J.K., and Johnson, A.T., Carbon nanotube composites for thermal management, *Appl. Phys. Lett.*, 80, 2767, 2002.
16. Bryning, M.B., Milkie, D.E., Islam, M.F., Kikkawa, J.M., and Yodh, A.G., Thermal conductivity and interfacial resistance in single-wall carbon nanotube epoxy composites, *Appl. Phys. Lett.*, 87, 161909, 2005.
17. Song, Y.S. and Youn, J.R., Influence of dispersion states of carbon nanotubes on physical properties of epoxy nanocomposites, *Carbon*, 43, 1378, 2005.
18. Du, F., Guthy, C., Kashiwagi, T., Fischer, J.E., and Winey, K.I., An infiltration method for preparing single wall nanotube/epoxy composite with improved thermal conductivity, *J. Polym. Sci., Part B: Polym. Phys.*, 44, 1513, 2006.
19. Gonnet, P., Liang, Z., Choi, E.S., Kadambala, R.S., Zhang, C., Brooks, J.S., Wang, B., and Kramer, L., Thermal conductivity of magnetically aligned carbon nanotube buckypapers and nanocomposites, *Curr. Appl. Phys.*, 6, 119, 2006.
20. Guthy, C., Du, F., Brand, S., Winey, K.I., and Fischer, J.E, Thermal conductivity of single-walled carbon nanotubes/PMMA nanocomposites, *J. Heat Transfer*, 129, 1096, 2007.
21. Haggenmueller, R., Guthy, C., Lukes, J.R., Fischer, J.E., and Winey, K.I., Single wall carbon nanotube/polyethylene nanocomposite: thermal and electrical conductivity, *Macromolecules*, 40, 2417, 2007.
22. Song, L., Zhang, H., Zhang, Z., and Xie, S., Processing and performance improvements of SWNT paper reinforced PEEK nanocomposites, *Composites: Part A*, 38, 388, 2007.
23. Kashiwagi, T., Grulke, E., Hilding, J., Groth, K., Harris, R., Butler, K., Shields, J., Kharchenko, S., and Douglas, J., Thermal and flammability properties of polypropylene/carbon nanotube nanocomposites, *Polymer*, 45, 4227, 2004.
24. Liu, C.H., Huang, H., and Fan, S.S., Thermal conductivity improvement of silicone elastomer with carbon nanotube loading, *Appl. Phys. Lett.*, 84, 4248, 2004.
25. Kim, S.W., Kim, J.K., Lee, S.H., Park, S.J., and Kang, K.H., Thermophysical properties of multi-walled carbon nanotube-reinforced polypropylene composites, *Int. J. Thermophys.*, 27, 152, 2006.
26. Kim, Y.A., Kamio, S., Tajiri, T., Hyashi, T., Song, S.M., Endo, M., Terrones, M., and Dresselhaus, M.S., Enhanced thermal conductivity of carbon fiber/phenolic resin composites by the introduction of carbon nanotubes, *Appl. Phys. Lett.*, 90, 093125, 2007.
27. Liu, C.H. and Fan, S.S., Effects of chemical modifications on the thermal conductivity of carbon nanotube composites, *Appl. Phys. Lett.*, 86, 123106, 2005.
28. Kim, Y.A., Hayashi, T., Endo, M., Gotoh, Y., Wada, N., and Seiyama, J., Fabrication of aligned carbon nanotube-filled rubber composite, *Scripta Materiala*, 54, 31, 2006.
29. Kuriger, R.J. and Alam, M.K., Thermal conductivity of thermoplastic composites with submicrometer carbon fibers, *Exp. Heat Transfer*, 15, 19, 2002.
30. Cummings, A.T., Shi, L., and Koo, J.H., Thermal conductivity measurements of Nylon11-carbon nanofiber nanocomposites, *Proceedings of IMECE2005*, November 5–11, 2005, Orlando, FL, p. 773.
31. Enomoto, K., Fujiwara, S., Yasuhara, T., and Murakami, H., Effect of filler orientation on thermal conductivity of polypropylene matrix carbon nanofiber composites, *Jpn. J. Appl. Phys.*, 44, L888, 2005.
32. Fukushima, H., Drzal, L.T., Rook, B.P., and Rich, M.J., Thermal conductivity of exfoliated graphite nanocomposites, *J. Therm. Anal. Calorim.*, 85, 235, 2006.
33. Hung, M.–T., Choi, O., Ju, Y.S., and Hahn, H.T., Heat conduction in graphite-nanoplatelet-reinforced polymer nanocomposites, *Appl. Phys. Lett.*, 89, 023117, 2006.
34. Mu, Q. and Feng, S., Thermal conductivity of graphite/silicone rubber prepared by solution intercalation, *Thermochim. Acta*, 462, 70, 2007.
35. Ghose, S., Working, D.C., Connell, J.W., Smith Jr., J.G., Watson, K.A., Delozier, D.M., Sun, Y.P., and Lin, Y., Thermal conductivity of Ultem/carbon nanofiller blends, *High Perform. Polym.*, 18, 961, 2006.

36. Gojny, F.H., Wichmann, M.H.G., Fiedler, B., Kinloch, I.A., Bauhofer, W., Windle, A.H. and Schulte, K., Evaluation and identification of electrical and thermal conduction mechanisms in carbon nanotube/epoxy composites, *Polymer,* 47, 2036, 2006.
37. Tyler, T., Shenderova, O., Cunningham, G., Walsh, J., Drobnik, J., and McGuire, G., Thermal transport properties of diamond-based nanofluids and nanocomposites, *Diamond and Related Materials,* 15, 2078 (2006).
38. Shenderova, O., Tyler, T., Cunningham, G., Ray, M., Walsh, J., Casulli, M., Hens, S., McGuire, G., Kuznetsov, V., and Lipa, S., Nanodiamond and onion-like carbon polymer nanocomposites, *Diamond and Related Materials,* 16, 1213, 2007.
39. Fan, L., Su, B., Qu, J., and Wong, C.P., Electrical and thermal conductivities of polymer composites containing nano-sized particles, *2004 IEEE Electronic Components and Technology Conference,* Las Vegas, NV, 148.
40. Mishra, S. and Mukherji, A., Studies of thermal conductivity of nano CaCO3/HIPS composites by unsteady state technique and simulation with Nielsen's model, *Polymer-Plast. Technol. Eng.,* 46, 239 (2007).
41. Seo, W.J., Sung, Y.T., Kim, S.B., Lee, Y.B., Choe, K.H., Choe, S.H., Sung, J.–Y., and Kim, W.N., Effects of ultrasound on the synthesis and properties of polyurethane foam/clay nanocomposites, *J. Appl. Polym. Sci.,* 102, 3764 (2006).
42. Caseri, W.R., Nanocomposites of polymers and inorganic particles: preparation, structure and properties, *Mater. Sci. Technol.,* 22, 807, 2006.
43. Tibbetts, G.G., Lake, M.L., Strong, K.L., and Rice, B.P., A review of the fabrication and properties of vapor-grown carbon nanofiber/polymer composites, *Compos. Sci. Technol.,* 67, 1709, 2007.
44. Lasance, C.J.M., The urgent need for widely accepted test methods for thermal interface materials, *Nineteenth Annual IEEE Semiconductor Thermal Measurement and Management Symposium,* 11–13 March, 2003, San Jose, CA, p. 123–128.
45. Jarrett, R.N., Merritt, C.K., Ross, J.P., and Hisert, J., Comparison of test methods for high performance thermal interface materials, *Annual IEEE Semiconductor Thermal Measurement and Management Symposium, SEMI-THERM Proceedings 23rd Annual IEEE Semiconductor Thermal Measurement and Management Symposium, SEMI-THERM,* March 18–22, 2007, San Jose, CA, p. 23.
46. Tzeng, J.J.-W., Weber, T.W., and Krassowski, D.W., Technical review on thermal conductivity measurement techniques for thin thermal interfaces, *Sixteenth Annual IEEE Semiconductor Thermal Measurement and Management Symposium. Proceedings,* March 2000, San Jose, CA, p. 174.
47. ASTM D5470-01, Standard Test Method for Thermal Transmission Properties of Thin Thermally Conductive Solid Electrical Insulation Materials, ASTM International.
48. ASTM E1530-06, Standard Test Method for Evaluating the Resistance of Thermal Transmission of Materials by the Guarded Heat Flow Meter Technique, ASTM International.
49. ASTM F433-02, Standard Practice for Evaluating Thermal Conductivity of Gasket Materials, ASTM International.
50. De Groot, J.J., Kestin, J., and Sookiazian, H., Instrument to measure the thermal conductivity of gases, *Physica,* 75, 454, 1974.
51. Healy, J.J., De Groot, J.J., and Kestin, J., The theory of the transient hot-wire method for measuring thermal conductivity, *Physica,* 82C, 392, 1976.
52. Trump, W.N., Luebke, H.W., Fowler, L., and Emery, E.M., Rapid measurement of liquid thermal conductivity by the transient hot-wire method, *Rev. Sci. Instrum.,* 48, 47, 1977.
53. Nagasaka, Y. and Nagashima, A., Absolute measurement of the thermal conductivity of electrically conducting liquids by the transient hot-wire method, *J. Phys. E: Sci. Instrum.,* 14, 1435, 1981.
54. Yu, Wenhua and Choi, S.U.S., Influence of insulation coating on thermal conductivity measurement by transient hot-wire method, *Rev. Sci. Instrum.,* 77, 076102, 2006.
55. Gustafsson, S.E., Karawacki, E., and Khan, M.N., Transient hot-strip method for simultaneously measuring thermal conductivity and thermal diffusivity of solids and liquids, *J. Phys. D: Appl. Phys.,* 12, 1411, 1978.

56. Gustavsson, M., Karawacki, E., and Gustafsson, S.E., Thermal conductivity, thermal diffusivity and specific heat of thin samples from transient measurements with hot disk sensors, *Rev. Sci. Instrum.*, 65, 3856, 1994.

57. Gustafsson, S.E., Karawacki, E., and Khan, M.N., Determination of the thermal conductivity tensor and the heat capacity of insulating solids with transient hot-strip method, *J. Appl. Phys.*, 52, 2596, 1981.

58. Parker, W.J., Jenkins, R.J., Butler, C.P., and Abbott, G.L., Flash method of determining thermal diffusivity, heat capacity and thermal conductivity, *J. Appl. Phys.*, 32, 1679, 1961.

59. ASTM E1461-01, Standard Test Method for Thermal Diffusivity by the Flash Method, ASTM International.

60. Miller, M.G., Keith, J.M, King, J.A., Hauser, R.A., and Moran, A.M., Comparison of the guarded heat flow and transient plane source methods for carbon filled nylon 6, 6 composites: experiments and modeling, *J. Appl. Polym. Sci.*, 99, 2144, 2006.

61. Progelhof, R.C., Throne, J.L., and Reutsch, R.R., Methods of predicting the thermal conductivity of composite systems: a review, *Polym. Eng. Sci.*, 9, 615, 1976.

62. Nielsen, L.E., The thermal and electrical conductivity of two-phase systems, *Ind. Eng. Chem. Fundam.*, 13, 17, 1974.

63. Prasher, R.S., Shipley, J., Prstic, S., Koning, P., and Wang, J.–L., Thermal resistance of particle laden polymeric thermal interface materials, *J. Heat Transfer*, 125, 1170, 2003.

64. Nan, C.–W., Birringer, R., Clarke, D.R., and Gleiter, H., Effective thermal conductivity of particulate composites with interfacial thermal resistance, *J. Appl. Phys.*, 81, 6692, 1997.

65. Shenogin, S., Xue, L., Ozisik, R., and Keblinski, P., Role of thermal boundary resistance on the heat flow in carbon-nanotube composites, *J, Appl. Phys.*, 95, 8136 (2004).

66. Nan, C.–W., Liu, G., Lin, Y., and Li, M., Interface effect on thermal conductivity of carbon nanotubes composites, *Appl. Phys. Lett.*, 85, 3549, 2004.

67. Every, A.G., Tzou, Y., Hasselman, D.P.H., and Raj, R., The effect of particle size on the thermal conductivity of ZnS/diamond composites, *Acta. Metall. Mater.*, 40, 123, 1992.

68. Foygel, M., Morris, R.D., Anez, D., French, S., and Sobolev, V.L., Theoretical and computational studies of carbon nanotube composites and suspensions: electrical and thermal conductivity, *Phys. Rev. B*, 71, 104201, 2005.

69. Agarwal, S., Khan, M.M.K., and Gupta, R.K., Thermal conductivity of polymer nanocomposites made with carbon nanofibers, submitted to *Polym. Eng. Sci.* for publication, 2007.

70. Shenogin, S., Bodapati, A., Xue, L., Ozisik, R., and Keblinski, P., Effect of chemical functionalization on thermal transport of carbon nanotube composites, *Appl. Phys. Lett.*, 85, 2229, 2004.

71. Clancy, T.C. and Gates, T.S., Modeling of interfacial modification effects on thermal conductivity of carbon nanotube composites, *Polymer*, 47, 5990, 2006.

72. Zhao, B., Hu, H., Bekyarova, E., Itkis, M.E., Niyogi, S., and Haddon, R., *Carbon Nanotubes: Chemistry, Dekker Encyclopedia of Nanoscience and Nanotechnology*, J.A. Schwarz, C.I. Contescu, and K. Putyera, Eds. Marcel Dekker Inc., NewYork, 2004.

73. Xie, X.–L., Mal, Y.–W., and Zhou, X.–P., Dispersion and alignment of carbon nanotubes in polymer matrix: a review, *Mat. Sci. Eng. R.*, 49, 89(2005).

74. Devpura, A., Phelan, P.E., and Prasher, R.S., Size effects on the thermal conductivity of polymer laden with highly conductive filler particles, *Microscale Thermophys. Eng.*, 5, 177, 2001.

75. Zhou, W., Qi, S., Tu, C., Zhao, H., Wang, C., and Kou, J., Effect of particle size of Al_2O_3 on the properties of filled heat-conductive silicone rubber, *J. Appl. Polym. Sci.*, 104, 1312, 2007.

76. Kim, D.H., Kim, M.H., Lee, J.H., Lim, J.H., Kim, K.M., Lee, B.C., Park, J.M., and Kim, S.R., Synergistic effect of hybrid filler contained composites on thermal conductivity, *Mater. Sci. Forum*, 544–545, 483, 2007.

77. Heiser, J. and King, J.A., Thermally conductive carbon filled Nylon 66, *Polym. Compos.*, 25, 186, 2004.

18

Bio-Based Nanocomposites from Functionalized Plant Oils

Chang Kook Hong, Jue Lu, and Richard P. Wool

CONTENTS

18.1 Introduction

Composite materials are widely used in various fields, such as the aerospace, automotive, marine, infrastructure, military, electronic, and industrial fields. Recently developed nanocomposites broaden their applications in many fields. With relatively low loading of nanometer-sized particles, they showed dramatically increased stiffness, hest distortion temperature (HDT), dimensional stability, gas barrier, electrical conductivity, and flame retardancy.[1, 2]

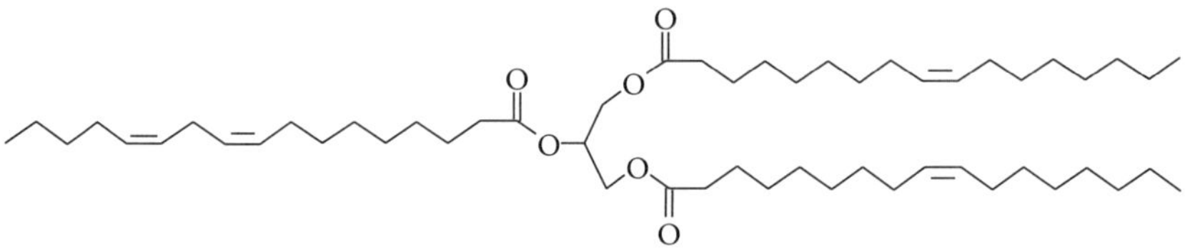

FIGURE 18.1
Molecular structure of triglyceride.

Developing affordable composite materials from renewable sources, such as plant oils, offers economic advantages and also environmental advantages.[3–9] The growing demands of polymer materials have increased dependence on petroleum-based resources. With the fast depletion of these resources, replacing some or all of these with readily available, renewable, and inexpensive natural resources, such as plant oils, is becoming important. The use of abundant renewable materials contributes to global sustainability and diminution of global warming gases; and as the number of applications of composite materials continues to increase, an alternative source of petroleum-based composites becomes important. Plant oils are found in abundance in all parts of the world, making them ideal alternatives of petroleum-based materials because of their unlimited resources and potential biodegradability. When bio-based resins derived from natural plant oils are combined with natural fibers, glass fibers, carbon nanotubes (CNTs), nanoclays, and lignin, new low-cost composites can be produced that will be economical in many applications.[10–20]

Plant oil resins are based on triglycerides, which are the major components of natural oils. Triglycerides, shown in Figure 18.1, are composed of three fatty acids joined at a glycerol juncture.[21] Plant oils have been widely used in coatings, inks, plasticizers, lubricants, and agrochemicals in addition to the food industry.[22–26] In recent years, extensive work has been done to develop polymers for engineering applications using plant oils as the main component.[27–33] Although unmodified triglycerides do not readily polymerize, the chemical functionality necessary to cause polymerization can be easily added to the triglycerides. The active sites on the modified triglycerides can be used to introduce polymerizable groups using the same synthetic techniques that have been applied in the synthesis of petrochemical-based polymers.[34–42] Multifunctional triglycerides are produced using active sites, such as the double bonds and the ester groups, which are amenable to chemical reaction; then styrene or other co-monomers are added as reactive diluents. Very detailed synthetic pathways and experimental methods can be found elsewhere.[43, 44] In this review, the focus is on three triglyceride monomers, as shown in Figure 18.2: (1) acrylated epoxidized soybean oil (AESO), (2) maleinized acrylated epoxidized soybean oil (MAESO), and (3) soybean oil pentaerythritol maleate (SOPERMA). These monomers, when used as a major component of a molding resin, have shown properties comparable to conventional polymers. The resins from soybean oil can be a substitute for liquid molding resins such as unsaturated polyester resins, vinyl esters, and epoxy resins. With suitable chemical functionalization and viscosity, the molding process of soybean resins is similar to that of conventional thermosetting liquid molding resins, using resin transfer molding (RTM), vacuum-assisted resin transfer molding (VARTM), sheet molding compound (SMC), etc. The triglyceride-based materials display the necessary rigidity and strength required for structural applications.

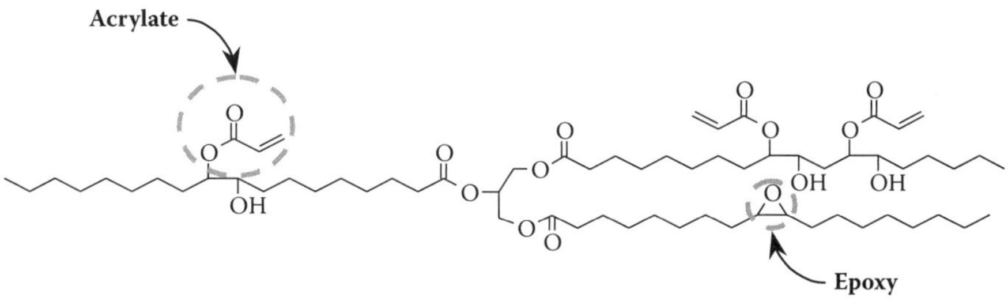

Acrylate

AESO

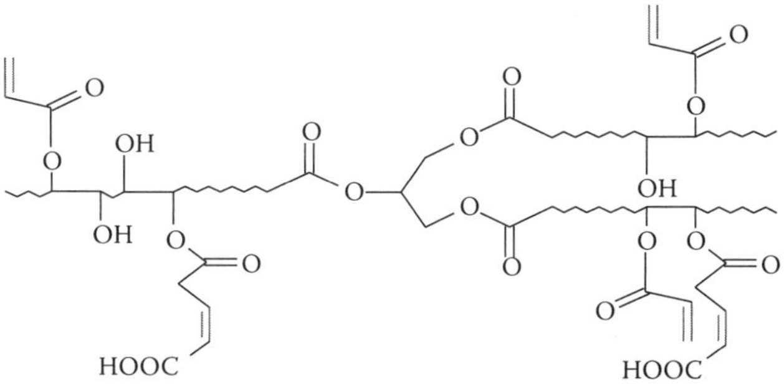

MAESO

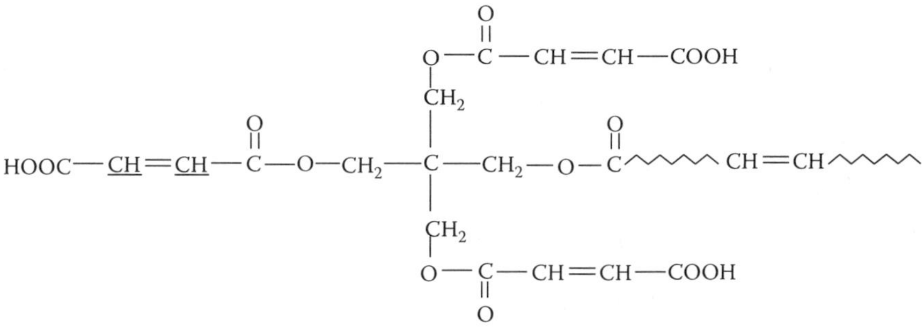

$$O - C - CH = CH - COOH$$

SOPERMA

FIGURE 18.2
Molecular structures of triglyceride-based monomers.

18.2 Bio-Based Nanocomposites from Functionalized Plant Oils and Nanoclays

Polymer-layered silicate nanocomposites have been successfully developed with most conventional polymers. The addition of nanoclays into plant-oil-based polymers to form nanocomposites can broaden the applications of these new bio-based materials by improving the mechanical properties of these materials. The extremely large surface area and high aspect ratio (between 30 and 2000) make it possible for the property improvements resulting from the formation of a nanocomposite. Nanoclays are first chemically treated to make them organophilic and compatible with the polymer matrix, and then combined with polymers to form nanocomposites. There are essentially three different approaches to synthesize polymer-clay nanocomposites: (1) melt intercalation, (2) solution, and (3) *in situ* polymerization. The synthesis of nanocomposites using triglyceride-based resins is included in the last category, as are epoxy resins and unsaturated polyester resins.[45, 46]The nanocomposites are prepared by first swelling the organo-modified nanoclay with the monomer, followed by cross-linking reactions. The miscibility of monomer or polymer with organo-modifier at the swelling stage can be measured by solubility parameters.[47–49] It is believed that the exfoliation of silicate layers is related to the polarity of the monomers.[50] Often, an intercalated structure is formed when nonpolar styrene is polymerized with nanoclay, but most epoxy-based nanocomposites show an exfoliated structure.[51–54] Several studies have shown the formation of a partially exfoliated morphology for unsaturated polyester-clay nanocomposites.[55–57] Because triglyceride-based monomers optionally have polar groups such as hydroxyl and carboxylic acid groups, these molecules are favored in the formation of a possible exfoliated structure. Triglycerides also offer the ability to control the number of polar groups through chemical modifications. In this review, the miscibility, morphology, and properties of these new triglyceride-based nanocomposites are reported.

18.2.1 Experimental

18.2.1.1 Synthesis of Maleinized Acrylated Epoxidized Soybean Oil (MAESO)

AESO was obtained in the form of Ebecry-860 from UCB Radcure Inc. This AESO is fully acrylated with approximately 3.4 acrylates per triglyceride and an average molecular weight of 1200 g/mol. Maleic anhydride, N,N-dimethylbenzylamine (BDMA), hydroquinone, and styrene were all obtained from Aldrich Chemical Co. and used as received. To synthesize MAESO, 50 g AESO and 0.05 g hydroquinone were first heated to 70°C, while being stirred. Then 8.17 g maleic anhydride was finely ground and added to the reaction at 70°C. The reaction was then heated up to 80 to 85°C, at which point the maleic anhydride dissolved, forming a homogeneous solution. The BDMA catalyst (1 g) was then added. The reaction was stopped after 6 hr before gelation occurred.

18.2.1.2 Synthesis of Soybean Oil Pentaerythritol Maleate (SOPERMA)

In the alcoholysis reaction, 400 g soybean oil (Aldrich) was mixed with 186.51 g pentaerythritol (Aldrich) and 5.87 g $Ca(OH)_2$ in a three-necked 1-L round-bottom flask equipped with a mechanical stirrer and a nitrogen gas inlet, calcium drier. The reaction mixture was heated to 230 to 240°C and agitated under N_2 atmosphere for 2 hr. The reaction product at room temperature was a light brown colored viscous liquid that separated into two layers with time. In the maleinization reaction, 540 g of the soybean oil pentaerythritol alcoholysis product and 311.86 g maleic anhydride were placed in a 1-L flask equipped with a thermometer and a mechanical stirrer. The mixture was heated in an oil bath with stirring until the maleic anhydride melted and then mixed with the soybean oil pentaerythritol alcoholysis product. Then, 8.52 g BDMA and 0.852 g hydroquinone were added and the reaction mixture was heated to 95°C. The mixture was agitated at this temperature for 2 hr. Both the FT-IR (Fourier transform infrared) and the H-NMR (proton nuclear magnetic resonance) spectroscopy of the product confirmed the consumption of maleic anhydride and the formation of maleate half-esters.

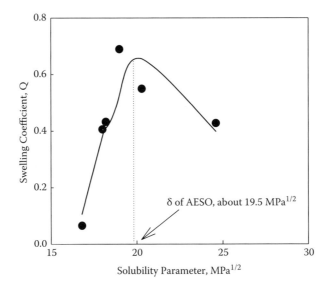

FIGURE 18.3
Representative plot of determining the solubility parameter of AESO.

18.2.1.3   Swelling Test

The solubility parameters of triglyceride-based polymers were determined by swelling experiments. The solvent with the closest solubility parameter swells the polymer the most. Several solvents of varying solubility parameter, such as cyclohexane, methyl methacrylate, toluene, benzene, chloroform, acetone, and acrylic acid, were selected, and the polymers were swelled to equilibrium in each of them. The solubility parameters of the triglyceride-based polymers were obtained from the maximum of the plot of swelling coefficient as a function of the solvent's solubility parameter. Figure 18.3 shows a representative plot of determining the solubility parameter of AESO.

18.2.1.4   Preparation of Clay Nanocomposites

Cloisite® 30B (C30B), a natural montmorillonite modified with methyl tallow bis-2-hydroxyethyl quaternary ammonium chloride, was obtained from Southern Clay Products. Several different clay concentrations in the triglycerides-based resins were investigated (3, 5, 7.5, and 10 wt% based on the total weight of the resin). The composition of the resin is 66.7 wt% functionalized triglyceride monomers and 33.3 wt% styrene. The desired amount of nanoclay was added to the resin and mechanically stirred for 24 to 48 hr with the flask well sealed to protect against the evaporation of styrene. Both room-temperature (RT) and high-temperature curing was used to cure the sample. For the high-temperature curing, 1.5 wt% *t*-butyl peroxy benzoate (TBP) radical initiator (Elf Atochem) was added and the mixture was cured at 110°C for 3 hr and post-cured at 150°C for 2 hr. For RT-curing, 3 wt% of trigonox 239A (Akzo Nobel) and 0.8 wt% of cobalt naphthalate were added. Samples were cured at room temperature for 24 hr and then post-cured at 150°C for 2 hr. To prevent oxygen free-radical inhibition, the resin was purged with nitrogen gas prior to curing.

18.2.1.5 Characterization and Mechanical Test

Wide-angle x-ray diffraction (XRD) measurements were performed on the solid samples with a Philips X'Pert diffractometer using Cu Kα radiation (40 kV, 40 mA) in the 2θ ranges of 0 to 10°. Transmission electron microscopy (TEM) was performed on a JEOL 2000 FX electron microscope on 100-nm-thick sections. The samples were microtomed from the cured samples. Samples for dynamic mechanical analysis were prepared to dimensions of 48 × 10 × 2.5 mm. Dynamic mechanical analysis was conducted in a three-point bending geometry on a Rheometrics Solids Analyzer II (Rheometric Scientific Inc.). Temperature scans were run from approximately 25°C to 180°C at a heating rate of 5°C/min with a frequency of 1 Hz and strain of 0.01%. Samples for flexural testing were prepared to dimensions of approximately 63.5 × 12.7 × 3.2 mm. Flexural strength and modulus were measured according to ASTM D 790-95a with a cross-head speed of 1.27 mm/min. A TA Instruments Thermogravimetric Analyzer Q500 was used to measure the weight loss of the cured sample in air. Approximately 4 mg of sample was heated from room temperature to 800°C at a heating rate of 20°C /min.

18.2.2 Results and Discussion

18.2.2.1 Miscibility

The solubility parameter approach to the prediction of nanocomposite morphology has been used with some degree of success in studies of polymer-clay modifier miscibility.[58, 59] For a binary mixture, Hildebrand has related the enthalpy contribution ΔH_m to the difference of δ_i, the solubility parameter of the polymers as

$$\Delta H_m = (\delta_1 - \delta_2)^2 V_m \phi_1 \phi_2 \tag{18.1}$$

where V_m is the volume of the mixture, ϕ_i is the volume fraction of i in the mixture, and δ_i is the solubility parameter of the i component. In the Flory-Huggins theories, the interaction parameter χ, which is responsible for the enthalpic contribution, is related to the solubility parameters via[60]

$$\chi = 0.34 + \frac{V_s}{RT}(\delta_1 - \delta_2)^2 \tag{18.2}$$

where V_s is the reference volume. For better miscibility, the polymers should have similar solubility parameters. This is in accordance with the general rule that chemical and structural similarity favors solubility. In this study, the solubility parameter values for each resin and clay modifier were calculated by the same theoretical model for comparison basis. The method used was the one proposed by Hoy, which is very similar to the Hansen and Hoftyzer and Van Krevelen's (H-VK) model in which they assume that the solubility parameter directly depends on dispersive, permanent dipole-dipole interaction (polarity) and hydrogen bonding forces:[61–64]

$$\delta_t = \delta_d + \delta_p + \delta_h \tag{18.3}$$

where δ_d, δ_p, and δ_h are dispersion, polar, and hydrogen bonding components, respectively. Using the group contribution method, the solubility parameter is given by[65]

$$\delta = \rho \frac{\sum\limits_i F_i}{M} \tag{18.4}$$

TABLE 18.1

Solubility Parameter Values for Monomers and Clay Organic Modifier

				SOPERMA			
	SO	AESO	MAESO	Mono-	Pen-	Styrene	C30B
Molecular weight (g/mol)	871	1167.9	1390.4	547.9	689.9	104	
δ (Hoy model) (MPa$^{1/2}$)	18.73	20.72	20.87	20.17	20.51	20.50	20.38
δ_d (MPa$^{1/2}$)	17.41	16.19	15.81	16.34	15.76	17.07	16.68
δ_p (MPa$^{1/2}$)	5.75	10.20	11.04	9.82	10.33	9.33	9.48
δ_h (MPa$^{1/2}$)	3.82	7.94	7.97	6.57	8.11	6.45	6.87
δ (Experimental) (MPa$^{1/2}$)		19.5	19.6	20.1		19.0	19.0[40]

where ρ is the density of the polymer, M is the molar mass of the polymer, and $F_i = (E_i V_i)^{1/2}$ where E_i and V_i are the cohesion energy and molar volume of the group considered, respectively. According to Table 18.1, the three functionalized triglycerides have very similar solubility parameters, and also the polarity δ_p values are very close. These calculated solubility parameters are relatively higher than the experimental values from the swelling test. With the same model, the solubility parameter of 30B modifier, which is methyl, tallow, bis-2-hydroxyethyl, quaternary ammonium was calculated to be 20.38 MPa$^{1/2}$. This value was very close to the experimental value found using a small-angle neutron scattering technique.[66] From calculation, all three triglyceride monomers would be expected to be miscible with the organo-modified clay. In the mixing stage, triglyceride molecules can diffuse between the clay layers and expand the spacing between the layers. In comparison to unmodified soybean oil (SO), the chemical modification of triglycerides increases the solubility parameters and polarity of the molecules. As seen in Table 18.1, both the values of δ_p and δ_h of modified triglycerides are doubled, resulting in the solubility parameters being very similar to the organic modifier parameters. Therefore, it can be used as an effective method to control the miscibility between these triglyceride-based molecules and the nanoclay surface. In the mixing stage, triglyceride molecules can diffuse between the clay layers and expand the spacing between the layers. This can be proved with x-ray diffraction (XRD) for the liquid samples during mixing.[67]

18.2.2.2 Structure and Morphology

The morphological state of cured nanocomposites has been investigated using XRD and TEM. The XRD scans of AESO-clay nanocomposites show that the morphology of nanocomposites does depend on the clay loading, as shown in Figure 18.4. From Figure 18.4a, the pure C30B shows a peak at 4.78°, corresponding to the (001) plane. Using the Bragg formula $\lambda = 2d \sin \theta$, where λ is the x-ray wavelength (1.5418 Å), the gallery spacing of C30B is calculated as 18.2 Å. This peak disappears in the scattering curves for the 3 wt% C30B-AESO nanocomposite. When the clay content increases to 5 wt%, there are two peaks; (1) one is a weak peak at about 4.64° and (2) the other is observed at 2.12°. These 2θ values correspond to interlayer spacings of 19.04 and 41.58 Å, respectively. The XRD data for both MAESO- and SOPERMA-clay nanocomposites have the same trend as shown in Figure 18.4. From XRD data, these triglyceride-based nanocomposites show an exfoliated

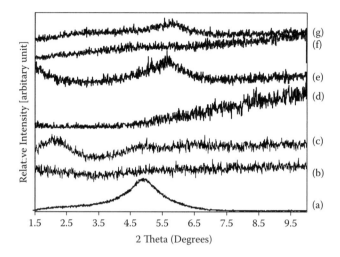

FIGURE 18.4
XRD data of triglyceride-based nanocomposite systems: (a) pure C30B, (b) AESO + 3 wt% C30B, (c) AESO + 5 wt% C30B, (d) MAESO + 3 wt% C30B, (e) MAESO + 5 wt% C30B, and (f) clay (C30B) nanocomposites. The scale bars represent 50 nm.

structure at 3 wt% nanoclay load, and an intercalated sheet structure at high clay load (≥5 wt%).

The morphology of triglyceride-based nanocomposites was further investigated by TEM of a thin section (100 nm). The results from TEM images of AESO-clay nanocomposites show a completely exfoliated structure at 3 wt% clay loading and a mix of intercalated and partially exfoliated structure above 5 wt%, as shown in Figure 18.5. For the 3 wt% clay nanocomposite, well-dispersed individual silicate layers are shown. However, for 5 wt% clay in an AESO matrix, the individual sheets of clay are separated by triglyceride molecules (intercalated structure), and some of the clay layers are partially exfoliated. Also, the micro-sized aggregates of clay sheets are observed at low magnification above 5 wt% clay loading. This may be attributed to the volume effect as the intercalation of monomer into the space between the silicate galleries depends on the number of clay layers available and the amount of monomer. Another possible reason might be insufficient mixing, as this process is very time dependent and mixing-strength dependent. The intercalated clays are being further investigated for their combined toughening and self-healing capabilities.

18.2.2.3 Thermomechanical Properties

Table 18.2 lists the storage moduli and glass transition temperatures for AESO, SOPERMA, and MAESO nanocomposites. The storage modulus of all triglyceride-based nanocomposites improved with the addition of clay. A change in the modulus indicates a change in rigidity and, hence, strength of the nanocomposites. SOPERMA-based nanocomposites had greater improvement in storage modulus than AESO and MAESO. Although SOPERMA has a similar solubility parameter as AESO and MAESO, the molecular weight of SOPERMA is much lower than the other two, as shown in Table 18.1. The diffusion of molecules into the clay interlayer is probably not only related to the miscibility, but also

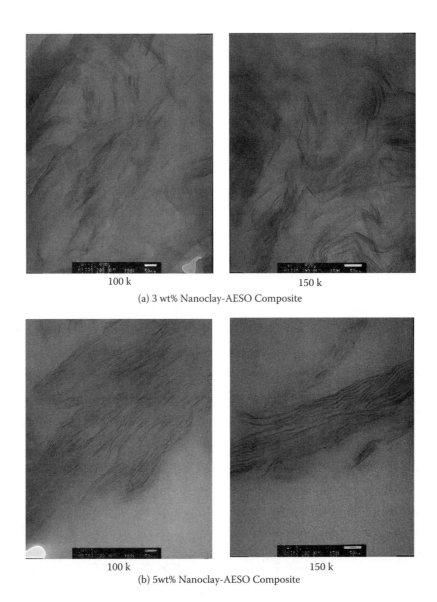

(a) 3 wt% Nanoclay-AESO Composite

(b) 5wt% Nanoclay-AESO Composite

FIGURE 18.5
TEM micrographs of AESO-clay (C30B) nanocomposites. The scale bars represent 50 nm.

to molecular flexibility. As SOPERMA molecules only have one fatty acid chain, these molecules have high flexibility that favors molecular intercalation, as shown in the XRD scattering curves. The higher exfoliated and intercalated yield resulted in the higher storage modulus increase.

The T_g was measured by dynamic mechanical analysis because the transition is too broad to determine accurately with DSC for these triglyceride-based polymers.[68] Generally, for most thermoset resins, the tan δ peak occurs at the glass transition temperature (T_g). Figure 18.6 shows the temperature dependence of tan δ, the ratio of viscous to elastic properties, for MAESO nanocomposites at various clay contents. The tan δ peak is shifted to

TABLE 18.2

Dynamic Mechanical Properties of Triglyceride-Based Nanocomposites

Resin	Clay Content (wt%)	Storage Modulus (Gpa)	T_g (°C)
AESO	0	1.258	71
AESO	3	1.321 (+5.0%)[a]	69
AESO	5	1.368 (+8.7%)[a]	65
SOPERMA	0	1.680	130
SOPERMA	3	2.009(+19.6%)[a]	126
MAESO	0	2.041	136
MAESO	3	2.306(+13.0%)[a]	130

[a] Percent improvement within parentheses.

lower temperature with increasing clay content. AESO and SOPERMA nanocomposites show the same behavior as shown in Table 18.2. The intensity of tan δ peak also diminishes with increasing clay content. This is expected because the formation of the nanostructure restricts the molecular motions, which cause the amount of energy that could dissipate throughout the nanocomposite to decrease dramatically. This could cause the T_g to shift to higher temperature. On the other hand, in this bi-component system, the styrene co-monomer can more easily penetrate into clay layers than triglyceride-based monomer, due to its small size. The aromatic nature of styrene imparts rigidity to the network, and the loss of styrene outside of the clay layer decreases the rigidity of the nanocomposite, which results in the decreased T_g. Indeed, Suh et al.[69] demonstrated that the loss of styrene outside the clay layers causes the decrease of T_g for unsaturated polyester-based nanocomposites. They proposed a sequential mixing method that successfully increased the T_g of the nanocomposites. In general, the loss factors (tan δ) showed the competition between two opposite effects, the formation of the nanostructure and loss of styrene outside the clay layers.

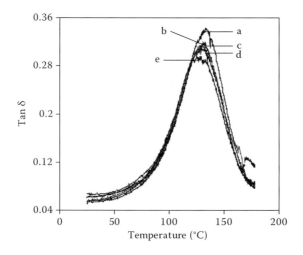

FIGURE 18.6
Tanδ of MAESO-clay nanocomposites at different clay contents as a function of temperature: (a) pure MAESO, (b) 3 wt% C30B, (c) 5 wt% C30B, (d) 7.5 wt% C30B, and (e) 10 wt% C30B.

TABLE 18.3

Flexural Properties of AESO-Clay Nanocomposites

Clay Content (wt%)	Flexural Modulus (Gpa)	Flexural Strength (Mpa)
0	0.896	34.807
3	1.124 (+25.4%)	38.870(+11.7%)
5	1.040 (+16.1%)	36.644(+5.3%)

[a] Percent improvement within parentheses.

18.2.2.4 *Flexural Properties*

The flexural properties of AESO nanocomposites at different clay content are shown in Table 18.3. The flexural modulus and flexural strength of triglyceride polymers were significantly enhanced with the addition of clay — 3 wt% nanocomposite shows better flexural properties than 5 wt%. From TEM results, a completely exfoliated structure was observed at 3 wt% clay loading, and a mix of intercalated and partially exfoliated structure was shown at 5 wt%. Therefore, the degree of exfoliation is a control mechanism to increase the flexural properties of clay nanocomposites. Figure 18.7 shows the increase of flexural modulus in MAESO-clay nanocomposites. The flexural modulus increases significantly with increasing clay load up to 7.5 wt%, which only corresponds to 4.0 vol%; the increase in modulus is 30%. After that, the modulus slightly decreases. In practice, the modulus for the conventional composite with flake-like inclusion can be estimated with simple empirical rules, such as the Halpin-Tsai equations, which predict a strong dependence of composite modulus on the filler aspect ratio.[70] Figure 18.8 suggests the experimental data did not follow the prediction, which was possibly due to the extreme difference of the modulus of silicate layer and polymer matrix. Silicate platelets have an in-plane modulus of 178 GPa.[71] Hue and Shia developed new equations for composites with aligned platelet inclusions.[72] The model was valid for the entire range of modulus and aspect ratios as the expressions

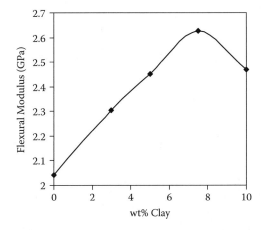

FIGURE 18.7
Flexural modulus for MAESO-C30B nanocomposites as a function of clay content.

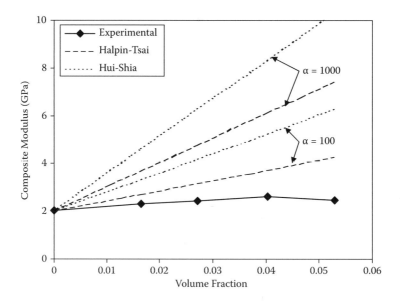

FIGURE 18.8
Experimentally measured modulus values and theoretical predictions by two different models: Halpin-Tsai and Hui-Shia (α is the aspect ratio).

were derived only by making the assumptions that the Poisson ratios of the inclusions and the matrix were the same and equal to 1/2.

$$\frac{E}{E_o} = \frac{1}{1 - \frac{\phi}{4}\left(\frac{1}{\xi} + \frac{3}{\xi + \Lambda}\right)} \tag{18.5}$$

$$\xi = \phi + \frac{E_o}{E_1 - E_0} + 3(1 - \phi)\left[\frac{(1 - g) - (g/2)\alpha^2}{1 - \alpha^2}\right] \tag{18.6}$$

$$\Lambda = (1 - \phi)\left[\frac{3(1 + 0.25\alpha^2) - 2}{1 - \alpha^2}\right] \tag{18.7}$$

$$g = \frac{\pi}{2\alpha} \tag{18.8}$$

where E, E_0, and E_1 are the modulus of the composite, matrix, and inclusion, respectively; ϕ is the volume fraction of the inclusion; and α is the aspect ratio of the inclusion, which was given by the ratio of the length over thickness of the inclusion. However, the experimentally measured modulus values were much lower than the values predicted by these equations for the aspect ratios in the range of 100 to 1000, as shown in Figure 18.8. This was possible because of the imperfect bonding between the clay surface and polymer matrix that reduced the reinforcing efficiency of the clay. Solving the Halpin-Tsai and Hui-Shia

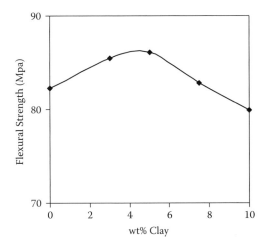

FIGURE 18.9
Flexural strength for MAESO-C30B nanocomposite as a function of clay content.

equations with experimental modulus data yielded effective aspect ratios in the range of 40 to 50 and 5 to 10, respectively, which were much lower than the real aspect ratio for silicate layers. The effective aspect ratio decreases with increasing content of clay, which is consistent with the morphology of nanocomposites, as shown in the TEM image in Figure 18.5. The flexural strength slightly increases at low clay content and then decreases above 5-wt% clay loading, as shown in Figure 18.9. Figure 18.10 indicates that the flexural strain decreases with increasing clay content because the formation of nanostructure restricted the flexibility of crosslink network.

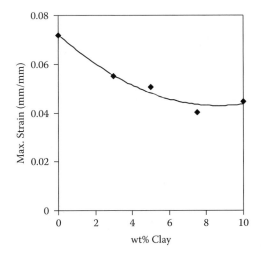

FIGURE 18.10
Flexural strain of MAESO-C30B nanocomposites as a function of clay content.

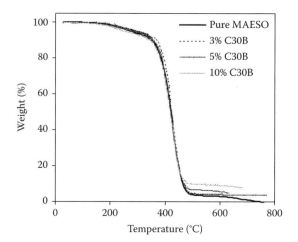

FIGURE 18.11
TGA thermal degradation profiles for MAESO-C30B nanocomposites at different clay concentrations.

18.2.2.5 *Thermal Stability*

Tyan et al.[73] found that the silicate layers can exert very effective retardation on the thermal degradation of polymers. Figure 18.11 shows the thermogravimetric analysis (TGA) trace for the cross-linked MAESO-clay nanocomposites. There is only a single degradation process under air atmosphere for all samples. The degradation occurs in the temperature range from 300 to 450°C, and is due to the degradation of the intercalated agent followed by the decomposition of the cross-linked polymer network. The 50% weight loss occurs at approximately 420°C. With the addition of 3 wt% clay, the onset of degradation is slightly slowed because of the constrained region in the nanocomposite. With increased clay content, the degradation is hastened because of the presence of the unfavorable hydroxyl groups in the organic modifier. After approximately 500°C, the curves all become flat and mainly the inorganic residue (i.e., Al_2O_3, MgO, SiO_2) remains.

18.2.3 Summary

Clay nanocomposites were successfully prepared from plant oil, one of the abundant, cheap, renewable resources. Organo-modified nanoclay was dispersed in functionalized triglyceride monomers and styrene, followed by copolymerization reactions. The monomers and organic modifier (methyl tallow bis-2-hydroxyethyl quaternary ammonium chloride) are proved to be miscible using their solubility parameters. The formation of nanostructure was confirmed both by XRD and TEM images. The morphology can be described as a mix of intercalated and partially exfoliated sheets. The exfoliated nanocomposites significantly increase the flexural properties. The increase in modulus is not consistent with the theoretical models for traditional flake-like inclusion composites. The glass transition temperature is lowered with increasing clay content due to styrene penetration. Thermal degradation is slightly hastened with increasing clay load.

18.3 Low Dielectric Constant Materials from Plant Oil Resins and Nanostructured Keratin Fibers

The continuing demand for higher-speed electronic devices has driven an extensive search for new low dielectric constant (k) materials that can replace silicon dioxide as the insulator, and polymers such as epoxy and polyimide used in the circuit boards. In a typical microchip, performance gain is mostly limited by the intra- and interlayer capacitance, dictated primarily by the dielectric constant of the insulators, known as dielectrics.[74] A decrease in the dielectric constant of the insulator containing the printed circuits increases the operating speed, minimizes the "cross-talk" effects between metal interconnects, and diminishes the power consumption.[75] Development of low-k dielectric materials is considered one of the main issues in modern high-speed microelectronics.

Developing a low-k material from renewable resources, such as hollow keratin fibers and plant oil from soybeans, is quite attractive from economic and environmental perspectives. They not only offer economic incentives through the use of cheaper materials, but also environmental advantages through the use of abundant renewable materials, which promote global sustainability. The new low-k material is a natural, bio-based, and environmentally friendly material. The hollow fiber, composed of the protein keratin with an α-helical structure at the molecular level, is highly nanocrystalline, and is tough enough to withstand both mechanical and thermal stress. The use of keratin fibers as reinforcing fibers in composite materials offers an environmentally benign solution for the feather industry. Due to the hollow nanostructure of the keratin fibers, a given volume of the fiber innately contains a significant volume of air. Air is an ideal dielectric material, representing the minimum dielectric constant of 1.0. Soybean-oil-based resin is a low-polarity, hydrophobic material. The triglyceride-based materials display the necessary rigidity and strength required for structural applications.

In this review, a new low-k material suited to electronic applications using nano-structured keratin fibers and chemically modified soybean oil is introduced. A low-polarity soybean resin is used as the composite matrix and hollow nano-crystalline keratin fibers are used as reinforcing fibers for taking advantage of the low k of air. It was important that the resin did not displace the air by filling the hollow fibers. The essential properties for electronic applications, such as dielectric and mechanical properties, and the thermal expansion coefficients, of the new composites are discussed.

18.3.1 Experimental

18.3.1.1 Sample Preparation

Chemically modified soybean oils such as acrylated epoxidized soybean oils (AESO, Ebecryl 860, UCB Chemical Company) were mixed with 33 wt% reactive diluents of styrene monomer as a resin. During mixing, 3.0 wt% of cumyl hydroperoxide (Trigonox 239A, Akzo Chemicals) was added as an initiator, and 0.8 wt% of cobalt naphthenate with 6% metal content (CoNap, Witco) was added as an accelerator. Various concentrations of nano-structured keratin fibers from feathers or the keratin fiber mats (supplied by Tyson Foods, Inc.) were physically mixed with the resin, or by resin infusion on the fiber mats, using a vacuum-assisted resin transfer molding (VARTM) process. The copolymerization reaction of the resin was carried out by redox decomposition of the free radical initiators using a metal promoter at room temperature to produce rigid composites. After room-temperature curing, the samples were post-cured at 120°C for 2 hr.

18.3.1.2 Characterization

The dielectric properties of the materials were measured on a Dielectric Analyzer (DEA 2970, TA Instruments) with a heating rate of 2°C/min at a frequency of 100 Hz. The dimension of a sample was 0.7 × 25.4 × 25.4 mm. In dielectric analysis, each sample was placed between two gold electrodes (parallel plate sensors, TA Instruments).

The storage modulus of the new low-k material was measured using a Dynamic Mechanical Analyzer (DMA 2988, TA Instruments) according to ASTM D5023. Fracture toughness was also measured to study the effect of HF content on the mechanical properties of the composites. The fracture toughness test was performed according to ASTM D5045 (three point bending mode) at room temperature. The crosshead speed of a tester (Instron 4201) was 1.27 mm/min.

The coefficients of thermal expansion (CTEs), in-plane direction, were measured using a Thermomechanical Analyzer (TMA 2940, TA Instruments) under a force of 0.05 N at a heating rate of 5°C/min. The CTE is defined as the fractional change in length per unit change in temperature.

18.3.2 Results and Discussion

Keratin fibers are light and hollow and can improve the structural properties of the composites. The density of keratin fiber is about 0.80 g/cm^3, and that of soybean resin is about 1.08 g/cm^3. The density of the new dielectric materials decreases with an increase of the fiber content and can be made to be less than 1 g/cm^3 at a fiber volume fraction of about 30%. This contrasts with a density of typical epoxy-glass fiber circuit boards. The low density of the keratin fiber composites makes them suited to automotive and aeronautical applications in addition to electronic materials.

Figure 18.12 shows the k values of the new composite materials developed from nanostructured keratin fibers and soybean oil resin, at a temperature of 25°C. The k values decrease from 2.7 to 1.7, with an increase of the fiber content approximately as

$$k = k(\text{resin}) - \text{WF}$$

where WF is in the weight fraction of fiber. The new keratin fiber composite material has a lower dielectric constant than conventional semiconductor insulators such as silicon dioxide (k = 3.8–4.2), epoxies, polyimides, and other dielectric materials. [77] In integrated circuits (ICs), a decrease in the k value of the insulator increases the operating speed. The delay time of the electronic signal is proportional to the square root of k, and values close to k = 1 are most desirable. The measured k value of the fiber mat itself was 1.7 because keratin fibers contain a significant volume of air. The ideal minimum k value is 1.0, as represented by air, and therefore porous or high-air-content materials may have dielectric constants in the ultra-low-k (<2.2) region. The dielectric constant of Teflon film is about 1.9. However, they are too soft and thermally unstable at high temperature. There are also serious concerns about the effect of fluorinated dielectrics on the interconnect metals and metal liners at elevated temperatures. Thermosetting epoxy resin has a k value of 4.1 and is used for electromagnetic components (EMCs), printed circuit board (PCBs), and the resin-encapsulating type semiconductor devices. The k value of soybean resin itself is

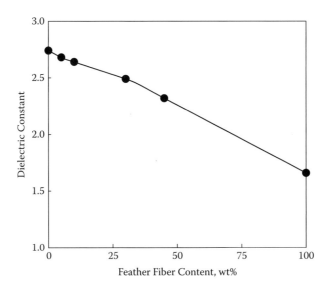

FIGURE 18.12
The dielectric constants of HF composites at a temperature of 25°C as a function of HF content.

2.7 and the molding process of soybean resin can follow the lead of conventional unsaturated polyester or vinyl ester resins. The low-viscosity resins are also capable of being spun into nano-size thin films. Thus, soybean resin alone can also be a substitute for petroleum-based resins in many electronic material applications.

The temperature dependence of the dielectric constant of the keratin fiber composites is shown in Figure 18.13. The dielectric constant of the composite slightly increased with

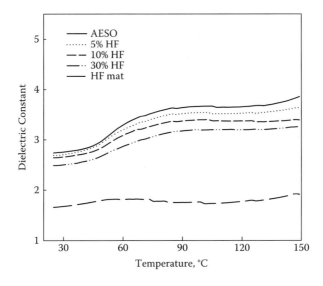

FIGURE 18.13
The dependency of the dielectric constants on temperature.

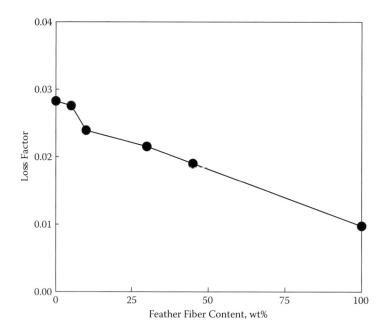

FIGURE 18.14
The loss factors of HF composites at a temperature of 25°C as a function of HF content.

increasing temperature, resulting from the alignment of the dipoles when the composite softened with temperature. However, because the effect of temperature on the k value of air is minimal, we expect relatively little change, as noted for the pure fiber mat in Figure 18.13.

In dielectric tests, the measured value is separated into dielectric constant (permittivity, e′) and loss factor (e″). The dependence of the loss factor upon fiber content is given in Figure 18.14. The loss factor represents the energy required to align dipoles and move ions. The loss factor decreases with increasing fiber content.

With the low k values, the mechanical strengths are also essential properties for the material to be used in electronic applications. The incorporation of keratin fiber in the soybean oil resin results in a considerable increase in the stiffness and strength of the soybean-oil-based composites. The storage modulus of soybean resin at 20°C is 1.87 GPa. With the addition of 30 wt% of the hollow keratin fibers, the storage modulus increases to 2.55 GPa. Also, the fracture toughness K_{1C} of the HF composite material was determined to be 0.86 to 0.97 MPa×m$^{1/2}$, which is higher than silica-based dielectric materials. In addition to their low density, the mechanical properties of the composites are significantly enhanced by the addition of the keratin fibers, suitable for electronic applications.

The coefficients of thermal expansion (CTEs) of the keratin fiber composites are shown in Table 18.4. The CTE decreased with increasing fiber content and these values are quite low, especially for the 30 wt% sample. In this case, the 30 wt% sample is a VARTM processed sample using the fiber mats and the hollow fibers having high in-plane orientation. The CTE of the 30 wt% composite below the glass transition temperature, 67.4 ppm/°C, is low enough for electronic applications, and similar to the value of silicon material (66 ppm/°C) or polyimides (~59 ppm/°C). However, the CTE of the AESO is higher and changes abruptly above its glass transition temperature (T_g). The glass transition temperature of

TABLE 18.4

The CTE Values of HF Composites

HF Content, wt%	α below T_{g-} (μm/m-°C)	α above T_{g-} (μm/m-°C)
0	127.2	205.8
5	106.1	200.6
10	100.4	196.3
20	93.1	193.2
30	67.4	69.6

AESO resin cured at room temperature was about 70°C and can be increased by chemical modification. From an atomic perspective, the CTE reflects an increase in the average distance between atoms with increasing temperature. The thermal expansion of AESO resin is compensated with the thermal contraction of keratin fibers in the longitudinal (oriented) direction. Interactions between resin and the fibers also act to resist changes in dimension with increasing temperature. The CTE can be further decreased by increasing the glass transition temperature and cross-linking density of the network structure, resulting in increasing chain stiffness.

To check the lithographic process potential, we printed a multi-chip module (MCM) and an optical element on the new keratin fiber dielectric materials using a semiconductor fabrication technique (Figure 18.15) in the Nano-Optics Fabrication facility. Incorporating low-*k* materials into viable sub-micron fabrication schemes will necessitate the use of chemical mechanical polishing processes to meet the critical dimension requirements of current ICs. Although detailed tests of the chip performance have not yet been carried out, preliminary results suggest that the new dielectric material from renewable resources is a promising alternative in a variety of electronic applications (e.g., high-speed circuits, high-performance capacitors, electrical cable insulation, and electronic packaging and components).

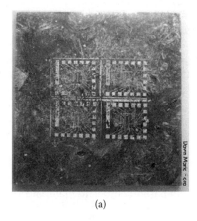 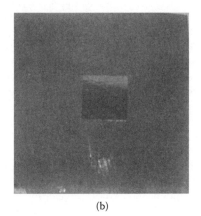

(a)   (b)

FIGURE 18.15

Pictures of the first chicken feather microchips printed using a semiconductor fabrication technique: (a) opto-electronic multi-chip module and (b) diffractive optical element.

18.4 Summary

A new low-k material was developed from renewable resources using nano-structured keratin fibers and chemically modified soybean oil in this study. The new low-k material was a natural, bio-based and environmentally friendly material. The k value was found to be in the range of 1.7 to 2.7, depending on the fiber fraction. The k values were lower than that of a conventional semiconductor insulator material such as silicon dioxide, epoxies, polyimides, and other dielectric materials. Also, the keratin fiber dielectric composite was lightweight and rigid, with good fracture toughness properties and approximates the shape and feel of a silicon dioxide insulator. The coefficient of thermal expansion of new low-k material (67.4 ppm/°C) was low enough for electronic applications, and similar to the value of silicon material or polyimides. Multi-chip-module circuit printing results suggest that the low-cost composite made with keratin fiber and plant oil has the potential to replace the dielectrics in microchips and circuit boards in the ever-growing electronic materials field.

References

1. M. Alexandre and P. Dubois, Polymer-layered silicate nanocomposites: preparation, properties and uses of a new class of materials, *Mater. Sci. Engr. R-Rpt.*, 28, 1, 2000.
2. P.C. LeBaron, Z. Wang, and T. Pinnavaia, Polymer-layered silicate nanocomposites: an overview, *J. Appl. Clay Sci.*, 15, 11, 1999.
3. S. Khot, J. LaScala, E. Can, S. Morye, G. Williams, G. Palmese, S. Kusefoglu, and R.P. Wool, Development and application of triglyceride-based polymers and composites, *J. Appl. Polym. Sci.*, 82, 703, 2001.
4. G.I. Williams and R.P. Wool, Composites from natural fibers and soy oil resins, *Appl. Compos. Mater.*, 7, 421, 2000.
5. E. Can, S. Kusefoglu, and R.P. Wool, J. Rigid, thermosetting liquid molding resins from renewable resources. I. Synthesis and polymerization of soy oil monoglyceride maleates, *Appl. Polym. Sci.*, 81, 69, 2001.
6. W. Thielemans, E. Can, S. Morye, and R.P. Wool, Novel applications of lignin in composite materials, *J. Appl. Polym. Sci.*, 83, 323, 2002.
7. E. Can, S. Kusefoglu, and R.P. Wool, Rigid thermosetting liquid molding resins from renewable resources. II. Copolymers of soybean oil monoglyceride maleates with neopentyl glycol and bisphenol A maleates, *J. Appl. Polym. Sci.*, 83, 972, 2002.
8. J. LaScala and R.P. Wool, The effect of fatty acid composition on the acrylation kinetics of epoxidized triacylglycerols, *J. Am. Oil Chem. Soc.*, 79 (1) 59, 2002.
9. J. LaScala and R.P. Wool, Effect of FA composition on epoxidation kinetics of TAG, *J. Am. Oil Chem. Soc.*, 79 (4) 373, 2002.
10. C.K. Hong and R. P. Wool, Low dielectric constant materials from hollow fibers and plant oil, *Journal of Natural Fibers*, 1 (2) 83, 2004.
11. C.K. Hong and R.P. Wool, Development of a bio-based composite material from soybean oil and keratin fibers, *J. Appl. Polym. Sci.*, 95(6), 1524, 2005.
12. M.A. Dweib, B. Hu, A. O'Donnell, H. Shenton, and R.P. Wool, All natural composite sandwich beams for structural applications, *Compos. Struct.*, 63(2), 147, 2004.
13. A. O'Donnell, M.A. Dweib, and R.P. Wool, Natural fiber composites with plant oil-based resin, *Compo. Sci. Technol.*, 64(9), 1135, 2004.

14. M.A. Dweib, B. Hu, H. Shenton, and R.P. Wool, Bio-based composite roof structure: manufacturing and processing issues, *Compos. Struct.*, 74(4), 379, 2006.
15. S.S. Morye and R.P. Wool, Mechanical properties of glass/flax hybrid composites based on a novel modified soybean oil matrix material, *Polymer Composites*, 26 4, 407, 2005.
16. W. Thielemans, I. McAninch, V. Barron, W. Blau, and R.P. Wool, Impure carbon nanotubes as reinforcements for acrylated epoxidized soy oil composites, *J. Appl. Polym. Sci.*, 98(3), 1325, 2005.
17. J. Lu, C.K. Hong, and R.P. Wool, Bio-based nanocomposites from functionalized plant oils and layered silicate, *J. Polym. Sci., B: Polym. Phys.*, 42,(8), 1441, 2004.
18. L. Zhu and R.P. Wool, Nanoclay reinforced bio-based elastomers: synthesis and characterization, *Polymer*, 47(24), 8106, 2006.
19. W. Thielemans and R.P. Wool, Kraft lignin as fiber treatment for natural fiber-reinforced composites, *Polym. Compos.*, 26(5), 695, 2005.
20. W. Thielemans and R.P. Wool, Lignin esters for use in unsaturated thermosets: lignin modification and solubility modeling, *Biomacromolecules*, 64, 1895, 2005.
21. K. Liu, *Soybeans: Chemistry, Technology, and Utilization*, Chapman & Hall, New York, 1997, p. 25.
22. A. Cunningham and A. Yapp, (Full text is not available for this patent), U.S. Patent 3,827,993, 1974.
23. G.W. Bussell and Inmont Corporation, Maleinized fatty acid esters of 9-oxatetracyclo-4.4.1 o, o undecan-4-ol, U.S. Patent 3,855,163, 1974.
24. L.E. Hodakowski, C.L. Osborn, E.B. Harris, and Union Carbide Corporation, Polymerizable epoxide-modified compositions, U.S. Patent 4,119,640, 1975.
25. D.K. Salunkhe, J.K. Chavan, R.N. Adsule, and S.S. Kadam, *World Oilseeds: Chemistry, Technology, and Utilization*, Van Nostrand Reinhold, New York, 1992.
26. D.J. Trecker, G.W. Borden, and O.W. Smith, Acrylated epoxidized soybean oil amine compositions and method, U.S. Patent 3,931,075, 1976.
27. R.P. Wool, S.H. Kusefoglu, G.R. Palmese, R. Zhao, and S.N. Khot, High modulus polymers and composites from plant oils, U.S. Patent 6,121,398, 2000.
28. A. Guo, D. Demydov, W. Zhang, and Z.S. Petrovic, Polyols and polyurethanes from hydroformylation of soybean oil, *J Polym. Environ.*, 10, 49, 2002.
29. F.K. Li and R.C. Larock, New soybean oil-styrene-divinylbenzene thermosetting copolymers. II. Dynamic mechanical properties, *J. Polym. Sci., B: Polym. Phys.*, 38, 2721, 2000.
30. F.K. Li and R.C. Larock, Thermosetting polymers from cationic copolymerization of tung oil: synthesis and characterization, *J. Appl. Polym. Sci.*, 78, 1044, 2000.
31. F.K. Li and R.C. Larock, New soybean oil-styrene-divinylbenzene thermosetting copolymers. III. Tensile stress-strain behavior, *J. Polym. Sci., B: Polym. Phys.*, 39, 60, 2001.
32. F.K. Li, A. Perrenoud, and R.C. Larock, Thermophysical and mechanical properties of novel polymers prepared by the cationic copolymerization of fish oils, styrene and divinylbenzene, *Polymer*, 42, 10133, 2001.
33. Z.S. Petrovic, W. Zhang, A. Zlatanic, C.C. Lava, and Ilavsky, Effect of OH/NCO molar ratio on properties of soy-based polyurethane networks, *J. Polym. Environ.*, 10, 5, 2002.
34. S. Khot, J. LaScala, E. Can, S. Morye, G. Williams, G. Palmese, S. Kusefoglu, and R.P. Wool, Development and application of triglyceride-based polymers and composites, *J. Appl. Polym. Sci.*, 82, 703, 2001.
35. E. Can, S.Kusefoglu, and R.P. Wool, Rigid, thermosetting liquid molding resins from renewable resources. I. Synthesis and polymerization of soy oil monoglyceride maleates, *Appl. Polym. Sci.*, 81, 69, 2001.
36. E. Can, S. Kusefoglu, and R.P. Wool, Rigid thermosetting liquid molding resins from renewable resources. II. Copolymers of soybean oil monoglyceride maleates with neopentyl glycol and bisphenol A maleates, *J. Appl. Polym. Sci.*, 83, 972, 2002.
37. J. LaScala and R.P. Wool, The effect of fatty acid composition on the acrylation kinetics of epoxidized triacylglycerols, *J. Am. Oil Chem. Soc.*, 79(1), 59, 2002.

38. J. LaScala and R.P. Wool, Effect of FA composition on epoxidation kinetics of TAG, *J. Am. Oil Chem. Soc.,* 79 4, 373, 2002.

39. E. Can, R.P. Wool, and S. Kusefoglu, Soybean- and castor-oil-based thermosetting polymers: mechanical properties, *J. Appl. Polym. Sci.,* 102(2), 1497, 2006.

40. E. Can, R.P. Wool, and S. Kusefoglu, Soybean and castor oil based monomers: Synthesis and copolymerization with styrene, *J. Appl. Polym. Sci.,* 102(3), 2433, 2006.

41. J. Lu, S. Khot, and R.P. Wool, New sheet molding compound resins from soybean oil. I. Synthesis and characterization, *Polymer,* 46(1), 71, 2005.

42. J. Lu and R.P. Wool, Novel thermosetting resins for SMC applications from linseed oil: Synthesis, characterization, and properties, *J. Appl. Polym. Sci.,* 99(5), 2481, 2006.

43. S. Khot, J. LaScala, E. Can, S. Morye, G. Williams, G. Palmese, S. Kusefoglu, and R. P. Wool, Development and application of triglyceride-based polymers and composites, *J. Appl. Polym. Sci.,* 82, 703, 2001.

44. R. P. Wool, S.H. Kusefoglu, G.R. Palmese, R. Zhao, and S.N. Khot, High modulus polymers and composites from plant oils, U.S. Patent 6,121,398 2000.

45. X. Kornmann, L.A. Berglund, and J. Sterte, Nanocomposites based on montmorillonite and unsaturated polyester, *Polym. Engr. Sci.,* 38, 1351, 1998.

46. X. Kornmann, H. Lindberg, and L.A. Berglund, Synthesis of epoxy-clay nanocomposites: influence of the nature of the clay on structure, *Polymer,* 42, 1303, 2001.

47. D.L. Ho, R.M. Briber, and C.J. Glinka, Characterization of organically modified clays using scattering and microscopy techniques, *Chem. Mater.,* 13, 1923, 2001.

48. D.L. Ho and C.J. Glinka, Effects of solvent solubility parameters on organoclay dispersions, *Chem. Mater.,* 15, 1309, 2003.

49. H. Ishida, S. Campbell, and J. Blackwell, General approach to nanocomposite preparation, *Chem. Mater.,* 12, 1260, 2000.

50. X. Kornmann, H. Lindberg, and L.A. Berglund, Synthesis of epoxy-clay nanocomposites. Influence of the nature of the curing agent on structure, *Polymer,* 42, 4493, 2001.

51. J.G. Doh and I. Cho, Synthesis and properties of polystyrene organoammonium montmorillonite hybrid, *Polym. Bull.,* 41, 511, 1998.

52. C.C. Zeng and L.J. Lee, Polymethyl methacrylate and polystyrene/clay nanocomposites prepared by in-situ polymerization, *Macromolecules,* 34, 4098, 2001.

53. X. Kornmann, H. Lindberg, and L.A. Berglund, Synthesis of epoxy-clay nanocomposites: influence of the nature of the clay on structure, *Polymer,* 42, 1303, 2001.

54. T. Lan and T.J. Pinnavaia, Clay-reinforced epoxy nanocomposites, *Chem. Mater.,* 6, 2216, 1994.

55. X. Kornmann, L.A. Berglund, and J. Sterte, Nanocomposites based on montmorillonite and unsaturated polyester, *Polym. Engr. Sci.,* 38, 1351, 1998.

56. R.K. Bharadwaj, A.R. Mehrabi, C. Hamilton, C. Trujillo, M. Murga, R. Fan, A. Chavira, and A.K. Thompson, Structure-property relationships in cross-linked polyester-clay nanocomposites, *Polymer,* 43, 3699, 2002.

57. D.J. Suh, Y.T. Lim, and O.O. Park, The property and formation mechanism of unsaturated polyester-layered silicate nanocomposite depending on the fabrication methods, *Polymer,* 41, 8557, 2000.

58. D.L. Ho and C.J. Glinka, Effects of solvent solubility parameters on organoclay dispersions, *Chem. Mater.,* 15, 1309, 2003.

59. H. Ishida, S. Campbell, and J. Blackwell, General approach to nanocomposite preparation, *Chem. Mater.,* 12, 1260, 2000.

60. J. E. Mark, Ed., *Physical Properties of Polymers Handbook,* AIP Press: New York, 1996.

61. K.L. Hoy, Solubility parameter as a design parameter for water borne polymers and coatings, *J. Ind. Textiles,* 19, 53, 1989.

62. C.M. Hansen, The three dimensional solubility parameter - key to paint component affinities I. Solvents, plasticizers, polymers, and resins, *J. Paint Technol.,* 39, 104, 1967.

63. C.M. Hansen, The three dimensional solubility parameter - key to paint component affinities II. Dyes, emulsifiers, mutual solubility and compatibility, and pigments, *J. Paint Technol.,* 39, 505, 1967.

64. D.W. Van Krevelen and P J. Hoftyzer, *Properties of Polymers*, Elsevier, New York, 1997.

65. J.E. Mark, Ed., *Physical Properties of Polymers Handbook*, AIP Press: New York, 1996.

66. D.L. Ho, R.M. Briber, and C.J. Glinka, Characterization of organically modified clays using scattering and microscopy techniques, *Chem. Mater.*, 13, 1923, 2001.

67. J. Lu, C.K. Hong, and R.P. Wool, Bio-based nanocomposites from functionalized plant oils and layered silicate, *J. Polym. Sci., B: Polym. Phys.*, 42(8), 1441, 2004.

68. S. Khot, J. LaScala, E. Can, S. Morye, G. Williams, G. Palmese, S. Kusefoglu, and R.P. Wool, Development and application of triglyceride-based polymers and composites, *J. Appl. Polym. Sci.*, 82, 703, 2001.

69. D J. Suh, Y.T. Lim, and O.O. Park, The property and formation mechanism of unsaturated polyester-layered silicate nanocomposite depending on the fabrication methods, *Polymer*, 41, 8557, 2000.

70. J.C. Halpin and J.L. Kardos, Halpin-tsai equations—review, *Polym. Engr.Sci.*, 16, 344, 1976.

71. C.Y. Hui and D. Shia, Simple formulae for the effective moduli of unidirectional aligned composites, *Polym. Engr. Sci.*, 38, 774, 1998.

72. C.Y. Hui and D. Shia, Simple formulae for the effective moduli of unidirectional aligned composites, *Polym. Engr. Sci.*, 38, 774, 1998.

73. H.L. Tyan, Y.C. Liu, and K.H. Wei, Thermally and mechanically enhanced clay/polyimide nanocomposite via reactive organoclay, *Chem. Mater.*, 11, 1942, 1999.

74. R.D. Miller, Device physics - In search of low-k dielectrics, *Science*, 286, 421, 1999.

75. H. Treichel, B. Withers, G. Ruhl, P. Ansmann, R. Wurl, Ch. Muller, M. Dietlmeier, and G. Maier, in *Handbook of Low and High Dielectric Constant Materials and Their Applications*, H.S. Nalwa, Ed., Vol. 1, chap. 1, Academic Press, San Diego, 1999.

76. W.F. Schmidt, in *Advanced Fibres, Plastics, Laminates and Composites*, F.T. Wallenberger, N.E. Weston, R. Ford, R.P. Wool, and K. Chawla, Eds., Materials Research Society, Pennsylvania, 25, 2002.

77. H. Treichel, B. Withers, G. Ruhl, P. Ansmann, R. Wurl, Ch. Muller, M. Dietlmeier, and G. Maier, in *Handbook of Low and High Dielectric Constant Materials and Their Applications*, H.S. Nalwa, Ed., Vol. 1, chap.1, Academic Press, San Diego, 1999.

78. R.D. Miller, Device physics-In search of low-k dielectrics, *Science*, 286, 421 1999.

79. H. Treichel, B. Withers, G. Ruhl, P. Ansmann, R. Wurl, Ch. Muller, M. Dietlmeier, and G. Maier, in *Handbook of Low and High Dielectric Constant Materials and Their Applications*, H.S. Nalwa, Ed., Vol. 1, chap.1, Academic Press, San Diego, 1999.

80. R.D. Miller, Device physics - In search of low-k dielectrics, *Science*, 286, 421, 1999.

81. I. Ogura, in *Handbook of Low and High Dielectric Constant Materials and Their Applications*, H. S. Nalwa, Ed., vol. 1, chap. 5, Academic Press, San Diego, 1999.

82. S.J. Martin, J.P. Godschalx, M.E. Mills, E.O. Shaffer II, and P. Townsend, Development of a low-dielectric-constant polymer for the fabrication of integrated circuit interconnect, *Adv. Mater.*, 12, 1769, 2000.

83. S.J. Martin, J.P. Godschalx, M.E. Mills, E.O. Shaffer II, and P. Townsend, Development of a low-dielectric-constant polymer for the fabrication of integrated circuit interconnect, *Adv. Mater.*, 12, 1769, 2000.

84. A.E. Eichstadt et al., Structure-property relationships for a series of amorphous partially aliphatic polyimides, *J. Polym. Sci., B: Polym. Phys.*, 40, 1503, 2002.

85. S. Khot, J. LaScala, E. Can, S. Morye, G. Williams, G. Palmese, S. Kusefoglu, and R.P. Wool, Development and application of triglyceride-based polymers and composites, *J. Appl. Polym. Sci.*, 82, 703, 2001.

86. W.D. Callister, *Material Science and Engineering: An Introduction*, 3rd ed., Wiley, New York, 1994.

Index